70079446

D1758851

Bosc

Bosch Professional Automotive Information is a definitive reference for automotive engineers. The series is compiled by one of the world´s largest automotive equipment suppliers. All topics are covered in a concise but descriptive way backed up by diagrams, graphs, photographs and tables enabling the reader to better comprehend the subject.

There is now greater detail on electronics and their application in the motor vehicle, including electrical energy management (EEM) and discusses the topic of intersystem networking within vehicle. The series will benefit automotive engineers and design engineers, automotive technicians in training and mechanics and technicians in garages.

Konrad Reif

Editor

Automotive Mechatronics

Automotive Networking, Driving Stability Systems, Electronics

 Springer Vieweg

Editor
Prof. Dr.-Ing. Konrad Reif
Duale Hochschule Baden-Württemberg
Friedrichshafen, Germany
reif@dhbw-ravensburg.de

ISBN 978-3-658-03974-5 ISBN 978-3-658-03975-2(eBook)
DOI 10.1007/978-3-658-03975-2

Library of Congress Control Number: 2014946887

Springer Vieweg
© Springer Fachmedien Wiesbaden 2015

Springer Vieweg is part of Springer Science+Business Media
www.springer-vieweg.de

▶ Foreword

As the complexity of automotive vehicles increases this book presents operational and practical issues of automotive mechatronics. It is a comprehensive introduction to controlled automotive systems and provides detailed information of sensors for travel, angle, engine speed, vehicle speed, acceleration, pressure, temperature, flow, gas concentration etc. The measurement principles of the different sensor groups are explained and examples to show the measurement principles applied in different types.

Complex technology of modern motor vehicles and increasing functions need a reliable source of information to understand the components or systems. The rapid and secure access to these informations in the field of Automotive Electrics and Electronics provides the book in the series "Bosch Professional Automotive Information" which contains necessary fundamentals, data and explanations clearly, systematically, currently and application-oriented. The series is intended for automotive professionals in practice and study which need to understand issues in their area of work. It provides simultaneously the theoretical tools for understanding as well as the applications.

Contents

Basics of mechatronics
Dipl.-Ing. Hans-Martin Heinkel,
Dr.-Ing. Klaus-Georg Bürger.

Architecture
Dr. phil. nat. Dieter Kraft,
Dipl.-Ing. Stefan Mischo.

Electronic control units
Dipl.-Ing. Martin Kaiser,
Dr. rer. nat. Ulrich Schaefer,
Dipl.-Ing. (FH) Gerhard Haaf.

Basic principles of networking
Automotive networking
Bus systems
Dipl.-Ing. Stefan Mischo,
Dipl.-Ing. (FH) Stefan Powolny,
Dipl.-Ing. Hanna Zündel,
Dipl.-Ing. (FH) Norbert Löchel,
Dipl.-Inform. Jörn Stuphorn,
 Universität Bielefeld,
Dr. Rainer Constapel, Daimler AG Sindelfingen,
Dipl.-Ing. Peter Häussermann,
 Daimler AG Sindelfingen,
Dr. rer. nat. Alexander Leonhardi,
 Daimler AG Sindelfingen,
Dipl.-Inform. Heiko Holtkamp,
 Universität Bielefeld.

Automotive sensors
Sensor measuring principles
Sensor types
Dr.-Ing. Erich Zabler,
Dr. rer. nat. Stefan Finkbeiner,
Dr. rer. nat. Wolfgang Welsch,
Dr. rer. nat. Hartmut Kittel,
Dr. rer. nat. Christian Bauer,
Dipl.-Ing. Günter Noetzel,
Dr.-Ing. Harald Emmerich,
Dipl.-Ing. (FH) Gerald Hopf,
Dr.-Ing. Uwe Konzelmann,
Dr. rer. nat. Thomas Wahl,
Dr.-Ing. Reinhard Neul,
Dr.-Ing. Wolfgang-Michael Müller,
Dr.-Ing. Claus Bischoff,
Dr. Christian Pfahler,
Dipl.-Ing. Peter Weiberle,
Dipl.-Ing. (FH) Ulrich Papert,

Dipl.-Ing. Christian Gerhardt,
Dipl.-Ing. Klaus Miekley,
Dipl.-Ing. Roger Frehoff,
Dipl.-Ing. Martin Mast,
Dipl.-Ing. (FH) Bernhard Bauer,
Dr. Michael Harder,
Dr.-Ing. Klaus Kasten,
Dipl.-Ing. Peter Brenner, ZF Lenksysteme GmbH,
 Schwäbisch Gmünd,
Dipl.-Ing. Frank Wolf,
Dr.-Ing. Johann Riegel.

Electric Actuators
Dr.-Ing. Rudolf Heinz,
Dr.-Ing. Robert Schenk.

Electrohydraulic Actuators
Electronic Transmission Control
Modules for Transmission Control
Dipl.-Ing. D. Fornoff,
D. Grauman,
E. Hendriks,
Dipl.-Ing. T. Laux,
Dipl.-Ing. T. Müller,
Dipl.-Ing. A. Schreiber,
Dipl.-Ing. S. Schumacher,
Dipl.-Ing. W. Stroh.

Antilock Braking System (ABS)
Traction Control System (TCS)
Electronic Stability Program (ESP)
Automatic brake functions
Hydraulic modulator
Dipl.-Ing. Friedrich Kost
(Basic Principles of Vehicle Dynamics),
Dipl.-Ing. Heinz-Jürgen Koch-Dücker
(Antilock Braking Systems, ABS),
Dr.-Ing. Frank Niewels and
Dipl.-Ing. Jürgen Schuh
(Traction Control Systems, TCS),
Dipl.-Ing. Thomas Ehret
(Electronic Stability Program, ESP),
Dipl.-Ing. (FH) Jochen Wagner
(Automatic Brake Functions),
Dipl.-Ing. (FH) Ulrich Papert
(Wheel-Speed Sensors),
Dr.-Ing. Frank Heinen and
Peter Eberspächer

Sensotronic brake control (SBC)
Dipl.-Ing. Bernhard Kant.

Overview of common-rail systems
High-pressure components of common-rail system
Electronic Diesel Control (EDC)
Dipl.-Ing. Felix Landhäußer,
Dr.-Ing. Ulrich Projahn,
Dipl.-Inform. Michael Heinzelmann,
Dr.-Ing. Ralf Wirth
(Common-rail system),
Ing. grad. Peter Schelhas,
Dipl.-Ing. Klaus Ortner
(Fuel-supply pumps),
Dipl.-Betriebsw. Meike Keller
(Fuel filters),
Dipl.-Ing. Sandro Soccol,
Dipl.-Ing. Werner Brühmann
(High-pressure pumps),
Ing. Herbert Strahberger,
Ing. Helmut Sattmann
(Fuel rail and add-on components),
Dipl.-Ing. Thilo Klam,
Dipl.-Ing. (FH) Andreas Rettich,
Dr. techn. David Holzer,
Dipl.-Ing. (FH) Andreas Koch
(Solenoid-valve injectors),
Dr.-Ing. Patrick Mattes
(Piezo-inline injectors),
Dipl.-Ing. Thomas Kügler
(Injection nozzles),
Dipl.-Ing. (FH) Mikel Lorente Susaeta,
Dipl.-Ing. Martin Grosser,
Dr.-Ing. Andreas Michalske
(Electronic diesel control),
Dr.-Ing. Günter Driedger,
Dr. rer. nat. Walter Lehle,
Dipl.-Ing. Wolfgang Schauer,
Rainer Heinzmann
(Diagnostics).

Active steering
Dipl.-Ing. (FH) Wolfgang Rieger,
 ZF Lenksysteme, Schwäbisch Gmünd.

Drive and adjustment systems
Dipl.-Ing. Rainer Kurzmann,
Dr.-Ing. Günter Hartz.

Heating, ventilation and air conditioning
Dipl.-Ing. Gebhard Schweizer,
 Behr GmbH & Co., Stuttgart.

Vehicle security systems
Dipl.-Ing. (FH) Jürgen Bowe,
Andreas Walther,
Dr.-Ing. B. Kordowski,
Dr.-Ing. Jan Lichtermann.

Electromagnetic compatibility
Dr.-Ing. Wolfgang Pfaff.

Fault diagnostics
Dr.-Ing. Matthias Knirsch,
Dipl.-Ing. Bernd Kesch,
Dr.-Ing. Matthias Tappe,
Dr,-Ing. Günter Driedger,
Dr. rer. nat. Walter Lehle.

and the editorial team in cooperation with the responsible in-house specialist departments of Robert Bosch GmbH.

Unless otherwise stated, the authors are all employees of Robert Bosch GmbH.

Basics

Basics of mechatronics

The term "mechatronics" came about as a made-up word from mechanics and electronics, where electronics means "hardware" and "software", and mechanics is the generic term for the disciplines of "mechanical engineering" and "hydraulics". It is not a question of replacing mechanical engineering by "electronification", but of a synergistic approach and design methodology. The aim is to achieve a synergistic optimization of mechanical engineering, electronic hardware and software in order to project more functions at lower cost, less weight and installation space, and better quality. The successful use of mechatronics in a problem solution is dependent upon an overall examination of disciplines that were previously kept separate.

Mechatronic systems and components

Applications
Mechatronic systems and components are now present throughout almost the entire vehicle: starting with engine-management systems and injection systems for gasoline and diesel engines to transmission control systems, electrical and thermal energy management systems, through to a wide variety of brake and driving dynamics systems. It even includes communication and information systems, with many different requirements when it comes to operability. Besides systems and components, mechatronics are also playing an increasingly vital role in the field of micromechanics.

Examples at system level
A general trend is emerging in the further development of systems for fully automatic vehicle handling and steering: more and more mechanical systems will be replaced by "X-by-wire" systems in future.

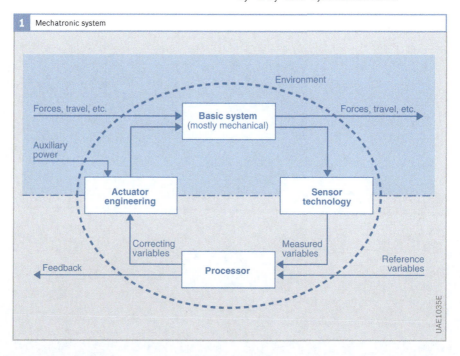

1 Mechatronic system

A system that was implemented long ago is the "Drive-by-wire" system, i.e. electronic throttle control.

"Brake-by-wire" replaces the hydrome-chanical connection between the brake pedal and the wheel brake. Sensors record the driver's braking request and transmit this information to an electronic control unit. The unit then generates the required braking effect at the wheels by means of actuators.

One implementation option for "Brake-by-wire" is the electrohydraulic brake (SBC, Sensotronic Brake Control). When the brake is operated or in the event of brake stabilization intervention by the electronic stability program (ESP), the SBC electronic control unit calculates the required brake pressure setpoints at the individual wheels. Since the unit calculates the required braking pressures separately for each wheel and collects the actual values separately, it can also regulate the brake pressure to each wheel via the wheel-pressure modulators. The four pressure modulators each consist of an inlet and an outlet valve controlled by electronic output stages which together produce a finely metered pressure regulation.

Pressure generation and injection are decoupled in the Common Rail System. A high-pressure rail, i.e. the common rail, serves as a high-pressure accumulator, constantly providing the fuel pressure required for each of the engine's operating states. A solenoid-controlled injector with a built-in injection nozzle injects fuel directly into the combustion chamber for each cylinder. The engine electronics request data on accelerator pedal position, rotational speed, operating temperature, fresh-air intake flow, and rail pressure in order to optimize the control of fuel metering as a function of the operating conditions.

Examples at component level

Fuel injectors are crucial components in determining the future potential of Diesel-engine technology. Common-rail injectors are an excellent example of the fact that an extremely high degree of functionality and, ultimately, customer utility can only be achieved by controlling all the physical domains (electrodynamics, mechanical engineering, fluid dynamics) to which these components are subjected.

In-vehicle CD drives are exposed to particularly tough conditions. Apart from wide temperature ranges, they must in particular withstand vibrations that have a critical impact on such precision-engineered systems.

In order to keep vehicle vibration away from the actual player during mobile deployment, the drives normally have a spring damping system. Considerations about reducing the weight and installation space of CD drives immediately raise questions concerning these spring-damper systems. In CD drives without a damper system, the emphasis is on designing a mechanical system with zero clearances and producing additional reinforcement for the focus and tracking controllers at high frequencies.

Only by combining both measures mechatronically is it possible to achieve good vibration resistance in driving mode. As well as reducing the weight by approx. 15 %, the overall height is also reduced by approx. 20 %.

The new mechatronic system for electrically operated refrigerant motors is based on brushless, electronically commutated DC motors (BLDC's). Initially, they are more expensive (motor with electronics) than previous DC motors equipped with brushes. However, the overall optimization approach brings benefits: BLDC motors can be used as "wet rotors" with a much simpler design. The number of separate parts is therefore reduced by approx. 60 %.

In terms of comparable cost, this more robust design doubles the service life, reduces the weight by almost half and reduces the overall length by approx. 40%.

Examples in the field of micromechanics

Another application for mechatronics is the area of micromechanical sensor systems, with noteworthy examples such as hot-film air-mass meters and yaw-rate sensors.

Because the subsystems are so closely coupled, microsystems design also requires an interdisciplinary procedure that takes the individual disciplines of mechanical components, electrostatics and possibly fluid dynamics and electronics into consideration.

Development methods

Simulation

The special challenges that designers face when developing mechatronic systems are the ever shorter development times and the increasing complexity of the systems. At the same time, it is vital to ensure that the developments will result in useful products.

Complex mechatronic systems consist of a large number of components from different physical domains: hydraulic components, mechanical components and electronic components. The interaction between these domains is a decisive factor governing the function and performance of the overall system. Simulation models are required to review key design decisions, especially in the early development stages when there is no prototype available.

2 Model library for a micromechanical yaw-rate sensor

Microsystem

Mechanical components

Electro-mechanical components

Rigid bodies

Elastic bodies

Comb-like structures

Detection electrodes

From segments of a circle

From segments of a rectangle

Bending beams

Segment of a circle

Undivided stator comb

Divided stator comb

From segments of a circle

From segments of a rectangle

UAE0942-1E

Basic issues can often be clarified by producing relatively simple models of the components. If more detail is required, more refined component models are needed. The detailed models focus mainly on a specific physical domain:

- This means that detailed hydraulic models of common rail injectors are available, for example. These can be simulated using special programs with numeric calculation methods that are exactly tailored to hydraulic systems. Cavitation phenomena have to be taken into consideration, among other things.
- Detailed models are also needed to design the power electronics that trigger the injector. Again, this involves the use of simulation tools which must be developed specifically to design electronic circuits.

The development and simulation of the software that controls the high-pressure pump and the power electronics in the control unit with the aid of the sensor signals also takes place using tools that are specially designed for this area of the overall system.

As the components in the overall system interact with each other, it is not sufficient to consider specific detailed models of the components in isolation. The optimum solution is also to take into account the models of other system components. In most cases, these components can be represented by simpler models. For example, the system simulation that is focussed on the hydraulic components only requires a simple model of the power electronics.

The application of various domain-specific simulation tools during the development of mechatronic systems is only efficient if there is some sort of support for exchanging models and parameters between the simulation tools. The direct exchange of models is highly problematic due to the specific languages used for describing the models of each of the tools.

However, an analysis of the typical components in mechatronic systems shows that they can be composed of a few simple elements specific to the domains. These standard elements are, for example:

- In the hydraulic system: throttle, valve or electric line
- In the electronic system: resistor, capacitor or transistor
- In the mechanical system: ground with friction, transmission or clutch (or the equivalent for micromechanics)

The preferable solution is that these elements should be stored in a central standard model library that is also decentrally accessible to product development. The essence of the standard model library is a documentation of all the standard elements. For each element, this comprises:
Description of physical behavior in words
The physical equations, parameters (e.g. conductivity or permeability), state variables (e.g. current, voltage, magnetic flux, pressure) and
The description of the associated interfaces

In addition, a major part of the environment is a reference model written in a modeling language that is independent of the tool. Overall, the library includes reference models from the mechanical, hydraulic, electronic, electrodynamic and software areas.

V model

The dependencies of the different product development phases are illustrated in the "V model": from requirement analysis to development, implementation, testing and system deployment. A project passes through three "top-down" levels during the development stage:

- Customer-specific functions
- Systems and
- Components

A requirements specification (what) must first be produced at each level in the form of specifications. This results in the design specification, which is drawn up on the basis of design decisions (the actual creative engineering work). The performance specifications describe how a requirement can be met. The performance specs form the basis for a model description which allows a review (i.e. validation) of the correctness of each design stage together with previously defined test cases. This procedure passes through each of three stages, and, depending on the technologies applied, for each of the associated domains (mechanical engineering, hydraulics, fluid dynamics, electrics, electronics, and software).

Recursions at each of the design levels shorten the development stages significantly. Simulations, rapid prototyping, and simultaneous engineering are tools that allow rapid verification, and they create the conditions for shortening product cycles.

Outlook

The major driving force behind mechatronics is continuous progress in the field of microelectronics. Mechatronics benefits from computer technology in the form of ever more powerful integrated computers in standard applications. Accordingly, there is a huge potential for further increases in safety and convenience in motor vehicles, accompanied by further reductions in exhaust emissions and fuel consumption. On the other hand, new

3 Recursion method at one level

UAE0943-1E

challenges are emerging with regard to the technical mastery of these systems.

However, future "X-by-wire" systems without the mechanical/hydraulic fall-back level must also provide the pre-scribed functionality in the event of a problem. The condition for their imple-mentation is a high-reliability and high-availability mechatronic architecture which requires a "simple" proof of safety. This affects both single components as well as energy and signal transmissions.

As well as "X-by-wire" systems, driver-as-sistance systems and the associated man/machine interfaces represent another area in which the consistent implementation of mechatronic systems could achieve signifi-cant progress for both users and vehicle manufacturers.

The design approaches of mechatronic systems should strive toward continuity in several aspects:

- Vertical:
 "Top-down" from system simulation, with the objective of overall optimiza-tion, through to finite element simula-tion to achieve a detailed understanding, and "bottom-up" design engineering from component testing through to system testing
- Horizontal:
 "Simultaneous engineering" across several disciplines in order to deal with all product-related aspects at the same time
- Beyond company boundaries:
 Step by step, the idea a "virtual sample" is nearing our grasp

Another challenge is training in order to further an interdisciplinary mindset and develop suitable SE processes and forms of organization and communication.

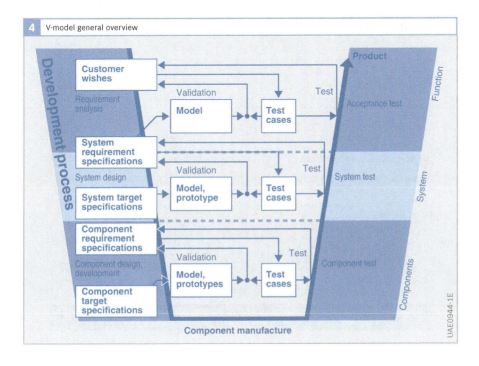

4 V-model general overview

Architecture

Over the last three decades, tremendous progress has been made in automotive engineering. Modern injection and exhaust-gas treatment systems drastically reduced pollutants in the exhaust gas, while occupant-protection and vehicle stabilization systems improved safety on the road. Much of this success is due to the introduction of electronically-controlled systems. The proportion of these systems used in cars increased continuously. The requirements of safety and environmental compatibility, but also the demand for comfort and convenience functions, will increase yet further and this will in no small part be achieved through the use of electronics. Up to around 90 % of innovations in the motor vehicle will be realized by electronics and microprocessor-controlled systems. The networking of these electronics creates the prerequisite for having this wide variety of electronic systems integrated within the complete vehicle system to form a whole. However, this results in a complexity that can only be overcome at considerable expense.

Overview

History
The on-board electrical network of a car around the year 1950 comprised approx. 40 lines. Essentially, cables were only required for the battery, starter, ignition and the lighting and signaling systems.

With the first electronic injection and ignition systems, cabling complexity began to increase fast. Sensors fitted in the engine compartment (e.g. speed sensor, engine-temperature sensor) had to deliver signals to the engine control unit, while the fuel injectors required their triggering signals from the electronic control unit.

A further increase in cabling complexity resulted from the introduction and rapid widespread adoption of the antilock brake system (ABS). Meanwhile, comfort and convenience systems, e.g. electrical power-window units, would also form part of the standard equipment. All these systems require additional connecting lines for the connection of sensors, control elements and actuators to the control unit.

Fig. 1
Source:
Mercer management
consulting

Technology of the present day

In the 1990s the cabling work in a luxury class vehicle amounted to around 3 km. This figure clearly demonstrates how complex the vehicle system has become. The growth of the proportion of electronics in the motor vehicle (Fig. 1) can mainly be attributed to the growth in microelectronics and sensor technology.

At first, many of the new systems were integrated into the vehicle by means of their own dedicated electronic control unit. For the most part, the individual electronic control units operated in mutual independence. All the same, connecting lines became increasingly necessary between electronic control units to enable the exchange of data by means of PWM signals, for example. Depending on the vehicle class, there are between 20 and 80 electronic control units fitted in today's vehicles. They control such equipment as the engine, antilock brake system or the airbags. The number of microcontrollers in the vehicle has therefore risen continuously in recent years (Fig. 2).

The components of the individual systems are optimally matched to each other. The systems may originate from different manufacturers that use previously agreed, albeit still their own, interfaces. The rain sensor, for example, "speaks" in a different way to the sensors for the engine management. The following example demonstrates just how networked the functions in a modern vehicle are: the radar sensor of the adaptive cruise control system (ACC) measures the distance to the vehicle traveling in front. If this distance is shorter than a specified minimum distance, the ACC electronic control unit sends this information to the engine management, the ESP electronic control unit and the airbag electronic control unit. The engine management reduces torque and thus driving speed. If this is not sufficient, the electronic stability program (ESP) must also generate brake pressure to decelerate the vehicle. If the distance continues to shorten, the airbag and seat-belt pretensioners are set to emergency standby. The communication between the electronic control units cannot take more than fractions of a second. The more electronic control units interact in the one complete system, the more difficult it becomes for them to communicate undisturbed.

With the number of electronic control units and the associated need for mutual communication, the costs of developing the systems rose as did the adaptation costs for making interfaces compatible. With the CAN bus (Controller Area Network) developed by Bosch, a powerful and widely used data bus system has become commonplace in vehicles for the first time. The data line of the CAN bus makes it possible for the electronic control units to exchange specific and relevant items of information with each other. At the start, the network only comprised a few electronic control units, such as the engine-management system, the electronic stability program and the transmission control. Gradually, further systems would expand this network, especially in the areas of comfort and convenience and infotainment. The CAN bus has gradually evolved into the standard for networking systems in the motor vehicle. Today it is the standard for communication between elec-

2 Number of microcontrollers in the motor vehicle

SVA0033E

tronic control units within different areas of the electronics (drivetrain, suspension, body electronics and infotainment) and forms a powerful backbone for networking these areas with each other. Additional bus systems (e.g. LIN bus, MOST bus) are used as subbuses or for transmitting at high data rates with comparatively low realtime requirements in the motor vehicle.

Development trends

The proportion of electrics and electronics in the motor vehicle will continue to increase. In the drivetrain, the number of components in the exhaust line (e.g. exhaust-gas sensors) is increasing due to stricter exhaust-emissions legislation. While the demands for reductions in fuel consumption can, for example, be fulfilled by means of new valve-gear concepts, even this requires additional electronic components. A further increase in the proportion of electronics results mainly from the growth of electronic systems in the areas of safety, comfort and convenience, and infotainment.

Objectives

Drivers demand a high level of reliability from a car. The vehicle manufacturer and the supplier of assemblies, meanwhile, are constrained by other requirements such as minimization of manufacturing costs, space restrictions and the weight of components. An opportunity to fulfill these requirements in the face of the increasing complexity of the "vehicle" system is seen in the shift of the traditional implementation technologies of mechanics, hydraulics and electrics towards microprocessor-controlled, electronic systems. For this reason, the development of software will continue to gain in importance in future.

The current situation in the electrical and electronic architecture of motor vehicles is characterized by an increase in functionality and an increasingly strained costs situation. To achieve both of these

objectives simultaneously, development partners are more frequently tapping into resources that are already available in subsystems. These can be sensors or actuators as well as realized functions that are available to different systems over the communications network. For new systems and functions, manufacturers strive to get by on a minimum of additional resources. In the meantime, engineers are faced with a new challenge in the form of "networked" thinking and subsystem integration, especially when the assemblies for the subsystems originate from different development partners (suppliers).

Complaints in the field (i.e. with series-production vehicles) due to electrical or electronic failures could be the consequence of not having taken the interactions of the subsystems into consideration. The causes – unmanageable behavior of functionality spread among networked systems, and their integration – are avoidable through the logical application of certified development processes as early as in the specification phase. Furthermore, modeling and tools for authoring a formal description of architectures are gaining ever more in importance.

Broadened requirements for a complete motor vehicle system in the future are leading to increased networking of vehicle components and subsystems. In this regard, new functions are being developed that go beyond the frontiers of traditional applications – and this is without additional expenditure on hardware wherever possible.

New development methods and technologies are required to make this achievable. With a top-down approach, new functions are viewed from the perspective of the complete vehicle. This means that, in accordance with the method of systems engineering, functional requirements and nonfunctional requirements (e.g. quality objectives, safety, costs, etc.) are set for the vehicle as a whole and derived as specifi-

cations for the subordinate subsystems. These requirements are formulated as a model and can thus be used as a specification for the subsystems and the creation of test cases. This is what is known as an "executable specification", which makes it possible to prove the completeness and the traceability of the requirements, for example, or to identify the requirements for interaction and communication between subsystems. In this way, it is possible to form an optimized architecture for the complete vehicle and its subsystems and components. The functional relationships between the complete motor vehicle system and the subordinate subsystems can be surveyed in different levels of detail and suitable interfaces can be defined for the functions. This approach supports an expanding networking of functions. Synergies are exploited between vehicle areas (domains such as the drivetrain, interior, infotainment) that were hitherto considered in isolation and resources are spared.

As an element of the development process that works in the opposite direction, the generation of new functions from available resources and existing systems (bottom up) should also be taken into consideration to minimize innovation risks. This is how new functions are integrated into existing systems, for example. Examples of this approach are measures to avert the consequences of an accident by "preparing" subsystems for an imminent crash (closing windows, closing the sliding sunroof, activating the airbag, etc.) or the assistance of the driver in emergency braking situations in ESP in future. In this way, it is possible to reduce the number of electronic control units and counteract rising system costs.

The development process described characterizes the CARTRONIC® concept that Bosch developed in the 1980s. The results of this concept are being incorporated into the Autosar Initiative (see Autosar Initiative)

Vehicle system architecture

Architecture

The architecture of a system represents its "construction plan". It describes the structural and dynamic system characteristics as a whole. The architecture is usually specified in a description language. Special draft mechanisms are used for specific requirements. With architecture being a construction plan for different realization technologies and a means of proving that functional and nonfunctional requirements have been fulfilled in the system draft, different views of the system architecture are required. Examples of this include:

- Hardware architecture
- Software architecture
- Network architecture in the area of realization technologies
- Cost and resource consumption in the area of economical analysis and
- For the area of social requirements, aspects such as safety, availability and legal conformity

The problems that arise in the integration of differently structured subsystems can be reduced by means of an architecture.

Functional structure

The domain of vehicle motion has the task of ensuring the controlled movement of the vehicle as well as its directional stability. This task can be subdivided into various levels (Fig. 3).

The navigation level is home to the planning tools for the driving route. These are merely informational in nature and have no interventional influence on vehicle motion.

At vehicle guidance level, the decisions of the driver are implemented by means of the steering wheel and accelerator pedal but also various assistance systems for vehicle handling (e.g. ACC, course stability

systems). At this level, the driver is able to overrule the assistance systems at any time.

At the stability level, there are the subsystems that are able to correct the decisions taken at handling level if these happen to be outside the range of safe reference variables (e.g. ABS, ESP). This may be the case when cornering or on wet road surfaces, for example.

At stabilization level, correcting variables for implementation by the vehicle's actuators are determined. Information about the environment (e.g. road condition, air temperature, rain sensor signal) is still required at the various levels for the implementation of the relevant tasks.

These tasks can be assigned to functional components, which are the architectural elements of the functional architecture. In this way, the driver information

functional component represents the tasks of the navigation level, which are to inform the driver of the driving route determined by means of a mapping system (Fig. 3). Vehicle guidance represents the guidance level, and stability intervention the tasks of the stabilization level. The vehicle motion coordinator determines the correcting variables for the actuators, e.g. of the drive and electronic stability program (ESP), from the information input by vehicle guidance and stability intervention.

Figure 4 shows how the functional components of guidance level, stabilization level and vehicle actuators are related in a hierarchical structure within vehicle motion. Communication relationships between the components and interactions with other domains, e.g. body and interior, are also featured in the model.

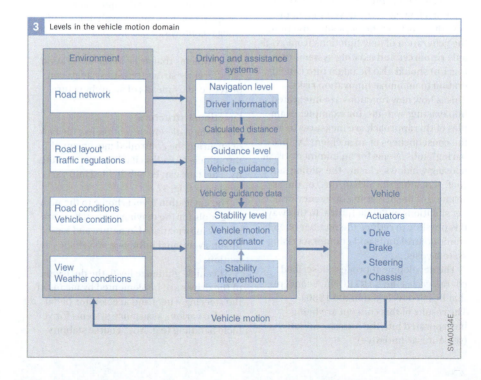

3 Levels in the vehicle motion domain

SVA0034E

In the same way as Vehicle motion is refined, these functional components require further detailing until the refined components represent manageable, clearly delimited tasks that make flexible, modular implementation possible through different realization technologies. Defined interfaces between the components enable communication and the exchange of data. For example, the transmission control issues a request through the engine-management system for a specific reduction in torque during a gearshift. This value is exchanged as a physical variable via the interface.

With its integration into a suitable procedural model, the functional structure is the starting point for subsequent stages in the development process.

Systematic creation of EE system architectures

The increasing amount of networking in traditional vehicle domains for the realization of new functions can be illustrated using the ACC (Adaptive Cruise Control) driver-assistance system as an example. Adaptive, same-lane driving is made possible by the networking of a combined cruise and distance control system with the engine-management system, brake system, transmission and cockpit. Here, subsystems from the drivetrain, chassis and infotainment (interaction with the driver) domains are used to realize the new function with minimal cost.

The decision as to whether a function (e.g. ACC) is realized in a dedicated logic close to the sensor or in one of the existing, subscriber electronic control units has no bearing on the function itself.

4 Example of a functional structure for the vehicle motion domain

SVA0035E

Rather, the decision is affected by non-functional requirements such as safety, availability, costs or resource availability. In addition to the functional requirements, these requirements mainly determine how the function is realized. The "how" is described by the architecture of the system. Different requirements result in different system architectures.

CARTRONIC® concept

With the CARTRONIC® architecture concept, all closed and open-loop control tasks in the vehicle have been structured in accordance with logical, functional viewpoints and modeled in the form of a functional architecture. Delimited functions (and their dependencies) that implement specific functional requirements have been represented by defined architectural elements. The functional structure, i.e. the structural description, represented a hierarchical decomposition of the subsystems down to manageable size. Interactions between elements of the functional structure have been described by communication relationships. Since the use of the architecture concept could have led to different functional structures, it was essential to reach agreement on the tasks and interfaces. It was necessary to choose interfaces that were based on physical variables and thus supported aspects such as reusability and interchangeability.

The motor vehicle system with all its open and closed-loop tasks was dismantled into subsystems that implement clearly defined tasks. These subsystems include the engine management, brake system, transmission control, ACC, lighting management, etc. Different levels of functional structure detail can be assigned to the system and subsystem levels (Fig. 5). It was therefore possible to create development frameworks for selected functional components and component groups on which to base implementation in the form of partial real-

ization stages. This required a decoupled development process and the exploitation of synergies between subsystems. The development frameworks took into consideration the dependencies and interface contents within the individual domains and with the rest of the vehicle, as is the case with a networked system such as ACC, for example.

Bosch introduced this concept to the Autosar Initiative (Workpackage 10.x).

Software architecture

The independence of the functional structure, or architecture, from the later realization stage results in a decoupling of functionality and technology and thus forms the first stage of a model-based development process. The functional structure can be used on several occasions and expanded as the foundation for drafting system architectures. This architecture is characterized by architecture drivers (specific criteria of the architecture) that are essentially the product of nonfunctional requirements (e.g. costs, quality, reusability, relocatability).

Further precision of the development frameworks devised from the functional structure, and of their interfaces in particular, is required if it is to be possible to evaluate an electronic control unit for the relocatability of functions and the integration of software, which is a contribution of various participants in the project. While retaining the realization-independent information from the functional structure – such as an agreed torque interface – the frameworks are supplemented by realization-specific information such as data type, quantization, runtime properties or resource requirements.

In the same way as hardware and disciplines such as mechanics or hydraulics, software can also be classified as a realization technology. Product-line or platform approaches have long been a familiar

feature of mechanical development or pro-
duction optimization and their application
is virtually universal. The trend towards
software-based system solutions and im-
provements, in conjunction with the vastly
expanding software scopes of electronic
systems, has given rise to the demand for
this strategy to also be transferred to soft-
ware-intensive systems.

The decisive challenge faced today is
less to do with technical feasibility but
more about how to develop methods fur-
ther and apply already developed methods
and processes in product developments
in a systematic and disciplined way, and
to anchor them within the organization.
The product-line approach in software de-
velopment was transferred to the motor
vehicle domain with the participation of
Bosch with methodological support from
the software engineering institute (SEI).
The method will be used systematically
in future Bosch product generations.

Network architecture
With the spreading of open standards
such as the CAN bus, the integration of
functions into application-specific elec-
tronic control units, and satellites linked
by subnetworks, network architecture has
become the synonym for the complexity
management of distributed systems.
Extensions and "attachment solutions"
are easily integrated until the limits of
network capacity are reached. If these
possibilities were to be exploited without
checking the system draft, this would
result in unmanageable increases in com-
plexities and integration conflicts. Biologi-
cal systems solve these unmanageable
complexities through specialization and
the creation of subnetworks with new
forms of organization. Their objectives
are stability and the ability to survive.
This model has, to a certain extent,
evolved on its own in motor vehicle sys-
tems through assignment to traditional

5 Functional architecture

System level

Subsystem level

Hardware architecture

Software architecture

Network architecture

SVA0036E

fields of application or domains and the comparatively slow growth of networking within these domains.

Bus systems for the individual domains are becoming more specialized due to their plainly different requirements. With the CAN in the drivetrain as the point of origin, new bus systems such as the LIN sub-bus have begun to infiltrate the area of body electronics or FlexRay in the case of safety-relevant x-by-wire systems. In the multimedia field, where demands for high data rates but low safety requirements prevail, bus systems such as Bluetooth have started to make an appearance.

Breaking through these traditional domains with ever more applications leads to known consequences, e.g. dramatic increase in complexity, high start-up costs, increasing integration times and costs, and more demanding work in customer service as a consequence of diagnostics no longer being manageable. A solution for these multidimensional optimization tasks has in the past been sought in the software field. In the case of technical systems in particular, the paradigm is still king, especially in software realizations, because the absence of physical boundaries supports unlimited growth.

Autosar Initiative

The Autosar Initiative (AUTomotive Open Systems ARchitecture) was founded in July 2003 by several vehicle manufacturers and suppliers – Bosch among them. Their global objective is the joint development of an open system architecture for future automotive applications. The aims of the partnership include the standardization of fundamental system functions (basic software) and function interfaces; they will replace the company-specific, individual solutions used to date. Model-based concepts and methods ought to reduce complexity in spite of an expanding range of functions. The demands for quality and

reliability are fulfilled by the multiple use of proven standards. Autosar concerns itself with all vehicle domains.

Based on the uniform electronics platform, which primarily consists of standard software modules, each vehicle manufacturer is then free to build its own specific content. They enable integration into the electronics network. These software functions permit differentiation between the competition.

Not only does software have to conform to the Autosar standard. The electronic control units must be built in such a way that the Autosar software is able to run on them. The Autosar members are hoping that the new development methods yield such benefits as shorter development times and lower development costs.

Until now, it was often the case that dedicated electronic control units would be developed and fitted for new functions (e.g. electronic transmission control, antilock brake system, air conditioning). The number of electronic control units fitted in the vehicle grew continuously; today's generation of vehicles are equipped with between 20 and 80 electronic control units. In future vehicle generations, it is intended that all functions be covered by a network of 10 to 20 electronic control units. Some of these will function a little like main computers that will bundle the important function groups together. These include the drivetrain, suspension management system, body and interior and the multimedia/telematics domain. On data buses, sensors with integrated electronics output processed and verified signals, while the buses carry the relevant control commands to actuators with integrated triggering electronics.

In future, new functions will often be able to use the existing computer architecture up to its performance limit and will be widely realized in the form of a software add-on. This would therefore render

unnecessary the additional electronic control unit that would have been required today. The system only needs to be supplemented by the sensors and actuators required.

Software will no longer be an inevitable component of hardware, but will increasingly become a stand-alone product. The first examples of business and collaborative models between supplier and manufacturer have already been put into practice at Bosch, e.g. in drivetrain management.

Examples

Individually-controlled drive components at each of the wheels with different wheel positions and wheel loads permit optimum use of tire force potential. This results in increased driving dynamics and safety while at the same time reducing consumption, wear and emissions. For this to be possible, all active elements in the drivetrain, suspension and steering must be networked.

One example of superordinate functions realized by networking is the ASIS (Active Shift Strategy) drive strategy for automatic transmissions in passenger cars. Based on the evaluation of various control elements (e.g. accelerator pedal) and conclusions drawn from the information of other systems (e.g. cornering detection from the wheel speeds), this strategy is able to control the gearshift in such a way as to meet the driver's real-time demand for agility and power through selection of the appropriate gear. In future, the telematics will be able to support additional, improved driving strategies, e.g. through the use of GPS signals for transmission control in terms of predictive driving.

Outlook

Increasingly greater demands for safety, reliability and availability are being placed on the network architecture of modern vehicles. This is where energy network architecture will play a key role. In the face of vehicle functions increasingly being realized electronically, this architecture makes for a reliable supply of power to systems and thus forms the basis for the reliability and safety of future systems. One possible future technology is the transmission of power and information on the supply line. The following benefits arise from the powerline communication (PLC) concept used in the public grid:

- Weight and cost reductions as well as space savings from the discontinuation of data lines
- Easier retrofitting for retrofit systems (spare parts trade)
- Reduction in complexity of the wiring harness in respect of manufacture and installation
- Increase in system safety, especially in mechanically stressed zones (e.g. door, mirror) that are characterized by premature aging of lines and increased risk of failure
- Powerline as a redundancy path for systems relevant to safety
- Simultaneous, or parallel, implementation of several bus systems or services, e.g. diagnostics

Electronic control unit

Digital technology furnishes an extensive array of options for open and closed-loop control of automotive electronic systems. A large number of parameters can be included in the process to support optimal operation of various systems. The control unit receives the electrical signals from the sensors, evaluates them, and then calculates the triggering signals for the actuators. The control program, the "software", is stored in a special memory and implemented by a microcontroller. The control unit and its components are referred to as hardware. The Motronic control unit contains all of the algorithms for open and closed-loop control needed to govern the engine-management processes (ignition, induction and mixture formation, etc.).

Operating conditions

The control unit is subjected to very high demands with respect to

- Extreme ambient temperatures (in normal vehicle operation from –40 to +60...+125 °C)
- Extreme temperature changes
- Indirect materials and supplies (oil, fuel etc.)
- The effects of moisture and
- Mechanical stress such as vibration from the engine

The control unit must operate reliably when the vehicle is started with a weak battery (e.g. cold start) and with high charge voltages (vehicle electrical system fluctuations).

Other requirements arise from the need for EMC (ElectroMagnetic Compatibility). The requirements regarding immunity to electromagnetic interference and limitation of high-frequency interference signal emission are extremely stringent.

Design

The printed circuit board with the electrical components (Fig. 1) is installed in a housing of plastic or metal. A multiple plug connects the control unit to the sensors, actuators and electrical power supply. The high-performance driver circuits that provide direct control of the actuators are specially integrated within the housing to ensure effective heat transfer to the housing and the surrounding air.

The majority of the electrical components are of the surface-mounted device technology type. This concept provides extremely efficient use of space in low-weight packages. Only a few power components and the connectors use push-through assembly technology.

Hybrid versions combining compact dimensions with extreme resistance to thermal attack are available for mounting directly on the engine.

Data processing

Input signals

In their role as peripheral components, the actuators and the sensors represent the interface between the vehicle and the control unit in its role as the processing unit. The electrical signals of the sensors are routed to the control unit via a wiring harness and the connector plug. These signals can be of the following type:

Analog input signals

Within a given range, analog input signals can assume practically any voltage value. Examples of physical quantities which are available as analog measured values are intake-air mass, battery voltage, intake-manifold and boost pressure, coolant and intake-air temperature. They are converted into digital values by an analog-digital converter in the microcontroller

of the control unit and used for calculations by the microcontroller CPU. The maximum resolution of these analog signals is 5 mV. This translates into roughly 1,000 incremental graduations based on an overall measuring range of 0 to 5 V.

Digital input signals
Digital input signals only have two states. They are either "high" or "low" (logical 1 and logical 0 respectively). Examples of digital input signals are on/off switching signals, or digital sensor signals such as the rotational-speed pulses from a Hall generator or a magnetoresistive sensor. Such signals are processed directly by the microcontroller.

Pulse-type input signals
The pulse-shaped input signals from inductive-type sensors containing information on rotational speed and reference mark are conditioned in their own control unit stage. Here, spurious pulses are suppressed and the pulse-shaped signals converted into digital rectangular signals.

Signal conditioning
Protective circuits limit the voltages of input signals to levels suitable for processing. Filters separate the useful signal from most interference signals. When necessary, the signals are then amplified to the input voltage required by the microcontroller (0 to 5 V).

Signal conditioning can take place completely or partially in the sensor depending upon the sensor's level of integration.

Signal processing
The control unit is the switching center governing all of the functions and sequences regulated by the engine-management system. The closed and open-loop control functions are executed in the microcontroller. The input signals that are provided by the sensors and the interfaces to other systems (such as the CAN bus) are used as input variables. and are subjected to a further plausibility check in the computer. The control unit program supports generation of the output signals used to control the actuators.

| 1 | Design of a control unit using the example of an ME Motronic (sectional view through housing cover) |

UAE0992Y

Output signals

The microcontroller uses the output signals to control output stages that usually provide enough power for connecting the actuators directly. It is also possible to actuate certain output stage relays for consumers that use up a great deal of power (e.g. motor fans).

The output stages are proof against short circuits to ground or battery voltage, as well as against destruction due to electrical or thermal overload. Such malfunctions, together with open-circuit lines or sensor faults are identified by the output-stage IC as an error and reported to the microcontroller.

Switching signals

Actuators can be switched on and off using the switching signals (e.g. motor fans).

PWM signals

Digital output signals can be in the form of PWM (Pulse-Width Modulated) signals. These are constant-frequency rectangular signals with variable on-times (Fig. 2), Various actuators can be moved to various operating positions using these signals (e.g. exhaust-gas recirculation valve, boost-pressure actuator).

Control unit-internal communication

In order to be able to support the microcontroller in its work, the peripheral components must communicate with it. This takes place using an address/data bus which, The microcontroller outputs the RAM address whose contents are to be read (for example) via the address bus. The data bus is then used to transmit the relevant data. For former automotive applications, an 8-bit bus topology sufficed. This meant that the data bus comprised 8 lines which together could transmit 256 values simultaneously. 65,536 addresses can be accessed using the 16-bit address bus in this system. Presently, more complex systems demand 16 bits, or even 32 bits, for the data bus. In order to save on pins at the components, the data and address buses can be combined in a multiplex system, i.e. addresses and data are dispatched through the same lines but offset from each other with respect to time.

Serial interfaces with just a single data line are used for data that does not have to transmitted extremely quickly (e.g. fault memory data).

EOL programming

The extensive variety of vehicle variants with differing control programs and data records, makes it imperative to have a system which reduces the number of control unit types needed by a given manufacturer. To this end, the Flash-EPROM's complete memory area can be programmed at the end of production with the program and the variant-specific data record (this is the so-called End-of-Line, or EoL, programming).

Another way of reducing the type diversity is to store several data variants (e.g. transmission variants) in the memory, which are then selected using coding at the end of the production line. This coding is stored in an EEPROM.

2 PWM signals

Signal voltage →

Time →

UAE0738E

Fig. 2
a Period duration (fixed or variable)
b Variable on-time

The performance of electronic control units goes hand-in-hand with advances achieved in the field of microelectronics. The first gasoline injection systems were still analog – with limited flexibility in the implementation of control functions. These functions were constrained by the hardware.

Progress advanced in quantum leaps with the arrival of digital technology and the microcontroller. The entire engine management system was taken over by the universally applicable semiconductor microchip. The actual control logic in microcontroller-controlled systems is in a programmable semiconductor memory.

From systems that initially simply controlled fuel injection, complex engine-management systems were then developed. They controlled not only fuel injection but also the ignition system including knock control, exhaust-gas recirculation and a whole variety of other systems. This continuous process of development is bound to continue in a similar vein over the next decade as well. The integration of functions and, above all, their complexity are constantly increasing. This pattern of development is only possible because the microcontrollers used are also undergoing a similar process of improvement.

Microcontrollers in the Intel 8051 family were used quite some time until they were replaced with 80515 derivatives with additional I/O facilities for timer-controlled signals and an integrated analog-digital converter at the end of the 1980's. It was then possible to create relatively powerful systems. Figure 3 shows a comparison between the performance of a fuel-injection system (LH3.2) and an ignition system (EZ129K) – equipped with 80C515 controllers – and that of the succeeding Motronic systems. The ME7 has approximately 40 times the performance capability of the LH/EZ combination with a clock frequency of 40 MHz. With the benefit of a new generation of microcontrollers and a further increase in clock frequency on the ME9, this figure will increase to a factor of well over 50.

In the foreseeable future microcontrollers will process more than just digital control sequences. Signal processors are integrated that can also directly process the signals provided by knock sensors, for example.

Advances in the development of semiconductor memory chips are also worthy of note. Complex control programs require an enormous amount of memory space. The capacity of memory chips at the start of the 1980s was still only 8 kilobytes. The ME7 now uses 1-megabyte chips and soon memory capacities of 2 megabytes will be required. Figure 3 shows this pattern of development and likely future trends.

Fig. 3
Chart illustrating
▶ Performance capability of engine-management systems
▶ Number of control unit connector pins
▶ Program memory capacity
▶ Data memory capacity (RAM)

By way of comparison: The performance capability of a state-of-the-art engine-management system far exceeds that of Apollo 13.

3 Development of electronic control units

SMK1930E

Digital modules in the control unit

Microcontroller

Structure

A microcontroller consists of the following interacting components (Fig. 1):

- *Central processing unit* (CPU): this contains the control unit and the arithmetic and logic unit. The control unit executes the instructions from the program memory, whereas the arithmetic and logic unit performs arithmetical and logical operations.
- *Input and output devices* (I/O, Input/Output), which handle the exchange of data with peripheral devices. Peripheral devices include input and output devices and external data storage media.
- *Program memory*, in which the operating program (user program) is permanently stored (ROM, PROM, EPROM or flash EPROM).
- *Data memory*, which is accessed for reading and writing (RAM). This contains the data that is currently being processed. Non-volatile memory (EEPROM) is used for data that must not be deleted when the supply voltage is switched off.
- The *bus system* connects the individual elements of the microcontroller.
- A *clock generator* (oscillator) ensures that all operations in the microcontroller take place within a defined timing pattern.
- *Logic circuits* are modules with specialized tasks such as program interrupts. They are integrated in individual I/O units.

The chief components of a microcomputer are generally separate modules connected to one another on a printed-circuit board. The microprocessor within such as system – the CPU – is not functional on its own: it is always part of a microcomputer.

In a microcomputer, however, the above-mentioned functions are integrated on a silicon wafer (system-on-a-chip). This is not functional on its own (standalone) and is therefore referred to as a single chip microcomputer.

The microcontroller is used to control self-regulating systems such as an engine-management system. Depending on the application, they may also have expansion modules connected to them (e.g. additional memory for data and program code).

The user program is fixed and is not replaced for different applications. This is the difference between a microcontroller system, for example, and a PC.

Programming

The only command form capable of direct interpretation by a microprocessor is a bit pattern, i.e. the binary representation of a number. Since, however, this form of instruction is not easy to work with for a programmer, and is therefore susceptible to errors, easily memorable abbreviations (mnemonics) are used. These are automatically translated by an assembler program into bit patterns (machine code) that can be understood by the microprocessor.

For more complex systems and programs, high-level programming languages such as C are needed, as otherwise it would be impossible to keep extensive programs manageable and free of errors. Such languages require sophisticated translation programs (compilers) which convert the text of the high-level language into a form that can be processed by the microcontroller.

The machine code is stored in the program memory, where it remains permanently. The CPU accesses these components via the bus system, reads the numerically coded commands and executes then.

1 Microcontroller

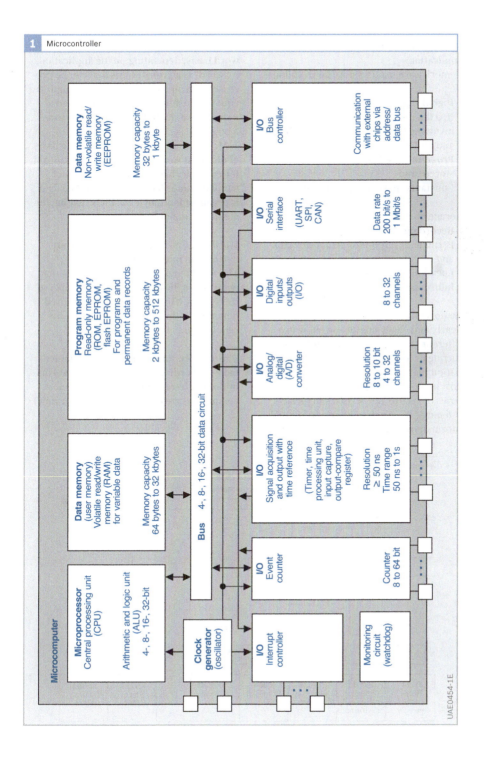

UAE0454-1E

Semiconductor memories
Applications
Memories are used to store large volumes of
- Digital signals representing data (I/O data, statuses, intermediate results involving frequent and rapid reading and writing)
- Program code (usually permanently stored) and
- Constants (permanently stored)

Storage involves
- Recording (writing)
- Permanent retention (actual storage) and
- Location and retrieval (reading) of information

Memories utilize physical effects that clearly and easily create and show two different states (e.g. conducting/non-conducting or loaded/not loaded). The information that is to be stored must therefore exist in binary form, i.e. encoded as a series of "yes or no" statuses (logical "1" or logical "0"). Such a "yes or no" unit of information is called a bit (binary digit).

Memory modules are organized on a bit or word basis, depending on the application. A "word" is a group of bits that can be processed as a single unit. The word length is equal to the number of bits processed as a single unit. Eight bits are referred to as a byte.

Memories can be organized on the basis of different word lengths. An 8 M x 8-RAM, for example, has a memory capacity of 8 million times 8 bits (64 Mbits). The data is organized into bytes (8 bits), making the memory capacity 8 Mbytes.

Word lengths of 4, 8, 16 and 32 bits are common in microcontroller systems. The word length is one of the factors that determines the performance capability of the system. The word length that is used depends on the performance capability requirements of the system.

The most important terms are explained below according to their standardized definitions, where applicable, or their most common usage (see Figure 2 for overview).

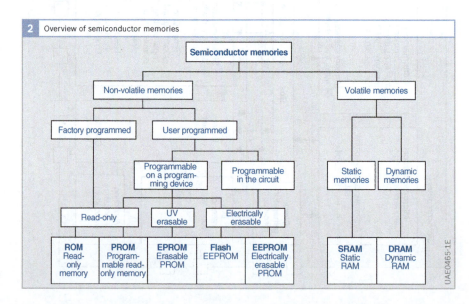

2 Overview of semiconductor memories

Random-access memory (RAM)

Random-access memory or RAM is a short-term memory that allows direct access to any storage location. Information can be read/written from/to the memory any number of times.

Static RAM (SRAM)

Static RAMs use bistable switching elements as the data storage cells. Their functionality is similar to that of a flip-flop, a simple circuit with two transistors, of which either the one (logical "1") or the other (logical "0") conducts at any one time. In SRAM, the information remains stored until the storage cell concerned is addressed and overwritten, or the operating voltage is switched off. SRAM is therefore volatile memory.

Dynamic RAM (DRAM)

Unlike SRAM, the information is stored as an electrical charge in the gate capacity of a CMOS transistor in dynamic RAM (DRAM). As such capacitors are susceptible to leakage, the charge is gradually lost. In order to retain the information, the charge has to be refreshed at regular intervals (every few ms).

Read-only memory

Read-only memory (ROM) is permanent-storage memory that allows any memory location to be accessed directly but - as the name indicates - allows the information only to be read and not modified.

A ROM is nonvolatile memory, i.e. the information it contains is retained even when the operating voltage is switched off. It is usually used to store program code (control programs) and fixed data (function tables, encoding rules, engine characteristic data maps) that need to be retrievable at any time. The information may be indelibly entered in the memory by the manufacturer or the user by means of appropriate programming of specially prepared memories (PROMs or programmable ROMs).

Erasable ROM

There are also ROMs whose contents can be erased and reprogrammed as outlined below.

EPROM (Erasable PROM)

This type of erasable read-only memory can have its contents completely wiped by irradiation with UV light and can then be reprogrammed using a programming device.

EEPROM (Electrical EPROM)

The EEPROM (also known as E^2PROM) can be electrically erased and reprogrammed. Every storage cell of an EEPROM be individually overwritten. For that reason, this type of memory module can also be used as nonvolatile data memory (e.g. for learned information in engine management systems).

Flash EEPROM

A more sophisticated variant of the EPROM and EEPROM is flash EEPROM. In this case, electrical flash pulses are used to erase specific storage areas or the entire contents of the memory. The erased areas can subsequently be reprogrammed.

The flash memory can be reprogrammed on a programming station. However, the advantage of flash EEPROM is that is can also be reprogrammed while still inside the sealed control unit.

Flash EEPROM is used in cases where relatively large quantities of data need to be stored, but must also be modifiable (e.g. program memory in vehicle control units).

Control unit software

Real-time capability

One of the requirements on electronic systems is real-time capability. This means that control procedures must react to input signals within an extremely short time. For example, a wheel that has a tendency to lock must be detected so quickly that the ABS control algorithm in the control unit can reduce the brake pressure via the hydraulic modulator quickly enough. Brake slip is therefore reduced before the wheel can lock. Engine management systems make considerable real-time capability demands so that crankshaft angles can be adhered to with extreme accuracy at fast engine speeds for injection and ignition timing purposes.

The complexity of an electronic system therefore makes extremely high demands of the software that is developed. The software structure is explained below on the basis of an example.

Software structure

The microcontroller in the control unit executes commands sequentially. The command code is obtained from the program memory. The time taken to read in and execute the command depends on the microcontroller that is used and the clock frequency. The microcontrollers that are currently used in vehicles can execute up to 1 million commands per second.

Because of the limited speed at which the program can be executed, a software structure with which time-critical functions can be processed with high priority is required.

The engine-management program has to react extremely quickly to signals from the speed sensor, which records the engine speed and the crankshaft position. These signals arrive at short intervals that can be a matter of milliseconds, depending on the engine speed. The control unit program has to evaluate these signals with high priority. Other functions such as reading in the engine temperature are not as urgent, since the physical variable only changes extremely slowly in this case.

Interrupt control

As soon as an event occurs that requires an extremely rapid response (e.g. speed sensor pulse), the program that is currently running must be interrupted. This can be done using the microcontroller's interrupt control facility. Events can trigger a program execution interrupt, whereupon the program jumps and executes the "interrupt routine". When this routine has been executed, the program resumes at the point at which it was interrupted (Fig. 1).

An interrupt can be triggered by an external signal, for example. Other interrupt sources are timers integrated in the microcontroller, with which timed output signals can be generated (e.g. ignition signal: microcontroller ignition output is switched at a point in time that is calculated beforehand). However, the timer can also generate internal time frames.

1 CPU power distribution principle

Tooth interrupt

Synchro interrupt

Ignition interrupt

Time frame

Background program

Time *t* ⟶

SAE1003E

Fig. 1
Depiction of several program levels on the example of the software from a Motronic system

The control unit program reacts to several of these interrupts. An interrupt source can therefore request an interrupt while another interrupt routine is currently being executed. Every interrupt source therefore has a fixed priority assigned to it. The priority controller decides which interrupt is allowed to interrupt another interrupt.

Tooth interrupt

The crankshaft is equipped with a pulse wheel (Fig. 2a) that has a certain number of teeth on its circumference. The teeth are scanned by the speed sensor. This allows the crankshaft position to be recorded. The typical distance between a pair of teeth on the crankshaft sensor wheel is 6°. In order to determine the crankshaft position, the control unit program must execute certain routines as each tooth is detected. At 6,000 rpm the detection time between two teeth is approximately 300 μs. Every command in these routines must be executed within this time. This requires a rapid response to the speed sensor signal. For this purpose, the engine-speed signal is connected to a microcontroller interrupt input. Every falling signal edge at this input interrupts the current calculations that are in progress and forces a branch to the interrupt routine. After executing the commands in the interrupt routine, the program continues execution at its point of origin.

In order to perform certain operations the control unit program requires the time taken for the crankshaft to travel between one tooth and the next. This calculation is performed by an internal timer. This is a freewheeling 16-bit counter (Fig. 3) that increments at a certain rate, depending on the microcontroller oscillator clock cycle. This time frame amounts to about 0.5 μs. When the falling tooth flank occurs, the current counter status is recorded. The difference (and therefore the tooth interval) is calculated using the stored counter from the previous tooth.

Example: crankshaft position calculation
The engine-management system (Motronic for gasoline engines, EDC for diesel engines) must know the crankshaft position at any given point in time. This is a prerequisite for injecting into the right cylinder at the right time and ensuring that ignition takes place at the calculated ignition angle (Motronic systems). In order to

2 Crankshaft sensor ring with speed sensor

a

b

Digital signal

SAE1004E

3 Time calculations via internal timer

Counter status

T_{n+1}
T_n
0

Time $t \longrightarrow$

Speed sensor signal

T_n: Counter status at tooth n
T_{n+1}: Counter status at tooth n+1

SAE1005E

Fig. 2
a Design
b Speed sensor signal

detect the engine position and the engine speed, the control unit evaluates the speed sensor signal (Fig. 2b).

There is gap in the crankshaft sensor wheel in which two teeth are missing. The tooth space has a defined position in relation to the top dead center (TDC) of cylinder no. 1. The control unit program has to synchronize itself with this tooth space. This is done by measuring the times between two consecutive falling tooth flanks. The time for the tooth space is considerably greater than the time before and after the gap. Following a "short – long – short" sequence the last thing to be scanned was the falling flank of the second tooth after the space.

The crankshaft has rotated by 6° for each falling tooth flank that has been detected by the control unit program. This is how the control unit program knows the crankshaft position within this time frame.

Since cylinder no. 1 is in the tooth space position in the vicinity of top dead center (TDC) or bottom dead center (BDC), an additional signal is required to determine the position. The camshaft sensor provides a different voltage level in both cases. The control unit is therefore able to uniquely assign the crankshaft and camshaft positions.

Combustion-synchronous interrupt

Some calculations have to be performed for every combustion cycle. For example, the ignition angle and the injection have to be recalculated synchronously with combustion for each cylinder. The program does this by branching to the "synchronization program" after certain teeth (Fig. 4). This interrupt takes place after every 30 teeth (ignition interval) for a four-cylinder engine, and after every 20 teeth for a six-cylinder engine.

The synchronization program is fixed to a certain tooth position and has to be executed with high priority. For this reason it is activated via an interrupt (triggered by a command in the tooth interrupt routine).

Since the synchronization program runs over several teeth at fast engine speeds, it has to be interrupted by the tooth interrupt. The tooth interrupt is given higher priority than the synchronization program.

Ignition interrupt

The ignition output takes place within a certain crankshaft range, depending on the value from the ignition map. Since the specified ignition angle has to be adhered to exactly, the ignition output is controlled by an interrupt. Like the synchronization program, the ignition interrupt is also called up once per combustion cycle.

The control unit program is aware of the crankshaft position in the 6° framework. However, this framework is not accurate enough for ignition angle output. For this reason, accurate ignition output between two teeth must take place as well as this approximate counting for the last 0 to 6 crankshaft degrees. This is done using a timer (Fig. 5). Ignition angle output that was purely timer-controller would lead to an ignition angle output error at high engine speed dynamics.

Firstly, the ignition coil must be enabled for a defined time (the so-called dwell period). In order to do this, the program calculates the switch-on time by calculating

4 Triggering the interrupt synchronously with combustion

Start synchronization program

Tooth counter elapsed
Synchronization program triggered

Second tooth after the gap: reference mark
Preset tooth counter for triggering the synchronization program

SAE1006E

backwards from the ignition angle at which the ignition coil has to be switched off. This makes it possible to calculate the tooth after which the ignition coil has to be switched off (approximate counting in 6° time frame). The remaining angle (detailed counting 0 to 6°) is converted into an output time using the current engine speed. As soon as the specified tooth position has been reached using approximate counting, a time is loaded with the output value from the detailed counting. When this time period expires, the timer triggers an interrupt. The commands that switch on the ignition coil are programmed in this interrupt routine. Then the timer is preset to the dwell period value, which causes an interrupt to be triggered when the timer elapses, switching off the ignition coil and therefore initiating ignition.

Time frame

Many control algorithms have to run within a certain time frame. Lambda control, for example, has to be processed within a fixed time frame (e.g. 10 ms) so that the correcting variables are calculated quickly enough.

5 Dwell and ignition time output

Ignition output

Switch on the ignition coil
Enter the time for ignition

Enter time for detailed counting

Synchronization program:
Calculate the dwell period and ignition angle
Preset the tooth counter for the dwell period
time (approximate counting)

SAE1007E

Background program

All other activities that do not run in an interrupt routine or a time frame are processed in the background program. At fast engine speeds, the synchronization program and the tooth interrupt are called frequently, leaving little CPU time for the background program. The time taken for a complete run-through of the background program therefore increases rapidly with the increasing engine speed. The background program must therefore only contain low-priority functions.

Software Development

A study of the current series projects together with the use of development capacities demonstrate that approximately 60% of the time spent on ECU development has to be devoted to creating the necessary software. For this reason, it is absolutely essential that modern tools and processes be used.

Development Process

Definition of Development Process

A depiction of the development steps in the form of a V-model (Figure 1) serves as the basis for all software-development activities. This model is used to detail the process steps which facilitate implementation within a product-development department.

Quality Assessment

Quality assessments are scheduled at defined points of the development process (Figure 2) for the purpose of process monitoring:

QA1
When: At the start of the project
What: Resource check (capacity, development environment, responsibility)

QA1F
When: Prior to function implementation
What: Function specification check

QA2F
When: After function implementation
What: Review of each individual function and checking of following documents:
- specification,
- function description,
- source code,
- data definitions, and
- test documentation.

QA2
When: Prior to software delivery
What: Review of all QA2F documents

QA3
When: Prior to start of series production
What: Series-production review for hardware and software

An essential part of the development process is also the distinction between specification and implementation. This separation allows programming by contract, whereby project teams use the software knowledge of ReUse teams who implement the functions (e.g. Keyword 2000 protocol) for various customers. To this end, the project teams write out function contracts which establish the boundary conditions of implementation. The scopes and levels of testing of the individual functions are determined in the project-specific quality plan (PQSP) with the customer requirements in mind.

1 Simplified V-model

Specification Test

Implementation

UTS0325E

2 Development process in detail

System initialization — System delivery
Function analysis specification — QA2
QA1
Function initialization — System integration/test
Function analysis specification — Function delivery
QA2F
QA1F — Function integration/test
Function development

UTS0326E

This also includes determining the QA x F scopes.

The PQSP is a central element of project implementation and should be fully discussed between vehicle manufacturer and supplier. It lays down, among others things, the responsibilities, the customer relationships, the development tools, the scopes of testing and documentation, etc.

The process in the ReUse team is set out as follows:

- The person in charge of the project formulates the task (if necessary, adoption of customer request).
- The function contract is drawn up (with details of task, project, desired date, scope, and reference documents) and given to the ReUse teams.

- The persons in charge of ReUse and the project discuss the application and establish the scopes and deadlines together.

- The person in charge of ReUse decides on the variant and version handling and implements the task.
- All documents are handed over to the person in charge of the project on completion.

In order to ensure that the created software can be reused to the greatest possible extent, there are C-programming guidelines binding an all programmers which are called up in the relevant reviews (e.g. QA2F).

Programming Guidelines

In any heterogeneous system of development that is spread over countries and continents, a standardized procedure for creating software is a vital and integral part of the time to market process. These guidelines address the following points and are binding for all programmers:

- general guidelines (terminology, vocabulary, variation handling),
- guidelines for software developments in C (templates, structure),
- definitions and declarations (include, defines, typedefs),
- check instructions (if, for, while, break, return ...),
- coding specifications and instructions (typecast, arithmetic, pointer),
- particular features when using variables (alignment, address),
- instructions on data consistency (preemptive), and
- instructions on resource relief.

These guidelines also serve as a source of knowledge for effective code configuration in order to counteract the limitations in relation to memory capacity and run time in the programming of microcontrollers.

Tools for Creating Software

As well as the formal aspects such as process and programming guidelines, it is crucially important to ensure that the tools are subject to constant support in the interests of product quality. Figure 3 provides an overview of the tools currently used for the various development phases. Significant features of this tool chain are:

- constant support throughout the entire development process and
- product-specific, optimized solutions with tools partly developed in-house.

As the wide variety of tools demonstrates, the process involved in creating the software for an ECU of the latest generation is highly complex. Figure 4 provides a simplified overview of the interplay between the individual tools from the specification through to the finished ECU program.

By way of example, two component parts of the tool chain will now be explained in closer detail:

- Design with ASCET-SD and
- Vehicle simulation with TCM-Simutec.

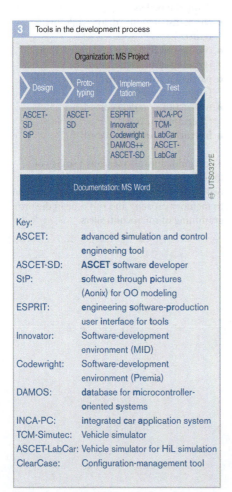

3　Tools in the development process

Organization: MS Project

Design | Proto-typing | Implemen-tation | Test

ASCET-SD StP	ASCET-SD	ESPRIT Innovator Codewright DAMOS++ ASCET-SD	INCA-PC TCM-LabCar ASCET-LabCar

Documentation: MS Word

Key:
ASCET:	advanced simulation and control engineering tool
ASCET-SD:	ASCET software developer
StP:	software through pictures (Aonix) for OO modeling
ESPRIT:	engineering software-production user interface for tools
Innovator:	Software-development environment (MID)
Codewright:	Software-development environment (Premia)
DAMOS:	database for microcontroller-oriented systems
INCA-PC:	integrated car application system
TCM-Simutec:	Vehicle simulator
ASCET-LabCar:	Vehicle simulator for HiL simulation
ClearCase:	Configuration-management tool

UTS0327E

4　Simplified process sequence

Design ASCET-SD	Function model Object model
Prototyping ASCET-SD	Source files
Code generation ESPRIT	ETC program
Test/application INCA/PC	

UTS0328E

5　Function design with ASCET-SD

UTS0329Y

Design with ASCET-SD

ASCET-SD (Figure 5) offers the following functions for designing software:
- interactive creation of function descriptions and function models,
- a graphical user interface,
- support of object-oriented design, data-flow-oriented design and state machines.

The ERCOSEK operating system is an integral part of the development environment, and facilitates real-time simulation of the function model.

ASCET-SD offers the following support for the rapid prototyping of functions:
- ASCET-SD operates as a bypass computer for the series ECU, i.e., individual ECU functions run on the PC, while the other functions continue to be executed by ETC (Figure 6).
- The connection is established via the CAN or the INCA probe.

The next step is the automatic C-code creation and the creation of the corresponding data files for the application from the models.

For further information, log on to

http://www.etas.de

Vehicle simulation with TCM-Simutec

To test the functions of a transmission-control system in the laboratory, there is a simulator for the vehicle and transmission environment which provides the input signals for ETC. One such simulator is shown in Figure 7.

The front panel of the simulator is equipped with assorted rotary potentiometers, switches, and pushbuttons which enable input variables such as output speed, selector-lever position, transmission temperature, etc., to be specified.

The top of the simulator accommodates a breakbox which permits access to every ECU pin. Measuring instruments can be easily connected to these sockets to enable, for example, a PWM signal of a pressure-regulator output to be viewed on an oscilloscope.

The laboratory car also contains computer cards which simulate the other ECUs in the vehicle network (e.g. ECUs for engine management, for ABS, etc.) and also their signals.

Process and Maturity Model

A clear definition of the development process and the corresponding implementation in the projects are made possible by a software development which can be evaluated with a maturity model such as CMM (capture maturity model).

| 6 | Test setup for ASCET-SD in bypass |

| 7 | TCM-Simutec (laboratory car) |

Fig. 6
1 ASCET-SD and INCA-PC
2 ASCET hardware (ETAS ES 1000.2)
3 ETK
4 ETC-Simutec (laboratory car)

8 Software layer model

Transmission software from
vehicle manufacturer or Bosch

Program
library

Component
driver

Operating system

Hardware

Software Structure

The software structure described in closer detail in the following is implemented within the transmission-control system. This layer model (Figure 8) comprises
- the application software (transmission software) with program library provided by the vehicle manufacturer or supplier (in this case by Bosch),
- the operating system,
- the component driver, and
- the hardware.

The separation of hardware and application software ensures that the software can be easily ported to new hardware platforms. Only the second layer, consisting of the operating system and component driver (BIOS), has to be adapted.
 The contents of the individual layers will now be broken down in the following:

The ECU hardware, the first layer of the software layer model, consists of the microcontroller (here, by way of example, the MPC555), the memory, the interfaces (SPI, CAN and UART), and the peripheral chips (ASICs):

CPU core timer and	SPI	TPU	MIOS	CAN (2)	UART (2)
Memory, hardware, driver, etc.					

The operating system with its services and the hardware-compatible software are implemented on this hardware:

ERCOSEK (OS)	EEPROM driver	Hardware Input/Output	KWP 2000 driver
	Device driver	Device driver	Device driver

The interface layer and program library for the application software contain:

Diagnosis handling	Diagnosis monitoring functions	Security software (SSK)
EEPROM handling	KWP2000 application	Shift by wire functions

The application software
(customized software) comprises:

e.g. ASIS (RB/ZF) AGS (BMW)	Transmission software
Software sharing (interface)	

Operating System

It is absolutely essential to use an OSEK-conforming operating system to fulfill the current real-time demands on an ECU. The ERCOSEK operating system from ETAS is used in Bosch transmission control units (available for all kinds of microcontroller).

An operating system is subdivided into processes and tasks (Figure 9):
 A *process* is a function which has no call or return parameters.
 A *task* consists of different processes and is characterized by
- the sequential execution of processes,
- the allocation of processes → task,
- each task being assigned a priority,
- tasks being assigned to a time base.

For task changing, there is either cooperative scheduling or preemptive scheduling (task management):

Cooperative Scheduling
In the case of cooperative scheduling, a task can only be interrupted between two processes by a higher-priority task (Figure 10).

The advantages of this procedure are low memory requirement (register banks, stack), simple management, and data consistency. The disadvantages are the limited response time (dependent on the process run time) and the jitter over the task period.

Preemptive Scheduling

Owing to the drawbacks of cooperative scheduling, preemptive scheduling is used in operating systems which operate as real-time systems.

With this form of scheduling, a higher-priority task can interrupt a lower-priority task at any time (Figure 11). The advantages of this procedure are the very short response times, the minimal jitter over the task period,

and a response time that is not dependent on process implementation. The disadvantages are increased memory requirement (stack, register banks) and data-consistency problems.

Mixed Scheduling

ERCOSEK offers the option of mixing both types of scheduling in *one* application. A combination of hardware and software scheduling serves this purpose. Figure 12 shows the distribution between cooperative and preemptive using the priorities assigned to the tasks.

A software call starts the operating system. It can support different application modes (e.g. different task sets for initialization, oper-

9 Processes and task

11 Preemptive task change

10 Cooperative task change

12 Priority distribution

13 Application-mode change

Application

Mode n Mode n+1

Mode

Init | Execution Init | Execution

Zeit *t* →

STS0336E

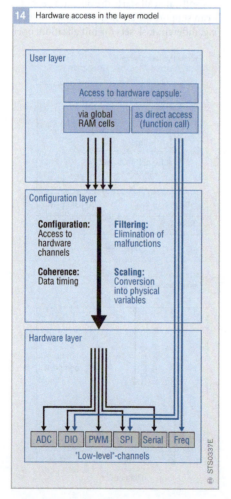

14 Hardware access in the layer model

User layer

Access to hardware capsule:

via global RAM cells | as direct access (function call)

Configuration layer

Configuration: Access to hardware channels

Filtering: Elimination of malfunctions

Coherence: Data timing

Scaling: Conversion into physical variables

Hardware layer

ADC | DIO | PWM | SPI | Serial | Freq

"Low-level"-channels

STS0337E

ation and ECU run-on, Figure 13). Each application mode consists of an initialization phase and an execution phase. Interrupts are prohibited during initialization of an application mode.

Further documents on the subject of ERCOS[EK] / OSEK can be found on the Internet at:

http://www.etas.de
http://www.osek-vdx.org

Acquisition of Input and Output Variables

Access to the hardware is obtained within the framework of the software layer model in accordance with three layers (Figure 14):
- user layer,
- configuration layer, and
- hardware layer.

The first implementation example to be featured is access (A) via global RAM cells. Table 1 describes the name of the RAM cell, the signal direction (input or output), the signal type (analog, digital, frequency, or PWM), the description, and the physical conversion.

The second example featured is access via (B) function interfaces. Table 2 is structured along the same lines as Table 1, but here the global RAM cell is replaced by a function call.

The objective in the configuration layer is to obtain independence from platform and project in the conversion of the hardware accesses into real software. This is achieved on the one hand by using tools which automatically create the C-code for access to the hardware, and on the other hand by using C-macros which are then resolved on a processor-specific basis.

1	Hardware access via global RAM cells				
RAM cell	**I/O**	**Type**		**Content**	**Scaling**
ugt_Batt	In	ANA	16 bit	Battery voltage	0...25 000 mV
CGT	In	ANA	8 bit	Oil temperature	−40...+215°C
ccu_Chip	In	ANA	8 bit	Substrate temperature	−40...+215°C
fgt_Fet	Out	DIG	8 bit	Status HSD-Fet	0.. 1 (On/Off)
fpo_L1	In	DIG	8 bit	Selector lever pos 1	0..1 (On/Off)
fpr_PinM	In	DIG	8 bit	M button	0..1 (On/Off)
NAB	In	FREQ	16 bit	Output speed	0...20 000 rpm
NAB32	In	FREQ	8 bit	Output speed/32	0...255 rpm/32
NTU	In	FREQ	16 bit	Turbine speed	0...20 000 rpm
NTU32	In	FREQ	8 bit	Turbine speed/32	0...255 rpm/32
hmv1	Out	PWM	16 bit	Solenoid-valve output	0...1000 per mil
idr1s	Out	ANA	16 bit	Nominal current pressure regulator	0..12 000 mA

Table 1

2	Hardware access via functions			
Software function	**I/O**	**Return value**	**Content**	**Scaling**
GetHWIO_U_IgnRunCrnk()	In	ANA 16 bit	Battery voltage	0...32 V
GetHWIO_T_TransOil()	In	ANA 16 bit	Oil temperature	−40...+215°C
GetHWIO_b_HSD()	In	ANA 8 bit	Status HSD-Fet	0...1 (on/off)
GetHWIO_e_TapUpDwnReq()	In	ENUM 8 bit	Tip (+/−) function	0 x 00...0 x 40
TsHWIO_PRNDL GetHWIO_s_PRNDL(void)	In	DIG 8 bit	Transmission control panel	0...1
TsHWIO_FreqParams GetHWIO_s_NTU(void)	In	FREQ struct	Turbine speed	Time stamp + counter value
TsHWIO_NAB_DualEdgeParams GetHWIO_s_NAB_DualEdge()	In	FREQ struct	Output speed, edge can be changed over	Time stamp + counter value + operating edge
SetHWIO_e_NAB_DualEdgeCptr Mode(BYTE)	Out	–	n_{out} Edge changeover	Rising, falling, both

Table 2

A further signal-acquisition component involves the exchange of signals via the communication interface. The CAN bus (controller area network) has gained acceptance in this field in the last few years.

CAN replaces the conventional wiring harness or the previously standard network of ECUs (Figure 15). The bus system must satisfy the following requirements here:
Real-time updating for
safety functions: 10 ms
Convenience functions: 10...100 ms
Maximum cable length: 40 m

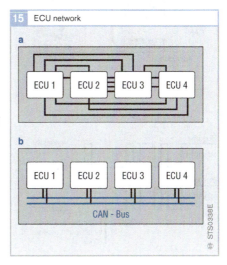

Fig. 15
a Conventional
b With CAN

3	Bit rates as a function of cable (bus) length	
Maximum bit rate		**Bus length**
kbit/s		**m**
1000		40
500		100
250		250
125		500
40		1000

Table 3

The system must also be resistant to temperature and moisture. The CAN bus has also gained acceptance in the field of automation technology. Table 3 lists the maximum possible data rates for different cable lengths.

Figure 16 shows the circuit-engineering implementation of the CAN interface in an ECU.

In the microcontroller itself, message handling is conducted via a dual-port RAM (Figure 17). Since this RAM chip, as the name suggests, can be described from two sides (CAN transceiver and microcontroller), the CPU workload is substantially relieved for signal transfer.

Complete arbitration (message organization, who sends what when) on the CAN bus is performed automatically by the CAN transceiver. It does not require any computing power in the microcontroller (Figure 18). A distinction is made within CAN messages between standard and extended data frames (Figures 19 and 20).

16 CAN interface

17 Use of dual-port RAM with the CAN bus

18 CAN-bus arbitration

19 CAN standard data frame

20 CAN extended data frame

The standard data frame is characterized by the following data (Figure 19):

Data capacity: 0...8 bytes
Identifier length: 11 bits
Message length: max. 130 bits

In contrast, the extended data frame (Figure 20) has the following characteristic data:

Data capacity: 0...8 bytes
Identifier length: 29 bits
Message length: max. 150 bits

Gear Selection and Adaptive Functions
Transmission control has undergone various phases or expansion stages within the framework of development.

The basic functions and the adaptive programs for shifting points and pressure control are now standard in the field of electronic transmission control (ETC). Where the various marques and vehicles differ is in the different strategies employed in the automatic adaptation of the transmission to the driving style and the traffic situation. This also represents an area of software which is being increasingly taken up by the vehicle manufacturer directly and which is no longer in the hands of a supplier. The adaptive functions for shifting-point control and pressure control have already been discussed in the chapter sections entitled Shifting-Sequence Control and Adaptive Pressure Control.

The following text will now deal with Bosch-specific implementations of automatic adaptations (learn functions). The adaptive shifting strategy determines a gear cyclically from the driver command, the vehicle status, and the driving situation. It is adaptive in relation to the driver type (sportiness) and also takes into account automatic or manual gear preselections (tip/nudge operation, as is familiar, for example, from Porsche Tiptronic© or BMW Steptronic©). The complete software package has been modeled with an object-oriented approach for optimum reuse.

Object-Oriented Approach
Vehicle Control
As an introduction, Figure 21 shows the 1-2 US and 2-1 DS shift curves for a driving program.

The shift curves shown in Figure 22 extend this system for different driving programs from super economy (XE) to super sport (XS). They clearly show that the upshift point in the sporty driving program moves towards higher vehicle speed or higher engine speed and thereby achieves optimum utilization of engine performance.

21 1-2 US shift curve

22 1-2 curve with several driving programs

Fig. 21 and 22
1 Upshift

To select a shift curve using the driver type and the total running resistance, it is necessary for them first to be recorded and evaluated once. The *overall structure of vehicle control* shown in Figure 23 serves this purpose.

On the one hand, there are the variables which determine the vehicle and its status:
● transmission control panel (TCP),
● the transmission itself,
● the engine,
● the accelerator-pedal position, and
● the vehicle variables (e.g. vehicle speed, wheel speed, etc.).

On the other hand, there are the variables which are crucial to shifting-point selection:
● driver type,
● driving situation, and
● driving program.

This is made clearer with the newly arranged graphic *overall structure of gear selection* in Figure 24. Each of these variables is then further divided into different sub-evaluations.

The *driver-type evaluation* specifies whether the current driving style is economical or sporty. The *driver-type determination* as featured in Figure 25 can be shown for this purpose.

The result of the driver-type determination is a *driver-type counter* (Figure 26) with an allocated driving program (XE to XXS).

Driver-type evaluation is followed by *hill recognition* (based on the vehicle running resistance), which distinguishes between different types of uphill and downhill driving with the following allocation (Figure 27):

B0 Downhill 2
B1 Downhill 1
B2 Level
B3 Uphill 1
B4 Uphill 2

23 Overall structure of vehicle control

24 Overall structure of gear selection

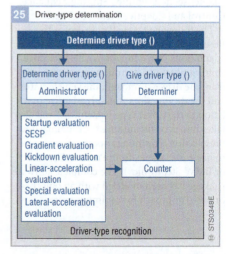

25 Driver-type determination

This is followed by determination of the *driving situation* (Figure 28).

Ratio Criteria (RC)

Driver type and driving situation form one of three ratio criteria (RC), all of which demonstrate different functions:

RC Driver Type

RC driver type makes a gear suggestion using a shift curve, depending on the relevant driver type. The driver thus serves as the *ratio supplier (RS)*.

RC FastOff

RC FastOff prevents upshifts if FastOff has been detected, i.e., a *shift prevention (USP, DSP) takes place.*

RC TCP

RC TCP changes the order of priority of the ratio supplier in accordance with operation of the transmission control panel (TCP). An RC selection adapted to the *shift situation (SS)* takes place.

Summary

The following features can be summarized for the object-oriented adaptive driving strategy:
- cyclical determination of the gear,
- consideration of driver command, vehicle status, driving situation,
- adaptive in relation to driver type (sportiness),
- division into static and dynamic driving programs,
- automatic and manual gear preselection (tip/nudge operation),
- new, flexible prioritization procedure, and
- object-oriented structure.

27 Hill recognition

26 Driver-type counter

28 Driving situation

Diagnostic Functions

As well as the actual transmission-control functions, monitoring functions are taking up an ever-increasing percentage of the software. At present, approximately 30% of the software stored in the ECU is assigned to diagnostics.

Full diagnostics serves to enhance driving safety and also with the aid of substitute functions to increase system availability. Diagnostic management must make the following functions available:

- management of the fault memory,
- CARB fault management, applicable to each individual fault code (only specific faults alter the vehicle's exhaust characteristics and are therefore relevant to CARB),
- filtering (time- and event-controlled),
- requirement of measures (substitute functions, limp-home),
- provision of data for the application of diagnostic management for each individual fault type,
- monitoring functions call up diagnostic management; this therefore only has a few calls in the task lists.

Fault Memory

The fault memory is divided into different subsections, which are handled differently in relation to the memory location and according to the input of the "ignition off" signal.

Primary Fault Memory (PFM)

The primary fault memory (PFM) exhibits the following features:

- saved in the non-volatile memory (EEPROM),
- typically 10 memory locations,
- contains fault code/type, environmental conditions, CARB/"warm-up" counter, and flags.

Secondary Fault Memory (SFM)

The secondary fault memory (SFM) exhibits the following features:

- one memory location for each fault code (only in the volatile RAM),
- contains filters, time stamps, and assorted flags,
- controls the internal program run via the flags.

Backup Fault Memory (BFM)

The backup fault memory (BFM) exhibits the following features:

- optional use,
- ring memory, saved in the non-volatile memory,
- typically 5 to 10 memory locations,
- contains fault entries which have been deleted from the PFM.

Snapshot Memory

The snapshot memory is optional and contains further environmental conditions for the first PFM entry.

The entries in this fault memory can be read out in the workshop with the diagnosis tester, and provide important information relating to vehicle repair.

Monitoring Functions

The following text deals with the most important monitoring functions in a transmission-control system:

Solenoid-Valve Monitoring

The following conditions apply to monitoring a solenoid valve in the transmission:

- activation in PWM mode,
- alternatively on/off function (100%/0% PWM),
- PWM with fundamental frequency 1 kHz,
- pulse-to-no-current ratio dependent on the supply voltage,
- jump from approx. 95% to 100% in PWM mode,
- analog feedback for diagnosis, separate for "on" and "off" states,
- general test condition: battery voltage ≥ threshold (7 V),
- fault detection as per Table 4.

4	Solenoid-valve monitoring			
PWM (%)	SCGO	SCG	O	SCP
0	–	HW	HW	–
0...5	–	SW	SW	–
5...95	–	SW	SW	SW
95...100	–	–	–	SW
100	HW	–	–	HW
Key:				
PWM	Pulse width modulation			
SCGO	Short circuit to ground or open circuit			
SCG	Short circuit to ground			
O	Open circuit			
SCP	Short circuit to positive			
HW	Diagnosis by hardware			
SW	Diagnosis by software			

A further method of solenoid-valve monitoring is ISIG evaluation (inductive **sig**nature). This method serves to monitor the voltage characteristic at the solenoid valve and evaluates the drop (U_{ISIG}) which occurs when the spool is moved (Figure 29). The objective here is to monitor the function of the on/off valve. Since the voltage drop is only very small and very short (t_{ISIG}), a special evaluator circuit must be used here.

Pressure-Regulator Monitoring
The pressure regulator must be permanently monitored because its function is crucial to the function of the transmission.

Program-Run Check (PRC)
The program-run check (PRC) ensures that the following incidents are detected:
● mix-up of components,
● double execution of code parts, and
● skipping of code parts.

Each module or each safety-relevant code part must have a check point at the beginning and the end in order to ensure a correct run with maximum high probability. The numbers represent the relevant module (function or process) and range from 0 to 9. The checksum algorithm (MISR procedure) ensures the sequence relationship.

Each reference (every 10, 20, and 30 ms reference) has its own enumerator and delivers its own part answer. The last checked point in

a PRC section must be followed by a "complete check" which generates the relevant PRC part answer with the aid of correction values. The part answers are XORed and form the complete part answer of the PRC to the question/answer communication. The current question serves as the input variable into the PRC (or, more precisely, into each of the PRC parts), i.e., the program run is monitored on a question-dependent basis. A dummy must simulate the PRC-monitored points during the initialization phase because these points are not run through during initialization.

If faults occur in the PRC, i.e. are detected, an incorrect complete part answer of the PRC is formed, which for its part causes an incorrect complete answer to be sent to the monitoring module (watchdog in the external ASIC, see also section entitled "ASIC"). This increases its fault counter by 1 (a fault-free complete answer results in a decrease to the minimum fault-counter content 0). The monitoring module shuts down the driver stages when the fault-counter content 5 is reached and initiates a reset at fault-counter content 7.

Table 4

29	ISIG signal characteristic

$t_{ISIG} \approx 10$ ms

$U_{ISIG} \approx 10$ mV

Voltage U

Time t →

STS0352E

Fig. 29
U_{ISIG} Voltage drop
t_{ISIG} Time interval for voltage drop

Basic principles of networking

With the tremendous speed at which computer technology is advancing, the number of electronic systems in use is increasing more and more. This growth is also continuing in automotive engineering. However, this also means that the complexity of an overall system (the vehicle in this case) is on the increase. Individual systems such as engine management have been improved over the last few years. However, innovations are mainly achieved by means of interaction between several individual systems. The individual components need to be networked so that the multitude of information that is managed by the individual systems can also be used elsewhere throughout the system. Different communication systems are used depending on requirements (e.g. transmission reliability, fault tolerances, costs).

Network topology

A network is understood to be a system in which a group of elements can exchange information via a transportation medium. If the elements are visualized as nodes and the communication relationships as lines, a picture is created of a network where many nodes are related to several other nodes. The nodes in a communication network are also often referred to as network subscribers or stations.

In motor vehicles, complex control units such as those for the engine management system (Motronic or electronic diesel control, EDC), the electronic stability program (ESP), the transmission control system or the door modules can be network subscribers (Fig. 1). However, a sensor with a conditioning circuit that merely prepares and digitizes a measured value can also act as a network subscriber and make the measured signals available to other net-

1 Vehicle system networking

UVA0014-1Y

work subscribers. The transport medium via which the communication takes place is referred to as a bus or a data bus.

A network topology is understood to be the structure consisting of network nodes and connections. This merely shows which nodes are interconnected, but does not depict underlying details such as the length of the connection. Every network subscriber must have at least one connection to another network subscriber in order to participate in network communication. Different requirements are made of the network topology for a variety of communication network applications, while the topology determines some of the characteristics of the overall network. All network topologies are based on the following four basic topologies
 Bus topology
 Star topology
 Ring topology and
 Mesh topology

Other structures (hybrid topologies) can be created by combining these basic topologies.

Bus topology

This network topology is also referred to as a linear bus. The core element of a bus topology is a single cable to which all nodes are connected via short connecting cables (Fig. 2). This topology makes it extremely easy to add other subscribers to the network. Information is transmitted by the individual bus subscribers in the form of so-called messages and distributed over the entire bus.

Nodes transmit and receive messages. If a node fails, the data that is expected from this node is no longer available to the other nodes on the network. However, the remaining nodes can continue to exchange information. However, a network with a bus topology fails completely if the main line is defective (due to a cable break, for example).

Star topology

The star topology consists of a main node (repeater, hub) to which all other nodes are coupled via a single connection (Fig. 3). A network with this topology is therefore easy to extend if free capacity is available (connections, cables).

2 Linear bus topology

SVA0002Y

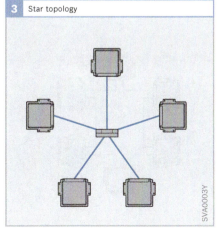

3 Star topology

SVA0003Y

In star topologies, data is exchanged between the individual node connections and the main node, whereby a distinction is made between active and passive star topologies. In active star topologies, the main node contains a computer that processes and relays information. The performance capability of a network is essentially determined by the performance capability of this computer. However, the main node does not have to have special control intelligence. In passive star systems, it merely connects the bus lines of the network subscribers together.

The following applies to active and passive stars: if a network subscriber fails or a connecting line to the main node is defective, the rest of the network continues to operate. However, if the main node fails the entire network is disabled.

In the automotive area, star structures are under discussion for safety and security systems such as brakes and steering. In this case, the risk of a complete network failure is prevented by designing the main node to be physically redundant. This means that several main nodes are used to which the nodes whose information is needed for safe operation of the vehicle can be connected in parallel.

Ring topology
In the ring topology, each node is connected to its two neighbors. This creates a closed ring (Fig. 4). A distinction can be made between single rings and double rings.

In a single ring, data transfers are unidirectional from one station to the next. The data is checked when it is received. If the data is not intended for this station it is repeated (repeater function), boosted and relayed to the next station. The data that is being transferred is therefore relayed from one station to the next in the ring until it has reached its destination or arrives back at its point of origin, where it will then be discarded.

As a soon as a station in a single ring fails, the data transfer is interrupted and the network breaks down completely.

Rings can also be set up in the form of a double ring (e.g. FTTI), in which the transfer of data is bidirectional. In this topology, the failure of a station or a connection between two stations can be overcome, since all data is still transferred to all operational stations in the ring.

However, if several stations or connections fail, the possibility of a malfunction cannot be ruled out.

4 Ring topology

SVA0004Y

5 Mesh topology

SVA0005Y

Mesh topology

In a mesh topology, each node is connected to one or more other nodes (Fig. 5). In a fully meshed network, each node is connected to every other node.

If a mode or connection fails it is possible for the data to be rerouted. This type of network therefore has a high degree of system stability. However, the cost of networking and transporting the message is high.

Radio networks form a type of mesh topology, since the transmissions from each station are received by every other station that is within range.

A mesh topology is bus-like as far as exchanging messages is concerned, and star-like regarding data transfers, since every station receives all transmissions from every other station, but connection failures can be overcome.

Hybrid topologies

Hybrid topologies are a combination of different network topologies. Examples of such combination are:

Star bus topology: the hubs of several star networks are interconnected as a linear bus (Fig. 6).

Star ring topology: the hubs of several star networks are connected to the main hub (Fig. 7). The hubs of the star network are connected in the form of a ring in this main hub.

6 Star bus topology

SVA0006Y

7 Star ring topology

SVA0007Y

Network organization

Addressing
In order to make it possible to transmit messages via a network and evaluate the contents thereof, the useful data (payload) that is transmitted is also accompanied by data transfer information. This can be explicitly contained within the transmission or implicitly defined using preset values. Addressing represents important information for data transfer information. It is needed in order for a message to be sent to the correct recipient. There are different ways of doing this.

Subscriber-oriented method
The data is exchanged on the basis of node addresses. The message sent by the transmitter contains the data to be transmitted and also the destination node address (Fig. 8a). All receivers compare the transmitter receiver address to their own address, and only the receiver with the correct address evaluates the message.

The majority of conventional communication systems (such as Ethernet) operate using the subscriber addressing principle.

Message-oriented method
In this method it is not the receiver node that is addressed, but the message itself (Fig. 8b). Depending on the content of the message, it is identified by a message identifier that has been predefined for this message type. In this method, the transmitter does not need to know anything about the destination of the message, since each individual receiver node decides whether or not to process the message. Of course, the message can be received and evaluated by several nodes.

Transmission-oriented method
Transmission characteristics can also be used to identify a message. If a message is always transmitted within a defined time window, it can be identified on the basis of this position. By way of a safeguard, this addressing is often combined with message or subscriber-oriented addressing.

Bus access method
A node must access the bus in order to transmit a message. In the bus access method a distinction is made between
 Predictable methods in which the bus access is determined by certain time-dependent network characteristics, whereby only one node can transmit at a time and
 Random methods whereby any node can attempt to transmit data if the bus appears to be free

In the predictable method the bus access right is determined before bus access. It can thereby be ensured that only one subscriber is using the bus at a time. Access collisions because of simultaneous bus usage will be prevented if all subscribers use this method.

8 Addressing modes

a

Adr 1 Adr 2 Adr 3 Adr 4

Adr 3

b

Id 1 Id 2 Id 2 Id 3
Id 3 Id 3 Id 4 Id 5
Id 6 Id 5 Id 6
 Id 7

Id 3

Fig. 8
a Subscriber-oriented
 method
b Message-oriented
 method

In the random method, the nodes can simultaneously attempt to use the bus as soon as it appears to be free. The timing of the bus access is therefore random. There is a risk of transmission collisions using this method, which will require attention. This can be dealt with by repeating transmissions after a collision has been detected (e.g. Ethernet), by giving the transmissions different codings (CDMA), controlling communication via a master or prioritizing message types or transmitters.

Time division multiple access (TDMA)

TDMA is a deterministic (predictive) access method. In this case each node is assigned a time window in which it is allowed to transmit (a priori). A fixed schedule is therefore required for the network. There is not usually a main communication subscriber controlling the communication procedure. However, concepts exist in which in which it is possible to switch between different schedules if necessary. The internal clocks of the different stations must run extremely synchronously with TDMA, since the transmit windows have to be adhered to with extreme precision.

Master-slave

In the master-slave system, one node on the network operates as the master. This node determines the communication frequency by interrogating its subordinate nodes (slaves). A slave only replies if it is spoken to by the master (Fig. 9). However, some master-slave protocols allow a slave to contact a master in order to transmit a message (e.g. transmit information about the position of the power-window unit to the door module).

Multimaster

In a Multimaster network, several nodes can access the transport medium independently without the assistance of another node. Bus access is uncontrolled. Every node can access the bus and transmit a message if the bus appears to be free. This means that each node is its own master, and that any node can start a message transfer with equal status. However, this also means that collision detection and handling methods have to be in place. For example, this may be in the form of a decision-making phase with prioritization or delayed transmission repeats. The use of priority control prevents a bus conflict if several nodes wish to use the bus at the same time, since the network node that has high priority or wishes to transmit a message with high priority forces its way through in the event of a conflict and transmits its message first. Normal message transmission resumes when the line is free again.

The Multimaster architecture has a positive effect on the availability of the system, since no individual node is in control of communication whose failure would lead to total communication breakdown.

9 Master-slave method

OSI reference model

Network protocols are usually defined
in layers, which combine properties and
tasks. The properties of the deeper-lying
layers are assumed in the next level up.
This has the advantage that individual
layers are exchangeable, provided that
the interfaces that are provided between
the layers remain unchanged.

The ISO OSI reference model (Open
Systems Interconnection) provides a basis
for describing and comparing many com-
munication protocols. This was developed
by the ISO (International Standardization
Organization) and led to the adoption of
international standards by ISO and IEEE
(Institute of Electrical and Electronic
Engineers).

In the OSI model, data communication
systems are depicted in different layers
(Fig. 10). The complex task of data commu-
nication is distributed among clearly ar-

ranged functional areas (layers). Not all of
the layers in the OSI model are needed in a
simple communication system. Layers can
also be combined for many applications.
Network protocols in the automotive area
are often divided up into
 Physical layer
 Communication layer and
 Application layer (user layer)

Physical layer
The electrical and procedural parameters
of the physical connection between the
network subscribers is defined in the
physical layer.

Signal level
In digital technology, data is represented
by sequences of the two binary statuses,
0 and 1. In order to transmit the data on
a bus, these statuses must be represented
on the transmission agent. It is particularly
important to avoid short-circuits on the

10 OSI reference model

11 Creating dominant and recessive levels

Control E	Status T	Bus level	
0	conducts	0 V	
1	blocks	5 V	

E1	E2	Bus level	
0	0	0 V	dominant
0	1	0 V	dominant
1	0	0 V	dominant
1	1	5 V	recessive

bus when one node is transmitting a status of 1 and another is transmitting a status of 0.

The binary statuses can be depicted in many different ways. The serial interface of the PC, for example, uses +12 V and -12 V, and CAN-B uses voltages of 0 V and 5 V. The voltages of the serial interface are unsuitable for a bus, since short-circuits can occur if several subscribers wish to transmit conflicting binary statuses simultaneously.

If the coding allows one level to overwrite another, the overwriting level is referred to as dominant, and the subordinate level as recessive.

It is also possible to depict dominant and recessive levels using visual media. A status of 1 (recessive) then corresponds to e.g. dark, and a status of 0 (dominant) corresponds to light. In an optical fiber, an individual node can override all of the others by feeding light into the conductor.

Bit stream
The application information cannot usually be transmitted directly. In order to make transmission possible, the information is first incorporated as a payload in the frame of a message that contains information to be transmitted. Since all protocols have been developed in accordance with different requirements, the frame format differs from protocol to protocol.

The frame needs to converted into a bit stream to actually transmit the information, i.e. a sequence of bits that can be transmitted via the transport medium as physical states.

Example of the UART interface
The microcontrollers that are used in control units have a simple interface (UART, Universal Asynchronous Receiver/ Transmitter) on the chip, via which they can communicate with the outside world (e.g. with a PC). The essential features of a data transmission are read out via this interface.

If no data is being exchanged, the bus level is 5 V (microcontroller operating voltage, Fig. 11). When the start bit is transmitted (dominant level), the other station connected to the bus (receiver) is notified that a data transfer is starting (Fig. 12). The length of the start bit determines a bit time that represents the basis for the entire data transfer. Every subsequent data bit has the same length. The reciprocal of this time corresponds to the data transfer rate, i.e. the number of bits that can be transmitted in one second in a continuous data stream. All participating stations must be set to the same data transfer rate.

After the start bit has been received, the transmission of an 8-bit data word commences (1 byte) with the lowest significant bit (LSB, Low Significant Bit). The receiver that has synchronized itself to the start bit scans the data bus between each data bit and therefore assembles the transferred data byte.

The eighth data bit is followed by the parity bit. This bit indicates whether the number of transmitted ones is odd or even. It therefore allows the receiver to perform a simple check for possible transmission errors. The sequence is completed with the stop bit, which is placed onto the bus

12 UART interface transmission frame

Start bit
8 data bits (1 byte)
Parity
Stop bit

SVA0009Y

with a dominant level. The next data transfer can then take place.

Communication layer

Control units can only interconnect and exchange data if they speak the same "language". This language determines the rules that are used to exchange data between the individual network subscribers.

The communication layer accepts data from the application layer, prepares it for transmission and forwards it to the physical layer.

The essential features of this protocol layer are:

Message frame format

Bus access control

Message addressing

Detection and handling of collisions

Network node synchronization

Checksum calculation

Application layer

The application layer consists of the application that processes and provides the information. The application layer is the only protocol layer to be affected by user or sensor input.

Control mechanisms

Event control

In an event-driven bus system, messages are transmitted as soon as an event that triggers the transmission of a message has occurred (Fig. 13a). Examples of such events are:

Pressing a button on the air conditioning system control panel

Operating the hazard warning flasher switch

Incoming message that requires a reaction (e.g. information from rpm sensor to speedometer needle motor)

Expiration of a fixed time period (time frame, e.g. 100 ms), after which messages are transmitted cyclically

Since the stations are not synchronized with each other, situations where several stations wish to access the bus simultaneously are unavoidable. In order to allow

13 Event control

a message to be transmitted without falsification, only one station at a time can transmit data on the bus. Collision avoidance mechanisms are available for preventing or solving bus conflicts.

If a node wishes to transmit a message whilst the bus is occupied, the transmission is delayed (Fig. 13b). A station that is ready to transmit must then wait until the transmission that is currently in progress has been completed.

Since bus access is subsequently renegotiated, the transmission may be delayed yet again. These delays become problematic if the bus becomes overloaded by a large number of network subscribers that wish to transmit messages. In this case messages may be lost if the transmitter abandons the transmission due to excessive delays.

Event-driven bus systems are suitable for reacting to asynchronous (unforeseen) events as quickly as possible. In an ideal case, they reduce the delay between the occurrence of the event and the message transmission (latency time) compared to time-driven systems. However, the latency time can vary considerably depending on the network loading.

Advantages
High level of flexibility and capability of retrofitting new nodes in the network
Good response time to asynchronous external events
Bus usage depending on event frequency in line with requirements
No network loading by unused events, since only events that have actually occurred trigger a transmission

Disadvantages
Static bus occupancy, non-deterministic (i.e. not possible to prove that a message was transmitted at the right time)

Timer control
In the most recent developments in dynamic driving systems such as brakes and steering, an increasing number of mechanical and hydraulic components are being replaced with electronic systems (x-by-wire). Mechanical connections such as the steering column are becoming superfluous, and the functionality thereof is being taken over by sensors and actuators. The reliability, safety and failure tolerance requirements of these systems are extremely high. This means:
Messages must be received on time
The latency time of critical messages must be extremely small
The system must have a redundant design
The failure of a node must affect the rest of the system as little as possible and
It must be possible to achieve a safe operating status from any fault situation

X-by-wire systems require close networking by the various components. The external increase in complexity places new demands on the safety, failure tolerance and availability of the communication system. The demands that are made of the electronic and network architecture therefore also increase. A reliable, fault-tolerant network architecture is required so that data is transmitted with guaranteed transmission characteristics, and electronic system malfunctions are handled in the most efficient way.

System architectures for real-time applications meet these requirements because their behavior is predictable and verifiable because of the way in which they are constructed. In these protocols, time windows within which a node is permitted to transmit are assigned to the control units in the communication network (nodes) during network planning (Fig. 14). In order to comply with the time window, the nodes must be synchronized as precisely as possible.

All transmissions are processed sequentially in accordance with the network planning (without collisions). Once each node has transmitted its message, the cycle restarts with the first transmitter. This makes it possible to determine how chronologically up-to-date the data is at any time. Since missing messages are detected immediately, time-triggered concepts are more reliable than event-driven systems.

If a fast data rate is required in a time-triggered system, the time delay between the occurrence of an event and the transmission of the data can be so small that the system complies with strict real-time requirements.

The bus can be protected from unauthorized access by a bus guardian. The bus guardian prevents a defective node from interfering with network communication by transmitting messages outside the relevant transmit window.

These characteristics make it possible to create redundant, fault tolerant systems in which transmission errors can be remedied and faults in the network can picked up by network nodes that can provide the functionality without errors.

Advantages
Deterministic system
Punctual data transmission
Reliable detection and isolation of defective network nodes

Disadvantages
Overall system must be planned for distributed developments
Capacity for expanding the communication system must be planned in
Good response time to asynchronous external events

Composability
If a communication system allows independently developed subsystems to be integrated in an overall system, it is said to support composability. An important criterion when doing this is that the properties that have been assured for the functionality of a subsystem are not adversely affected by adding other subsystems. If this has been ensured, the checking of system functionality is restricted to subsystem checking that can be carried out by the constructor of the subsystem.

If a communication system supports composability, changes can be made to a control unit without affecting the functionality of other control units. It is therefore not necessary to recheck the entire system after integrating a modified control unit – it is sufficient to check that the individual subsystems are operating reliably. Composability therefore reduces the time and cost of integrating new subsystems. This is the only way to increase the complexity of the electronics in the vehicle.

14 Timer control

Occurrence of events

1

0

Time slots

Max. output delay

Time →

SVA0011Y

► Overview of bus systems used in vehicles

	CAN-C high-speed CAN	CAN-B low-speed CAN	LIN	TTP
Definition	Controller area network	Controller area network	Local interconnect network	Time-triggered protocol
Bus type	Conventional bus	Conventional bus	Conventional bus	Conventional and optical bus
Domains	Drivetrain	Comfort/convenience	Comfort/convenience	Safety-related networking
Applications	Engine management, transmission control and ABS/ESP networking	Body and comfort and convenience electronics networking	Low-cost expansion of CAN bus for simple applications in the comfort and convenience electronics area	Networking in safety-related environments such as brakes, steering, railway signal boxes or aircraft landing gear
Most frequently used topology	Linear bus	Linear bus	Linear bus	Star topology
Data transfer rate	10 kbit/s to 1Mbit/s	Max. 125 kbit/s	Max. 20 kbit/s	Unspecified, typ. 10 Mbit/s
Max. number of nodes	10	24	16	Unspecified
Control mechanism	Event-driven	Event-driven	Time-driven	Time-driven
Bus lines	Copper conductors (twisted pair)	Copper conductors (twisted pair)	Copper conductor (single wire)	Copper conductors (twisted pair)
Deployment	in all vehicles	in all vehicles	in all vehicles	Premium class vehicles, aircraft, rail control systems
Standard	ISO 1198	ISO 11519-2	LIN consortium	TTA group
SAE classification	Class C	Class B	Class A	Drive-by-wire

	MOST Bus	Bluetooth	Flexray
Definition	Media oriented systems transport	Proprietary name (Danish king)	Proprietary name
Bus type	Optical bus	Wireless	Conventional and optical bus
Domains	Multimedia and Infotainment	Multimedia and Infotainment	Deployment across all domains
Applications	Transmission of control, audio and video information	Data transfers over short distances, e.g. mobile phone integration in the infotainment system	A network system for use in safety-related and simple applications
Most frequently used topology	Ring topology	Network topology (radio)	Star topology
Data transfer rate	Max. 22.5 Mbit/s	Max. 3 Mbit/s (v2.0) Max. 723 kbit/s (v1.2)	Typ. 10 Mbit/s Max. 20 Mbit/s
Max. number of nodes	64	8 active (up to 256 passive)	Theoretically up to 2,048 Max. 22 per passive bus/star
Control mechanism	Time and event-driven	Event-driven	Time and event-driven
Bus lines	Plastic or glass optical waveguides	Electromagnetic radio waves	Copper conductors (twisted pair)
Deployment	Premium class vehicles made by European manufacturers	All vehicles, connection between multimedia equipment and infotainment system	Pilot application
Standard	MOST cooperation	Bluetooth SIG	Flexray consortium
SAE classification	Mobile Media	Wireless	Drive-by-wire

Table 1

Automotive networking

Electrical and electronic systems in motor vehicles are often not independent of each other but influence and complement each other. For this reason, signal lines were used in previous injection and ignition systems in order to simplify communication between these two systems. However, the increasing number of electronic systems rapidly increased the demand for and the scope of the information that was being exchanged. The number of signal lines and plug connections that is required increased accordingly, meaning that the technology that has so far been used was approaching the limit of its capability.

The solution was provided by the development of serial bus systems, with which large volumes of data from different sources can be transferred. A serial bus system was first used in a vehicle in 1991, when the CAN bus was used in the Mercedes-Benz 500E.

The demand for additional driving safety, convenience, economy and stricter legal requirements on the environmental compatibility of motor vehicles can only be achieved with the aid of additional electronics. The number of electronic systems in vehicles is therefore increasing all the time (Fig. 2).

Cross-system functions

If you examine the signals that are processed in the individual systems, it becomes evident that many signals are needed in several control units. For example, the driving speed is evaluated in the electronic stability program (ESP) for the vehicle dynamics control, in the engine management for the automatic speed control (cruise control) and in the car sound system for the speed-dependent volume control.

The preparation of these variables from sensor signals requires computing power and therefore hardware and software resources. It is therefore advisable for these variables always to be calculated in a control unit and transmitted to other control units via a communication network.

However, in order to implement cross-system functions, electronic systems can also exchange information and therefore influence each other. Intelligent sensors are also considered to be electronic systems that prepare the sensor signal in an evaluation circuit and put the information on the data bus via a bus interface. Pre-crash sensors detect a pending collision, for example; the airbag control unit then sends the door modules and the overhead control panel a request to close the windows

1 Automotive networking

SVA0037Y

and the sliding roof. This protects the occupants from penetrating objects.

Another example of a system-encompassing function is the adaptive cruise control (ACC), in which the radar sensor, the engine management, the electronic stability program (ESP) and the transmission control communicate with each other. Distance control from the vehicle in front that is adapted to the flow of traffic is made possible in this system by means of engine torque adjustments, automatic brake system intervention and gear selection.

Coordination between the individual systems is therefore required for cross-system functions. Large volumes of data must be exchanged to do this. As well as powerful components, a powerful communication system is also required, with a low-cost network that is suitable for automotive vehicles. Special serial databus systems have been developed for this purpose.

The use of bus systems has the following advantages in comparison to a solution that uses conventional wiring:

Reduced costs with less weight and installation space because of fewer cables in the wiring harness

Better reliability and functional reliability due to fewer plug-in connections

Simplification of vehicle assembly during production

Multiple use of sensor signals

Simple connection of system components to a bus

Easier handling of equipment and special equipment variants in a vehicle

Requirements for bus systems

Both financial (e.g. cable costs, component costs) and technical constraints must be taken into consideration when a bus system is being selected. The most important technical selection criteria are explained in the following.

Data transfer rate

This variable specifies the volume of data that is transmitted during a time unit. The smallest unit of data is the bit, and the data transfer rate is usually specified in bits/seconds. Alternative names for this expression are transfer rate, data rate or bit rate.

The required data rate is dependent on the application. A slower transfer rate is required to switch the air-conditioning compressor on and off than to transfer audio signals, for example.

Interference immunity

Ideally, the data should be transferred without interference. However, this cannot be guaranteed in a motor vehicle because of electromagnetic effects. The interference immunity requirements that are made depend on the safety relevance of the electronic systems concerned. Lesser requirements are made of comfort and

2 Number of control units in the Mercedes S-class that are networked via CAN

convenience systems than the antilock brake system (ABS), for example.

In order to meet these requirements, mechanisms that detect transmission errors are incorporated in the network protocols. A simple check can be carried out using the parity bit, which is calculated in the transmitter and is transmitted together with the useful data. This specifies whether the number of 1's in the transferred byte is even or off. This information is checked by the receiver. Single errors can be detected using this method.

Another method is the checksum check. If several data bytes are being transmitted, the transmitter calculates a checksum from the individual data bytes using a predefined formula and transmits this value. The receiver also calculates the checksum of the data bytes that have been received and compares it with the checksum that has been received. If a data transmission error is detected, the received data is not used and a repeat transmission is requested.

Real-time capability
A real-time system guarantees that its results are calculated within a fixed time interval. The duration of the time interval depends on the application. The antilock brake system (ABS) must react to the incipient locking of a wheel within a few milliseconds (wheel speed reduction), whereas response times of 100 ms are adequate for actuating the power-window motor. Human beings cannot perceive delay periods of less than 100 ms.

Different demands are made of real-time behavior depending on the application:

Soft real-time requirement: the system generally adheres to the specified response time, and if these times are occasionally exceeded, it does not produce any serious effects (e.g. image jerking during picture transmission).

Hard real-time requirement: the time specification must be strictly adhered to. If the specified response time was exceeded, the calculated result would not be able to be used. This can lead to serious problems in safety-critical systems.

For example, if time allowances were exceeded in the ABS system, the incipient locking of the wheels would not be detected soon enough and the pressure in the master cylinder would not be reduced in time. This would result in locked wheels.

The time allowances must also be strictly adhered to for many engine-management system functions. Delays in transmitting injection and ignition signals could lead to engine judder and even misfiring. These reactions must be avoided, since they represent a potential danger. Hard real-time requirements must therefore be made of these systems.

However, this does not necessarily mean that the transmission of data via a bus system also has to be subject to these hard real-time requirements. Adherence to soft real-time requirements is usually sufficient. If signals from other control units are needed for functions (e.g. a torque reduction request during a shift operation), the bus system must transmit the data at a faster data transfer speed and with a smaller time delay so that the overall system complies with the specified real-time requirements.

Number of network nodes
The maximum number of nodes to be integrated varies for different areas of vehicle operation. The number of nodes for comfort and convenience systems may be high due to servomotor networking (e.g. seat adjustment) and intelligent sensors (e.g. rain sensors). Several identical busses can be used if necessary.

Classification of bus systems

Because of differing requirements, bus systems can be subdivided into the following classes.

Class A	
Transfer rates	Low data rates (up to 10 kBit/s.)
Applications	Actuator and sensor networking
Representative	LIN

Class B	
Transfer rates	Average data rates (up to 125 kBit/s.)
Applications	Complex mechanisms for error handling, control unit networking in the comfort functions
Representative	Low speed CAN

Class C	
Transfer rates	High data rates (up to 1 MBit/s.)
Applications	Real-time requirements, control unit networking in the drive and running gear functions
Representative	High speed CAN

Class C+	
Transfer rates	Extremely high data rates (up to 10 MBit/s.)
Applications	Real-time requirements, control unit networking in the drive and running gear functions
Representative	FlexRay

Class D	
Transfer rates	Extremely high data rates (> 10 MBit/s.)
Applications	Control unit networking in the telematics and multimedia functions
Representative	MOST

Applications in the vehicle

The overall vehicle system can be divided into four domains or functional areas from the point of view or electrics/electronics:
Drivetrain
Chassis
Interior and
Telematics

In the drivetrain and chassis domains, the emphasis is primarily on real-time applications. In the interior domain, the main focus is on multiplex aspects in net-working. Mainly multimedia and infotainment applications are networked in the telematics domain.

Real-time applications
The networking of these systems makes considerable demands of the performance capability of the communication system. Crankshaft-synchronous processes or processes within a fixed time frame with cycle times of a few milliseconds are typical. If the system response times are adequate for the task in hand, it is described as having real-time capability (e.g. rapid ignition-timing advance in the Motronic after a request from the traction-control system for reducing torque and therefore preventing the wheel from spinning).

3 Domains in the overall vehicle system

Chassis Drivetrain

Interior Telematics

SVC0013-1Y

The drivetrain and chassis systems are assigned to class C. These require fast transfer rates in order to ensure the real-time behavior that is required for these applications. They also make considerable fault tolerance demands. These requirements are met by the event-driven CAN bus with a transfer rate of 500 kBaud (high-speed CAN).

Examples:
Engine-management system (Motronic or electronic diesel control, EDC)
Transmission control
Antilock brake systems, ABS
Vehicle dynamics control (e.g. electronic stability program, ESP)
Chassis control systems (e.g. active body control, ABC)
Support systems (e.g. adaptive cruise control, ACC)

Multiplex applications
The multiplex application is suitable for controlling and regulating components in the body and comfort and convenience electronics area (class B), such as
Displays
Lighting
Access authorization with anti-theft warning device
Air-conditioning
Seat and mirror adjustment
Door module (power-window unit, door-mirror adjustment)
Windshield wipers
Headlamp adjustment

The transfer rate requirements are not as high for class B systems as they are for class C systems. For this reason, low-speed CAN with a transfer rate of 125 kBit/s or single-wire CAN with 33 kBit/s. can be used.
If the transfer rate requirements drop to less than 20 kBit/s, the low-cost LIN is more frequently used. Applications are mainly in the mechatronics area; examples being the transfer of switch information or the activation of actuators.

Multimedia networking
Mobile communication applications combine components such as
Car sound system
CD changer
Navigation system
Driver-information systems
Telephone
Video system
Voice input
Internet, E-mail
Back-up camera

The networking of these components makes it possible to have a centrally located display and control unit for several applications. Operating procedures can be standardized in this way, and status information can be summarized. Driver distraction is therefore minimized.

A distinction must be made between control data and audio/video data in multimedia networking. Transfer rates of up to 125 kBit/s are sufficient for control tasks (such as CD changer control), meaning that the low-speed CAN bus can be used, for example. The direct transmission of audio or video data requires extremely high transfer rates of more than 10 MBit/s. The MOST bus is used for this purpose, for example.

Coupling of networks

The network topologies and protocols that are most suitable for requirements are used for the different applications. However, the different network protocols are incompatible, meaning that data cannot be simply exchanged between networks.

In this case, help is provided by a gateway. A gateway can be compared to an interpreter that receives the "data" from a discussion partner, translates it and passes it to another discussion partner. Technically speaking a gateway is a computer that reads in the data that is transmitted by the networks and converts it into another format. The use of gateways therefore makes it possible to exchange information between different networks.

A central gateway (Fig. 4a) or several distributed gateways can be used (Fig. 4b) to interconnect the bus systems. All bus lines are routed to the central gateway. In the other case, one gateway connects two or more busses.

Examples of networked vehicles

Topology
The topologies of the communication networks can differ considerably depending on the vehicle equipment. Figure 5 shows examples of how the network can be structured for different vehicle classes. In some cases, different car manufacturers use different bus systems for communication.

Signal transmission
Signal types
A wide variety of information can be transmitted in a communication network in a vehicle. Some examples are:

Engine operating conditions (e.g. engine temperature, engine speed, engine load)
Physical measurements recorded by sensors (e.g. outside temperature)
Control signals for activating servo-motors (e.g. power-window units)
Control-element switch positions (e.g. for the windshield wiper)
Multimedia data (audio and video) for transmitting music and speech (e.g. from radio stations or when handsfree talking with a cellular phone) and moving pictures (e.g. when playing a DVD or the displays from a reversing camera)

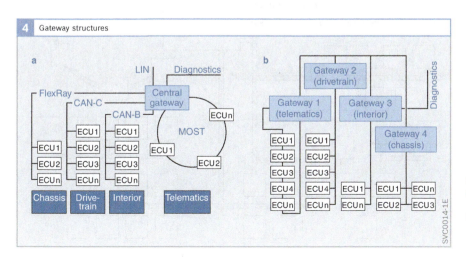

4 Gateway structures

Fig. 4
a Network with central gateway
b Network with several distributed gateways

5 Network topologies

a

Sensor CAN

Diagnostics CAN

Drive CAN

Instrument cluster CAN

LIN

Comfort CAN

Infotainment CAN

LIN

Comfort CAN

LIN

b

Distance control CAN

Bluetooth™

Instrument cluster CAN

Diagnostics CAN

MOST

Drive CAN

LIN

LIN

LIN

LIN

Comfort CAN

SVA0028E

Fig. 5
Typical network topologies for new-generation vehicles
a Compact class
b Luxury class

Resolution

The signals must be available with a suitable resolution. Switch positions can simply be shown as a 1-bit value (0 for switch open, 1 for switch closed). Other signals such as digitized analog voltages from the engine-temperature sensor or the calculated engine speed are shown as 1-byte or 2-byte values, for example, depending on resolution requirements. One byte can be used to represent 256 values, and two bytes can be used to represent 65,536 values ($= 256 \cdot 256$). The resolution for sensor signals with a voltage range of 0...5 V is approximately 20 mV with a 1 byte representation ($= 5$ V/256). A resolution of 5 mV requires a 10-bit data representation.

The conversion between the binary value and the physical value must be uniform so that the transferred signals represent the same physical value in all systems. A resolution of 30 rpm is sufficient for accessing the ignition map as far as the engine speed n is concerned. A value range of 0 to $255 \cdot 30$ rpm ($= 7,650$ rpm) and therefore the entire speed range can consequently be represented by one byte (8 bits). An incrementation of 30 rpm is too little for idle-speed control, on the other hand. More bits are required to represent the signal with a higher resolution, provided that the same measuring range is being recorded.

Output

In event-driven systems such as the ones that are primarily used in the automotive area, the signals can be transferred on the data bus when an event occurs. Examples of such events are the operation of switches for switching on the air-conditioning system or the windshield wiper. Signals that represent the operating state of the engine, for example, are not necessarily associated with an event. The engine temperature, which only changes slowly, is cyclically measured in a fixed time frame by the engine control unit (e.g. 1 second). The engine speed, on the other hand, can change extremely rapidly. The times at which the engine management carries out measurements and calculations depends on the crankshaft position and is performed once per combustion cycle. At fast engine speeds, this corresponds to a time interval of a few milliseconds, i.e. approx. 3.3 ms at an engine speed of 6,000 rpm for a 6-cylinder engine. However, not every system that needs the speed information for its control and regulation functions depends on this availability. The engine control unit therefore does not need to output the speed information on the data bus as soon as it is calculated. In this case too, data transmission takes place according to a cyclic time frame. A time frame of 10 ms is normal in the engine control area. This means that the speed information is transmitted on the bus 100 times per second.

Multimedia data

A multimedia-compatible digital bus system is frequently used in luxury-class vehicles to transfer audio data – as an alternative to using analog cables. In addition to improved audio quality, a bus system of this type offers the advantage that various audio streams and the associated check commands can be transferred in parallel. In the automotive industry, the optical MOST bus (Media Oriented Systems Transport) has proven successful as a the multimedia bus system.

The transfer of audio signals requires on the one hand synchronized transfer between the transmitter and one or more receivers and on the other hand a high data rate. In view of the fact that data transfer on the MOST bus takes place synchronously with a fixed clock-pulse rate, synchronization is already assured by its transfer mechanism.

The digital transfer of CD-quality audio signals, i.e. with a resolution of 16 bits and a clock-pulse rate of 44.1 kHz, requires a constant data rate of 1.35 Mbit/s for one stereo

channel. Up to 15 stereo channels can be transferred in parallel with the current version of the MOST bus (MOST25 with a total transfer rate of up to 24.8 Mbit/s).

Data transfer: examples
The following examples show which signals are measured and evaluated in which systems.

Driving speed
The ESP control unit calculates the driving speed from the wheel-speed sensors. This variable is transmitted on the CAN-C bus (drive CAN). The engine-management system needs this value for the cruise control, among other things, and the transmission control unit determines gear changes from the driving speed. The adaptive speed control (ACC, Adaptive Cruise Control) needs the current driving speed to calculate the necessary distance from the vehicle in front and use it as a setpoint value.

A gateway transmits the speed information via another CAN bus (instrument cluster CAN) to the instrument cluster, which displays the value via a needle instrument.

The CAN-B bus (comfort CAN) is also connected to the network via the gateway. Some luxury class vehicles are equipped with dynamic seats. The padding of the seats is inflated depending on the speed and the acceleration, counteracting the centrifugal force of the driver. This increases comfort considerably when cornering.

The speed information is sent to the Infotainment CAN via the gateway and relayed to the car sound system. This allows the volume to be adapted to the driving speed. The navigation system needs the speed to calculate the position if the GPS signal is missing (e.g. in a tunnel).

The diagnosis interface is directly connected to the engine and transmission control unit via the serial K-line. All other control units are connected to the diagnosis interface via a virtual K-line that is simulated on the CAN bus. This allows the

driving speed to be read out in the workshop via the connected diagnostic tester (example: the correct assignment of the wheel-speed sensors must be checked for the ABS functional test).

Engine speed
The injection and (with a gasoline engine) the ignition timing are output with a resolution of less than 1° of the crankshaft angle. In order to ensure real-time behavior, the crankshaft position must be recorded in the engine control unit. The engine-speed sensor scans the crankshaft trigger wheel and relays the signal to the control unit, which calculates both the crankshaft position and the engine speed. This variable is used to calculate the injection time and the ignition angle, for example.

The engine speed is a variable that is needed in many other systems. The engine control unit therefore outputs it on the data bus. The shifting points are defined in the transmission control unit depending on the speed. The engine speed is needed for the ASR function (acceleration slip control) in the Electronic Stability Program (ESP) – ASR intervention (torque reduction) must not make the engine stall.

As in the previous example, the engine speed is transmitted to the diagnosis interface and the instrument cluster (display on the rev counter).

Turn signaling
The driver operates the turn-signal lever (Fig. 6, Item 1). A signal is relayed to the steering column control unit via a discrete line (2) depending on whether the driver is indicating a right or left turn. This may be a resistance-coded signal, for example. The control unit evaluates the signal and detects that the driver is indicating a left turn, for example.

The comfort CAN relays this information to the vehicle power supply control unit (3). The indication direction is defined on the basis of the received information (normal flash frequency, increased flash

frequency in the event of bulb failure). The front left and rear left turn-signal lights are then actuated via discrete lines (4, 5). The vehicle power supply control unit also transmits the "left turn signaling" information on the comfort CAN. The gateway (6) relays the information to the instrument cluster CAN. The indicator lamp then flashes on the instrument cluster (9). If the vehicle has a trailer hitch, the information goes to the trailer-recognition control unit (7) via the comfort CAN. This actuates the turn-signal lamp on the trailer (8) via cables.

Wiper stage 1
The wiper switch (Fig. 7, Item 1) transmits a signal via a discrete line to the steering column control unit (2), which evaluates the information (e.g. wiper stage 1). The control unit transmits this information on the comfort CAN bus. The vehicle power supply control unit (3) picks up the information and relays it to the wiper motor (4) via the LIN bus. The vehicle power supply control unit acts as a gateway between the comfort CAN and the wiper LIN.

Load management
At low revs (idle speed) and when a considerable amount of power is being used up by the electrical consumers that are switched on, the battery or alternator voltage can drop to a low value. The vehicle power supply control unit calculates the current status of the vehicle power supply from the current battery voltage, the DF signal from the alternator (alternator utilization) and the information about heavy current consumers that are switched on with a short switch-on duration. The vehicle power supply control unit requests an idle speed increase via the CAN bus if the vehicle power supply is insufficient at idle speed. The engine control unit implements the request. If this action does not solve the problem, the vehicle power supply control unit switches off specific consumers such as the heated rear windshield, the seat heating or the heated outside mirror. These consumers are connected to the vehicle power supply control unit via discrete lines.

6 Data transfer during turn signaling

7 Data transfer during windshield-wiper operation

Fig. 6

1 Turn-signal lever
2 Steering column
 control unit
3 Electrical-system
 control unit
4, 5 Turn-signal lights
6 Gateway
7 Trailer-recognition
 control unit
8 Turn-signal light
 on trailer
9 Instrument cluster

Fig. 7

1 Windshield-wiper
 lever
2 Steering column
 control unit
3 Electrical-system
 control unit
4 Wiper motor

Multimedia application

The signal sequences during the output of a radio signal on the amplifier is described in the following as an example of the transfer of signals and multimedia data in a luxury-class vehicle infotainment system with MOST bus (Fig. 5b).

The functions of the vehicle infotainment system are controlled via the central operating unit, the head unit (Fig. 8, Item 1). In current systems, this unit usually has a screen in the instrument panel and a rotary/pushbutton controller in the center console, by means of which the user interacts with the system.

To output a radio station, the head unit, which in most systems is responsible for managing the audio channels of the infotainment system, establishes on the MOST bus a stereo channel of the required audio quality between the radio tuner (2) and the amplifier (3). It now tunes in the requested station at the radio tuner via corresponding check commands and if necessary makes further settings. Finally, the head unit connects the tuner output with the previously created stereo channel. For the information to be displayed in the user interface, the head unit receives from the radio tuner corresponding data relating to its current reception status (the current station name, for example) which are updated where necessary.

At the amplifier, the head unit connects the stereo channel with a selected input and uses corresponding check commands to set the properties for outputting the audio signal, e.g. the set volume. When all the settings have been made, it instructs the amplifier to fade in the audio signal of the radio signal, which is then output through the infotainment system's speakers.

Before this, the head unit has if necessary ensured that the output of a previously output audio signal has been faded out, the output of the associated device stopped, and the associated audio channel removed.

In parallel the CAN/MOST gateway (4) constantly transmits the driving-speed data, which it receives via the comfort CAN, via the MOST bus to the amplifier, which has requested a notification for this information when the system was started. The amplifier can use the speed together with further vehicle information to calculate additional settings for outputting the audio signal, e.g. adapting the volume depending on the current speed.

8 Multimedia networking

Fig. 8
1 Head unit
2 Radio tuner
3 Amplifier
4 CAN/MOST gateway
5 Screen
6 Control element
7 Speakers

Control of an automatic steel folding roof

Vehicles with automatic steel folding roofs are currently very much in vogue. These roofs are automatically opened and closed. The technical realization of this seemingly simple mechanical function poses a huge challenge for the networking of the electronics and control units.

The networking depicted in Figure 9 shows a typical present-day mid-size vehicle with the additional function of an automatic steel folding roof. At its maximum equipment specification, this vehicle has over 35 electronic control units (ECUs), which communicated with each other via four CAN buses and a further four LIN sub-buses. In this example, a total of 13 control units are involved in the control of the automatic steel folding roof. The text below describes which functions are performed by the individual control units to move the roof.

Roof control unit
Power activation of the motors for the steel folding roof is performed by this control unit. This control unit also assumes the role of complete monitoring of the movement process. Proximity-type sensors which monitor, record and evaluate the

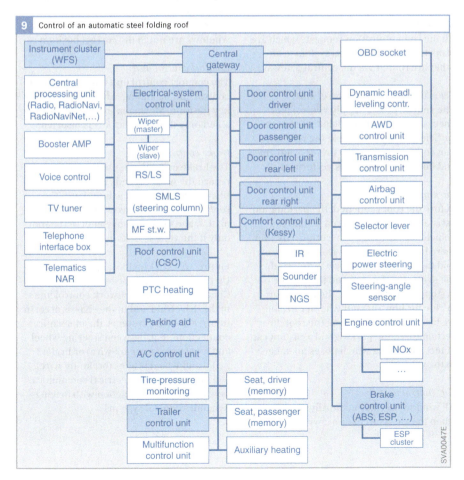

9 Control of an automatic steel folding roof

SVA0047E

Fig. 9
Systems in blue boxes are involved in the control of the steel folding roof

movement of the roof are used for this purpose. In order to inform the driver of the current status while the roof is being opened and closed and to prevent any damage to the vehicle or the surroundings, the roof control unit receives numerous further parameters from other control units in the vehicle.

Instrument cluster
The instrument cluster receives via the gateway the current status from the roof control unit and from other participating control units, and informs the driver accordingly. Example: "Obstacle behind the vehicle! Roof movement stopped".

Gateway
All the communication requests which are transmitted from one bus system to another are routed via the central gateway.

Vehicle power supply control unit
The vehicle power supply control unit checks whether the vehicle battery has sufficient charge to enable the roof to be moved. If necessary, the driver is informed and no roof movement is performed.

Parking-aid assistant
Before the roof is opened, the parking-aid assistant monitors the area behind the vehicle for obstacles to ascertain whether there is sufficient space available to unfold the roof. If an obstacle is detected, the driver is informed and movement is stopped. In this event, the driver can decide for him-/herself whether he/she wishes to continue moving the roof. In certain cases the parking-aid assistant can detect something which poses no danger to the vehicle.

A/C control unit
All windows are automatically closed when the air-recirculation switch is actu-

ated. This also applies to the sliding sunroof, which is activated by the roof control unit.

Trailer control unit
The trailer control unit informs the roof control unit whether a trailer is hitched. If this is the case, opening and closing of the roof is disabled.

Door control units
The status of the windows and doors is interrogated by all four door control units and transmitted to the roof control unit. Opening or closing of the roof is only begun if all the doors are closed and the windows are in the correct positions. This setting is automatically corrected if the windows are not in the correct positions.

Comfort control unit
The comfort control unit informs the roof control unit of, among other things, the key position. Movement of the roof is enabled only if the correct key is inserted.

Brake control unit
The brake control unit uses sensors to record the wheel speeds and thereby identify whether the vehicle is moving. The roof control unit receives the speed information and enables movement of the roof only if the vehicle is stationary.

Future areas of FlexRay application
Up to now the high-speed CAN bus (CAN C) has been used to network control units in the drivetrain and in the chassis area. In future X-by-wire systems, the mechanical connections, e.g. between steering wheel and front axle (steer-by-wire) or brake pedal and wheel brakes (brake-by-wire), will be replaced by electrical communication systems in conjunction with driving-dynamics control systems.

Furthermore, vehicle architectures have up to new often used several CAN buses which are linked to each other via gateways to distribute the high data volume. Future architectures will use, for a fast, powerful connection between several master computers (which assume, for example, central functions in the safety and driver-assistance areas), a backbone bus with a high data rate to which in each case sub-buses (e.g. drivetrain CAN and sensor CAN in the chassis area) are connected.

The new FlexRay bus system satisfies the demands which will be placed in future on the vehicle architecture, such as, for example

high data rates and guaranteed real-time capabilities in the drive and chassis areas,
large date volume in the backbone, and
high failure safety of safety-relevant applications (e.g. X-by-wire)

through properties such as
high availability and redundancy by means of two physically independent channels,
high data rates with up to 10 Mbit/s per channel,
data transfer with guaranteed latency, and
synchronicity of all communication users by means of a global time base.

FlexRay is currently still in the development stage, although some initial systems are already in volume production. The following example is a current approach to a FlexRay topology.

Excerpt from chassis domain
Concepts for global driving-dynamics control are being developed to facilitate control which is as cooperative and thus as effective as possible. Components of these concepts are ESP (Electronic Stabil-

ity Program), AFS (Active Front Steering), and ARC (Active Roll Control). Higher-level driving-dynamics control controls the individual systems without limiting their functionality and prevents negative interactions. Figure 10 shows the interconnection of these systems with bus system and sensors.

The yaw sensor transmits the vehicle's acceleration values and yaw rate to ESP. If a critical driving situation is detected by ESP using these data (e.g. vehicle oversteer), situation-conditioned brake pressures and engine-management interventions are calculated. Higher-level global driving-dynamics control also evaluates the driver-command steering angle from the steering-angle sensor and calculates a supplementary steering angle, which is converted by AFS. More effective and comfort-enhancing control can be achieved by this interaction of the individual systems. In this way, the vehicle can, for example, be stabilized by a corrective steering movement already at a very early stage such that braking interventions can be partially or even completely avoided.

10 FlexRay

Global driving-dynamics control

CAN — Driving dynamics — FlexRay — ESP (active braking)

Sensor technology — AFS (Active Front Steering)

Further compo-nents — ARC (Active Roll Control)

SVA0048E

Bus systems

CAN bus

In 1991 the CAN bus (Controller Area Network) was the first bus system to be introduced to a motor vehicle in mass production. It has since established itself as the standard system in the automotive sector, but the CAN bus is also commonly used as a field bus in automation engineering in general. In imitation of other network types, such as the local area network (LAN), wide area network (WAN) or personal area network (PAN), this bus system was given the name, CAN.

Applications

The CAN bus is used in various domains in the motor vehicle. These domains differ in the requirements they demand of the network. Due to the fast processes involved in the area of engine management, information is required much faster here than in the area of comfort/convenience where the controlled systems are located further apart and as such lines are more prone

to damage. As a result of these different requirements, buses with different data rates are used that offer an optimum cost-benefit ratio for the field of application concerned. A distinction is made between high-speed and low-speed CAN buses.

High-speed CAN (CAN-C)

CAN-C is defined in ISO Standard 11898-2 and operates at bit rates of 125 kBit/s to 1 MBit/s. The data transfer is therefore able to meet the real-time requirements of the drivetrain.

CAN-C buses are used for networking the following systems:
- Engine-management system (Motronic
- for gasoline engines or EDC for diesel engines)
- Electronic transmission control
- Vehicle stabilization systems (e.g. ESP) Instrument cluster

Low-speed CAN (CAN-B)

CAN-B is defined in ISO Standard 11898-3 and operates at a bit rate of 5 to 125 kBit/s. For many applications in the comfort/convenience and body area, this speed is sufficient to meet the real-time requirements demanded in this area. Examples of such applications are:
- Control of the air-conditioning system
- Seat adjustment
- Power-window unit
- Sliding-sunroof control
- Mirror adjuster
- Lighting system
- Control of the navigation system

The CAN bus is finding ever more use in vehicle diagnostics. Here, the electronic control unit is connected directly to the CAN bus and thus receives the information it needs for diagnostics immediately. Previous diagnosis interfaces (e.g. KWP2000) are becoming less important.

1 Networking of electronic control units by CAN

ECU 1 ECU 2 ECU 3 ECU 4

CAN-C

Gate-way

Diagnosis interface

ECU 5 ECU 6 ECU 7 ECU 8

CAN-B

ECU 9 ECU 10 ECU 11

SVC0016E

Fig. 1
ECU Control unit
ECU 1 to ECU 4
 On the CAN-C
 high-speed bus
ECU 5 to ECU 11
 On the CAN-B
 low-speed bus

Topology
Bus topology
In the development of the CAN, particular focus was placed on eliminating the need for a central control element for communication. This approach is most effectively supported by a bus topology in which all network nodes are connected to a bus and each node is able to receive all information sent on the bus. The bus topology is most commonly selected during the conceptual design of the communications system.

In addition to offering favorable electrical properties, the linear bus topology has the advantage that the failure of one station would not affect the functionality of the data transmission system. Furthermore, additional stations can be connected to the system with little extra effort.

Star topology
The use of a central coupler makes it possible to build star topologies. Active as well as passive couplers can be used. The use of a star topology achieves a high level of flexibility in adapting to the networking task.

The coupler is used to build a star topology and simply forwards the messages to the individual segments. Since the signal transit times remain unchanged, the circuit length of the star topology is the same as that of a bus topology.

Data transmission system
Network nodes
A network node (Fig. 2) comprises the microcontroller for the application software, the CAN controller and the CAN transceiver (bus driver). The CAN controller is responsible for the transmit and receive modes. It generates the bit stream for data communication from the binary data to be transmitted and forwards it to the transceiver on the TxD line. This amplifies the signals, generates the voltage level required for differential data transfer and transmits the processed bit stream serially on the bus line (CAN_H and CAN_L).

Incoming messages are processed by the transceiver and sent to the CAN controller on the RxD line.

The microcontroller, which runs the application program (e.g. Motronic), controls the CAN controller, prepares the data to be sent and evaluates the data received.

Logic bus states and coding
CAN uses two states for communication, dominant and recessive, with which the information bits are transmitted. The dominant state represents a binary "0", the recessive a binary "1". NRZ (Non-Return to Zero) is used as the encoding method for the data transmission. With this method, there is no compulsory return to zero between two transmission states of the same value.

When it receives messages, the CAN transceiver converts the signal level back to logical states. In the process, a differential amplifier subtracts the CAN_L level from the CAN_H level (Fig. 3). If lines become twisted, disturbance pulses (e.g. from the ignition system) have the same effect on both lines. Differential data transfer therefore makes it possible to filter out interference on the line.

2 Network nodes in the CAN

Some transceivers also evaluate the voltage level on the CAN_H and CAN_L line separately. It would then be possible for operation to continue in single-line mode if one of the two bus lines were to fail as a consequence of a short-circuit or cable break. However, the bus subscribers would have to share a common ground that would assume the function of the failed line.

Transmission agent and bus coupling

Two-wire line

For the CAN, any transmission agent on which dominant and recessive states can be transmitted is worthy of consideration. An untwisted or twisted pair of wires, depending on the ambient conditions, is usually used, and the wires are either galvanically coupled or decoupled. The two bus lines are designated CAN_H and CAN_L. The two-wire line supports symmetrical data transfer whereby the bits are sent on both bus lines and represented by different voltages. This reduces sensitivity to in-phase interference because the interference affects both lines equally and can be filtered out. Additional shielding of the lines reduces their own radiation emissions, especially at high baud rates.

Single-wire line

The single-wire line is a means of reducing manufacturing costs by dispensing with the second line. For this to be possible, however, all bus subscribers must share a common ground that would assume the function of the second line. The single-wire version of the CAN bus is therefore only an option for a communications system of limited spatial dimensions.

The data transfer on the single-wire line is more prone to interference radiation because it is not possible to filter out disturbance pulses as it is with the two-wire line. For this reason, a higher level increase is required on the bus line to improve the signal-to-noise ratio. This in turn has a negative effect on interference radiation. The flank steepness of the bus signals must therefore be reduced in comparison to the two-wire line. This is accompanied by a lower data transfer rate. Consequently, the single-wire line is only used for the low-speed CAN in the body electronics and comfort/convenience electronics. Thanks to this feature, a low-speed CAN with two-wire line would still remain functional in the event of a line failure (CAN_H or CAN_L).

3 Filtering out interference on the CAN bus

a V

3.5
2.5
1.5
0

CAN high level (CAN_H)
CAN low level (CAN_L)

a V

2.0
0

Differential signal

SVC0018E

Voltage level

The CAN transceiver converts the logical states 0 and 1 received by the CAN controller into voltage levels that are fed to the CAN_H and CAN_L bus lines.

The high-speed and low-speed CANs use different voltage levels for the transmission of dominant and recessive states. The voltage levels of the low-speed CAN are shown in Figure 4a, those of the high-speed CAN in Figure 4b.

In the recessive state, the high-speed CAN uses a voltage of 2.5 V on both lines. In the dominant state, a voltage of 3.5 V is present on CAN_H and a voltage of 1.5 V is present on CAN_L.

On the low-speed CAN, a voltage of 0 V is present on CAN_H in the recessive state, and 5 V on CAN_L. In the dominant state, voltages of 3.6 V and 1.4 V are present on the CAN_H and CAN_L respectively.

Reflection-free termination

Reflections of the electrical signals at open ends of lines would interfere with communication. To dampen these reflections, the bus lines are terminated at each end with a resistor of 120 Ω.

Alternatively, the terminating resistors may be integrated into the electronic control units themselves.

Emission limits

To permit correct evaluation of bit transmission, the signal must arrive at each node within the bit time without interference, and yet still at the relevant sampling point. Delays arise from the signal transit time on the data bus. The maximum permissible bit rate thus depends on the total length of the bus. ISO 11898 specifies the bit rate for a defined circuit length. The following recommendations exist for longer lines.

- 1 MBits/s for 40 m (specified)
- 500 kBits/s up to 100 m (recommendation)
- 250 kBits/s up to 250 m
- 125 kBits/s up to 500 m
- 40 kBits/s up to 1,000 m

It is possible to connect at least 30 network nodes to the bus without the need for additional measures.

4 Voltage level

5 Data transmission system

Fig. 4
a Voltage level of the low-speed CAN (CAN-B)
b Voltage level of the high-speed CAN (CAN-C)

CAN protocol

Protocol layers

For communications protocols, it is standard practice to arrange thematically coherent tasks as a set of layers, which affords a high level of flexibility in the implementation of a bus system. With the CAN, both the CAN hardware and the CAN software are subdivided into several layers (Fig. 6).

Application layer

The application layer represents the information in the form of data structures used by the application. These sets of data which are to be transmitted are forwarded to the object layer for this purpose.

Object layer

The task of the object layer is to manage the messages. The functions of this layer are used to decide which message should be sent at which time. For incoming messages, this layer is responsible for message filtering.

Transport layer

The transport layer furnishes the object layer with received messages and processes the messages prepared by the object layer for sending in such a way that the physical layer is able to transmit this

information. To fulfill this purpose, the transport layer is responsible for such functions as arbitration or fault detection and signaling.

Physical layer

The physical layer is the lowest level in the transport stack. It consists of the physical components of the network, such as the wiring and the voltages used to send the information.

Multimaster principle

The CAN protocol supports communication between network nodes without the need for a central control unit. Each node may attempt to send messages at any time. Whether this attempt is successful or not essentially depends on two factors:
- Is the bus free before the start of transmission?
- Has the arbitration phase been passed successfully?

This design ensures that, even if any number of nodes were to fail, it can still be determined whether a node is authorized to send.

Content-based addressing

Unlike other networks, the CAN does not address the individual network nodes but the messages that have been sent. Each message has a unique marker, or identifier. The identifier classifies the content of the message (e.g. engine speed or power-window position). A station is therefore able to broadcast a message to all other stations (multicast or broadcast method). These stations read only those messages whose identifiers are stored in their acceptance list (message filtering, Fig. 7). In this way, each station decides for itself whether or not it needs a message sent on the bus. The identifier has 11 bits (standard format, CAN 2.0 A) or 29 bits (extended format, CAN 2.0 B). With 11 bits in the standard format, it is possible to distinguish between 2,048 different CAN messages;

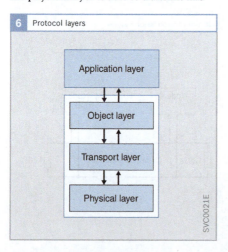

6 Protocol layers

Application layer

Object layer

Transport layer

Physical layer

SVC0021E

in the extended format, this number rises to over 536 million.

The advantage of this addressing method is that the network nodes do not require any information about system configuration and are thus free to operate fully independently of each other. This results in a highly flexible complete system, which makes it easier to manage equipment variants. If one of the ECUs requires new information which is already on the bus, all it needs to do is call it up from the bus. It is possible to integrate additional stations into the system (provided they are receivers) without having to modify the existing stations.

Controlling bus access
Arbitration phase
If the bus is unoccupied (recessive state) and messages are available for sending, each station is free to initiate the sending of its message. The message begins with a dominant bit (start-of-frame bit), followed by the identifier. When several stations start to transmit simultaneously, the system responds by employing "wired-and" arbitration (arbiter = logical AND operator) to resolve the resulting conflicts over bus access. The message with the highest pri-

ority (lowest binary value of the identifier) is assigned first access, without any data loss or delay (non-destructive protocol).

The arbitration principle permits the dominant bits transmitted by a given station to overwrite the recessive bits of the other stations (Fig. 8). Each station outputs the identifier of its message onto the bus bit by bit, with the most significant bit first. During this arbitration phase, each station wishing to send data compares the level present on the bus with the level it actually possesses. Each station that attempts to send a recessive bit but encounters a dominant bit loses the arbitration process. The station with the lowest identifier, i.e. the highest priority, makes its way onto the bus without having to repeat the message (non-destructive access control). The transmitters of lower-priority messages automatically become recipients of the message just sent by another station. They repeat their attempt to send as soon as the bus is free again.

Without this access control, bus collisions would result in faults. To guarantee unequivocal bus arbitration, therefore, it is not permissible for more than one node to send a message with the same identifier.

7 Addressing and message filtering (acceptance check)

8 Bit arbitration

Fig. 7
Station 2 transmits
Stations 1 and 4 accept
the data

Fig. 8
Station 2 gains first
access (signal on the
bus = signal from
station 2)

0 Dominant level
1 Recessive level

With this access method, maximum-priority messages only have to wait for the transfer of the message currently being sent and, with bit times of 130 (CAN 2.0 A) or 150 (CAN 2.0 B), they have the lowest latency. At a data transfer rate of 500 kBits/s, this equates to 260 µs or 300 µs. The higher the load on the bus, the greater the data transfer's temporal offset becomes for messages of lower priority, and thus the greater the uncertainty over when a message to be sent will arrive at the recipient.

In order that all messages have a chance of accessing the bus, the data transfer rate must be matched to the number of bus subscribers.

Priority assignments
The direct consequence of the arbitration process is that the identifier also has the role of prioritizing the frame during transmission in addition to identifying the frame content. An identifier corresponding to a low binary number has high priority and vice versa. Message priorities are derived from the speed at which the message content changes or from the importance of the message to safety considerations, for example. It must not be possible for messages to have the same priority.

Message format
The message transfer on the CAN bus is based on four different frame formats:
- *Data frame*:
 The transmitted message contains data (e.g. current engine speed) that is provided by the transmitting station (data source).
- *Remote frame*:
 Stations can request the data they need from the data source (example: the windshield wiper requests how wet the windshield is from the rain sensor). The data source responds by sending the relevant data frame.

- *Error frame*:
 If a station detects a fault or error, it communicates this to the other stations using an error frame.
- *Overload frame*:
 This can be used to create a delay between a preceding and subsequent data frame or remote frame. The transmitting node reports that it cannot currently process another frame.

For data transfer on the data bus, a message frame is created. It contains the information to be transmitted arranged in a defined sequence. CAN supports two different formats of frame, which are specified in CAN 2.0 A and CAN 2.0 B. The most important difference between these two frame formats is the length of the identifier. A CAN 2.0 A frame has an 11-bit identifier, while a CAN 2.0 B frame has a 29-bit identifier divided into two parts (11 bit and 18 bit).

Both formats are compatible with each other and can be used together in a network. The frames compliant with CAN 2.0 A and B are shown in Figure 9. They have a maximum length of 130 bits (standard format) or 150 bits (extended format).

9 CAN message format

Start of frame
Arbitration field
Control field
Data field
CRC field
ACK field
End of frame
Inter-frame space

IDLE | 1* | 12* | 6* | 0...64* | 16* | 2* | 7* | 3* | IDLE

Data frame
Message frame

UAE0285-3E

Fig. 9
0 Dominant level
1 Recessive level

In CAN 2.0 A and CAN 2.0 B data frames, the sequence of transferred information is arranged in the same way. The start-of-frame bit is followed by the arbitration, control, data and CRC fields. A frame is completed by the ACK field and the end-of-frame.

Start of frame
The bus is recessive in idle state. The start of frame, represented by a dominant bit, indicates the start of a transmission and, together with the process of bit stuffing, serves to synchronize all stations.

Arbitration field
With a CAN 2.0 A frame, the arbitration field comprises the 11-bit identifier and a control bit, the RTR bit (Remote Transmission Request).

With a CAN 2.0 B frame, the arbitration field comprises the 11-bit identifier, followed by the SRR bit (Substitute Remote Request) and the IDE (Identifier Extension Bit). Both bits are sent recessively and thereby ensure that a CAN 2.0 A frame always takes priority over a CAN 2.0 B frame if it has the same 11-bit identifier. The second 18-bit identifier follows. The RTR bit completes the arbitration field.

The RTR bit indicates whether the transmitted frame is a data or remote frame. The RTR bit is dominant in the data frame and recessive in the remote frame. If station A, for example, happens to send a message by data frame and station B requests this message by remote frame at the same time, the arbitration conflict cannot be resolved in this situation by means of the message identifier: the RTR bit is the decisive factor for access authorization. First of all, station A wins arbitration with the sending of the dominant RTR bit and continues to transmit the message. Station B, which has requested precisely this message, prepares to receive and is able to read the other data sent by station A.

Control field
In a CAN 2.0 A frame, the control field comprises the IDE bit (Identifier Extension Bit), which is always sent as dominant here, followed by a reserved bit for future extensions, which is sent recessively. The remaining four bits in this field define the number of data bytes in the next data field. This enables the receiver to determine whether all data has been received.

The structure of a CAN 2.0 B frame is practically identical. However, since the IDE bit already belongs to the arbitration field, a further reserved bit for future extensions takes its place and this reserved bit is sent recessively.

Data field
The data field contains the actual message information comprised of between 0 and 8 bytes. A data field in which the length of the data is expressed in 0 results in the shortest possible data frame with a length of 44 or 64 bits. A frame like this can be used to synchronize distributed processes. A number of signals can be transmitted in a single message (e.g. engine temperature and engine speed).

CRC field
The CRC field (Cyclic Redundancy Checksum) contains a 15-bit checksum (frame check word) across the preceding frame from the start bit to the final bit of the data field. The 16th bit (CRC delimiter) in this field is recessive and closes the checksum. The checksum is a means of detecting possible transmission interference.

ACK field
Unlike all preceding fields, the ACK field (acknowledgment) is not set by the sender of the frame but by a different node that is able to acknowledge receipt of the frame directly after the data field. This field comprises the ACK slot and the recessive ACK delimiter. The ACK slot is also transmitted recessively by the sender and over-

written as dominant by a receiver upon the message being correctly received. Here, it is irrelevant whether the message is of any significance for the particular receiver in the sense of an acceptance check (message filtering). Only correct receipt is confirmed. This signals to the sender that no malfunction occurred during data transfer.

End of frame
The end of frame marks the end of the message and comprises seven recessive bits. With seven bits of the same value in succession, the stuffing rule is deliberately broken (see Error detection).

Interframe space
The interframe space comprises a succession of three recessive bits which serve to separate successive messages. After a total of 10 of these recessive bits, the stations are permitted to begin transmitting within the network again. Until then, the bus remains in idle state.

Only data and remote frames need to include the interframe space. Error and overload frames can be sent immediately after the last frame. This enables immediate signaling of errors and problems.

Transmitter initiative
The transmitter will usually initiate a data transmission by sending a data frame. It is also possible for a receiving station to call in data from a sending station by depositing a remote frame. The network node that is able to deliver the information requested makes the information available in response to this query.

Error detection
The bus line may suffer interference, e.g. electromagnetic interference. To eliminate the risk of faulty behavior, the transmitted data must be checked for correctness. A number of control mechanisms for detecting errors are integrated in the CAN protocol.

Cyclic redundancy check
For every transmitted message, the transmitter calculates a check sequence from the start-of-frame, arbitration, control and data field. A 15-bit checksum is derived from the bit sequence by means of a generator polynomial. The checksum is then used to detect errors in the data transfer.

With CRC generation, a defined generator polynomial is used to carry out polynomial division across a given frame range. The remainder forms the checksum.

Once the CRC field has been received, the receiver is able to check that the frame has been transferred correctly by carrying out its own polynomial division of the received frame range using the generator polynomial and checking whether the received checksum matches the calculated remainder.

Frame check
The frame check involves all bus subscribers, senders as well as receivers, checking the sent/received data frame for compliance with the predefined frame structure. The CAN protocol contains a number of fixed-format fields (start and end-of-frame, delimiter) which are checked by all stations.

ACK check
With the ACK check, a receiver acknowledges that the frame has been received correctly by sending a dominant bit in the ACK slot. The sender of the frame can then verify whether a message was transferred correctly. An absence of this bit indicates that a transmission error has been detected.

Monitoring
The sender of a frame continuously monitors the bus level. It is able to detect a transmission error by comparing the sent bit and the sampled bit.

Bit stuffing

Compliance with bit stuffing is checked by means of the code check. The stuffing convention stipulates that in every data frame or remote frame a maximum of five successive bits of equal state may be sent between the start of frame and the end of the CRC field. As soon as five identical bit states have been transmitted in succession, the sender includes a bit with the opposite state. The receiving station clears all of these inserted bits after the message has been received (destuffing).

This measure permits detection of line faults, e.g. short-circuit or burst interference. Rare signal changes would affect the possibility for synchronization in the nodes.

Error handling

If a CAN controller detects a fault or format error, it interrupts the current transmission by sending an error frame comprising six successive dominant bits. This breaks the stuffing rule that prohibits this type of bit sequence. If the sender detects that its message has been interrupted by an error frame, it stops transmitting and makes another attempt at a later time. This effect prevents other stations from accepting the erroneous message and thereby ensures consistency of data across the entire system.

Fault confinement in the event of failures

Defective stations could weigh heavily on bus traffic if they were to frequently send erroneous messages or interrupt the transmission of correct messages by repeatedly sending an error frame. The CAN protocol localizes station failures by means of statististical error analysis. A station recognizes the probability of its own malfunction by how often it aborts messages before other stations send an error frame. The protocol's first measure in this case is to prevent a station such as this from continuing to abort transmissions. In an emergency, the station shuts down automatically.

Hardware
CAN controller of a subscriber
The CAN controller of a subscriber generates a frame from the data to be sent containing all the fields required for compliance with the CAN protocol. It then converts the frame into a bit stream.

CAN only specifies the physical, transportation and object layers. The interfaces that are provided in the application layer can be arranged differently to suit the field of application. In order to provide proper CPU support (electronic control unit microcontroller for application software) for a wide range of different requirements, semiconductor manufacturers have brought onto the market products that provide a variety of capabilities. They differ neither in the frame format they produce, nor in their error-handling methods, but solely in the type of CPU support required for message administration.

If messages on the bus are managed in extended format (29-bit identifier), the CAN controllers must be compatible with the CAN 2.0 B specification. There are controllers that only support CAN 2.0 A and send and receive messages in the standard format, but that generate errors with messages in the extended format. Other controllers are able to tolerate the extended format without generating an error. These modules can be used in a CAN together with controllers that administer the sending and receiving of messages in the extended format.

Basic CAN
In modules with basic-CAN implementation, only the basic function of the CAN protocol for the generation of the bit stream is implemented in the hardware. For the management of the messages to be sent and received, there is an intermediate

10 Basic-CAN module

11 Full-CAN module

buffer to which the local computer (application software microcontroller) has access (Fig. 10). Since the buffer capacity is limited, the computer must read the received data before new messages are received. Message filtering also takes place in this computer. A part of the computer's capacity is therefore used for CAN management. Since the computers do not usually have sufficient processing capacity, modules with basic CAN are primarily suitable for low bit rates, or for the transmission of fewer messages but at higher bit rates.

The advantage of these modules, in comparison to modules with full CAN, is the smaller chip surface and the lower manufacturing costs.

Full CAN

Full CAN implementation is the protocol of preference in cases where a station has to manage several messages at high bit rates and the local computer has no free capacity for communication tasks. They contain several "communication objects", each of which contains the identifier and the data of a particular message. During the initialization of the CAN module by the local computer, it is decided which messages the CAN controller should send and which received messages it should process further. Received messages are only accepted (message filtering) if the identifier matches one of the communication objects.

CAN controllers with full-CAN implementation relieve the burden on the local computer by performing all of the communication including message filtering in the controller (Fig. 11).

The CAN controller can be coupled to the microcontroller in the electronic control unit as a stand-alone module by the address/data bus. Powerful microcontrollers have the CAN controller integrated on-chip. This type of bus coupling is the more cost-effective and thus the more common solution.

Modules without local computer

A further category of CAN module is one that is supplied without a local computer. These SLIOs (Serial Linked Input/Output) are able to input and output data via ports. They are therefore suitable for making sensors and actuators bus-compatible at low cost, but they do need a master that controls them.

Transceiver

The bit stream generated by the CAN controller is made up of binary signals. They do not yet correspond to the required voltage levels of the CAN bus. The CAN-bus interface module, or transceiver, generates the differential signals CAN_H and CAN_L and the reference voltage U_{ref} from the binary data stream.

Sleep mode

The CAN comfort bus must remain ready for operation even with the ignition switched off so that functions such as the radio, power windows or parking lamp may continue to operate. The bus subscribers must therefore be supplied by terminal 30 (permanent positive). After terminal 15 has been switched off (ignition off), a CAN node may enter sleep mode (standby) to relieve the vehicle electrical system of as much load as possible. The transmitter part of the transceiver module is switched off in this condition to minimize the power consumption in this mode of operation. However, the receiver part remains active and checks whether messages are being sent on the bus. In this way, the CAN controller, which also enters standby mode, is able to react to a wake-up message and fully activate the CAN node.

Data transfer sequence

Using the transfer of engine speed as an example, it shall be demonstrated how data is transferred on a CAN with full-CAN module.

Transmission

The engine management's application software calculates the engine speed from the signal from the engine-speed sensor. This value is calculated once for every combustion cycle. The measured value enters both the receive buffer and the transmit buffer of the microcontroller (Fig. 12a).

The engine-management microcontroller is coupled to the CAN controller by a parallel interface. The contents of the transmit buffer are cyclically passed (e.g. every 10 ms) to the transmit buffer of the CAN controller. A flag notifies the CAN controller that a message is ready to be sent. With this transmit instruction, the engine management has completed its part of the task.

Once the engine-speed information has been stored in the transmit buffer, the CAN controller begins to compile the CAN frame. Over the RxD line, the CAN controller now monitors the bus. If the bus is free, it immediately starts to generate the bit stream and directs it to the transceiver on the TxD line; if the bus is occupied, it waits. From the bit stream, the transceiver creates a signal with the requisite voltage level.

Reception

The message originally sent by the engine management and forwarded on by the transceiver is received by all stations connected to the bus (Fig. 12b). The message reaches the CAN controller on the RxD lines in the form of a bit stream.

At the first stage (monitoring layer), the CAN controller checks the incoming message for errors using the CRC checksum. If the message is free of errors, each

12 Data transfer sequence

station responds with an acknowledge (ACK check).

At the second stage, or acceptance layer, the message undergoes message filtering. Each station checks whether the received identifier is addressed to the particular station and whether the message is required in the application software. If not, the message is rejected. Otherwise, it makes its way to the receive buffer. A flag notifies the application software that a new message is ready for processing.

The instrument cluster, for example, calls up the available message, processes the engine-speed information and calculates triggering signals for the actuator of the rev counter.

Standardization

The International Organization for Standardization (ISO) and SAE (Society of Automotive Engineers) have issued CAN standards for data exchange in automotive applications:

- For low-speed applications up to 125 kBit/s: ISO 11 519-2 and 11 898-3
- For high-speed applications faster than 125 kBit/s: ISO 11 898-2 and SAE J 22 584 (passenger cars) or SAE J 1 939 (commercial vehicles)
- An ISO Standard for diagnosis via CAN has also been published as ISO 15 765

Standardization makes it possible for components of different manufacturers to function together. No adaptations are required.

CAN is also widely used in industrial automation. These applications are supported by an alliance of companies in the "CAN in Automation" users group (CiA).

Bosch has concluded contracts with its licensees that guarantee that any CAN implementations will be able to communicate with each other. Users will be able to rely on the interaction of any CAN modules.

Characteristics

- Standardized in accordance with ISO 11898
- Prioritized communication
- Data transfer rates: up to 1 MBits/s
- Data capacity: up to 8 bytes per message
- Real-time response: the data protocol is sufficient for the real-time requirements in the motor vehicle
- Non-destructive bus-access method
- Low power consumption
- Flexibility of configuration
- Simple and economical design with twisted line pairs
- Very high reliability of data transfer
- Fault detection and signaling
- Localizing of failed stations
- Handling of intermittent and permanent faults
- Short-circuit resistance
- The number of nodes is theoretically unlimited. However, a limit arises in practice from the capacitive load of the bus and the increasing latencies of messages when a high number need to be sent.

LIN bus

Overview

The increasing use of mechatronic systems in the motor vehicle gave rise to the idea of designing a cost-effective bus system as an alternative to the low-speed CAN. In 1998, several automotive manufacturers founded a consortium with the aim of developing a specification for a serial bus for the networking of sensors and actuators in the body electronics area.

It was believed that a bus system with simple bus protocol and a simple sequence control would make it possible to use even low-capability microcontrollers without additional hardware for the communication interface.

The workgroup's resulting LIN bus specification was introduced into mass production with the Mercedes-Benz SL as early as 2001.

The name, LIN (Local Interconnect Network), is derived from the fact that all electronic control units are located within a demarcated installation space (e.g. in the door). The LIN, therefore, is a local subsystem for supporting the vehicle network by means of superordinate CAN networks.

The LIN bus is suitable for low data rates of up to 20 kBit/s and is typically limited to a maximum of 16 bus subscribers.

The electrical interface can be created easily and cost-effectively in the network nodes. As far as the nodes are concerned, a distinction is made between the master, which is generally an electronic control unit connected to a superordinate bus system, and the slaves. These are intelligent actuators, intelligent sensors or, quite simply, just switches with additional hardware for the LIN-bus interface.

The bus subscribers are usually arranged in a linear bus topology and connected to each other by a single-wire line. This topology, however, is not explicitly specified.

Communication on the LIN bus takes place in a time-synchronous manner, whereby the master defines the time grid. Consequently, there arises a strictly deterministic LIN bus response.

Figure 1 shows an example of a LIN network as a subbus in the roof/wiper area of the motor vehicle. Here, the bus comprises a central electronic control unit, as the master, and the four slaves: mirror, garage-door opener, rain/light sensor and wiper actuator. The master also functions as a gateway to the Chassis CAN, the Body CAN and the Diagnostics CAN.

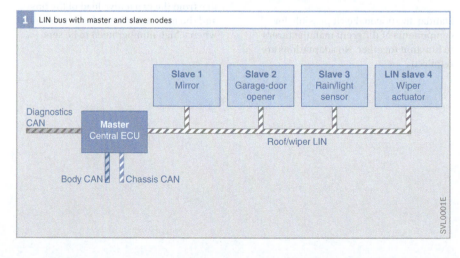

1 LIN bus with master and slave nodes

Applications

The LIN bus as a means of networking me-
chatronic systems can be used for many
applications in the motor vehicle for which
the bit rates and variability of the CAN bus
are not essential. Examples of LIN applica-
tions:
- Door module with door lock, power-win-
 dow drive and door-mirror adjustment
- Control of the power-sunroof drive unit
- Control of the wiper motor for the wind-
 shield wiper
- Sensor for rain and light detection
- Air-conditioning system (transmission
 of signals from the control element,
 activation of the fresh-air blower)
- Headlight electronics
- Control of motors for seat adjustment
- Theft deterrence
- Garage-door opener

Data transmission system

The LIN bus is designed as an unshielded
single-wire line. The bus may adopt two
logical states:
- The dominant level corresponds to the
 electrical voltage of approx. 0 V and
 represents logical 0
- The recessive level corresponds to
 battery voltage U_{Bat} and represents
 the logical 1 state

The recessive level is characterized by a
resistance to battery voltage of 1kΩ in the
master and 30kΩ in the slaves.

Due to circuitry variations, there may be
differences in the voltage levels. A stable
data transfer is ensured by the defining of
tolerances for sending and receiving in the
areas of the recessive and dominant levels.
The tolerance zones on the reception side
are wider (Fig. 2) to make it possible for
valid signals to be received despite inter-
ference radiation.

The data rate of the LIN bus is limited
to a maximum of 20 kBit/s. This is the out-
come of a compromise between the de-
mand for high edge steepness for easy syn-
chronization of the slaves on the one hand,
and the demand for lower edge steepness
for improved EMC characteristics on the
other. The standard bit rates of 2,400 Bit/s,
9,600 Bit/s and 19,200 Bit/s are recom-
mended. The minimum permissible value
for the bit rate is 1 kBit/s to prevent time-
out conflicts. The edge steepness itself
is defined in the LIN specification as
1 to 3 V/µs.

The maximum number of nodes is not
specified in the LIN specification. It is
theoretically limited by the number of
available identifiers. In practice, the num-
ber of subscribers is restricted to 16 due
to the maximum permissible total capacity
of the bus system.

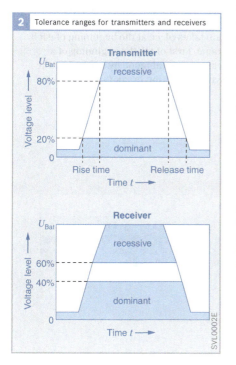

2 Tolerance ranges for transmitters and receivers

Bus access

Access to the LIN bus is determined by the master-slave principle. Each message is initiated by the master. The slave has the possibility to respond. The messages are exchanged between the master and one, several or all of the slaves (point-to-point, multicast, broadcast).

The following relationships are possible in the communication between master and slave:

- Message with slave response: the master sends a message to one or more slaves and asks for data (e.g. switch states or measured values)
- Message with master instruction: the master gives a slave a control instruction (e.g. switch on a servomotor)
- The master initiates communication between two slaves

No arbitration or conflict management is required because, with the master-slave access control, the master alone controls access to the data line.

LIN protocol

Frame

The information transferred on the LIN bus is embedded in a defined frame (Fig. 3). A message initiated by the master always begins with a header. The response (message field) contains different information depending on the type of message. If the master wishes to transmit control instructions for a slave, it will populate the response with data to be used by the slave. If the master is transmitting a data request, the slave that it is addressing will populate the response with the data requested by the master.

Header

The head is composed of the following parts:

- The synchronization break
- The synchronization field
- The identifier field

Synchronization

To guarantee consistent data transfer between master and slaves, a synchronization takes place at the beginning of each frame. First of all, the beginning of a frame is unambiguously marked by the synchronization break (SynchBreak).

3 LIN message format

	LIN frame										
Header			Response								
Synch Break	Synch Field	Ident Field	Data field 0	Data field 1	Data field 2	Data field 3	Data field 4	Data field 5	Data field 6	Data field 7	Check-sum

SVL0003E

The SynchBreak comprises at least 13 consecutive dominant levels and one recessive level.

At the end of the synchronization break, the master sends the synchronization field (SynchField) consisting of the bit sequence 01010101. The slaves are then able to adjust themselves to the time basis of the master and thus synchronize.

The synchronization method described permits a loose specification of the timing of the bus subscribers. The clocking of the master should not deviate more than ±0.5 % from the nominal value. The clocking of the slaves is permitted to deviate by up to 15 % before synchronization as long as the synchronization reduces the deviation to a maximum of 2 % before the end of the message.

In this way, the slaves can be built without an expensive quartz oscillator, e.g. using a simple RC circuit.

Identifier

The third byte in the header is used as the LIN identifier. As with the CAN bus, a content-based addressing method is used – the identifier therefore provides information about the content of a message (e.g. engine speed). Based on this information, all nodes connected to the bus decide whether they would like to receive and process the message further or simply ignore it. This process is known as acceptance filtering.

Six of the eight bits of the identifier field determine the identifier itself. Their permutations give rise to a possible 64 different identifiers (ID). They have the following meanings:
- ID = 0 to 59: transmission of signals.
 ID = 60: master request for the commands and diagnosis
- ID = 61: slave response to ID 60
- ID = 62: reserved for manufacturer-specific communication
- ID = 63: reserved for future extensions to the protocol

Of the 64 possible messages, 32 may only contain two data bytes, 16 may only contain four data bytes, and the remaining 16 eight data bytes each.

The last bits in the identifier field contain two checksums, which are used to check the identifier for transmission errors and any resulting incorrect message assignments.

Data field

Once the header sent by the master node has been transmitted, it is time for the transfer of the actual data to begin. The slaves know from the transmitted identifier whether or not they are being addressed and, if they are, they reply with their response in the data field.

Several signals can be packed into one frame. In this case, each signal has precisely one generator, i.e. it is always written by the same node of the network.

During the data transfer of the bytes, it is always the least significant bit (LSB) that is output first. Each byte (8 bits) is preceded by a start bit and followed by a stop bit, which means that each byte involves the transmission of ten bits. The purpose of the start and stop bits is to resynchronize the nodes and thereby prevent transmission errors.

The data response of the slaves is verified by means of a checksum.

LIN description file

The configuration of the LIN bus, in other words the specification of network subscribers, signals and frames is managed in the LIN description file, or ldf. For this purpose, the LIN specification provides for an appropriate configuration language.

From the ldf, a set of C codes or header files is automatically generated using suitable tools. These are used as the basis for implementing the master and slave functions in the electronic control units connected to the bus.

The ldf is therefore a means of configuring the entire LIN network. It represents a common interface between the vehicle manufacturer and the supplier of the master or slave modules.

Message scheduling

The scheduling table in the ldf defines the sequence and time grid in which messages are sent. Frequently needed information is sent more often. Once the table has been worked through, the master begins with the first message again. The sequence in which the table is worked through may change depending on the operating status (e.g. diagnostics active/inactive, ignition on/off).

The transfer grid of each message is therefore known. This deterministic behavior is guaranteed by the fact that, with the master-slave access control principle, all messages are initiated by the master.

Network management

The nodes of an LIN network can be forced into sleep mode to minimize the no-load current of the entire electronics and electrical system in the vehicle. Sleep mode can be achieved in two ways:

• The master sends the go-to-sleep command with the reserved identifier 60
• The slaves enter sleep mode of their own accord if no data transfer has taken place on the bus for a relatively long period (4 seconds)

Both the master and the slaves are able to wake up the network. To do so, they must send the wake-up signal. This comprises data byte 128. After a break of 4 to 64 bit times (wake-up delimiter), all nodes must have been initialized and be ready to respond to the master.

Example: air-conditioning control

The LIN bus is typically used in the control of the air-conditioning system. The operating and display unit functions as the bus master. This is where the software for the open and control algorithms are stored.

One of its tasks is to adjust the speed of the fresh-air blower. The controlled variable for this is the actual temperature in the passenger cell and the desired temperature set by the driver. The electronic control unit receives the interior temperature from a temperature sensor located in an appropriate place in the interior. The desired temperature is set at the control unit, e.g. using a sensor ring.

From the input variables, the electronic control unit calculates that the blower speed needs to be increased to a value of 200 rpm, for example. The electronic control unit transmits a message containing the master instruction to the LIN bus at a defined time interval. The identifier in this example would be "set the blower speed". In this subbus, this instruction corresponds to identifier 25, for example. Following the header containing this identifier, the master transmits a numerical value in the data field that equates to the physical value of 200 rpm. Each slave possesses a list of identifiers that are relevant for this node. The fresh-air blower is the one and only slave that responds to the "set blower speed" identifier and executes the master's request.

At speeds below idle, the fresh-air blower must be switched off. At this low speed, the heavy load could cause the engine to stall. The LIN bus is coupled to the CAN bus via the gateway and thus receives the current engine speed on a continuous basis. If the speed falls below the specified engine-speed threshold, the LIN master sends the message containing the "set blower speed" identifier where the data field contains the value 0. The fresh-air blower switches off in response.

4 LIN networking (example)

Intelligent actuators can send up-to-date operating-status information to the actuating unit. The fresh-air blower records the speed by means of a sensor and sends it back on the LIN bus as a numerical value. Due to the master-slave access method, the value is only contained in a message with slave response initiated by the master. The feedback signal with the current engine speed makes it possible to control, and therefore accurately maintain, the desired value.

Summary

The essential characteristics of the LIN bus are:

- Single-master/multiple-slave concept
- Master/slave access control
- Independent synchronization of the slave possible even without quartz
- Deterministic signal transmission
- Communication in the form of very short messages
- Character-based transmission (UART)
- Bit rate max. 20 kBit/s
- Data transfer over an unshielded single-wire line
- The reference potentials of the data line are battery voltage and ground
- Maximum bus length 40 m
- Maximum number of nodes 16, typically fewer than 12

Bluetooth

Overview

Mobile communication is gaining ever more importance in all domains. For convenient communication between various devices, a wireless link is indispensable. Infrared connections, as often used in the past, tended to be manufacturer-specific, they required a direct line of sight and imposed constraints on the area of movement. Only with a standardized, wireless link could mobile devices of different manufacturers communicate with each other without problems.

The development of the Bluetooth standard began in 1994 at the telecommunications company, Ericsson. As part of an initial study, Ericsson investigated the possibilities for completely replacing the cable connections between the cell phone and additional devices. In cooperation with other industry partners, the Bluetooth SIG (Special Interest Group) was formed in 1998 to create a uniform, internationally-accepted standard. The SIG set itself the aim of specifying a wireless technology with low manufacturing costs, low energy consumption and robust resistance to interference. The wireless interface had to be suitable for transmitting data for multimedia applications.

In the meantime, the Bluetooth SIG has expanded to involve around 2,000 companies from the telecommunications, data processing and automotive engineering fields.

Applications

Bluetooth is an industry standard for the networking of mobile multimedia devices, such as a car sound system, cell phone, headset, PDA (Personal Digital Assistant), PC and peripheral equipment (Fig. 1). Bluetooth stands for the simple exchange of data between portable terminals and the wireless transmission of audio and video signals for entertainment and information.

The short-range wireless connectivity of Bluetooth eliminates the need for connection cables between the cell phone and the hands-free system: no more tangling of cables when making a call. It is in this area that Bluetooth has been the most well received.

Meanwhile, a large number of consumer electronics devices, be they cell phones, PDAs, notebooks or car phones, now offer an address book function.

1 Bluetooth networking of mobile multimedia devices (example)

Fig. 1

1 Cellular phone
2 Headset
3 Personal digital assistant (PDA)
4 Laptop computer
5 Car sound system with navigation unit

With Bluetooth, it is possible to conveniently synchronize the entries stored in the various address books. The car driver is spared the laborious task of typing the phone number stored in the cell phone or PDA or inputting an address into the navigation system.

Bluetooth also makes it possible to listen to the radio through a headset in the rear of the vehicle and to control the car sound system using a PDA.

Bluetooth is predominantly used for the transfer of data in the multimedia domain. However, this technology can quite easily be used for diagnostic and service purposes. Workshops or breakdown services would be able to conveniently read vehicle information during fault diagnosis and record the condition of the vehicle.

Bluetooth versions

The first specification (Bluetooth 1.0) of the Bluetooth standard was adopted in July 1999. In December 1999 this specification was replaced by Bluetooth 1.0b, which contained improvements and clarifications for interoperability. February 2001 saw the release of the Bluetooth 1.1 specification, which fixed fundamental interoperability problems between the first devices and chips and is used in most Bluetooth devices today. Three years later, in November 2003, version 1.2 of the specification was published. This used the IEEE (Institute of Electrical and Electronics Engineers) language definition for technical documentation and introduced adaptive frequency hopping (AFH) as a technical advancement that improves resistance to interference in the frequency band. The most significant modification was introduced with version 2.0 of the specification in November 2004. With the enhanced data rate (EDR), it was possible to triple the bit rate from 1 MBit/s to 3 MBit/s and reduce power consumption. With each new version of

the specification, great emphasis was placed on backward compatibility with the previous versions.

Transmission technology
Frequency band
Bluetooth operates in the worldwide-license-free 2.4 GHz ISM band (Industrial Scientific Medicine). This is available practically worldwide without approval - apart from certain national limitations - and is reserved for applications in industry, science and medicine. For this reason, the possibility of interference from garage-door openers, microwave ovens or other appliances which operate using the same frequency band has to be accepted. To minimize interference, the best possible use should be made of the frequency spectrum. Devices that use the ISM band therefore implement spread-spectrum techniques.

Frequency-hopping method
Bluetooth divides the 2.4 GHz band into 79 channels at intervals of 1 MHz ($f = 2{,}402 + k$, $k = 0...78$). The band is terminated at both the upper and lower end by two guard bands with a width of 3.5 MHz (upper guard band) and 2 MHz (lower guard band). The transmission

2 Frequency-hopping method

Frequency f

Time slots Time t ⟶

SVB0002E

uses a combined frequency-hopping/time division duplex (FH/TDD) method: the frequency-hopping spread spectrum (FHSS, Fig. 2). The channels are switched 1,600 times per second (i.e. in 625 μs slots). The slots are assigned in accordance with the TDD method, i.e. the transmitter and receiver are authorized to send alternately. The frequency-hopping method guarantees optimal and uniform use of the ISM band and makes Bluetooth resistant to interference from other transmitters in the same frequency band (e.g. WLAN). In addition, it also offers some security because a hacker would not know the destination of a frequency switch or indeed the time frame for which a particular frequency would be valid. This makes interception of a connection much more difficult.

The introduction of adaptive frequency hopping (AFH) in the Bluetooth 1.2 specification further increased Bluetooth's resistance to interference and reduced the effect of Bluetooth itself as a source of interference on other devices in the 2.4 GHz ISM band. The number of effective channels is dynamically adaptable to the conditions in the frequency band. Frequencies used by other transmitters are then removed from the list of 79 possible channels.

Modulation method

Bluetooth uses the Gaussian frequency-shift keying (GFSK) method of modulation. This is a special variant of frequency-shift keying (FSK) supplemented with a Gaussian low-pass filter. GFSK varies the frequency and delivers 1 bit per keying interval. Bluetooth has a symbol rate of 1 MBit/s. With the enhanced data rate (EDR) introduced with the Bluetooth 2.0 specification, it has been possible to increase this data rate to 3 MBit/s. This is made possible by the use of two variants of phase-shift keying (PSK): the π/4 differential quaternary phase-shift keying (π/4DQPSK) and the

8 differential phase-shift keying (8DPSK), which modulate the phase of the carrier.

π/4DQPSK represents one of four combinations of a bit pair (00, 01, 10 and 11) and so transfers a maximum of two bits; 8DPSK can represent any combination of three bits. This is used to increase the bit rate from 1 MBit/s (net 727 kBit/s) to 2 MBit/s (net 1,446 kBit/s) with GFSK, or 3 MBit/s (net 2,169 kBit/s) with EDR.

Power classes

Bluetooth devices are categorized into different power classes. Manufacturers are free to decide which power class they implement. The present specifications define three power classes:
- Class 1 with 100 mW (20 dBm) transmission output for a range of 100 to 150 m
- Class 2 with 2.5 mW (4 dBm) transmission output for a range of 10 to 25 m
- Class 3 with 1 mW (0 dBm) transmission output for a range of 10 m

Topology

Bluetooth supports the automatic configuration of ad-hoc networks. This means that two or more devices are able to spontaneously form a network with no prior knowledge of each other. Bluetooth is based on the master-slave principle. In principle, any Bluetooth device may be a master or a slave. The master is a special device that coordinates communication between the devices.

Exported content

Piconet

A network of Bluetooth devices is known as a piconet (Fig. 3). For the devices within a piconet, there are two roles: master or slave. Which device assumes which role is not decided until communication is established. A device sets up a piconet and transmits its device ID and the value of its internal clock. This device becomes the master.

Any device that is not participating in a piconet but is not switched off remains in standby, which is marked by a small current draw.

The first stage in the creation of a piconet involves the potential master searching for other devices in its range (inquiry scan). Special frequency-hopping sequences (inquiry sequences) are available for this purpose. Devices that wish to enter into communication with other devices respond with an inquiry response message. By the end of the inquiry sequence, the information about all devices willing to communicate is available. The potential master is now able to address a specific device by means of a special paging sequence. Paging is a process whereby a fixed connection is established between a master and a slave. The slave contains all the information it needs to synchronize with the master.

The piconet is defined by the frequency-hopping sequence, i.e. the sequence of switching between the 79 possible channels. The hopping sequence is specified by the master and is calculated from the master's device ID (a worldwide unique 48-bit identifier) and its internal clock. The slaves synchronize with the master's clock and its hopping sequence.

A master is able to manage up to seven active slaves. Each active slave receives a 3-bit active member address (AMA).

Active devices can send data or even simply remain connected. By deregistering, a device can switch back to standby. A Bluetooth device may also switch to any of three energy saving modes: hold mode, sniff mode and park mode. The devices continue to be synchronized with the hopping sequence of the piconet. The lowest energy saving mode is the park mode, where the device relinquishes its AMA and receives in its place an 8-bit parked member address (PMA). With the PMA, the device is still a subscriber in the piconet but makes way for another, active device. Parked devices remain synchronized with the hopping sequence of the piconet and can be addressed and reactivated by the master using the PMA.

Scatternet

In a piconet, only one slave is ever able to communicate with the master simultaneously. When more slaves enter the network, the data rate per device falls very quickly. This is where scatternets provide a solution. A scatternet consists of up to ten piconets in which the individual piconets partly overlap. In this way, it is possible to set up even larger networks in which all network subscribers may connect to one another. The scatternet in Figure 4 comprises two piconets. One of the Bluetooth devices is a member of both piconets. Through this device, the two networks are able to exchange data. However, this device, which is assigned

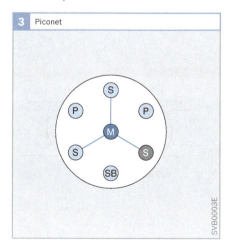

3 | Piconet

Fig. 3
M Master
S Slave
P Park mode
SB Standby

to more than one piconet, can only function as master in one of these networks.

A device that is assigned to two piconets in a scatternet must synchronize with the network with which it wishes to communicate. It cannot remain connected to both networks at the same time. Before the device leaves the old piconet, it must inform its master that it will be unavailable for a certain period of time. The remaining network subscribers can continue to communicate.

A master, too, may leave its own piconet and become a slave in another. All communication in its old network, however, is broken until it returns and assumes its master function again.

A master cannot become master of a second piconet. If this were to happen, both piconets would behave in an identical manner and form a single network.

Physical data channel

A data channel is represented by a hopping sequence between the 79 possible frequencies in the ISM band. Any device that is actively participating in a piconet must also hop at the same carrier frequency (frequency f_i). As soon as a master has sent data at frequency f_k, a slave can respond at frequency f_{k+1}. This procedure is illustrated in Figure 5.

It is conventional for the transmissions of the master to begin in the even time slots and those of the slaves to begin in the odd time slots. This characteristic conforms to the regular time-division multiplexing method (TDD here) whereby the master uses one half of the time slots and the slaves the other half.

In addition to packets that may occupy one time slot, Bluetooth provides for packets that may occupy three or five time slots (multi-slot packets). As soon as a master or slave sends a packet three or five time slots long, this transmitter will remain on the same frequency. There is no frequency change within a packet (Fig. 6).

Once the packet has been transferred, the frequency hops as determined by the hopping sequence (independently of the transmission process). In the illustration, for example, the frequency hops from f_k to f_{k+3} after the transmission of a 3-slot packet). The reason for this behavior is that not every station may have received the transmission and cannot therefore respond specifically to the transfer of data in several time slots. For this reason, all stations that are not involved in the transmission always continue to hop in the sequence specified by the master.

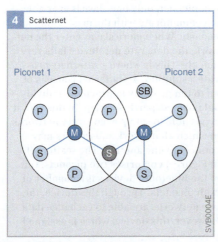

4 Scatternet

Piconet 1 Piconet 2

Fig. 4
M Master
S Slave
P Park mode
SB Standby

SVB0004E

5 Timing of data transfer

$f(k)$ $f(k+1)$ $f(k+2)$

Master

Slave

625 µs

SVB0005E

Physical connections

Bluetooth supports both circuit-switched and packet-switched data channels. For circuit-switched, synchronous communication, Bluetooth offers a synchronous connection-oriented link (SCO), and an ACL (Asynchronous Connectionless Link) for packet-switched, asynchronous communication. Between two Bluetooth devices, there may only ever be one ACL link but there can be up to three SCO links. The ACL links are the foundation for the data transfer. The SCO links are purely voice channels with a data rate of 64 kBit/s. The voice is transmitted by continuously variable slope delta modulation (CVSD) or pulse-code modulation (PCM). A Bluetooth master can maintain several different data channels between its slaves even at the same time.

Logical data channels

Logical channels refer to different types of channels streamed over a physical connection. The data transmitted over a physical channel has different logical meanings. Bluetooth distinguishes between two categories of channel: link channels, which are used for the exchange of control information between master and slave; and user channels, which transfer data from the user level – the applications.

Data packets

A Bluetooth data packet generally comprises three fields:
- Access code
- Header and
- Payload

A packet may simply comprise the access code, or the access code and a header, or the access code, header and payload (Fig. 7).

A packet begins with an access code. If a header follows, the access code has a length of 72 bits, otherwise it is 68 bits long. The access code is used for the synchronization and identification of a piconet, for device polling and for device calls. There are three different types of access code:
- Channel access code (CAC), the purpose of which is to synchronize and identify a piconet. The CAC is sent with each packet transmitted over the piconet.
- Device access code (DAC), which is used to transfer specific identifiers during the device call (e.g. when paging).
- Inquiry access code (IAC), which is used for device polling.

6	Frequency hopping with multi-slot packets

7	Structure of a packet

Access code 72 (68) bits	Header 54 bits	Payload 0 to 2,745 bits
4 preamble	3 AMA	
64 synchron-ization	4 type	
	1 flow	
(4) appendix	1 ARQN	
	1 SEQN	
	8 checksum	

SVB0007E

The header of a Bluetooth packet is 54 bits long and contains information about the link (link control information). It is made up of the address, packet type, flow control, error monitoring and checksum fields. The actual length of the header (i.e. the sum of the bits in the header) is 18 bits. The header is protected by forward error correction (FEC). The bits of the header are sent three times (1/3 FEC), giving a header length of 54 bits. A receiver can then simply take the majority decision; each bit triplet is resolved to the value with majority in the triplet.

The payload of a Bluetooth packet can be up to 341 bytes in size (1,023 bytes with EDR). The data field of the payload consists of up to three segments: the payload header, the payload itself and, in certain circumstances, a checksum.

Bluetooth uses 16 different packet types, the common feature of all being the access code and the header. The structure of the payload depends on the packet type concerned. Along with packets for ACL and SCO links, there are also packet types for polling the slaves, synchronizing the hopping sequence and acknowledging data transfers.

Bluetooth device addresses
A Bluetooth device is assigned a worldwide unique Bluetooth device address. The address is derived from the IEEE-802 standard and implemented in the device by the manufacturer. The 48-bit addresses are subdivided into three parts
- LAP (Lower Address Part, containing 24 bits)
- UAP (Upper Address Part, containing 8 bits)
- NAP (Non-significant Address Part, containing 16 bits)

The LAP and UAP fields form the significant part of a Bluetooth device address.

Bluetooth architecture
The Bluetooth architecture is complex. This is inherently associated with the idea behind general cable emulation. The architecture is designed to enable new protocols and applications to be adapted for the use of Bluetooth. Accordingly, the Bluetooth protocol stack is very extensive (Fig. 8 shows a simplified version of the protocol stack).

The Bluetooth technology can be subdivided into three logical sections:
- Hardware
- Host stack
- Applications

The hardware layer and the host stack are defined in the Bluetooth Core Specification; the applications are part of the Bluetooth Profile Specification.

Hardware-side protocols
The hardware is represented by the Bluetooth chip. This is where the radio layer, baseband and link management are contained. The radio layer deals with the radio transmission, used frequencies, modulations and transmission output. The baseband represents the link-establishment mechanisms, the packet structure and the timing. The link management establishes and manages the link between

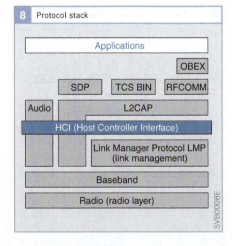

8 Protocol stack

Applications

OBEX

SDP · TCS BIN · RFCOMM

Audio · L2CAP

HCI (Host Controller Interface)

Link Manager Protocol LMP (link management)

Baseband

Radio (radio layer)

SVB0008E

two devices and also implements security and authentication functions.

The host controller interface (HCI) is the interface between the hardware of a Bluetooth module and the host-side protocols. The HCI is a hardware abstraction and realizes various interfaces for controlling the Bluetooth hardware and transferring data (e.g. via USB or UART).

Host stack

The host stack uses the HCI to control the Bluetooth hardware and transfer data. The format of this data is determined by the upper protocol layers. The logical link control and adaptation protocol (L2CAP) implements the abstraction of the hardware's properties, adapts the upper layers of the protocol stack to the capabilities of the baseband and hides transmission details such as the connectionless or connection-oriented transmission type. The L2CAP layer essentially has three main functions:

- It can receive packets with a length of up to 64 kB from the upper layers and decomposes them into smaller data packets (segments) for processing in the lower layers if necessary. At the other end, the segments are recomposed back into packets.
- It manages the multiplexing and demultiplexing of several packet sources. If a packet is being recomposed, the L2CAP layer determines to which protocol of the upper layers the packet is forwarded.
- L2CAP offers functions for negotiating quality of service and configuration parameters. This means, for example, that the maximum size of the payload can be negotiated so that a device with limited resources would not be overwhelmed by overly large packets. Using the configuration parameter for the quality of service, it is possible to define the properties of the data transfer: best effort (best attempt but with no guarantee of data transfer) or guaranteed.

In addition to L2CAP, there are also audio and control protocols that govern the handling of audio data and control data. Audio applications, for example, can make direct use of the baseband layer once the audio signals have been coded accordingly.

The service discovery protocol (SDP) is another important host-side protocol. It serves to identify and scan for services with specific properties and to describe services within range of a Bluetooth device. Bluetooth devices are meant to be able to interact on an ad hoc basis with other devices in different environments. It is therefore necessary to know which services are made available by which devices within range. All devices wishing to provide services must use an SDP server; for all other devices, an SDP client is sufficient. The SDP prepares the services available on a device in a service database. The service information in the possession of the SDP server is stored in a service record. The service record consists of a list with service attributes that describe the properties of the service more precisely and is identified by a 32-bit service record handle.

Two further protocols in the protocol stack form the fundamental basis for the interoperability of Bluetooth devices.

The RFCOMM (Radio Frequency Communication) cable emulation protocol above the L2CAP layer simulates up to 60 virtual serial interfaces derived in accordance with the ETSI 07.10 standard. As a result, almost any software that previously expected a serial interface can work with Bluetooth.

TCS BIN (Telephony Control Protocol Specification - Binary) is employed as the means of controlling telephone and telephony functions. This is a bit-oriented protocol for establishing voice and data connections between Bluetooth devices. The use of appropriate, indus-

try-wide standards has made it possible, here too, to ensure wide-ranging compatibility with legacy applications.

Applications
Many more protocols have been adapted for the Bluetooth standard (adapted protocols) and can be found in the protocol stack. Internet applications can, for example, continue to use TCP/IP through the point-to-point protocol (PPP) or the Bluetooth network encapsulation protocol (BNEP). For the exchange of vCalendars and vCards, it is possible to use the object exchange protocol (OBEX) covered by the IrDA standard.

Profiles
The uppermost layer contains the applications and profiles. Profiles represent "standard solutions" for a particular usage scenario. The Bluetooth specification currently brings together 13 different applications described as profiles. Each profile uses a certain choice of protocol; in principle, application profiles make a different protocol stack available for each application. Profiles describe the vertical slice through the Bluetooth protocol stack; the protocols represent the horizontal layers (Fig. 9). Within the profiles, the required and optional functions of the layers are defined. These standardized profiles make it possible to ensure interoperability between different devices.

The four fundamental profiles of the Bluetooth specification are:
- Generic access profile (GAP)
- Service discovery application profile (SDAP)
- Serial port profile (SPP) and
- Generic object exchange profile (GOEP)

The Bluetooth profiles form the hierarchy (Fig. 10). The GAP forms the basis and describes all the essential functions that a Bluetooth device must fulfill at the lowermost level. These include, for example, the functions for establishing and managing the link, the supported operating modes and the security of a link.

The SDAP defines the access interface for the service discovery protocol (SDP) with which devices can discover or poll the services offered by other devices. SDP builds on the GAP. All Bluetooth devices must implement these two profiles.

The SPP is used by most other Bluetooth profiles. One exception to this is the telephony control profile. This is always used when Bluetooth is being used for cable emulation or if a serial data connection is to be used. The SPP builds on the GAP and uses the RFCOMM protocol.

The GOEP defines the fundamental functions necessary for the exchange of complex objects. It defines a client-server relationship for the exchange of data. Like the SPP, the GOEP provides the foundation for further profiles.

9 Profiles and protocols

10 Profile hierarchy

SVB0010E

▶ Origin of the name, Bluetooth

The name, Bluetooth, originates from
King Harald Gormsen, a Danish Viking.
Harald had the nickname, Blatand.
In the Middle Ages, Bla actually meant
dark, while Blatand made reference
to the dark figure of Harald Gormsen.
Bluetooth, therefore, is not a direct
translation of Blatand.

In the 10th Century, Harald Gorm-
sen united large parts of Denmark
and Norway. The Bluetooth wireless
system developed 1,000 years later
unified a wide range of different infor-
mation, data processing and wireless
mobile devices. This is why it was
named after King Harald.

MOST bus

Introduction

The MOST bus (Media Oriented Systems Transport) was specifically developed for the networking of infotainment systems in motor vehicles. In addition to traditional entertainment functions, such as radio tuners or CD players, these infotainment systems (which are becoming an increasingly common feature of modern vehicles, and premium class vehicles in particular) also offer video functions (DVD and TV), route guidance capabilities and access to mobile communications and information.

Requirements

The functions of an infotainment system place high demands on a bus system. The transmission of multimedia data, both audio and video, requires a high data rate and a synchronization of the data transfer between source and sink as well as between sinks themselves.

For the transmission of informational data, e.g. detailed information about the music tracks on an mp3 player, and the sending of software updates, it is necessary for the data transfer to be flexible with varying and, at certain times, equally very fast data rates.

Likewise, the requirements for use in a motor vehicle, and the associated requirements of electromagnetic compatibility (EMC), must also be fulfilled.

MOST Cooperation

It was with these requirements in mind that the MOST bus was developed by the MOST Cooperation. This was founded in 1998 by BMW, DaimlerChrysler, Harman/Becker and Oasis SiliconSystems (now part of SMSC). By 2006, the MOST Cooperation had 16 automotive manufacturers and 67 suppliers and tool makers among its membership.

The MOST Cooperation creates and administers the specifications on which the MOST bus is based. Furthermore, it defines the requirements for implementation of MOST devices and provides appropriate compliance tests through accredited test houses.

Type of use

At present, the MOST bus is almost exclusively used for the networking of infotainment systems in motor vehicles. It can currently be found in over 35 vehicle models of various manufacturers, especially in the premium and mid-range classes (as at 2006).

Features of the MOST bus

The MOST bus supports the logical networking of up to 64 devices, at which point constraints associated with the chipsets may be encountered.

In its current version, the MOST bus offers a data rate of 24.8 MBit/s (MOST 25). Versions with higher data rates of 50 MBit/s (MOST 50) and 150 MBit/s (MOST 150) are already available as a development model.

Data transmission channels

For the transmission of data, the MOST bus supports the following channels and their different attributes:

- The control channel is used for the simple transmission of control commands, for the signaling of device statuses and for the sending of messages necessary to system management. With MOST 25, the control channel has a gross bandwidth of 705.6 kBit/s.
- For the transmission of multimedia data, the MOST bus has a flexible number of synchronous channels that are able to carry both audio and video data. In MOST 25, there are a maximum of 15 audio channels available in stereo quality.

- On the asynchronous channel, data is sent in packets. It is therefore suitable for transferring information that does not require a fixed data rate but does need a high data rate at certain times. Examples could be the transfer of track information of an mp3 player or a software update. With MOST 25, the asynchronous channel has a gross bandwidth of up to 12.7 MBit/s.

The available bandwidth can be flexibly distributed between the asynchronous channel and the synchronous channels by means of the "boundary descriptor", which can also be shifted to the transit period if relevant preconditions are met.

Topology

A MOST system is arranged in a ring structure whereby a device is connected to its predecessor or successor in the ring by an input or output respectively (Fig. 1). One of the devices acts as the "timing master" and generates the data frames for data transfer with which the other devices synchronize.

Each device has a bypass. When this is closed, the device forwards the signal directly and is therefore invisible to the MOST system. This service is useful, for example, when the system is starting and a device requires more time for initialization, or in the event of a temperature-dependent shutoff.

There are also other possible structures for a MOST system, e.g. a star topology. Internally, however, these must always be arranged to form a (logical) ring. With a star structure, this can be achieved through use of a central hub, for example.

Device model

The MOST standard defines the model shown in Figure 2 for a MOST device for which the following elements are required:

At the lowermost level, the physical layer provides access to the transmission agent. Departing from the original definition that provided for optical transmission by POF cable (Plastic Optical Fiber), there are today various physical layers available with optical and electrical transmission as well as various speeds of transfer.

1 Ring structure of the MOST system

2 Model for a MOST device

The MOST network interface controller (NIC) is a hardware controller that is responsible for controlling the physical layer and implements the basic transfer services. While in the original version of the NIC many control tasks had to be carried out by the main processor of the device, the current version (Intelligent Network Interface Controller, INIC) is, for the most part, able to handle tasks autonomously and already implements substantial parts of the network service. Typical controllers for the MOST are the OS8104 (NIC) and the OS81050 (INIC), manufactured by SMSC.

The MOST network service is the driver layer through which applications and system services have access to the NIC. The lower layer of the network service (layer 1) offers basic communication and management functions, while the layer above (layer 2) offers functions for supporting the development of applications, such as a translator for the application protocol or support for the notification service.

The applications of the device are implemented on top of the network service. In a MOST system, the interface of an application is realized as a function block (FBlock). Each device must at least implement a special FBlock, the NetBlock, which is required for management functions within the MOST system. In addition, a device will usually implement one or more applications that can be used as FBlocks of applications on other devices. A device of a vehicle infotainment system is often integrated with several functions, e.g. those of a radio tuner and those of an amplifier, which can be represented in the MOST system as independent FBlocks.

Transmission agent

The MOST bus was originally defined as an optical transmission agent that used a plastic optical fiber (POF). The optical signal used here with a wavelength of 650 nm (in the red wavelength range) is generated by an LED on the transmit side, which is designated Tx-FOT (FOT = Fiber Optic Transceiver). On the receive side, Rx-FOT, the optical signal is converted back into an electrical signal by a PIN photodiode.

The POF cables used for automotive applications (Fig. 3) consist of a 980 µm thick optical core insulated by 20 µm thick optical cladding with a low refractive index. In total, the optical conductor therefore has a diameter of 1 mm. The optical fiber is insulated with a black buffer, which is in turn surrounded by a protective cable sheathing. This gives the cable a total diameter of 2.3 mm.

POF lines offer the following benefits when used in the motor vehicle:
- No electromagnetic interference radiation
- Insensitivity to interference irradiation
- Lower weight than equivalent shielded electrical lines
- More flexible routing in comparison to equivalent shielded electrical lines

3 POF cable

Cable sheathing

Buffer

2.3 mm

1 mm 1.5 mm Optical fiber

UVM0003Y

To date, modern vehicles are equipped exclusively with MOST 25, with POF data transmission via POF connections. In addition to this, however, further transmission techniques have been defined for the MOST:
- The optical transmission by glass fiber cable (PCS) with laser diodes (VCSEL), which offers a greater damping reserve, supports higher speeds and is less sensitive to high temperatures.
- An electrical transmission by copper cable, which is also less sensitive to temperature and is more economical by comparison, requires additional shielding measures at higher bandwidths. This has consequences for costs and cabling.

Data transfer
Data transfer on the MOST bus is organized into data frames, which are generated by the timing master with a fixed data rate and passed on by subsequent devices in the ring.

Data frames
The timing master normally generates data frames with a cycle of 44.1 kHz, and in rarer cases 48 kHz. The cycle is defined by the system manufacturer (i.e. the vehicle manufacturer in conjunction with its device suppliers) to suit the predominant media formats in the system. The size of the data frames, by extension, determines the bus speed of a MOST system. With MOST 25, the size of a data frame is 512 bits.

A data frame is made up of the following ranges (Fig. 4):
- Areas for administrative information at the start and end of the data frame, the tasks of which include the synchronization of the data transfer, determination of the devices' ring positions and the transmission of the boundary descriptor

- An area for the control channel (with MOST 25, two bytes per data frame)
- An area for the synchronous channels, whereby one or more bytes are always assigned to a specific synchronous channel
- An area for the asynchronous channel

In the case of MOST 25, the synchronous and asynchronous area demand a 60-byte share of the data frame. The distribution between the synchronous channels and the asynchronous channel is determined by the value of the boundary descriptor with a resolution of 4 bytes. The synchronous area must have at least 24 bytes (6 stereo channels). This means that 24 to 60 bytes are permitted for the synchronous area and 0 to 36 bytes are permitted for the asynchronous area.

4 Data frame

Administrative area: 1 byte
(preambel, boundary descriptor)

Administrative area: 1 byte
(frame control, status and parity)

Synchronous channels ← → Asynchronous channel → Control channel: 2 bytes

← Total of 60 bytes: distribution by boundary descriptor →

← 1 data frame for MOST 25: 64 bytes →

UVM0004E

System events

The fundamental system events described below indicate the current status of the data transfer in the MOST system. These are monitored by the NIC and forwarded by the network service to the applications, which must then respond accordingly.

- Lock, stable lock: a lock occurs as soon as communication is established in the ring, i.e. the correct device receives the correct data frames. A stable lock is set when the system has been in lock status for a specific period.
- Unlock, critical unlock: an unlock denotes the loss of the lock, i.e. that no correct data frames are being received. If the system is in the unlock state for a certain period, the status switches to critical unlock.
- Network change event: if the number of devices in a MOST system changes because one device has opened or closed its bypass, this is signaled by a network change event.

Addressing

Devices on the MOST bus are addressed by means of a 16-bit address. The following different addressing types are supported:

- Logical addresses: each device has a logical address that is unique in the system and is set and distributed when the system starts. The logical addresses can be set dynamically based on the ring position, for which, beginning with the timing master, the hexadecimal range 0x100 to 0x13F is reserved. Or they can be defined statically by the system manufacturer in a range reserved specifically for this purpose.
- Physical addresses: for system administration services, e.g. determination of logical addresses, each device also has a physical address that is derived from its current ring position. The physical addresses are assigned the range 0x400 to 0x43F.

- Group addresses: each device may also be assigned a group address in the 0x300 to 0x3FF range, making it possible to address all devices in a given group at the same time.
- Broadcast address: the MOST additionally supports a special broadcast address, hexadecimal 0x3C8, which makes it possible to address all devices in the MOST system.

Transmission of control commands

The control channel permits the transmission of control messages with a maximum length of 32 bytes. These have to be shared among consecutive data frames. With MOST 25, the 16 data frames required here are also known as a block. The format of the control messages is defined by the protocol described in the next section.

The sending of messages on the control channel is supported by the NIC. It provides the necessary services for data protection and channel arbitration, i.e. for determining the next opportunity for a message to be sent. The reliable transmission of control messages is safeguarded by the use of a cyclic redundancy code (CRC) and an automatic repeat request.

The control channel's gross bandwidth of 705.6 kBit/s is reduced to a net data rate of 406 kBit/s following the deduction of approximately 3 % for system administration overheads (two of 64 control messages) and less the extra overhead to account for arbitration and data protection. Since arbitration only allows a particular device to populate every third available control message, an individual device has a maximum data rate of 135 kBit/s at its disposal for sending.

To enable the sending of messages of more than 32 bytes, the network service supports the segmentation and desegmentation of application messages up to 65,556 bytes in size on the control channel. This service, also known as the appli-

cation message service (AMS), differs from the control message service (CMS), which only allows 32-byte control messages to be sent.

Transmission of multimedia data
In the MOST system, the transmission of multimedia data takes place on the synchronous channels. The data transfer is controlled by relevant control commands on the control channel. A synchronous channel can be assigned a specific bandwidth, which is achieved with a resolution of one byte of a data frame. A stereo audio channel with a resolution of 16 bits requires four bytes, for example. In the case of MOST 25, a maximum of 60 bytes (depending on the value of the boundary descriptor) are available for the synchronous channels, which equates to 15 stereo audio channels.

The sinks and sources for multimedia data are each assigned to an FBlock, which provides the necessary functions for their management. An FBlock may therefore contain several sources and sinks, which are numbered consecutively with a source and sink number respectively.

An FBlock has functions that supply information about the quantity and type of sources and sinks that it makes available (SyncDataInfo, SourceInfo and SinkInfo). Furthermore, each FBlock with a source has an "Allocate" function with which it requests a synchronous channel and associates its source to it. Its "Deallocate" function, on the other hand, is used to free up the channel again.

Similarly, an FBlock with a sink has a "Connect" function for connecting the sink to a particular synchronous channel, and a "DisConnect" function for its disconnection. Several sinks may also be connected to a particular channel at the same time.

With MOST 25, the channel assignment for synchronous channels is managed by the timing master. A source wishing to occupy a synchronous channel sends an appropriate system message request (ResourceAllocate) to the timing master. A corresponding message (ResourceDeallocate) is used to free the channel.

Transmission of packet data
On the asynchronous channel, data is sent in the form of packets. The asynchronous channel currently supports two modes: a slower 48-byte mode, where 48 bytes are available in each packet for the net data transfer but which places less heavy demands on device implementation, and a 1,014-byte mode that is more complex to implement.

With MOST 25, between 0 and a maximum of 36 bytes of a data frame can be assigned to the asynchronous channel, which corresponds to a maximum gross bandwidth of 12.7 MBit/s. For the 1,014-byte mode, this produces a maximum net bandwidth of almost 11 MBit/s, and a data rate of almost 3 MBit/s for the 48-byte mode. In practice, however, substantially lower data rates are achieved mainly due to restrictions in device implementations.

The transmission of data packets on the asynchronous channel is also directly supported by the NIC. Likewise, a packet on the asynchronous channel contains a CRC to permit detection of simple transmission errors. Unlike the control channel, however, there is no issuing of automatic repeat requests on the data link layer.

To ensure reliable transfer and flow control in the large data volumes typical of the asynchronous channel, it is conventional to use yet another transport protocol that is implemented in a higher driver layer. This may either be the MOST high protocol (MHP), developed specifically for the MOST, or the popular TCP/IP protocol, which is set up on an appropriate adaptation layer known as the MOST asynchronous medium access control (MAMAC).

Administrative functions

The MOST standard defines the following administrative services that are required for the operation of a MOST system.

Configuration status

A valid configuration status in the MOST system is the prerequisite for communication in the application layer. The configuration of the MOST system is managed by a single device, the network master. An applicative communication is only permitted once this device has signaled a conflict-free configuration status.

- OK configuration status: the network master transmits the OK configuration status as soon as a conflict-free configuration is achieved.
- NotOK configuration status: if a conflict arises, e.g. because two devices have the same address, the network master transmits its NotOK configuration status. In response to this, all devices initialize their addresses and communication settings.
- New and Invalid configuration statuses: these configuration statuses indicate that a new application (FBlock) has been registered in the system or that a previously registered application has dropped out.

NetBlock

It is mandatory for each device to implement an FBlock called the NetBlock, which covers a range of different administrative functions (e.g. for address initialization) and supplies information about the device and the FBlocks that it implements.

Network master

The network master is implemented by a special device in a MOST system and is responsible for system configuration. In present-day systems, the network master is usually realized by the headunit (i.e. the control element) of the infotainment system. This device often assumes the role of timing master too, but this does

not necessarily have to be the case. The other devices in the MOST system in this relationship are known as network slaves.

The network master manages an image of the current configuration of a MOST system in the central registry. In addition to the addresses of all devices, this also contains all FBlocks implemented by them (down to the system FBlocks such as the NetBlock). A network slave searching for the device address for a particular FBlock may request it from the network master.

If a network slave frequently accesses other devices, it may, for faster access, also store a local copy of the central registry – a decentral registry – which it must update accordingly in the event of any change in the configuration.

To build the central registry, the network master carries out a network scan when the system starts and as soon as a network change event has occurred. During this scan, it queries the NetBlocks of all devices for the FBlocks implemented by this device. If it detects a conflict while doing so, e.g. duplicate device address or duplicate instance of an FBlock, it takes measures to resolve the conflict. To do this, the network master may initiate a recalculation of addresses, convert the InstId of an FBlock or, if necessary, ignore a device.

Connection master

The connection master manages the synchronous connections existent in the MOST system at a given time. It makes available a table containing information about all the connections currently present in the MOST system (SyncConnectionTable). It also makes it possible to query the bandwidth remaining for further synchronous connections (AvailableChannels).

On request, the connection master builds connections between a specific source and sink (BuildSyncConnection) or removes them (RemoveSyncConnection) by sending the necessary commands to the associated FBlocks.

It is even possible to dispense with the connection master completely if a central entity in a MOST system, e.g. the headunit, is responsible for the administration of the synchronous channels.

MOST application layer

The MOST standard defines a suitable protocol in the application layer for the transmission of control commands, status information and events. This protocol makes it possible to address a specific function of an application interface (i.e. of an FBlock) that is provided by any device within the MOST system. For example, it is possible to start playback of a CD on a separate CD player or to query the number of the track currently playing in it. While the MOST protocol is mainly applied to the control channel, it can be transferred across the asynchronous channel if necessary.

Overview

The protocol for MOST control messages provides for the following elements of a control message (Table 1):

- The address of a device in the MOST system (DeviceID)
- An identifier for an FBlock implemented by this device (FBlockID) and its instance in the MOST system (InstID)
- The identifier for the requested function within the FBlock (FktID)
- The type of operation (OpType) that should be applied to this function, e.g. the setting or querying of a property of the FBlock
- A data range containing the parameters of the function call (Data) and a corresponding length value (Length)

1 Elements of a control message

Field	Size	Description
DeviceID	16 bits	Device address
FBlockID	8 bits	FBlock identifier
InstID	8 bits	Instance of the FBlock
FktID	12 bits	Function identifier
OPType	4 bits	Type of operation
Length	16 bits	Length of the data field
Data	0 to 65,535 bytes	Data field

Table 1

Function block (FBlock)

A function block (FBlock) defines the interface of a specific application or system service. An FBlock is assigned an address comprising an 8-bit FBlockID, which specifies the type of FBlock, and an additional 8-bit InstID. The latter is a means of distinguishing between several instances of FBlocks of the same type in a MOST system.

An FBlock that is controlled by an associated application, and the device that implements the FBlock, are known as the slave in this relationship (e.g. an application that addresses an external CD player through the associated FBlock). The application that controls the FBlock is called the controller.

The functions that an FBlock possesses are defined by the function catalog. In the case of system services, the function catalogs are defined by the MOST standard. The MOST standard even defines the interfaces for common applications from the area of vehicle infotainment systems (e.g. for an amplifier or a CD player/changer). However, these interfaces tend to be extended by the system manufacturer. Other proprietary FBlocks are defined in their entirety by the system or device manufacturer.

Functions and operations

A function block is made up of several functions that are addressed with a 12-bit FktID. Depending on the type of function, the function may be assigned differently predefined operations, which are described by a 4-bit OpType identifier. In the case of functions, a further distinction is made between methods and properties.

A method describes an action that an FBlock is able to execute, e.g. starting the station scan of a radio tuner. A method can be initiated by the controller using a Start or StartResult operation. In response, the slave returns a corresponding result message if applicable (for StartResult). If this result is not made available within a specific period, the controller is notified that the method is still being processed by means of a processing message. In the end, a method may even be terminated by the sending of an abort message.

A property describes a particular attribute of the FBlock, such as the number of the track currently playing. A property such as this can be requested (Get operations) or set (Set or SetGet operations) by the controller. As the result of such a query, the slave responds with a status message, if applicable (for Get and SetGet), containing the current contents of the property.

Furthermore, the MOST standard provides for a notification service with which a controller may register with a slave for certain properties. If the value of one of these properties changes, all controllers that have registered for notification are informed of the change by a status message.

Data field

The data field of the control message contains the parameter values for a function call or its results. The data field is interpreted as specified in the definition of the function concerned and may contain one or more parameter values.

The MOST standard defines the following parameter types:
- Boolean and BitField for individual boolean values (1 byte) or sequence of individual bits (1, 2 and 4 bytes).
- Enum for enumerations.
- Unsigned and signed byte, word and long for integer values with or without sign and a size of 1, 2 or 4 bytes.
- String for character strings. These are zero terminated and contain a description of the character coding in the first byte.
- Stream, ClassifiedStream and ShortStream for various types of byte sequences of any length.

A parameter type for real numbers does not exist. The MOST standard provides for real numbers represented as fixed-point values by the integer parameter types. More complex data types are not defined in the standard either. These have to be realized by function classes instead.

Function classes

To standardize the way in which functions are defined, the MOST standard specifies a series of function classes for properties. These determine which properties the function has and which operations are permissible.

In addition to simple function classes (with a single parameter for numerical values or texts), there also exist complex, composite function classes. These are used to represent composite data structures such as records and arrays. They can also be nested at a specific nesting depth, which makes it possible to create two-dimensional arrays or arrays of records, for example. These nested data structures are, for example, used to represent the phone book of a mobile phone.

With multidimensional function classes, the two parameters PosX and PosY are used to address a particular position in the first or second dimension, where 0 always stands for all elements of this dimension. This makes it possible to access not only a specific element but also an entire line of a twodimensional array.

Applications

As well as defining the lower layers necessary for data transfer, the MOST standard defines the interfaces for typical applications from the area of vehicle infotainment systems, e.g. a CD changer, amplifier or radio tuner.

The FBlocks defined by the MOST Cooperation are summarized in a function catalog and listed with their current versions in the next section. The function catalog defines in detail all functions for the FBlocks contained in them as well as the permissible operations and their parameters.

For the function catalog, there is a machine-readable description available in XML format that can be used to import the catalog into various MOST tools.

In addition to the function catalog's description of all functions of an FBlock, developments are underway to specify the dynamic behavior for the use of an FBlock. This involves the use of message sequence charts (MSCs) compliant with the MSC 2000 standard but with minor, MOST-specific enhancements.

Functions required by various applications, e.g. the aforementioned functions for managing sources for multimedia data, are defined in the GeneralFBlock. This is used as a function compilation, i.e. when a new FBlock is defined, all necessary functions are copied from the GeneralFBlock to the new FBlock.

Standardization

The MOST standard is maintained by the MOST Cooperation, which also publishes the corresponding specifications. The specifications are available through the home page of the MOST Cooperation (www.mostcooperation.com).

The following versions of the MOST specifications are currently valid (as at 2006):

- MOST Specification, Version 2.4.
- MOST Dynamic Specification, Version 1.2.
- MOST Specification of Optical Physical Layer, Version 1.1.
- MOST Specification of Advanced Optical Physical Layer, Version 1.0.
- MOST Specification of Electrical Physical Layer, Version 1V1.
- MOST High Protocol Specification, Version 2.2.
- MAMAC Specification, Version 1.1.

The MOST Cooperation defines within the framework of the compliance process requirements placed on MOST devices which must be satisfied by all devices in order to be able to carry the MOST logo. MOST compliance is tested and awarded by test institutes which have received appropriate accreditation from the MOST Cooperation. The compliance requirements are governed by the following specifications, which describe the compliance process itself and the compliance tests for the physical layer, for system mechanisms (Core) and for application interfaces (Profile):

- MOST Compliance Requirements, Version 2.0.
- MOST Compliance Test of Physical Layer Specification, Version 1.0.
- MOST Core Compliance Test Specification, Version 1.1.1.
- MOST Profile Compliance Test Specification, Version 1.0.

The following application interfaces are standardized by the MOST Cooperation. The following versions are currently valid:

- AudioAmplifier (FBlockID: 0x22), Version 2.4.2: a simple amplifier.
- AuxIn (FBlockID: 0x24), Version 2.4: an interface for connecting MP3 players or data carriers with pieces of music.
- MicrophoneInput (FBlockID: 0x26), Version 2.3.1: a microphone input.
- AudioTapePlayer (FBlockID: 0x30), Version 2.3.1: a cassette player.
- AudioDiskPlayer (FBlockID: 0x31), Version 2.4: a CD player or CD changer.
- DVDVideoPlayer (FBlockID: 0x34), Version 2.4.1: a DVD player or DVD changer.
- AmFmTuner (FBlockID: 0x40), Version 2.4.2: a radio receiver for FM/AM.
- TMCTuner (FBlockID: 0x41), Version 2.3.1: a special receiver for traffic-message signals (TMC).
- TVTuner (FBlockID: 0x42), Version 2.3.2: a television receiver.
- DABTuner (FBlockID: 0x42), Version 4.0: a receiver for digital radio (DAB).
- SDARS (FBlockID: 0x44), Version 2.4: a receiver for satellite radio.
- Telephone (FBlockID: 0x50), Version 2.3.2: a telephone module or a connection to a cellular phone.
- GeneralPhonebook (FBlockID: 0x51), Version 2.3.1: access to a phonebook, e.g. that of a cellular phone.
- NavigationSystem (FBlockID: 0x52), Version 1.11: an interface for a navigation system; but only unofficially approved by the MOST Cooperation.
- GraphicDisplay (FBlockID: 0x60), Version 2.3: an independent display screen.

TTP/C

Overview

The Time-Triggered Protocol for SAE Class C networks (TTP/C) is a time-controlled protocol in which the network users transfer data in special time intervals determined in advance. In other words, all network users have a global time definition which is determined by means of a time-synchronization protocol.

TTP/C was developed specially to satisfy real-time requirements in distributed, fault-tolerant systems. It was developed specifically with a view to its use in the automotive field, but has in the meantime also come to be used in other fields, such aviation and railroads.

According to the requirements of the Society of Automotive Engineers (SAE) for Class C protocols, TTP/C is a protocol for high-speed networks (>125 kbit/s bit rate) which can be used for real-time communication in check systems. For this reason, particular emphasis was placed during development on the fields of fault detection and tolerance, robustness against malfunctions, composability, and guaranteed latencies.

TTP/C is based on the principle of Time-Triggered Architecture (TTA), a framework for developing highly reliable, distributed real-time systems and the applications that run on them.

The development of TTP/C dates back to the MARS project (Maintainable Real-Time System), which was conducted in 1979 at the Technical University of Berlin. Further development followed within the framework of further projects, such as TTA (Time-Triggered Architecture) or the Brite-EuRam III project "Safety Related Fault Tolerant Systems in Vehicles (X-by-Wire)", in which systems for activating safety-relevant automotive components, such as brakes or steering systems, via electronic networks without a mechanical fallback level were examined.

Areas of application

TTP/C was originally developed for use in safety-relevant systems in the automotive field. Because the core functions of TTP/C were formally verified and a TTP/C network can be put together economically compared with other protocols with similar reliability properties, TTP/C has also in the meantime come to be used in aircraft, a field which is subject to very stringent requirements with regard to safeguarding functionality. TTP/C is also used in train control systems.

Fault-tolerance strategy

The fault-tolerance strategy is used to describe which faults can be detected and handled. TTP/C is able to detect and handle each individually occurring fault (Single Fault Hypothesis). In addition, a series of multiple fault scenarios is handled.

However, some important multiple fault scenarios, such as a complete temporary breakdown of communication for example, cannot be detected on the node level. Additional methods of fault detection and handling are required on the application level for this purpose.

The TTP/C network must consist of at least four real member nodes to ensure that all single faults are detected.

The network ensures on the architecture level that a faulty network node cannot prevent any properly functioning network node from transmitting its message. This is done by bus guardians, which permit bus access only at defined points in time.

Time-Triggered Architecture

Time-Triggered Architecture (TTA) forms the basis on which the TTP/C protocol was specified (Fig. 1). The core element of TTA is the network node, which comprises a host processor and a TTP/C controller, referred to in the following as the controller. The network node is the "Smallest Replaceable Unit" (SRU) in a TTP/C system. The host processor processes inputs from

sensors and activates actuators. The Communication Network Interface (CNI) forms the interface between the host processor and the controller and provides memory areas via which the host processor and the controller can exchange information.

The controller is in the end connected to the TTP/C bus, by way of which it connects the node with other nodes. The group of all the nodes connected to a TTP/C bus and the bus forms a cluster.

Host Processor

The host processor executes the application, the actual function of the network node. In order to support the application, an operating system, which also for the most part provides the FT-COM layer, runs on the host processor as well as the application. The FT-COM layer introduces a further abstraction level of the technical realization of communication. The application can provide data for transfer via the interfaces of the FT-COM layer, thanks to which the application designer does not have to deal with deeper concepts, such as the memory areas in the CNI.

Communication Network Interface (CNI)

The CNI functions as a temporary firewall because it makes available a memory area via which data received and to be sent are exchanged between the host processor and the controller. This interface represents the host processor's sole possibility of transferring information via the network. It is therefore not possible for the host processor to influence the communication sequence. In particular, the point in time at which information is provided by the host processor has no influence on the network's communication sequence.

TTP/C Controller

The controller (Fig. 2) operates entirely independently of the host processor. Its most important components are the protocol processor, a local bus guardian, and the Message Descriptor List (MEDL).

The function of the protocol processor is to prepare information provided by the host processor in the CNI in such a way that they can be transferred as TTP/C frames.

The controller is connected via two interfaces with the remaining components of the network node. The connections to the host processor and to the transceivers (drivers) are established via the CNI and the Logical Line Interface respectively.

1 TTA network

SVT0001E

Message Descriptor List (MEDL)

The MEDL contains all the check information required by the controller to send and receive data. It is a list adapted for each controller containing, for example, information on the position and size of transmit and receive slots.

The information stored in the MEDL includes the points in time at which the controller frame may be transmitted and at which frames of other controllers are expected which are processed by the host processor.

CNI memory positions in which information from the host processor is to be provided for transfer and in which received information from the controller is to be stored are also defined in the MEDL.

The MEDL also contains further information needed for the operation. For example, the identification of frames used to synchronize the internal clocks and the synchronization times.

Strictly speaking, the designation Message Descriptor List is incorrect. Information blocks within a frame are designated as messages in TTP/C (Fig. 3). However, this structure exists only from the perspective of the host processor, not from the perspective of the controller, which only processes frames. The designations has persisted, however, because it was introduced in earlier specifications.

Local and Central Bus Guardians

The primary function of the bus guardian is to limit transmissions of a node to the defined time windows. For this purpose, it accesses the information of the MEDL in which the transmit slots of the node are stored.

There are two different types of bus guardians. Local bus guardians are connected between the controller and the TTP/C bus and prevent bus access outside defined transmit slots. Central bus guardians operate in a central coupler. If exclusively central bus guardians are used, however, only faults are excluded for the nodes which are directly connected to the couplers. Connected bus topologies, on the other hand, cannot be completely protected.

2 TTP/C node

3 Message components

In addition to monitoring bus access, the central bus guardian can also assume the function of detecting and containing SOS faults (Slightly-Off Specification). These are faults in which a node continually violates slightly one or more communication parameters, such as the voltages used or the timing of the transmit slots. The upshot of this is that some network nodes which were designed with greater tolerances process these messages, but other reject them as faulty. As a result of SOS faults, it becomes problematic to establish whether data have been correctly transmitted or incorrectly received.

Logical Line Interface

The Logical Line Interface (LLI) is the interface between the controller and the transceiver (driver). The LLI establishes the structure of the logical information in which the driver must accept information to be transmitted and in which it must provide received information. This makes it easier to adapt controllers to a new transmission medium since only the drivers have to be modified; however, the format of the interface to the rest of the system remains unchanged.

Driver

The driver is the transceiver for the transmission medium. It converts the physical states of the TTP/C bus into logic states, which can be processed by the controller.

Network

Network Size

TTP/C is designed for networks of up to 64 nodes. A TTP/C network can consist of a minimum of two nodes, but at least four nodes are required to satisfy the Single Fault Hypothesis. A problem in the data transfer lies in the danger that a transfer is not received uniformly by all the network nodes. If there are at least four real member nodes (nodes with sole access to one transmit slot) in a TTP/C network, TTP/C can identify reliably whether the message transmission was successful or whether faults occurred during transmission or reception.

Topology

All TTP/C networks have two independent physical channels via which a network node simultaneously transmits information redundantly. This configuration is required in order that the Single Fault Hypothesis, i.e. the handling of the failure of any component, can be satisfied. However, it is also possible for nodes which do not fulfill any safety-relevant functions to be connected only to one of the two channels.

Since a controller always forwards messages via the LLI to its drivers, it is possible to adapt TTP/C to very different topologies. Bus, star and hybrid topologies from any combinations of bus and star topologies are described (Fig. 4) in the TTP/C Specification (Version 1.1).

Bus topology
In a bus, all the network nodes are connected in series to central cables. If a bus topology is used, the structure dictates that no central bus guardians can be used. However, the overall length of the cabling can be reduced to a minimum for this purpose.

Star topology
If a star topology is used, the network nodes are connected to each other via central couplers. Here, each TTP/C channel is brought together in its own coupler.

Star topologies facilitate the use of central bus guardians, which are able to provide additional protection against faults.

Hybrid topologies of bus and star
Even hybrid technologies composed of bus and star topologies can be built. For example, it is possible to cascade several star topologies into a multi-star or to connect a sub-bus to a star topology.

Transmission Media
Within a TTP/C network, communication takes place via two channels. The channels are designated Channel 0 (Ch0) and Channel 1 (Ch1), or Channel A (ChA) and Channel B (ChB).

Because the TTP/C Specification is formulated very openly with regard to the transmission medium, it is possible to build TTP/C networks using very different media. Both electrical and optical media can be used here. In particular, it is also possible to design one channel to be optical and the other to be electrical. Differences in the latencies of the different media can be compensated for by way of the configuration in the MEDL.

The achievable bit rate also varies together with this flexibility; however, the bit rate itself is fixed during operation within a network and cannot be altered.

It is possible with current implementations to achieve bit rates of 5 Mbit/s for asynchronous data transfer during system starting and of up to 25 Mbit/s for synchronous data transfer during normal operation. Prototypes in the laboratory environment has achieved bit rates of up to 1 Gbit/s.

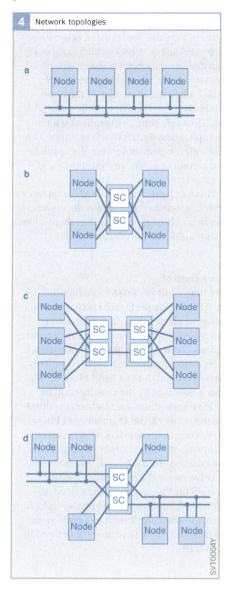

4 Network topologies

a

b

c

d

SVT0004Y

Fig. 4
a Bus
b Star
c Multi-star
d Star/bus
 combination
SC Coupler

Communication cycle

TTP/C controls bus access by means of a TDMA process (Time Division Multiple Access, Fig. 5). Here, the available bandwidth is divided into time windows (slots) in which in each case an established network node can utilize the network's full bandwidth.

TTP/C distinguishes between rounds and cycles in the sequence of slots. All rounds contain the same sequence of slots whose size can be independently established. A cycle is a defined sequence of rounds in which the individual rounds differ only in the content of the individual slots, but not in the sequence of size of the slots.

In round, each member node of the cluster transmits one frame precisely in one slot. A distinction is made here between two types of member node: real and virtual. Real member nodes are nodes which have their own transmit slot which only this node can access. Virtual member nodes consist of a group of nodes which use a common transmit slot. This structure is used to build up redundant structures (Fault Tolerant Unit, FTU) which make the failure of nodes more tolerable.

FTUs are used so that safety-relevant information can also be provided in the event of the failure or one or more network nodes. Here, all the nodes combined in a FTU are able to process a task and use in a cycle alternately the transmit slot of a round allocated to the FTU.

Passive Nodes

A node transfers data only when current data are available for transfer, otherwise it remains silent. A node which does not transfer any data in its transmit slot is a passive node. This is the standard behavior when the controller of the node establishes that its host has not updated the "Host Life Sign", by means of which an application signals that it is active. The controller then assumes that the application cannot generate any valid data and does not transmit.

This mechanism is also used for monitor nodes, which merely monitor the communication of the other nodes, but themselves do not influence the communication.

Time control

The time window of a slot can be subdivided using the performed activities into different phases (Fig. 6). Before a node can transmit information, it needs time in order to read in the transmission parameters or to prepare the transceivers. This phase is called the Pre-Send Phase (PSP). This is followed by the time interval in which information is actually transmitted via the network, the Transmission Phase (TP). After a message has been received, a node needs time in order to prepare the information. This takes pace in the Post-Receive Phase (PRP). Because the time needed for PSP and PRP is not fixed, the PRP is followed by the Idle Phase, a phase through which the length of a node slots is stretched to the time interval defined in the MEDL.

5 Schematic bus-access diagram

A Node C Frame Ch0 ⬜ Frame
B Round slot D Frame Ch1

| A | 0 | 1 | 2 | 3 | 0 | 1 | 2 | 3 | 0 |
| B | 0 | 1 | 2 | 3 | 4 | 5 | 6 | 7 | 0 |

Node slot

TDMA round TDMA round

Cluster cycle

Time

SVT0005E

A TDMA slot is fixed at points in time which can be observed using the transmission. But because this is not applicable to all the activities of a controller during a slot, a distinction is made between node slot and TDMA slot.

For a node, the slot begins with the PSP. But because no signal is transmitted in this phase, external observers lack a precise point in time at which the node slot begins.

The first precise point in time which can be observed from an external source is the start of the TP. This point is also called the Action Time (AT). A TDMA slot therefore comprises the time interval which passes between two ATs. Therefore, there exists in a TDMA slot only the TP and the Interframe Gap (IFG), in which the PRP, Idle Phase and PSP are combined.

A node slot, on the other hand, begins with the PSP, which is followed by the TP, the PRP and the Idle Phase.

TTP/C Protocol

TTP/C systems use the TTA framework (Time-Triggered Architecture) as the basis for which - with the exception of the application - finished solutions can be used.

The TTA protocol stack (Fig. 7) can be divided hierarchically into three areas. The top layer is the host layer, the executed application. Under this is the FT-

COM layer, also an application, which is processed on the host processor.

The lower layers are implemented in the controller, where data communication takes place between the host processor and the controller via the CNI interface.

In the controller, the top layer is the Protocol Service Layer; under this is the Data Link Layer followed at the lowest level by the Physical Layer.

Physical Layer

The Physical Layer comprises the requirements which are placed on the physical transmission of information. Often the connections to be used, the cable types or the voltage levels are established for this purpose in a protocol specification.

Efforts are indeed underway to formulate a specification for a TTP/C physical layer, but the TTP/C protocol specification merely establishes requirements which must be fulfilled by the physical layer used. This has the advantage that the protocol is very flexible with regard to the networking and bit rate. It is thus possible, for example, for networking to use both optical and electrical media, which can also be combined with each other.

However, the system designer is faced with the problem of initially having to identify and evaluate possible solutions.

6	Slot timing

(Figure 6: Slot timing diagram showing Node slot i-1, Node slot i, Node slot i+1 with Δslot, ΔTP, ΔPRP, ΔIdle, ΔPSP; PRP Idle PSP TP PRP Idle PSP TP; IFG, Action Time (AT); TDMA slot i-1, TDMA slot i, TDMA slot i+1. SVT0006Y)

7	Protocol stack

Figure 7: Protocol stack — Executed application (Host Layer); FT-COM CNI; Redundancy management, Message permanence (FT-COM Layer); TTP CNI; Communication services, Safety services, Higher-layer services (Protocol Service Layer); Frame structure, Endianess of bus (Data Link Layer); Bit synchronization, Bit coding/decoding (Physical Layer). Layers processed by the host processor / Layers processed by the controller. SVT0007E

Essentially, three requirement are placed on the physical layer:
- Two independent physical transmission channels must exist.
- The transmission channel must permit distributed broadcast transmissions, i.e. transmissions which are received by all the network users.
- The delay which occurs during the transmission of information via a medium (Propagation Delay) must be known.

Data Link Layer
The Data Link Layer contains the functions which are needed to transmit frames between nodes. These include the structures which are required for access to the transmission channels and for the transmission of frames. In particular, the Data Link Layer establishes the format of the message frames.

Protocol Service Layer
The Protocol Service Layer is responsible for implementing the protocol services offered. Here, a distinction is made between communication services, safety services and higher-level services.

The functions of the communication services are the safeguarding of reliable data transfer, startup of a cluster, reintegration of member nodes in the cluster, transmission acknowledgment, and fault-tolerant clock synchronization.

The safety services include the management of node membership, avoidance of cliques, and the independent bus guardian.

The higher-level services include functions for delayed mode changes, distributed alarm, external clock synchronization, or reconfiguration of a node.

FT-COM Layer
The FT-COM Layer offers functions for realizing replicated, distributed applications. These include redundancy control and the decision as to whether replicated messages of other nodes are correct.

In order to allow applications standardized access to communication data, the FT-COM Layer offers an interface similar to CNI in which the data for the Host Layer applications are provided.

Host Layer
The Host Layer comprises the applications which run on the host processor. In addition to the actual applications, these include the operating system and the check structures (control loops).

In order to give the communication system as modular a structure as possible, the applications of this layer should access the FT-COM interface and not the CNI. This makes it possible, for example, for an application which previously ran on a real member node to be switched with minimal complexity to a series of virtual member nodes.

Protocol services
Cluster Startup
All the nodes of a cluster act without synchronization during cluster startup; bus access is therefore not yet controlled via TDMA. Because there is always a transmission delay caused by the medium used when messages are distributed in a network, two unsynchronized nodes of a cluster can start simultaneously with the message transmission if the transmission delay is greater than the duration of the startframe. This can occur during system startup if several nodes are authorized to initiate synchronization (coldstart nodes).

The situation can arise where different cliques arise within one cluster of which only the nodes of one clique are synchronized, but not all the nodes of the cluster.

In order to avoid this, all the nodes reject the first correctly received coldstart frame. The coldstart nodes wait for a time interval (startup timeout) before trying again to transmit a coldstart frame. Because the size of the startup timeout must be different for all the coldstart nodes in the cluster and differ at least by the size of the transmission delay, collisions can no longer occur during the second cluster startup. This process is also known as "Big Bang".

Synchronization
In a communication system which controls bus access by means of TDMA, it is important for the individual users to have as precise an idea as possible of a global time.

Synchronization in a TTP/C system uses for the operation time only the offset correction by means of which the onset of a point in time is synchronized to all the internal clocks of the nodes of a cluster. For this purpose, TTP/C uses a variant of the Welch-Lynch algorithm [1] which was described by Kopetz and Ochsenreiter [2].

The internal clocks of the network nodes of a cluster can run with different levels of precision. Only nodes which have precise clocks should be used for time synchronization. These nodes receive the flag as "Master Clocks" in the MEDL, whereby the SYF (Synchronization Frame Flag) is set for their frames. Nodes with more imprecise internal clocks act as "Slave Clocks" and use the clock information of the master clocks. There must be at least four master clocks in a cluster to ensure that clock synchronization is Byzantine-fault-tolerant. Byzantine faults are faults in which different nodes interpret the same transmission differently.

The nodes of a cluster are synchronized by measuring the Action Time (AT) in each transmitted slot. If the SYF flag of a frame is set and the received frame was faultlessly received, the AT of the frame is stored in a stack four deep. In other words, the last four correction values determined are stored in the stack.

The actual clock synchronization takes place in frames, for which the ClkSyn flag (Clock Synchronization) is set in the MEDL. If this is the case, the mean value is formed from the stack values, where the highest and lowest values are ignored. The internal clock is now corrected by the calculated mean value.

TTP/C also supports the inclusion of external reference clocks (e.g. GPS) in the system. Time Gateway Hosts are nodes of this type which offer an external time reference. These nodes transmit the external correction value as application data in a frame. The remaining nodes of the cluster store this value in the CNI and include it in the calculation of the AT correction values.

[1]
Jennifer Lundelius-Welch,
Nancy A. Lynch

[2]
Hermann Kopetz, Wilhelm Ochsenreiter: "Clock Synchronization in Distributed Real-Time Systems", IEEE Transactions on Computers, 36(8): 933-940, August 1987.

Membership Service

TTP/C uses Membership Service to ensure that all the nodes of a cluster at each point in time have the same consistent view as to which nodes are present in the network and which are not. For this purpose, each node carries a member vector in which it notes whether the last received transmission of a node has run correctly. If a node receives a message incorrectly, it changes the entry of the message sender to "incorrect".

By comparing the member vectors of the following frames, a calibration is made with the remaining nodes of the cluster.

Clique Avoidance

In order to achieve correspondence within the cluster with regard to the current state of the cluster, it is necessary to prevent the formation within the cluster of subgroups or cliques whose view of the current state differs from the real state of the cluster.

To prevent the formation of cliques, each node logs how many transmissions were successfully and how many were incorrect.

Before it can transmit a frame itself, the Clique Avoidance Algorithm is used. The controller checks whether more successful transmissions were received than incorrect transmissions. Only if this is the case does the node set both counters to zero and transmit its frame. If this is not the case, the controller switches to the "Freeze" state, in which the controller is deactivated and no data are transmitted. From this state, it can be placed by the host processor in the initialization mode again.

Transmission acknowledgment

TTP/C offers an implicit transmission acknowledgment. Because the member vector of the next sender in the round is also transmitted as part of the frame, the original sender can check whether its transmission was correctly received.

But because the successor may be faulty, the sender also observes the transmission of the second successor. If the latter identifies in its member vector that it has also received the transmission incorrectly, this means for the sender that a transmit fault has occurred at its end. If the second successor has received the message correctly, the first successor detects that a receive fault has occurred.

Measures can now be taken to handle both fault types.

Controller State (C-State)

In order for the applications executed on the host processors to be able to deliver consistent results, it is necessary that all the nodes of a cluster share common knowledge of the state of the cluster. This knowledge is summarized in the C-state.

Each node may indeed determine its C-state independently, but the cluster agrees by comparison with the other nodes to the view of the majority of the users.

The controller state is influenced by the values for global time, current position in the TDMA round (MEDL position), current cluster mode, requested mode change (DMC), and the member vector.

Cluster Mode

In many real-time systems, operation and check functions can be divided into phases which are mutually exclusive. For example, the operation of a motor vehicle could be divided into the phases "Driving" and "Stationary". In order to support the different requirements placed on the exchange of information in the different phases, it is possible to define in a TTP/C network modes in which the transmitted information in a frame can vary. However, the order of the slot in a TDMA round is the same in all modes.

For the purpose of correct interpretation, it is therefore necessary when modes are used for all the nodes of a cluster to be in the same mode.

Deferred Pending Mode Change (DMC)

A change of cluster mode can be initiated by applications if the controller is permitted in the MEDL to change to the requested new mode.

Because the change of cluster mode comes into force only at the beginning of the following cycle, the information for the deferred pending mode change is recorded in the C-state.

If another change of cluster mode is requested by another application later in the cluster cycle, this request overwrites the previous request. A special case is CPM (Clear Pending Mode change), in which an application prevents the requested change of cluster mode.

Reintegration of a node

A node must synchronize itself with the cluster if it has failed due to an incorrect transmission from the member vector or if it has been re-initialized. To this end, it waits for a frame in which an explicit C-state was transmitted, e.g. an initialization frame. The controller adopts this C-state and uses it for its transmission.

In order for this to be able to function, a frame with an explicit C-state must be transmitted every two rounds. If this is not the case, the controller assumes that there has been a communication fault and begins with a cluster startup.

Frame format

Because TTP/C regulates bus access via TDMA, all messages are transmitted in broadcast mode and a priori knowledge of the communication sequence and its content is present in the MEDL, it is possible to keep very low the overhead of the protocol, i.e. the volume of information needed to process the frame.

Two different frame types are used in the operation of a cluster (Fig. 8). Cold-start Frames, also known as Initialization Frames or I-Frames, occur exclusively during system startup. Normal Frames, also known as N-Frames, are used for communication during normal operation.

8 Frame format

Header

| I/N Frame 1 bit | Mode Change Flags 3 bits |

I-frame

| Header | C-state | CRC |

N-frame (explicit C-state)

| Header | C-state | Application data up to 240 bytes | CRC |

N-frame (implicit C-state)

| Header | Application data up to 240 bytes | CRC |

SVT0008E

I-Frame

An I-frame begins with the TTP/C header. This is followed by the current controller C-state and finally the CRC value of the frame.

N-Frame

To ensure that Membership Service can function, it is necessary for the C-state to be included in the frame structure. This can occur explicitly or implicitly in the case of N-frames.

If the C-State is explicitly transmitted, the header is followed by the controller C-state. This is followed by application data and then the CRC value of the frame.

If the C-State is implicitly included in the frame, the frame consists of the header, application data and finally the CRC value of the frame, the determination of which was initialized with the current C-state.

Download frames are a special form of the N-frame; these are accepted by all the receiving nodes, regardless of cluster association and C-state. To facilitate this behavior, a constant value which is known to all the nodes is used for these frames to determine the CRC.

Frame Fields

Header

The header of all the TTP/C frames is identical. Firstly, one bit signals the frame type, whether an I-frame or an N-frame is transmitted.

This is followed by three bits, by means of which a change of current cluster mode can be requested. These bits are set to zero if no change has been requested.

Application data

The information of the application is transmitted in this area. The size of this field can be determined for each slot of a round individually in the MEDL and must not exceed 240 bytes per slot.

C-state

The C-state consists of the global time, information on the MEDL and the member vector of the cluster. The MEDL position corresponding to the Round Slot Position in the current cluster mode, the current cluster mode and the requested cluster mode change (DMC) is transmitted as information on the MEDL.

An explicit C-state has a size of 96 bits (six words of two bytes each).

CRC checksum

TTP/C uses the CRC checksum only to identify transmission faults. Transmission faults are not corrected. Transmission faults are detected by a process whereby each node determines a CRC checksum for the received message and compares this with the transmitted checksum.

A polynomial is used to determine the checksum. The TTP/C Specification in this respect does not stipulate a specific polynomial; instead, it requires the polynomial to facilitate a Hamming distance of at least 6. In other words, it must be possible within one transmission for at least five incorrectly transmitted bits to be detected.

The CRC calculation is initialized by the Schedule ID, where the ID is split into two parts. One part is used for initialization for Channel 0 and the other for Channel 1. This prevents a node which does not have the correct MEDL or which was connected crosswise to the channels from transmitting or receiving successfully.

The C-state for initializing the CRC calculation is also used in the transmission of frames with an implicit C-state. Thus, only those frames whose CRC checksum was determined with the same C-state are accepted as valid.

The calculated CRC value is transmitted at the end of a frame in a 3-byte field.

Composability

Networking technology has changed greatly over the years. Where initially stand-alone systems (e.g. activation of a turn-signal lamp via switches and relays) were used, the changeover was quickly made to cooperative systems in which individual components exchange information with each other.

However, the interaction of components is increasingly making it difficult to ensure the correct fulfillment of a function. For this reason, the aspect of composability, i.e. the possibility of combining separately developed subsystems into an overall system without having to verify the function of the overall system again, is becoming increasingly important.

In order to ensure that this functions, it is necessary to ensure that a subsystem autonomously fulfills its functions, but also that this is the case when the subsystem is integrated in an overall system.

TTP/C facilitates the implementation of composability through the communication parameters which are stored in the MEDL.

Verification of the Specification

The functionality of TTP/C has been thoroughly analyzed. The core algorithms of the protocol specification have been formally verified with regard to certain aspects of consistency, stability and safety.

Furthermore, the function of TTP/C systems has been tested using failure-injection experiments. This involved the use of both physical and software faults to check the fault tolerance and fault-detection properties.

Standardization

The development of TTA stretches back over a period of 20 years. Today, development of the Specification is coordinated by the TTAgroup, a cross-industry consortium which was founded by the companies Airbus, Audi, Delphi, Honeywell, PSA Peugeot Citroën, Renault, and TTTech.

The Specification of TTP/C has been influenced in particular by Hermann Kopetz, who between 1979 and 1982 was Professor of Computer Process Control at the Technical University of Berlin and since 1982 has run the Institute for Technical Information Technology at the Technical University of Vienna. Professor Kopetz also directed the MARS project and was substantially involved in each further development of the protocol.

Characteristics

- Support of communication for safety-relevant functions.
- Guaranteed transmission properties.
- Guaranteed detection of all single faults.
- Fault-tolerant communication units (FTU) for fault-tolerant provision of data.
- Local and central bus guardians possible.
- Low protocol overhead.
- Data efficiency of 85 % for transmission.
- Supports composability with regard to time response and ranges of values.
- Functions for the most part formally verified.
- Consistent view by all nodes of the cluster state thanks to Membership Service and Clique Avoidance Algorithm.
- Clear structuring of the protocol interfaces.
- No restriction of the bit rate from the Specification. TTP/C components were tested with a bit rate of up to 1 Gbit/s.
- Experience from product use.

FlexRay

Overview

FlexRay is a field that was designed to support open and closed-loop control technologies in the automotive sector. Development focused on the suitability for use in active safety systems in particular. FlexRay therefore offers transfers with guaranteed compliance with transfer properties, high bit rates and a fault-tolerant design.

The development of FlexRay, managed by the FlexRay Consortium, goes back to cooperation between German automotive manufacturers, BMW and DaimlerChrysler. In 1999 these two companies began to compile requirements for a new communication system. In the process, development would incorporate experience gained from the byteflight system of BMW and the prototype development of DaimlerChrysler.

The aim of FlexRay is to provide a system with high transfer rates that will work in a deterministic and fault-tolerant manner while being as flexible as possible to use and expand. The main fields of application of the FlexRay are drivetrain systems (drive) and active safety systems with no mechanical fall-back level (x-by-wire). However, the areas of passive safety systems and comfort/convenience and body electronics are also supported.

To support these very different domains, FlexRay uses two different types of bus access. For deterministic transfer properties, it offers time-controlled bus access. In the case of applications whose transfers place less demanding requirements on transfer properties, it is desirable to make as effective use of the available transfer capacity as possible. To combine these two approaches, communication takes place in cycles. In each cycle, there is firstly a static transmission component in which bus access is controlled by TDMA (Time Division Multiple Access). There then follows a dynamic transmission component in which bus access is controlled by the use of minislots. Minislots are small time windows in which a defined message can be transferred. This access method is also known as FTDMA (Flexible Time Division Multiple Access).

With messages transmitted in the static component of the cycle, the transfer properties can be assured in accordance with known methods. In the dynamic component, messages are sent as required. In this case, messages are prioritized based on the message ID. The division between the static and dynamic component is freely configurable but it cannot be modified later during system operation. The same applies to the slot lengths in the static range, which, while configurable, must remain constant once operation is underway.

FlexRay systems can be equipped with two transmission channels, with each channel having separate lines. This makes it possible to create a redundant data transfer or increase the available bandwidth for specific applications. The latter can be used both for the parallel transmission of information of two network nodes and for the faster transfer of information of one node.

FlexRay operates at a maximum bit rate of 10 MBit/s, which is achievable in optimal environmental conditions. A rate of up to 20 MBit/s is achievable if two channels are used without redundant access. In addition to the bit rate of 10 MBit/s, which is the only bit rate defined in Specification 2.1 dated 15th December 2005, there are endeavors to support lower bit rates of 2.5 or 5 MBit/s.

Areas of application

The areas of application of the FlexRay are very diverse. By offering redundant, time-controlled and fault-tolerant transfers, not only is FlexRay suitable for use in active safety systems with no mechanical fall-back level and in the area of the drive-train, but its optional, dynamic transmission range means that it is able to support application in the areas of passive safety systems and networking in the body and comfort/convenience domains.

With the high bandwidth of up to 20 MBit/s for non-redundant transfers, it is even conceivable, in theory, that FlexRay could be used for the transfer of audio or highly compressed video.

Topology

Point-to-point

The simplest configuration of a FlexRay system is the direct link between two network nodes. The maximum distance between the two nodes is 24 m in this configuration.

Bus topology

Between 4 and 22 nodes can be connected in a bus topology. The maximum distance between any two nodes in a bus topology may be 24 m.

Star topology

A star topology is another architectural variant of the FlexRay network. Both active and passive couplers are supported.

Passive star topology

3 to 22 nodes can be connected in a passive star. Here, too, the distance between any two nodes cannot be greater than 24 m.

Active star topology

An active star topology can be viewed as a point-to-point link between a number of nodes and an active coupler. In this topology, therefore, only the distance from a node to the coupler is of importance. This cannot be more than 24 m.

Apart from the connection lengths and the number of connectable nodes, the most important difference between an active and a passive star topology is in the achievable bit rates. With the use of passive couplers, the achievable data rate (approx. 1 MBit/s) is below the rate that can be achieved with active couplers (up to 10 MBit/s).

Cascaded star topologies

Cascaded star topologies are the product of several active star topologies linked together. In a FlexRay system, a topology like this is limited to three cascaded star topologies.

Hybrid topologies

Hybrid topologies are a blend of topologies from bus and star topologies, e.g. the connection of several bus topologies to a star topology instead of individual nodes.

Two-channel topologies

Since it is possible to implement both channels of a FlexRay system independently, different topologies can be used for each channel. For example, one channel could be realized as an active star topology, the other as a bus topology.

Hardware

FlexRay controller of a subscriber

A FlexRay node (Fig. 1) comprises the host processor, a communication controller (CC) and one bus driver (BD) for each channel to which the node is connected. Optionally, each node may contain a bus guardian (BG) for monitoring each bus driver and the bus guardian may receive information through an additional host processor.

It is optional whether a network node is connected to one or both channels; here, at the very most, the requirements from the field of application of the network node take effect.

Transmission agents

The transmission agent for a FlexRay system is of a twisted-pair cabling design, where shielded (STP) or unshielded (UTP) cabling may be used.

Each of the two FlexRay channels consists of two wires designated bus plus (BP) and bus minus (BM).

Host processor

The host processor gathers information from the sensors, which is forwarded to the communication controller for trans-

mission. The information received by the communication controller is forwarded to the actuators responsible for processing.

Communication controller

The communication controller (CC) realizes all aspects of the FlexRay system relating to the protocol. Its tasks include scheduling, synchronizing with other network nodes, creating a macrotick signal, creating a bit stream from the information of the host or controlling bus access.

Bus driver

The bus driver (BD) assumes the role of a transceiver. It converts logical information into physical voltages that are carried on copper wires, and vice versa. The role of the BD also includes a protective function against electrostatic discharge (ESD).

Furthermore, the BD monitors the states of the BP and BM and is thereby able to detect whether the bus is affected by physical faults.

The BD also processes wakeup signals and, optionally, is able to control the power management of the network node because functions of the node can be switched off.

1 Network node architecture

UVF0001E

Bus guardian

The bus guardian (BG) is a device that will only permit transmissions if the connected network node is authorized to send a message. With this system, it is theoretically possible for faulty network nodes to interfere at the very most with their own transmission but not with the transmissions of other network nodes.

The FlexRay protocol only provides the framework for use of a BG: its use is not a mandatory specification. Specification 2.0 only described a local bus guardian that would operate inside the network node and monitor access to one of the channels. A critical aspect of this concept was that correct operation of the bus guardian in a faulty network node could not be guaranteed as a consequence of the integration of the bus guardian in the node and the permission for the bus guardian to access the oscillator of the network node.

Specification 2.1 contains a refined concept for the local bus guardian and introduces the concept of a central bus guardian.

Local bus guardian

The local bus guardian (node-local bus guardian) is implemented as a stand-alone CC inside the network node that monitors operation of a CC (Fig. 2). Its synchronization process is independent of that of the CC's; in particular, the CC and BG have separate oscillators.

In addition, it is also possible for the BG and CC to be controlled by dedicated host processors that communicate with each other. This is, however, not mandatory.

As a consequence of direct control lines (bus-guardian enable, BGE) between the BG, CC and BD, information can only be transferred from the CC to the BD if the BG permits this transfer to take place.

Central bus guardian

The central bus guardian (CBG, Fig. 3) was a new feature introduced with Specification 2.1. It operates in the coupler of a star topology and monitors the operation of all connected nodes or networks. Faults, such as short-circuits on buses or network nodes, that transmit information like a "babbling idiot" with no consideration of the transmission window, and thus interfere with orderly communication, can be detected by a CBG and suitable counter-

2 Local bus guardian

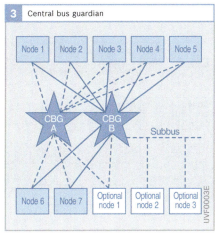

3 Central bus guardian

measures may then be taken. As a result, the effects of these faults can be confined to small areas of the network.

Operating modes

In addition to normal operation, a FlexRay system also supports the complete initialization of the network from sleep to standby (Fig. 4). In order for this system start to function, the BDs of all network nodes have the capability to detect wakeup signals on the bus and to start the remaining other components of the dedicated network node.

Special network nodes that may be connected to the starter motor of a vehicle, for example, generate the wakeup signal.

The system start is split into the wakeup and startup phases. The wakeup phase involves the activation of the network nodes by the wakeup signal. In the startup phase, the network nodes are initialized and synchronized.

If the function of a node is not required for the time being, the node can be set to a standby mode in which all operations of the coding and decoding process are stopped. This reduces the energy consumption of the node. In sleep mode, the power consumption of a network node is at its minimum. In this mode, all functions of the network node are deactivated: only

the bus driver looks out for the transmission of a wakeup signal, in response to which the node is set to wakeup mode.

Sending and receiving

A bus driver connects a FlexRay network node to the channel that contains a receiver and transmitter. If a controller is connected to both channels, it follows that there must be two receivers and transmitters available in the network node.

Protocol

Protocol layers

The FlexRay protocol is built on five core mechanisms:
- Coding and decoding
- Control of bus access (media access control, MAC)
- Processing of frames and symbols
- Clock synchronization and
- Schedule monitoring by the bus guardian

Unlike in other network protocols, with FlexRay there are interfaces between all core mechanisms. This requires a process that coordinates and synchronizes the changes in the core mechanisms. This is the task of the protocol operation control (POC).

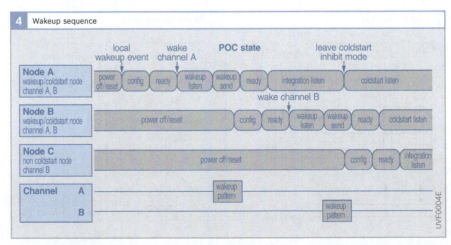

4 Wakeup sequence

FlexRay gives the host system the capability of influencing all five core mechanisms directly. To arrange this, communication takes place through an interface - the controller host interface (CHI) - between the network node and host.

Concentrating on the main interfaces, the task areas can be arranged as a protocol stack (Fig. 5). At the top level is the application, which forwards its commands to the CHI. Under the CHI is the POC, which in turn has access to the MAC, clock synchronization and frame & symbol processing. In the next lower layer are the processes for coding and decoding. The lowermost level is represented by the transfer characteristics of the physical layer.

The bottom three layers are in nodes that are connected to both channels: double presence because these functions have to be fulfilled for each channel separately. More precisely speaking, this means that for one node connected to both FlexRay channels there is double availability of synchronization, MAC, frame & symbol processing, coding and decoding, and physical layer.

Coding on the physical layer

For coding, FlexRay uses a NRZ method (Non-Return to Zero) in which the two identical transmission states are not divided by a range in which the voltage falls back to a zero value. This type of coding makes it necessary to have mechanisms with which a network node can divide the transmitted states. To this end, FlexRay adds a sequence behind each transmitted byte: the byte start sequence (BSS). From this sequence, each node can detect when a byte has been transmitted and, with this information, encode the individual bits of the transmission.

By applying different voltages to the two wires of one channel, it is possible to create four bus states, which are designated Idle_LP, Idle, Data_0 and Data_1 (Fig. 6).

A bus state is identified by measuring the differential voltage. Here, the bus voltage ($uBus$) is made equal to the difference between uBM and uBP (voltages at BP and BM).

$$uBus = uBP - uBM$$

5 Protocol stack

Application

Controller host interface

Protocol operation control

| Media access control | Clock synchronization | Frame & symbol processing |

Coding and decoding

Physical bus

UVF0005E

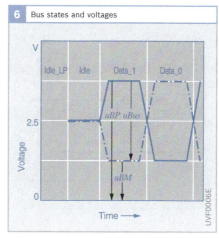

6 Bus states and voltages

UVF0006E

With this method, the data transfer is protected against external electromagnetic interference because these act equally on both wires and are canceled out in the difference.

Idle_LP (LP = low power) is the state in which a very low voltage of -200 mV and 200 mV is present at BP and BM. This state, for example, is used to identify the start of a transmission.

In Idle state, a voltage of 2.5 V with a tolerance of 500 mV is present at BP and BM.

To set the channel to Data_0 state, at least one transmitting node must apply a negative differential voltage of -600 mV to the channel.

For the channel to be set to Data_1 state, at least one transmitting node must apply a positive differential voltage of 600 mV to the channel.

If the transmission of an information signal is neither blocked by the bus guardian nor the communication controller, a HIGH bit is signaled by the Data_1 channel state and a LOW bit by Data_0.

Generation of a frame bit stream

Before a node can transmit a frame containing the data of the host, the frame is converted into a bit stream. To this end, the frame is first decomposed into individual bytes. The start of a frame is populated with a transmission start sequence (TSS) followed by a frame start sequence (FSS) (Fig. 7). From the bytes of the frames, an extended byte sequence (EBS) is then generated whereby each frame byte is preceded with a byte start sequence (BSS).

The 24-bit checksum (CRC) for this bit sequence (TSS+FSS+EBS) is now calculated and appended to the bit sequence. To finish the bit stream, a frame end sequence (FES) is appended.

If the frame belongs to the dynamic segment, a dynamic trailing sequence (DTS) can be additionally appended to prevent another node from beginning its transmission on the channel prematurely.

Communication cycle

Each FlexRay cycle (Fig. 8) contains a static segment that is transmitted as the first part of the cycle. The static segment contains a fixed number of transmission ranges, the "static slots".

Optionally, there may be a dynamic segment in the FlexRay cycle and this is transferred in second place. Each dynamic segment contains a fixed, freely configurable number of "minislots".

Also optional is the symbol window, which is transmitted as the third element of the cycle. It can be used for the transmission of an individual symbol and has the same size as a static slot.

To terminate the cycle, there is a phase – network idle time (NIT) – in which the bus is set to idle state. In general, the length of the NIT corresponds to the remaining macroticks not used by the static and dynamic segment or symbol window. This is not the case if it was detected during synchronization that an offset correction is required through which the length of the NIT can be increased or reduced.

7 Bit stream (dynamic segment)

Static segment

The static segment comprises a fixed number of equally sized transmission windows (static slots), which are sent simultaneously on both channels. Exactly one frame can be sent in each static slot. Both the size and quantity of the static slots in a static segment are configured during integration.

Bus access in the static segment is controlled by a TDMA method whereby, in each slot, the frame with the corresponding frame ID is sent. Frame ID 1 is therefore sent in slot 1, frame ID 2 in slot 2 and thus frame ID n in slot n.

Dynamic segment

Unlike in the static segment, the transmissions in the dynamic segment are not of fixed length. To control bus access, "minislots" are used that hold the same fixed number of macroticks for all nodes connected to the network. Macroticks are ranges in which exactly one frame can be transmitted.

"Dynamic slots" (transmission windows of variable size) are now created based on the minislots. The maximum size of a dynamic slot is limited by a fixed number of minislots specified during configuration.

To identify the end of a transmission, a phase in which the channel is idle (dynamic slot idle phase) is added to the end of a dynamic slot. The phase in which the transmission takes place is called the dynamic slot transmission phase.

Another difference between the dynamic and static segment is that the counters of slots transferred are incremented synchronously on both channels in the dynamic segment. In the dynamic segment, the counter of both channels is incremented independently in line with the current transmission status (Fig. 9).

9 Slot counter

8 Communication cycle

Symbol window

The symbol window makes it possible to send a collision avoidance symbol (CAS) or a media access test symbol (MTS). These two symbols are identical and help to prevent collisions during the system-startup phase.

The third symbol defined in the FlexRay protocol – the wakeup symbol (WUS) – is not permitted to be sent in the symbol window. It serves only to generate a wakeup pattern (WUP), which is used during the system startup.

Network idle time (NIT)

During the NIT, a network node corrects the ascertained time deviations of its internal clock. It also gives the network node the opportunity in this phase to make implementation-dependent adjustments and changes to settings relating to the communication cycle.

Clock control

In both the static and the dynamic part of the communication cycle, FlexRay reverts to the identifiers of the transmission window in which a node may transmit. If this method is to be successful, it is necessary for all nodes connected to the network to have synchronous time information.

For this purpose, selected messages from the static part of the cycle are used (sync frames) and are transmitted by the network nodes that are connected to both channels.

In a distributed communication system, each network node generally needs its own internal clock for setting transmission timing correctly. As a consequence of temperature and voltage fluctuations and production-related precision tolerances, these internal clocks can soon deviate from an abstract, system-wide global time. In the case of a system such as FlexRay, which controls bus access by means of time slots, the synchronization of all network nodes plays an especially important role because, without a standardized, global time preset for all nodes, the specified transmission windows cannot be maintained and may break down.

Clock control hierarchy

In a FlexRay system, clocking is controlled on three levels (Fig. 10). The bottom level is represented by the microticks, which are derived directly from the oscillator clock of the network node. A microtick is therefore not a system-wide variable, but only relevant to the node concerned.

At the next level up, the clock is controlled by means of macroticks – ranges in which exactly one frame can be sent. Within fixed tolerances, macroticks form the lowermost system-wide time unit. Each network node determines the duration of a macrotick in microticks. The number of microticks per macrotick does not have to be the same for all macroticks.

The top level is characterized by a clock control based on the communication cycle, which always contains the same, fixed number of macroticks.

10 Timing

Time synchronization

A synchronization method is required in order to guarantee a standardized time across the entire system. The basic parameters for timing are the number of microticks per communication cycle (*pMicroPerCycle*), the duration of a communication cycle (*gdCycle*) and the duration of a microtick (*pdMicrotick*), where the relationship between each is determined by the following equation:

$$pMicroPerCycle = round(\frac{gdCycle}{pdMicrotick})$$

During operation of a FlexRay network, the clocks of the oscillators of various network nodes will not be equal or even simply remain constant. They will fluctuate in response to outside influences. This deviation is balanced by two corrective measures: the gradient correction and the inclusion of an offset in the communication cycle.

Fault-tolerant average algorithm

Since an individual node can only observe the differences between the expected time at which a transmission should have been actioned and the actual time at which the transmission took place, a fault-tolerant average algorithm (FTM), as described by J. L. Welch and N. A. Lynch, is used in both corrective measures. The measured time differences of all transmitted sync frames are sorted by their value. If more than seven sync frames were received, the two highest and two lowest values are removed from the list. If between three and seven sync frames were received, the highest and the lowest value are removed. For fewer than three sync frames, the average is derived from all measured differences.

Gradient correction

With gradient correction, deviations in the transmission frequency of a node are continuously corrected during the communication cycle. The node here checks the transmission frequencies of all other nodes. By means of the FTAA, these values are used to adjust the node's own transmission frequency.

Offset correction

Offset correction is a way to correct deviations in the communication-cycle phase by including an offset in the NIT. With this method, the node similarly employs the FTAA to determine its phase difference in relation to the other nodes of the network and uses it to determine the offset required to shorten or extend the NIT phase. Since all nodes use this approach, all nodes can initiate the transmission of a cycle in a synchronous manner.

Timing

Transmissions in the FlexRay system are timed by action points, which are specially-marked macrotick boundaries. A transmission in a static slot begins once a static-slot action point has been reached. The position of this action point in a macrotick is defined system-wide.

In a dynamic slot, a transmission begins once a minislot action point has been reached and ends at a subsequent minislot action point. This method ensures that there is always an idle period between individual transmissions.

Welch, Lynch, "A New Fault-Tolerant Algorithm for Clock Synchronisation", Information and Computation, vol.77, No.1, April 1988

Frame format

FlexRay uses the same format in both the static and dynamic part of the frame. The format can be broken down into three parts: header, payload segment and trailer segment (Fig. 11).

Header segment

The header segment contains a total of five bytes, at the start of which is a series of indicators.

The reserved bit is intended for future modifications to the protocol and is sent as a logical "0".

The payload preamble indicator indicates whether or not the payload segment contains a network-management vector (NMVector). An NMVector enables the host processor to transmit data directly without first being processed and prepared by the CC.

The null frame indicator indicates a null frame that contains no usable information.

The sync frame indicator indicates that this frame is intended to be used for system synchronization. This indicator may only be set by network nodes that operate as sync nodes.

The startup frame indicator reveals that this frame is a startup frame. Startup frames are used in the network's starting phase and may only be sent by special coldstart nodes.

The indicators are followed by the frame ID, payload length, header CRC and cycle count.

The frame ID has a length of 11 bits and corresponds to the number of the slot in which the frame is sent. In one cycle, each frame ID exists only once on each channel.

The payload length expresses the size of the information sent in the payload segment. Since the payload in a FlexRay frame is always sent in 2-byte words, a 62 in the payload length field indicates, for example, that 124 bytes of payload are being transmitted in the payload segment.

In the static segment, the payload length field always contains the same value because the size of information here is constant for all frames. Of course, this property does not apply to frames in the dynamic segment.

In the header CRC, an 11-bit CRC value is transmitted. This is formed from the polynomial

$$x^{11} + x^9 + x^8 + x^7 + x^2 + 1.$$

Since the information in the header segment is constant for all frames in the static segment, the header CRC value can be calculated offline and assigned to the network node through configuration settings.

If a frame in the dynamic segment is always of the same size, it is possible here also for the CRC value to be calculated offline.

The final field in the header segment is the cycle count. This field contains the count number of the cycle in which the sending network node is currently active.

11 Frame format

- Reserved bit
- Payload preamble indicator
- Null frame indicator
- Sync frame indicator
- Stratup frame indicator

Header CRC
Covered Area

| Frame ID | Pay-load length | Header CRC | Cycle count | Data 0 | Data 1 | Data 2 | Data n | CRC | CRC | CRC |

| 11 bits | 7 bits | 11 bits | 6 bits | 0...254 bytes | | | | 24 bits | | |

Header Segment | **Payload Segment** | **Trailer Segment**

FlexRay Frame 5 + (0 ... 254) + 3 bytes

UVF0011E

Payload segment

The payload segment has a maximum length of 254 bytes, which are transmitted in 2-byte words.

The payload segment customarily transfers the payload that is further processed by the host processors. Optionally, however, an NMVector may also be sent in the payload segment, or a 16-bit message ID.

With the assignment of message IDs, it is possible to send several data blocks in a frame. For this purpose, the message ID is placed at the front of the other data by the host processor as application data.

Trailer segment

The trailer segment contains a single field in which a 24-bit CRC checksum (frame CRC) is sent. The polynomial used to determine the frame CRC is:

$$x^{24} + x^{22} + x^{19} + x^{18} + x^{16} + x^{15} + x^{14} + x^{13} + x^{11} + x^{10} + x^8 + x^7 + x^6 + x^3 + x + 1$$

Standardization

FlexRay Group

FlexRay is a protocol specified by the FlexRay Group. The FlexRay Group was founded in 2000 by the companies, BMW, DaimlerChrysler, Motorola and Philips Semiconductors (www.flexray-group.com).

The members of the FlexRay Group are organized in a hierarchy. They also have different levels of influence on the further development of FlexRay, depending on the obligations that they have accepted. The uppermost group is represented by the core partners, which first and foremost include the founding members. Alongside the core partners are the premium associates and associate memberships.

By 2004 General Motors, Bosch and Volkswagen had joined the FlexRay Group as core partners. In 2004 Freescale took over Motorola's position in the group of core partners.

Characteristics

- Deterministic transfers possible
- Optional bus guardian monitors network access and protects against faults
- Differential signal transmission
- Collision-free transmission while operation is in progress
- Redundant transmission of information on two channels
- High transmission speed of up to 10 MBit/s; up to 20 MBit/s with parallel transmission on two channels
- Startup and initialization from an inactive sleep state are directly supported by the protocol
- Support for diverse fields of application
- Event and time-controlled transmission of information possible
- Published standardization by the FlexRay Group

Diagnosis interfaces

Diagnosis legislation (e.g. CARB, California Air Resources Board) demands constant monitoring of emission-related components in various electronic systems (OBD, On-Board Diagnosis). Faults (e.g. electrical short-circuits of sensors, implausible operating states) are stored in a fault memory in the control unit. These faults can be read using a scan tool (official testing station) or workshop tester. For this purpose, there is a socket in the footwell, dashboard or center console of the vehicle to which the tester is connected by means of a standard connector (ISO 15031-3) (Fig. 1).

Using a workshop tester, the after-sales service is able to read the entire fault memory: emission-related faults and vehicle-specific faults (e.g. from Motronic, ABS). With the information stored in the fault memory, it is possible to diagnose a fault and repair it efficiently. The tester can also be used to clear the fault memory. Furthermore, measured values and control-unit identification values can be evaluated. Using the workshop tester, it is also possible to control special diagnostic functions. With actuator diagnosis, for example, individual actuators can be targeted for activation to check them for correct operation. At the end of the line (EOL), the diagnosis interface is used to test the electronic control unit and make changes to configurations (e.g. immobilizer, transmission type). It is even possible to program the entire flash EPROM using end-of-line programming.

Diagnostics requires an interface for connecting the tester to the electronic control units. For communication, there are basically two options available at the present time:
- K line and
- CAN bus

The K-line network is not a bus in the true sense, which means that collisions may occur. Since most electronic control units have a CAN interface for the exchange of data between electronic systems, this option for communication with the tester is gaining ever more in importance.

Various protocols are implemented on the interfaces. These tend to be the CARB or official protocol and the manufacturer-specific communication, which can be closely related in their properties and functions. It is even possible to use both interfaces: CARB-relevant functions run on the CAN, manufacturer-specific functions on the K line.

Figure 2 provides an overview of the standards and the various layers of the OSI reference model.

1 Diagnosis connector (diagnosis interface)

Pin 2: Bus line (high) from SAE J1850
Pin 10: Bus line (low) from SAE J1850
Pin 7: K line in accordance with ISO 9141-2
 and ISO 14230-4
Pin 15: K line in accordance with ISO 9141-2
 and ISO 14230-4
Pin 6: CAN_H from ISO 15765-4
Pin 14: CAN_L from ISO 15765-4
Pin 1, 3, 8, 9, 11, 12, 13
 Not used by OBD
Pin 4: Vehicle ground
Pin 5: Signal ground
Pin 16: Battery positive

UWT0106-1E

Diagnostic protocol

The communication method in diagnostics is characterized by the fact that the tester (client) addresses one or several electronic control units (servers) and requests data output (e.g. from the fault memory) or an action (e.g. actuator diagnosis). These services and the transmission agent are defined in the various protocols.

KWP 71

This protocol was developed in collaboration with Bosch as the first standardized serial interface and is based on a 5-baud initialization and byte handshake. For addressing, the protocol provides for the use of a unidirectional line (L line). The line for data exchange (K line) is bidirectional. Alternatively, if the L line is not used at all in the system architecture, the K line can also be used for addressing because addressing and data exchange are two chronologically sequential states. The way in which communication is established is predefined, the services are standardized.

KWP 2000 (ISO 14230: 1-3)

This diagnostic protocol connects a tester to the electronic control units using a K line. In the KWP 2000 standard, there are various possibilities for the way in which communication is established (fast initialization and 5-baud initialization). The communication services are specified but there is some freedom for customer-specific adaptations.

ISO 15765: 1-3 (CAN)

This standard reproduces the existing protocols, ISO 14230: 1-4, on the CAN. It defines a way to leave untouched the fundamental format of messages exchanged between tester and electronic control unit and to transmit them on the CAN bus.

CARB

The Californian environmental authority (CARB) introduced guidelines to promote support for a standardized interface. The CARB protocol is built on existing standards and specifies the data flow and timing (transport layer). This is established in ISO 14230-4/ISO 9141-2 for the K line and in ISO 15765-4 for the CAN. The actual communication messages (services) in the application layer are precisely defined in ISO 15031-5/SAE J1979 and permit no manufacturer-specific deviations.

In California, every newly registered vehicle must support at least one of these protocols with all emission-related electronic control units. The European counterpart is the EOBD standard.

2 Standards for diagnosis communication

a

Layer	CARB			
	K line			CAN
7	ISO 15031-5	ISO 15031-5	ISO 15031-5	ISO 15031-5
6				
5				ISO 15765-4
4				
3				ISO 15765-2 / ISO 15765-4
2	ISO 9141-2	ISO 14230-2 / ISO 14230-4	SAE J1850	ISO 11898 / ISO 15765-4
1	ISO 9141-2	ISO 14230-1 / ISO 14230-4	SAE J1850	ISO 11898 / ISO 15765-4

b

Layer	Manufacturer-specific	
	K line: KWP 2000	CAN / UDS
7	ISO 14230-3	ISO 15765-3 / ISO 14229-1
6		
5		ISO 15765-3
4		
3		ISO 15765-2
2	ISO 14230-2	ISO 11898-1
1	ISO 14230-1	ISO 11898

SVA0018E

Fig. 2
a CARB
 communication
b Customer-specific
 communication

Layers of the
OSI reference model
7 Application
6 Presentation
5 Session
4 Transport
3 Network
2 Data link
1 Physical

UDS Unified Diagnostic
 Services

Application protocols

The increasing complexity of control-unit functions places high demands on the tuning and optimization of these systems to suit the types of engine and vehicle concerned. This process is known as application engineering. For this purpose, an application system is connected to the electronic control unit. The application protocol for the K line is the McMess protocol, and the CCP protocol for the CAN.

McMess

With the McMess protocol, parameters in the electronic control unit can be modified and operands recorded. In a special application mode, the protocol describes the method and contents of communication between application system and electronic control unit for the K line. With McMess, for example, the measuring device is able to read the contents of the variable memory (e.g. sensor values, measured values) quickly.

CCP (CAN Calibration Protocol)

The CCP protocol describes the contents of communication between an application system and an electronic control unit if these communicate on the CAN bus. CCP allows electronic control units in the network to function at higher payload rates than on the K line.

Communication on the K line

Physical layout

The diagnostic tester connects to one or several electronic control units using the K-line system (Fig. 3). All electronic control units have equal rights. Communication flows bidirectionally between the tester and the electronic control units. This means that data can be received and sent by all subscribers. However, only one subscriber is ever able to send at a given time.

The communication is an asynchronous process. There is no additional line for the transmission of a clock signal. The baud rate on the K line is between 1,200 and 10,400 baud in accordance with ISO standards. In special applications (e.g. manufacturer-specific flash programming), it could even be as high as 250 kBd.

Data transfer on the K line is realized as follows: the resistor in the tester increases the potential of the line – provided no electronic control unit is sending on it – to battery-voltage level (Fig. 4). If an electronic control unit is activated, it connects its K-line connection to the shared ground and therefore switches the potential of the K line to ground. Each of the connected electronic control units is able to detect that one of the bus subscribers is actively accessing the K line.

3 System architecture of bus systems

CAN bus

ECU 1 ECU 2 ECU 3

K line

Diagnosis interface

Tester

SVA0017E

4 Circuit diagram of an electronic-control-unit network

ECU 1 ECU 2 ECU n

U_{Bat}

Tester

SVA0019E

Message format

The message format of the KWP 2000 protocol comprises three parts (Fig. 5):
- Header
- Data bytes and
- Checksum

Header

The header is made up of no more than four bytes.
- The format byte (Fmt) contains information about the form and composition of the message. Two bits (A0 and A1) provide address information, while six bits (L0 to L5) provide the quantity of data bytes. With this coding, there can be up to 63 data bytes. If L0 to L5 are set to zero, the number of payload bytes must be coded using the length byte (Len).
- The target address (Tgt) indicates the communication partner for which the information is destined.
- The source address (Src) indicates the communication partner that sent the information.
- The length byte (Len) specifies the quantity of payload bytes (max. 255 bytes). Up to a value of 63, the coding can take place in Fmt. For more than 63, the length must be specified using this byte.

The existence of a target and source address as well as the length byte is dependent on the parameters from the format byte.

Data bytes

Up to 255 bytes of payload can be sent per message. The first byte of the payload is always a service-identification byte (SId). The subsequent data bytes are data that differ depending on the service concerned.

Checksum

The 1-byte checksum is always positioned at the end of a message. It contains the value of the modulo 256 sum across all bytes in the message – apart from the checksum itself.

Initialization

To be able to send messages, each electronic control unit must support a subset of the possible message formats. At the beginning of communication, the format supported by the electronic control unit is communicated to the tester by means of key bytes as part of initialization.

The ISO standards specify which of the possible initialization procedures are customary for ECU diagnostics. The following initialization procedures are supported:
- 5-baud initialization and
- Fast initialization

5 Message format of the KWP 2000 protocol

Header	Data bytes	Check-sum
Fmt Tgt Src Len SId		
max. 4 bytes	max. 255 bytes	1 byte

Checksum calculation

A1	A0	L5	L4	L3	L2	L1	L0

A1	A0	
0	0	No address information
0	1	Exception mode (CARB)
1	0	With address information, physical addressing
1	1	With address information, functional addressing

SVA0020E

These different initialization methods are used either for CARB diagnostics or for after-sales diagnostics. Furthermore, a distinction can be made between physical initialization and functional initialization, which introduces the topic of "point-to-point" or "point-to-multi" communication. With functional initialization, a group of electronic control units is addressed and initialized. A function initialization can only be successful if all the electronic control units of this group support the same baud rate, the same transmission timing and the same protocol. By contrast, only one electronic control unit is initialized in the case of physical initialization.

5-baud initialization
Communication is initiated by means of a 5-baud address sent by the tester. It is thus possible for each electronic control unit to be addressed individually and addressing is clearly separate from the data transfer.

The tester outputs the address onto the K line at a transfer rate of 5 baud (Fig. 6). 10 bits (eight data bits, one start and one stop bit) take approx. 2 seconds.

On receiving the address, the electronic control unit responds with the baud-rate recognition pattern. From the bit sequence 01010101 (binary format), the tester determines the baud rate of the electronic control unit, which may be between 1,200 and 10,400 baud. The electronic control unit then transmits the two key bytes, which communicate the header format and timing method that the electronic control unit supports. As acknowledgment that communication has been successfully established, the tester sends the complement of the second key byte back to the electronic control unit and, in return, receives from the electronic control unit the complement of the address. Initialization is then complete and the regular data transfer may begin.

Fast initialization
Fast initialization establishes communication in a shorter time than 5-baud initialization. With fast initialization, the tester sends a wakeup pattern (WuP) comprising a low and high phase, each lasting 25 ms (Fig. 7). At the end of the WuP, the tester sends the StartCommunication Service at a rate of 10,400 baud. The electronic control unit returns a positive reply containing the key bytes. Initialization is then complete and regular communication may begin.

6 5-baud initialization

☐ Tester ■ Control unit

5-baud initialization

Address (2s)

5 baud 0x55 KB1 KB2 $\overline{KB2}$ $\overline{Address}$

W5 W1 W2 W3 W4 W4

1,200 to 10,400 baud

Synchronization
Baud-rate recognition

W1	60 to 300ms
W2	5 to 20ms
W3	0 to 20ms
W4	25 to 50ms
W5	300ms

SVA0021E

7 Fast initialization

☐ Tester Control unit ■

Fast initialization

T_{WuP}

T_{idle}

T_{iniL} 10,400 baud

T_{iniL}	25±1ms
T_{WuP}	50±1ms
T_{idle}	300ms

SVA0022E

CARB initialization
CARB initialization is a special case of
5-baud initialization, although CARB may
also be activated by fast initialization.
A functional initialization of all connected
electronic control units is achieved with
the fixed address 0x33. The baud rate is
permanently set at 10,400 Bd. The require-
ments are described in more detail in
ISO 14230-4 and ISO 9141-2.

One of the outcomes of these standards
is that all emission-related electronic con-
trol units fitted in a vehicle may only sup-
port one protocol: ISO 14230-4, ISO 9141-2
or SAE J1850. A mixture of protocols in the
same vehicle is not permitted.

Key bytes
The electronic control unit uses key bytes
to inform the tester of which header bytes,
length bytes and timing method are sup-
ported. The key-byte decoding process
is defined in ISO 9141 and ISO 14230-2.

Arbitration
Arbitration is a means of preventing or de-
tecting collisions of data sent by different
electronic control units at the same time.
Arbitration is necessary if a correct response
to the tester request is to be received in
the case of functional communication
(e.g. reading the fault memory of all emis-
sion-related electronic control units using
a scan tool). Collisions can be expected
because more than one electronic control
unit is permitted to send following initial-
ization by the tester.

Arbitration is only relevant to data
transfer on the K line.

Collision prevention
To prevent collisions following a tester re-
quest, each electronic control unit checks
the K line for a falling edge after a time of
P2min. A falling edge may be interference,
or another electronic control unit has
already been sending data.

If a falling edge was detected, the elec-
tronic control unit delays sending its
own data until the bus has returned to
idle state and the P2min time period has
again elapsed. Only then may it attempt
to send its data again.

Collision detection
A collision cannot be prevented if two
electronic control units send at the same
time. In this event, the collision still has
to be detected. This is possible by monitor-
ing the K line. To this end, the electronic
control units reread their own transmitted
byte through the serial interface.

In Figure 8, two electronic control units
(ECU 1 and ECU 2) send a start bit at the
same time. ECU 2 then sends a 0 (dominant
bit) and ECU 1 a 1 (recessive bit). The 0
appears on the K line. With bit 2, the exact
opposite is true. Here, ECU 1 persists with
the dominant level. When the complete
byte has been transferred, the electronic
control units compare the byte sent on the
K line with the byte they have read. In the
example, a recessive level was overwritten
by a dominant level for both electronic
control units. From this, both electronic
control units deem the transfer to have
been faulty and retreat from sending.

8 Sequence of unsuccessful communication

To prevent cyclical collisions, the electronic control unit reports after a time of P2min, which is calculated in accordance with a specific algorithm.

Let us assume that ECU 2 in the previous example sent a 1 as its first data bit. ECU 1 would then receive nothing from the data transfer of ECU 2 and would continue to establish communication. ECU 2, however, detects the collision in bit 2 and retreats. The collision in this event is non-destructive.

Communication on the CAN
Physical layout
A CAN interface containing the necessary hardware already exists in many electronic control units. This bus system can also be used for diagnostics. It is possible to communicate with the electronic control units using a diagnostic tester connected to the bus by a diagnosis connector.

Baud rates
The typical baud rates associated with CAN are 500 kBd or 1 MBd.

Addressing and message types
Tester communication on the CAN bus is defined by ISO 15765. The communication services of ISO 15765-3 or 14229-1 are defined in a similar way to those of ISO 14230-3 (Fig. 2).

The fundamental difference between these protocols is in the format of messages and how they are transmitted. While up to 255 data bytes can be transmitted in a message on the K line, only eight data bytes are possible with the CAN. To reproduce the service messages on the CAN, the contents (data bytes) of the header and the checksum are separated and embedded in a similar, new message format. The address of the electronic control unit is now the CAN identifier itself. With different CAN identifiers, it is now possible to support functional or physical addressing.

Different addressing methods exist: normal and extended addressing, whereby normal addressing is the regular addressing method and is analyzed below. Longer messages that do not fit into a CAN frame (seven bytes with normal addressing, six with extended addressing) are segmented into several CAN frames and recomposed by the recipient.

Unsegmented messages
If the quantity of data bytes to be transmitted is no greater than seven bytes (for normal addressing), a single frame is transferred. The first byte to be sent is the PCI (Protocol Control Information). It identifies the frame in the high nibble (top four bits of a byte) as being a single frame and indicates the data length (DL) in the low nibble (Fig. 9).

The data is sent by the transmitter to the receiver in a single message (Fig. 10a).

Segmented messages
If more than seven data bytes (for normal addressing) are to be delivered, the diagnosis tester (client) sends a first frame on the CAN data bus (Fig. 10b) first of all. The electronic control unit (server) acknowledges with a flow-control frame. The other data is subsequently sent in consecutive frames.

9 Format of an unsegmented message

0	1	2	3	4	5	6	7
PCI				Data			

7	6	5	4	3	2	1	0
0	0	0	0	DL			

Single frame

SVA0024E

The first frame contains the PCI, an additional length byte (DL, Data Length) and the first six data bytes (Fig. 11). In the high nibble, the PCI contains the information for identifying the frames as the first frame. The extended data length (XDL) is stored in the low nibble. Together with the DL, a 12-bit data word is formed with which it is possible to express values from 0 to 4,095.

The transmitter sends the first frame and waits for an acknowledgment of receipt from the receiver in the form of a flow-control frame. This message again contains the PCI, the high nibble of which identifies the frame as the flow-control frame. The low nibble contains the flow status (FS), which can be used to authorize or delay the sending of further frames (consecutive frames).

BlockSize (BS) indicates how many consecutive frames can be received without the need for a further flow-control frame to be sent. The values expressed in ST_{min}

(Separation Time) specifies the interval to be maintained between consecutive frames.

The flow-control frame is followed by the consecutive frames. Again, the PCI in the high nibble contains the identifier, while the low nibble contains the sequence number (SN). With the first consecutive frame to be sent, the SN is set to 1. The SN is incremented with each subsequent consecutive frame; after 15, the SN restarts from 0. By evaluating the sequence number, the receiver is able to detect whether or not all the frames have arrived.

10 Diagram of message transfer

a Transmitter Receiver

Single frame

b Transmitter Receiver

First frame

Flow control

Consecutive frame

ST_{min} Consecutive frame Block size (BS)

Consecutive frame

Flow control

Consecutive frame

Consecutive frame Last block (≤BS)

SVA0025E

11 Format of a segmented message

a

0	1	2	3	4	5	6	7
PCI	DL			Data			

7	6	5	4	3	2	1	0
0	0	0	1		XDL		

First frame

3	2	1	0	7	6	5	4	3	2	1	0
XDL				DL							

b

0	1	2	3	4	5	6	7
PCI	BS	ST_{min}		Data			

7	6	5	4	3	2	1	0
0	0	1	1		FS		

Flow control

| | | | | 0 | 0 | 0 | 0 | Clear to send (CTS) |
| | | | | 0 | 0 | 0 | 1 | Wait (WT), receiver not ready |

c

0	1	2	3	4	5	6	7
PCI		Data					

7	6	5	4	3	2	1	0
0	0	1	0		SN		

Consecutive frame

SVA0026E

Fig. 10

a Message transfer in a single frame (unsegmented message)

b Segmented message transfer

Fig. 11

a First frame

b Flow control

c Consecutive frame

Automotive sensors

The term sensor has become common, as in the past 20 to 40 years measuring gages have also come into use in consumer applications (e.g. motor vehicle and domestic appliance technology). Sensors - another term for measuring detectors or measuring sensors - convert a physical or chemical (generally non-electrical) variable Φ into an electrical variable E; this process often also takes place over further, non-electrical intermediate stages.

Table 1 summarizes and compares the various fields of application for sensors. Figure 1 provides an overview of the abundance of electronic vehicle systems already on the market. Undoubtedly, this number will increase immensely in the years to come.

Basics and overview

Sensor, term/definition
The electrical sensor outputs are not only provided in the form of current and voltage, but are also available as current or voltage amplitudes, frequency, phases, pulse widths, and cycles or periods of an electrical oscillation, or as the electrical parameters, resistance, capacitance, and inductance. A sensor can be defined using the following equation:

(1) $E = f(\Phi, Y_1, Y_2...)$
 Sensor output signal
(2) $\Phi = g(E, Y_1, Y_2...)$
 Required measured variable

If the functions f or g are known, then they represent a sensor model with the help of which the measured variable sought may be mathematically calculated from output

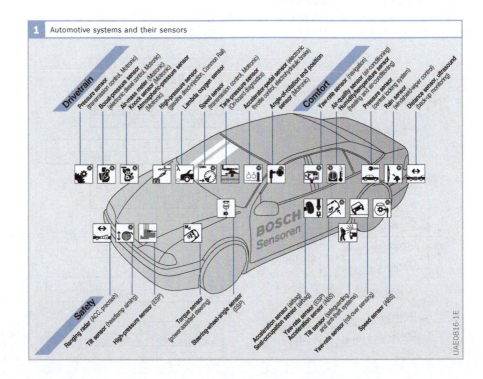

1 Automotive systems and their sensors

Drivetrain
Pressure sensor (transmission control, Motronic)
Boost-pressure sensor (electronic diesel control, Motronic)
Air-mass meter (Motronic)
Knock sensor (Motronic)
Atmospheric-pressure sensor (Motronic)
High-pressure sensor (gasoline direct-injection, Common Rail)
Lambda oxygen sensor
Speed sensor (transmission control, Motronic)
Tank-pressure sensor (On-board diagnostics)
Accelerator-pedal sensor (electronic throttle control, electrohydraulic brake)
Angle-of-rotation and position sensor (Motronic)

Comfort
Yaw-rate sensor (navigation)
Air-quality sensor (air-conditioning)
Humidity/temperature sensor (heating and air conditioning)
Pressure sensor (central locking system)
Rain sensor (windshield-wiper control)
Distance sensor, ultrasound (back-up monitoring)

Safety
Ranging radar (ACC, precrash)
Tilt sensor (headlamp aiming)
High-pressure sensor (ESP)
Torque sensor (power-assisted steering)
Steering-wheel-angle sensor (ESP)
Acceleration sensor (airbag)
Seat-occupation sensor (airbag)
Yaw-rate sensor (ESP)
Acceleration sensor (ABS)
Tilt sensor (safeguarding and anti-theft systems)
Yaw-rate sensor (roll-over sensing)
Speed sensor (ABS)

BOSCH Sensoren

UAE0816-1E

1	Areas in which sensors are used				
Typical features	**Primary standards**	**Precision measurement**	**Industrial measurement**	**Consumer technology**	
Accuracy	10^{-11} to 10^{-7}	2 to $5 \cdot 10^{-4}$	2 to $5 \cdot 10^{-3}$	2 to $5 \cdot 10^{-2}$	
Costs	€ 100 k to € 1 m	€ thousands	€ hundreds	€ 1 to 10	
Units/year	single figures	approximately 10	100 to 1 k	10 k to 10 m	
Use	– Research – Testing of secondary normals	– Calibration	– Process instrumentation – In-process measurement	– Auto electronics – Building services	

Table 1

signal E and the influencing variables Y_i practically without error (intelligent or smart sensors).

Adjustment

In the real world, the sensor model always includes some free parameters, with which the model can be adapted to the actual properties of the individual sensor type in a kind of calibration process (Fig. 4a).

These model parameters are generally stored in a programmable, nonvolatile memory component (PROM) in the digital processes for conditioning sensor signals, which have come to dominate. By contrast with the conventional analog compensation of influencing variables, it is not only roughly linearly acting influences that may be corrected, but also strongly nonlinear processes that can be corrected successfully. Another significant advantage of this type of calibration, which takes place over a purely electrical connection, is that each sensor may easily be maintained under operating conditions during the calibration phase.

2	Sensor symbol

3	Basic sensor function

▶ Concept of the smart sensor

In a somewhat more general form, intelligent sensors (smart sensors) may be defined as follows: Intelligent sensors, sometimes also known as integrated sensors or sensors with specific local electronics, allow the (static or dynamic) accuracy inherent in a sensor to be exploited by means of (generally digital) microelectronics to a far greater degree than conventional sensors. The sensor information, in particular the complex information from multisensor structures, can be compressed here by local further processing, that is to say, it is brought to a higher level (than the sensor alone is capable of providing), without a multiplicity of external connections being required for this.

There is no clear determination as to
whether sensors can already incorporate
a part of the signal processing or not; it is,
however, recommended, for example, that
no distinction be made between elemen-
tary sensor, sensor cell, or the like and
an integrated sensor.

Calibration process

A smart sensor is generally programmed
or calibrated, in a procedure correspond-
ing to the calibration of conventional ana-
log sensors, in three steps with the help of
an external computer (host) (Fig. 4):

Recording the actual value
The host computer systematically varies
both the measured variable x_e and the
influencing variable(s) y, thus setting a
specific number of relevant and represen-
tative operating points. The smart sensor
here passes the, still uncorrected, raw sig-
nals x_a^* to it. The host, however, is simulta-
neously also receiving the "true" values x_e
and y from substantially more accurate
reference sensors. The host computes the
correction values necessary by comparing
these two values, and interpolates these
corrections over the full measuring range.

Saving the correction parameters
The host computer calculates the model
parameters for the specific sensor from
the data previously obtained, e.g. for a lin-
ear characteristic curve course and saves
these to the smart sensor's PROM. In a
check process, these parameters may also
first be emulated in the host computer's
RAM before they are finally burnt to the
nonvolatile memory of the smart sensor.
If characteristic curves are approximated
with higher order polynomials, program
maps (look-up tables) may also be stored
in the smart sensor to avoid protracted
computing processes. The solution of sav-
ing a coarse-grid program map in conjunc-
tion with a simple linear interpolation
between the data points (example shown
in Fig. 5) has also proved very useful.

Operating phase
The smart sensor is now disconnected
from the host computer and is in a position
itself to calculate the measured variable x_e
with very low error thanks to the stored
model data. It is able to pass the value to a
connected control unit in digital, bit-serial
or even analog form (e.g. pulse-width
modulated), for instance. The measured
variable can also be distributed digitally
to further control units via a bus interface.

4 Trimming/calibration of a smart sensor

SAE1014-1E

This process can, unlike conventional laser calibration, in principle also be repeated if an erasable PROM is used. This is a particular advantage in the development phase of sensors.

5 Interpolation of measured values using a data point program map

Example: two-dimensional data point-program map s (T_n, ϑ_m) of a smart sensor for the measurement of a travel s:

For a highly precise evaluation of the sensor, which operates using variable inductance, its natural characteristic curve and its temperature sensitivity are each approximated using 5th order polynomials. As a frequency-determining element of a very simple oscillator circuit, it outputs the period duration T as an uncorrected output signal. As a sensor model for the measuring travel s, an outline program map embracing only 32 x 64 = 2,048 example-specific values $s_{n,m}$ is stored (in the PROM) along with a simple interpolation algorithm (in the ROM), instead of a total of 36 polynomial coeffients and a protracted polynomial evaluation. Should a signal T occur between these data points T_n and T_{n+1} in conjunction with a temperature ϑ between data points ϑ_m and ϑ_{m+1}, the sensor interpolates in two dimensions in accordance with the mapping between the benchmark figures stored "error-free" $s_1,...s_4$ and thus the measured value $s(T, \vartheta)$ sought is determined as the result of interpolation.

Automotive applications

With increasing demands on all functions in the vehicle, the control and regulation functions that were previously implemented by mechanical means have, in the last 40 years, successively been replaced by electronic devices (ECU, Electronic Control Unit). This has necessarily caused high demand for sensors and actuators with which these electronic control units can, on the one hand determine the relevant states of the vehicle, and on the other hand, actually influence these states. Over this period, the vehicle industry has become one of the, previously unprecedented, drivers of the development of sensors that could be manufactured in large numbers.

Whereas at the outset they still were of electromechanical or macromechanical form of some sort, the trend beginning in the 80s was for miniaturization, with

6 Milestones in the development of sensors for automotive applications

Year	Development
1950	Lambda oxygen sensor
1960	Electromechanical pressure sensor
	Piezoelectric knock sensor
1970	First integrated Hall sensor
	Strain-gage acceleration sensor for airbag
	First pressure sensor on silicone base
1980	Hot-wire air mass meter
	Thick-film air mass meter
	Integrated pressure sensor
1990	Micromechanical acceleration sensor for airbag
	Piezoelectric yaw-rate sensor for ESP
	Micromechanical air-mass meter
	Micromechanical yaw-rate sensor
2000	Yaw-rate sensor for roll-over sensing

semiconductor methods (batch process-ing) for the manufacture of high use sen-sors. For a while, sensors originating in hybrid technology using thick-film tech-niques also played a not insubstantial role. These are still used in isolated applications today, e.g. in the wafer-shaped Lambda oxygen sensors and high temperature sensors for the exhaust line.

Where temperature and magnetic field sensors could initially still be designed in circuit-like structures and be produced in batches, this trend increased as it became possible also to micromachine silicon in many various ways in two or three dimen-sions, and also to connect them, even in multiple layers, very soundly by very efficient methods.

Where the technologies of the electronic semiconductor circuits were practically exclusively based on silicon as the base material, quite different materials and technologies played a not insignificant role in sensors. Thus, for instance, quartz can also be micromachined using anisotro-pic etching techniques, and unlike silicon also has very advantageous piezoelectric properties. III-V semiconductors such as

gallium arsenide (GaAs) have a substan-tially greater operating temperature range than silicon, which can be very advanta-geous, particularly at some points in the vehicle. Thin metallic layers are very well suited to the manufacture of precise strain-gage resistors, accurate tempera-ture sensors and magnetic field-dependent resistors.

With silicon it is possible also to inte-grate the electronics monolithically with the sensor. This technique has lost its im-portance – with a few exceptions (e.g. Hall-effect ICs) – because of the generally very different number and type of process steps and also because of the inflexibility associ-ated with this. Hybrid integration technol-ogies in the tightest of spaces as a rule lead to substantially more cost effective, but functionally equivalent solutions (Fig. 7).

While the development of sensors was ini-tially concentrated almost exclusively on systems inside the detail in the drivetrain, the suspension and the body and driving safety, the direction of sensing of newer developments is increasingly named to the outside and to the area close to and further from the vehicle:

7 Hybrid integration of sensor and electronics: surface micromechanical acceleration sensor on a microhybrid circuit

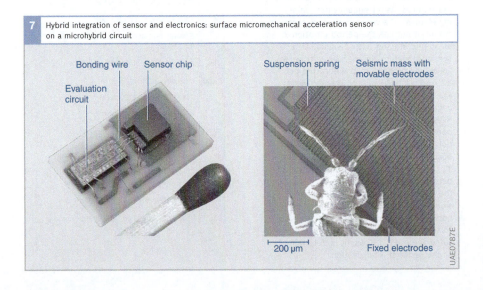

Bonding wire Sensor chip Suspension spring Seismic mass with
 movable electrodes
Evaluation
circuit

200 μm Fixed electrodes

UAE0787E

- Ultrasound sensors detect obstacles on parking and will even allow automatic parking in the foreseeable future, perhaps in combination with other sensors.
- Near-range radar scans the area around the vehicle to detect objects that probably could cause a collision, to gain time and to prime safety systems before a collision occurs (precrash sensors).
- Imaging sensors can not only detect traffic signs and send them to the driver's display, but also detect the edge of the carriageway, warn the driver of any hazardous deviation and, where required, in the long term also permit automatic driving. In combination with infrared beams and a screen in the driver's field of vision, IR-sensitive imaging sensors could permit long distance observation of the carriageway, even at night (night vision) or in foggy conditions.
- Long-range radar sensors observe the carriageway for 150 m in front of the vehicle, even in poor visibility, to adapt the driving speed to vehicles ahead and, in the longer term, also to support automatic driving.

As part of the vehicle's peripheral equipment, the sensors and actuators form the vehicle's interface to its complex drive, braking, chassis, and bodywork functions, as well as to the vehicle guidance and navigation functions and the (usually digital) ECUs which operate as the processing units (Fig. 8). An adapter circuit is generally used to convert the sensor's signals into the standardized form (measuring chain, measured-data registration system) required by the control unit.

8 Sensors in the multilayered process that is the motor vehicle

Fig. 8
1 Measuring sensors
2 Adapter circuit
3 Driver circuit
4 Actuators
AK Actuator
AZ Display
SA Operating switch
SE Sensors
ECU Control unit
Φ Physical variable
E Electrical variable
Y_i Disturbances

These customer-specific adapter circuits are tailor-made for specific sensors and are available in integrated design and in a wide variety of versions. They represent a quite substantial and very valuable addition to the sensors described here, without which their use would not be possible, and the measuring accuracy of which is, properly said, only defined in conjunction with these.

The vehicle can be regarded as a highly complex process, or control loop, which can be influenced by the sensor information from other processing units (control units), as well as from the driver using his/her controls. Display units keep the driver informed about the status and the process as a whole.

Details of the sensor market

The proportion of value added in the vehicle by the electrical and electronics systems is currently around 26 %. Nowadays, practically every second sensor is fitted into a vehicle, with annual growth rates that are, in some cases, still in double digits. Since the start of the 90s, micromechanical and microsystem-technical sensors have taken a rapidly increasing share, which amounted to about a third in 2005.

Unlike in the general sensor market, Europe has a current market share of 41 % of the vehicle sensors sector and Bosch is the world market leader; this has clearly surpassed America with a share of just 34 %. The total market for sensors for vehicle applications is expected to grow from US$ 8.88 billion in 2005 to around US$ 11.35 billion in 2010, that is by a total of 28 % (Fig. 9).

There are typically three groups of companies offering sensors for use in the vehicle:
• The semiconductor industry: sensors from semiconductor manufacturing originated here with the application of a few process steps. They serve the entire sensor market including the automotive industry and have an efficiently-operating sales and distribution system.

Micromechanical processes for the production of sensors are continually being further developed in tandem with the semiconductor processes. These companies do not, however, have any specific know-how in the area of specification, testing and packaging appropriately for the vehicle.
• Special, generally medium-sized manufacturers of sensors, who do not produce semiconductor circuits, but have generally chosen just a few sensor types as their product to supply the sensor market overall or just preferred branches such as the vehicle market.
• Large vehicle suppliers and system manufacturers (e.g. Bosch) or large subsidiaries of vehicle manufacturer who have specialized in the needs and support of their mother groups. Here too, since the introduction of electronics in the vehicle, experience has been gained with the production of semiconductor and hybrid circuits in close collaboration with semiconductor manufacturers (process development, licensing). Thanks to system knowledge, it has been possible to build up comprehensive know-how in the field of specification, testing and packaging techniques appropriate for the vehicle.

9 Sensor market for automotive applications

Billion US $

12 — 11.05 11.35
 10.67
 10.24
10 — 9.66
 8.88
 8.22
8 — 7.51
 7.21
6.71
6 —
4 —
 2001 2005 2010

UAE1046E

Fig. 9
Source: Bosch

Features of vehicle sensors

While general sensors are being developed for as broad a user group as possible and in multilevel measuring ranges, often without the manufacturer being aware of the application, vehicle sensors are generally specified and optimized for a special application. They are a part of a system and are often not freely available in the trade. Their development generally takes longer than that of commercial sensors, not just because of the enhanced demands. It is more associated with the development of the system and generally lasts as long as this does, as the sensor specification can still change until system development is concluded.

The high thrust for innovation in systems for the automobile sector very often compels the development of new sensor technologies or significant extensions of their specifications. Figure 10 shows the typical development phases that vehicle sensors pass through at the supplier's works.
 The development process begins naturally with the system idea originating with the vehicle manufacturer or the supplier. The initial task is to make a selection of the measured variables required, whether or not this is feasible. At this stage, the sensors' function is simulated as part of the systems engineers' usual system simulation and a first specification is drawn up. If it is possible to use a sensor technology that has already been introduced, the wishes in respect of the sensors are passed directly to product development or the producing area. If no technology is immediately available, sensor and technology experts are gradually increasingly included in research and advanced development. Here, the first laboratory prototypes can be created using known technologies and often with the help of external partners. These prototypes can then be passed to product development for initial tests.

If no sensor principles are known for the requirements set, research will be started into new procedures and methods for the measurement of the variables required. This phase is one of fundamental research which ultimately also supplies new initial technology prototypes. This procedure can repeat itself recursively until a promising solution which can then be sent to product development is found. It is not unusual for this development loop to be repeated again in full or for a different measured variable to be selected.

Five distinct phases are generally identified in the development of a completely new sensor, just as with other electronic products (Table 2). While prototypes and A-samples mostly still come from advance development or research, B- and C-samples

10 Genesis of a functional Bosch sensor

Implementation phase Design phase

■ Technology issues
■ Application issues

Technical assistance Technical clarification

──────── Research ────────
(Technology pool) Process engineering
 New methods
 Key problems

──── Advance development ────
Prototype, A-sample Sensor principle
Laboratory test/ Sensor technology
measuring technique Packaging

──── Product development ────
Vehicle test, trial System definition
B-sample/C-sample Measurement selection
Production model Sensor target
 specifications

Product specification Vehicle parameters
──────── Customer ────────

Product: Customer specification
Vehicle system/
Components Customer requirements:
 Vehicle system

SAE1012-1E

are products of product development. If recursions are necessary in difficult developments, there can easily also be a number of B- or C-sample phases (B1, B2, C1, C2).

2	Sensor prototyping phases leading to series production		
Prototype phase	**Function/ specifications**	**Manufacture**	
Prototype	restricted	Prototype construction without tools	
A	restricted	Prototype construction without tools	
B (possibly B1, B2)	full	Prototype construction without tools (construction identical to C)	
C (possibly C1, C2)	full	Prototype construction with series production tools	
D	full	Pilot series, partially manual	
Series production	full	Automated	

Table 2

11 Characteristic curve types

Fig. 11
S Output signal
X Measured variable

a Continuous, linear
b Continuous, nonlinear
c Discontinuous, multi-step
d Discontinuous, two-step (with hysteresis)

UAE0286-1Y

Sensor classification

Sensors can be classified and grouped according to very different points of view. They can be classified with regard to their use in the vehicle as follows:

Assignment and application
- Functional sensors (pressure, air-mass flow) mainly used for open and closed-loop control assignments
- Sensors for safety (passenger protection: airbag, ESP) and security (theft-deterrence feature)
- Sensors for vehicle monitoring (on-board diagnosis (OBD), fuel-consumption and wear parameters) and for driver/passenger information

Characteristic curve type
- Continuous linear curves (Fig. 11a) are used mainly for control assignments covering a wide measuring range. Linear curves are also distinguished by uncomplicated testing and calibration.
- Continuous nonlinear curves (Fig. 11b) are often used for the closed-loop control of a measured variable across a very restricted measuring range (e.g. exhaust-gas control to $\lambda = 1$, vehicle spring-deflection level). When, for instance, a constant permissible deviation relative to the measured value is demanded throughout the complete measuring range (e.g. HFM air-mass meter), curves which feature both pronounced non-linearity and a special shape (e.g. logarithmic) are at an advantage.
- Such two-step curves (possibly even featuring hysteresis, Fig. 11d) are used for limit-value monitoring in such cases where remedial measures are easy to apply when the limits are reached. If remedial measures are more difficult, then multi-step curves (Fig. 11c) can be used for an earlier warning.

Type of output signal

Sensors may also be distinguished by the type of output signals (Fig. 13):

Analog signals
- Current/voltage, or a corresponding amplitude
- Frequency/period duration
- Pulse duration/pulse duty factor

Discrete output signal
- Two-step (binary coded).
- Multi-step, with irregular steps (analog coded).
- Multi-step equidistant (analog or digital coded).
- A further distinction must be made, as is represented in Figure 12 in a systematic overview of the determined, i.e. nonrandom (stochastic) signal – as to whether the signal is permanently available at the sensor output (continuously) or only at discrete intervals (discontinuously).

For instance, the signal is bound to be discontinuous if it is digital and issued in bit-serial form.

13 Signal shapes

Fig. 13
a Output signal U, information parameter: frequency f
b Output signal U, information parameter: pulse duration T_P

12 Classification of the determined signals according to the information parameter (IP) with examples

Error types and tolerance requirements

The deviation of the actual characteristic curve of a sensor from its reference characteristic curve is designated its error e. It is best stated in relation to the input range y (measured variable) and not to the output range x (output signal):

(3) $e = y_{indicated} - y_{true}$

$y_{indicated}$ = indicated value for the measured variable

y_{true} = ideal value, setpoint value for the measured variable (is determined with a measuring sensor that is more accurate than the sensor being examined by at least 1 class)

The amount of the deviation represents, as shown in Figure 14, the absolute error e_{abs} (unit as measured variable). Related to the true measured value y_{true}, this becomes the relative error (% of read-ing), related to the measuring-range final value it becomes the percentage error of the range.

Assuming a generally desirable linear characteristic, the absolute deviation e_{abs} can be classified into three categories (Fig. 15):
• Zero offset (offset error) e_{zero}
• Gradient deviation (gain error) e_{gain}
• Linearity deviation e_{lin}

The causes of these errors are predominantly found in
• The production scatter of the characteristic curve

15 Subdivision of the total error

a

b

c

14 Characteristic curves and error graph for a sensor

a

b

Fig. 14

a Actual and
reference
characteristic

b Error graph

y Measured variable
x Output signal
Δx Measuring range
e Error (deviation)

Fig. 15

a Zero error

b Gradient error

c Linearity error

y Measured variable
x Output signal
Δx Measuring range
e Error

- The temperature sensitivity of the characteristic curve and
- The production scatter of the temperature sensitivity

The deviations referred to are exclusively systematic or terministic errors which, by contrast with random (stochastic) errors, are well defined, foreseeable and, as a matter of principle, correctable, and are also for the most part more or less precisely corrected.

The non-correctable, stochastic errors include, for example,
- Drift (low and higher frequency background noise) and
- The effects of aging

In the specification of a sensor, the total error in the new state and after aging is prescribed in the product specifications by means of a tolerance graph (Fig. 16). Sometimes, however the permissible individual error shares such as the offset, lead and linear deviations are additionally specified.

Strict metrology teaching states that, in the case of systematic errors, the sum of the amounts of the individual errors must be assumed as the total error (these may add up in the worst case). In the case of stochastic errors, statistical addition, which calculates the total error as the root of the sum of squares of the individual errors, is permitted. As statistical addition leads to a smaller total error, it is, however, often also applied to the systematic errors in a less strict configuration:

(4) $e_{total} = \sum_{1}^{n} |e_i|$ Sum of n systematic errors

(5) $e_{total} = \sqrt{\sum_{1}^{n} e_i^2}$ Sum of n stochastic errors (statistical addition)

Reliability

Failure rate

The operational reliability of a sensor is a purely statistical quantity and, as with any component, is characterized by its failure rate λ, which is stated in $1/h$, $\%/h$ or ppm/h. Here λ is determined with a large number of parts. If it is wished to determine approximately the failure rate with a number N (< 40) of sensors that is not too great, it is possible to observe the failure behavior of this random sample under operating conditions until, after a finite time, all the parts have failed. If this observation is begun at time t_0 and if the inventory of parts remaining at a later point t_i is identified as $B(t_i)$, then a good approximation for the failure rate λ is the failure quota q:

(6) $q(\Delta t_i, t_i) = \dfrac{B(t_i) - B(t_{i+1})}{\Delta t_i \cdot B(t_i)}$ Failure quota

where $\Delta t_i = t_{i+1} - t_i$

t_i are the times at which individual parts or several parts fail (Fig. 17). The ratio of the current inventory of parts to the initial inventory of parts is known as the relative inventory B_R:

16 Tolerance graph for a sensor

Temperature T

UAE1050E

Fig. 16
Tolerance graph illustrated on the example of a resistance temperature sensor

(7) $B_R(t_i, t_0) = \dfrac{B(t_i)}{B(t_0)}$

With a very large number of parts ($N \to \infty$), the survival probability $R(t)$ at the, here now continuously variable, time t corresponds to this. The failure rate is calculated here for a large number (in practice $N \approx 2{,}000$) of sensors as a percentage variation on the survival probability R per unit of time dt as:

(8) $\lambda(t) = -\dfrac{1}{R(t)} \cdot \dfrac{dR}{dt}$ Failure rate

Reliability is the inverse value of the failure rate:

(9) $z = \dfrac{1}{\lambda}$ Reliability

The definition of the failure rate λ demands a failure criterion:
• Complete failure
• Partial failure
• Sudden failure (sudden change in a feature)
• Drift failure (gradual change in a feature)

Furthermore, it is absolutely necessary to specify under what operating conditions the failure rate defined in this way is to be understood. Particularly in the case of electrical components such as sensors, for example, it is necessary to distinguish between true, active operating time (switched on state) and service life in the sense of pure storage time. Any statement of a failure rate without this additional information is worthless.

Failure rates are normally determined using time-acceleration methods. The acceleration factors are achieved here by exposing the sensors to enhanced operating conditions. A high degree of experience is required in order to apply time acceleration methods that truly reflect reality.

The concept of the mean service life T_M is also used to characterize the reliability of a sensor. A good approximation of this can be calculated from a random sample from the sum of the individual service lives T_i:

(10) $T_M \approx \dfrac{1}{N} \cdot \displaystyle\sum_{1}^{N} T_i$ or

(11) $T_M = \displaystyle\int_{0}^{\infty} R(t)\, dt$ for a very large number of parts

Fig. 17
Observation for a
random sample of
$N = 20$ sensors;
mean service
life $T_M = 4{,}965$ h.

17 Inventory and relative inventory of a random sample

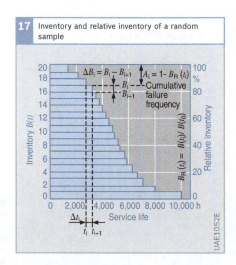

18 Distribution of the failure rate $\lambda(t)$ over time

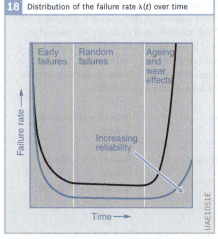

The failure rate of a product against time shows a typical bathtub pattern (Fig. 18). At the start, the failure rate is somewhat high because of early failures, it then passes through a longer, relatively low, horizontal range, then to rise again drastically towards the end of the service life (aging and wear effects). In the case of sensors for which high operational reliability is demanded, an attempt is made to avoid the elevated failure rate at the start by sorting out the early failures by pre-aging, e.g. by conditioning at higher temperatures ("burn in"). Early failures are basically nothing other than manufacturing defects that have not been recognized.

Table 3 gives some examples of permissible failure rates λ applicable for the vehicle. The ppm values given relate to a period of 10 years, or alternatively to 150,000 km covered, if this is completed in a shorter time. Where the sensors are given a flat rate of <10 ppm, it means that in 10 years, only fewer than 10 of 1 million sensors may fail. This value is, however, substantially lower for sensors in passenger protection systems.

Measures for increasing reliability

The best method for ensuring high quality is to engineer and design in reliability. This means that as soon as the sensor is designed, correspondingly durable materials must be selected and solid protective measures against anticipated mechanical,

chemical and electrical ambient influences must be provided (packaging, passivation). On the other hand, it is costly to create operational reliability by testing alone, i.e. to exclude early failures through premature aging.

It is sensible to provide monitoring and diagnostic capabilities for sensors in complex systems (e.g. signal range check, etc.), so that any failures occurring may be detected in good time. In an emergency, a sensor's function can here be temporarily replaced by other measured variables or sensible fixed values (limp-home mode, back up). A sensor-less, often purely mechanical, limp-home mode can be provided. For instance, should the accelerator-pedal sensor failure in a diesel vehicle, the vehicle could also be driven slowly back for repair under regulated idle speed alone (limp home).

Where reliability must be guaranteed with a probability bordering on certainty (e.g. for sensors in electronic braking and steering systems), redundancy, i.e. provision of multiple parts, is generally the means used. Dual redundancy of a sensor of the same type here only permits failure detection for the case of widely different indications, whereas triple redundancy with 2 from 3 analysis additionally still supplies a correct measured value. Here, however, care must be taken to provide not only redundancy of the sensors, but other essential parts such as the power supply, signal evaluation and transmission means must also be correspondingly redundant, as otherwise the probability of simultaneous failure increases. It is often also advisable to provide sensor redundancy in different technologies.

3	Reliability requirements on vehicle systems
Warranty target: 150,000 km/10 years	
→ ECU failure rate (field)	< 50 ppm
→ ECU failure rate (0 km)	< 15 ppm
→ **Failure rate of modules and sensors**	**< 10 ppm**
→ ASIC failure rate	< 3 ppm
→ IC failure rate	<< 1 ppm
→ Failure rate of discrete components	< 0.5 ppm
→ For comparison Mobile phone	~ 5,000 ppm

Table 3

Main requirements, trends

The vehicle sensors customized for the requirements of special electronic systems in the vehicle, unlike the universal sensors conventionally found in the market, are subject to five strict requirements (Fig. 19) which must be fulfilled by development and to which the most important development trends also correspond.

Low manufacturing costs

Electronic systems in modern vehicles easily contain up to 150 sensors. Compared to other sectors of sensor application, this abundance forces a radical reduction in costs. The target costs – typically in the range of € 1 to 30 – are here often less than one-hundredth of those for conventional sensors of the same capacity. Naturally, particularly with the introduction of a new technique or technology, the costs are generally on a falling learning curve, starting from a higher level.

Development trends

Automated manufacturing processes (Fig. 20) are largely used, working to a high yield. This means that each process step is always executed for a large number of sensors simultaneously. For example, semiconductor sensors are manufactured using "batch processing" in which 100 to 1,000 sensors are manufactured simultaneously on a single silicon wafer. On the other hand, such manufacturing equipment is only an economic proposition when correspondingly large numbers of sensors are produced. These quantities sometimes exceed an automotive-industry supplier's own in-house requirements, and can typically often be between 1 and 10 million per year. Here, the high numbers of sensors needed by the automotive industry has played an unprecedented and revolutionary role, and set completely new standards.

High reliability

In accordance with their assignments, vehicle sensors are subdivided into the following reliability classes, given in descending order of severity:
- Steering, brake, passenger protection
- Engine/drivetrain, chassis/tires
- Comfort and convenience, diagnosis, information, and theft deterrence

19 Principal requirements on vehicle sensors

Vehicle sensor	Requirement
Rational mass production	⇐ Low costs
Robust, proven technology	⇐ High reliability
Appropriate packaging technology	⇐ Extremely severe operating conditions
Appropriate miniaturization technologies	⇐ Low space requirement
On-site error compensation	⇐ High-level accuracy

SAE1013-1E

20 Relationship of costs, markets, technologies

UAE1053E

The requirements on the highest class here easily match the high reliability values familiar from aerospace applications. To some extent, they demand similar measures, such as the use of the best materials, redundant componentry, self-monitoring, (short-term) backup power supply, multiple programming of critical decision-making algorithms.

Development trends

Appropriate design measures guarantee built-in reliability. For instance, this necessitates the use of reliable, top-quality components and materials, coupled with rugged and well-proven techniques and engineering. And furthermore, an effort is made to achieve a consistent integration of the systems to avoid connection points which may be broken and are at risk of failure. This is possible, for instance, with radio-scanned sensors based on the antenna-coupled SAW (Surface Acoustic Wave) elements which do without wiring completely. If necessary, redundant sensor systems are fitted.

Severe operating conditions

Like practically no other breed, vehicle sensors are exposed to extreme stresses because of where they are fitted, and must withstand all kinds of aggression:
• Mechanical (vibration, shock)
• Climatic (temperature, moisture)
• Chemical (e.g. spray, salt mist, fuel, engine oil, electrolyte)
• Electromagnetic (incident radiation, line-conducted interference, excess voltages, polarity reversal)

It is this trend of using sensors immediately at the site where the measurement is taken in order to exploit the benefits associated with this which itself leads to the requirements being considerably cranked up.

Figure 21 illustrates this problem on a sensor which might be a temperature sensor, speed sensor, flow-rate sensor or concentration sensor, for instance. The sensor cannot in the slightest always be surrounded by a hermetic protective sleeve. Although this sleeve could offer rough protection for the measurement of temperature, flow rates and concentration, it must permit more or less direct contact between the sensor and the generally very aggressive monitored medium (an exception to this is the inertia sensor). Sometimes, thin, but very resistant, passivation layers are permissible on the sensor.

The form of the plug-in sensor demands a permanently tight seat of the mount in the associated wall (inner packaging, outer packaging). The sensor may be connected to the electronic control units via a fixed housing plug or also, as illustrated, via a plug flexibly fastened with a length of cable (e.g. ABS wheel-speed sensor). Here, too, there are three critical electrical connection points that absolutely must be protected against conductive fluids and corrosion: the inner connection on the sensor element, the connection of the cable tail and finally the connection of the partially shielded cable to the external plug connec-

21 Technologies accompanying sensors

Monitored medium

External environment

Fig. 21
1 Sensor
2 Protective sleeve (coating)
3 Seal
4 Mount
5 Seal, mounting
6 Data point
7 Seal, strain relief
8 Insulation (flexible)
9 Contacts
10 Plug housing
11 Plug contact

tion. If the seal is inadequate, it is only a question of time until corrosive liquids penetrate from the external connector into the internal sensor terminal.

The plug-in connection itself must be sufficiently tight overall, that no shunts form at the sensor output. The cable itself must retain its flexibility and tightness after many years of operation under the most adverse conditions.

Separable plug-in connections in the vehicle unfortunately still represent one of the most frequent causes of failure. Wireless signal connections (e.g. infrared light or radio) could break down this problem in the longer view, especially if the sensors were even to have a wireless energy supply (autonomous sensors).

The costs of these indispensable technologies complementary to the sensors often exceed those of the actual sensor element many times over. They constitute the actual value of an vehicle sensor, not only in terms of cost, but also in terms of function.

Development trends

Protective measures against the stresses referred to demand a very high degree of specific know-how in the field of sensor packaging.
- Passivation and connecting techniques
- Sealing and joining techniques
- EMC protective measures
- Low-vibration installation
- Service life test and simulation methods
- The use of resistant materials, etc.

Furthermore, detailed knowledge of the loading to which the sensor will be subjected at the particular installation point is required. It is often forgotten that the quality of a sensor stands and falls with total competence with these protective measures.

Fiber-optic sensors in which the light guided in the optical fibers (glass, plastic) can be modified as a function of the measured variable are particularly immune to electromagnetic interference. This is true up to the point at which the optical signals are converted back into electrical signals. If these are to be used in the future, there will have to be some development work the provision of low-priced measuring elements and the accompanying technologies. There are, for example, some very interesting applications for this in the field of force measurement, e.g. in providing finger protection for electric power windows and sliding sunroofs (Fig. 22). Sensors of this type have also been tested very successfully as very early responding, distributed sensors in the door and frontal area of the vehicle for triggering passenger and pedestrian protective systems.

22 Microbending effect

Microbending effect

$F = 0$ $F > 0$
Damping ≈ 0 Damping > 0

Door frame
Sensor
Sealing rubber

Example of development: fiber-optic finger protection in power windows based on the microbending effect:
The light in an optical fiber is attenuated proportionally to the corrugated distortion on the application of a force F transverse to the fiber (large measuring effect), independently of where the force is applied or whether it is acting at a point or is distributed (distributed sensors).

Low-volume design
On the one side, the number of electronic systems in the vehicle continues to climb steadily. On the other, today's vehicles are becoming more and more compact. These facts, together with the need to retain the high level of passenger-compartment comfort forces development to concentrate on an extreme space-saving designs. Furthermore, the increasing demand for further improvements in fuel economy mean that minimization of the vehicle's weight is of prime importance.

Development trends
Widespread use is made of the familiar technologies applied in circuit engineering for the miniaturization of electronic components:
• Film and hybrid technologies (deformation-dependent resistors, thermistors, and magnetoresistors); nanotecholology
• Semiconductor technology (Hall-effect and temperature sensors)
• Surface and bulk micromechanical techniques (silicon pressure, acceleration and yaw-rate sensors)
• Microsystem technology (combination of two and more microtechnologies such as microelectronics and micro-mechanics, Fig. 23)

Micromechanical manufacturing refers on the one hand to dimensions in the µm-range and tolerances in the sub-µ-range which cannot be achieved with conventional machining methods. On the other hand, sensors which may have dimensions in the mm-range, but which are manufactured with micromechanical methods, are also considered micromechanical (Fig. 24). We will discuss only the most well-known and important method here, anisotropic etching of silicon. This is the most important method, because silicon is manufactured to high precision and in large numbers at low cost, and is the most thoroughly researched and best known material. In addition, it offers the capability of the monolithic integration of sensor and evaluation electronics. The etching speeds, which can be very different depending on the crystal axis (1:100), with which suitable etching fluids such as KOH attack are exploited in this method (Fig. 25); some crystal faces, for instance, remain practically untouched, while other faces are rapidly etched to some depth.

In conjunction with appropriate etching stop methods (doping, depletion layer) practically three-dimensional structures of the smallest dimensions may be produced to the highest precision in this way

23 Microsystem technology

Packaging

Micro-mechanics:
– Sensor micro-technology
– Actuator micro-technology

Microsystem technology

Micro-electronics

Micro-optics

LAE0915E

24 Micromechanics and mechanical structures

Basic mechanics	Advanced mechanics	Micro-mechanics	Nano-mechanics

1,000µm 100µm 10µm 1µm 0.1µm

Traditional material machining

(turning, milling, etc.)

Special machining methods

(e.g. extruder, laser, etc.)

Semiconductor technology methods

(lithography + etching/layer separation)

Nanolithography

UAE1055E

25 Test structure for the determination of the etching rate in Si monocrystal wafers

UAE1056Y

26 Anisotropic etching of (100) silicon

UAE1057E

Fig. 26
a Top view
b Cross-section
 at AA'

27 First attempts at three-dimensional forming of silicon

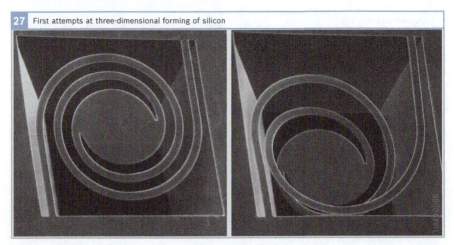

Fig. 27
Source:
Prof. Heuberger,
Fhg Berlin

(Figs. 26, 27). The problem of simultaneously integrating the circuit can be considered largely resolved, even if currently the associated circuit for most sensors is still made separately for reasons of a better yield and higher flexibility.

By contrast with bulk micromechanics, anisotropic, often wafer-depth etching has no role to play in surface micromechanics (SMM). SMM sensor structures are mostly built up additively on the surface of a silicon substrate (Figs. 28 and 29). Even if the dimensions of bulk silicon sensors are still mostly in the mm-range, those of SMM structures are an order of magnitude smaller (typically 100 μm), as a rule.

Frequently, mechanically indispensable parts are used simultaneously for enclosing and even cooling (associated) sensors and electronics (e.g. microhybrid control unit as an add-on control unit to the VP44 diesel distributor-type injection pump). This fusion of mechanical and electronic components known as mechatronics is coming more and more to the forefront in the search for cost and space savings. In the foreseeable future, practically all systems will operate on this basis.

Another example of a mechatronic sensor system is the Hall-effect speed sensor built into the wheel bearing. The addition of magnetic particles to the indispensable shaft seal in the bearing can even give it the function of the rotor or a pole wheel. The sensor benefits from excellent protection and encapsulation along with the high precision of the roller bearing and no longer needs any bias magnets. Finally, the Hall-effect technology allows installation in the very tight space within the bearing.

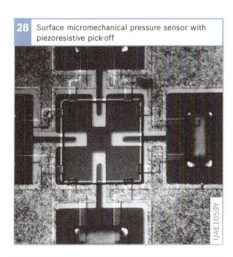

28 Surface micromechanical pressure sensor with piezoresistive pick-off

UAE1059Y

29 Surface micromechanical acceleration sensor with capacitive pick-off

Sensor element

UAE1060E

High accuracy

By comparison with the probes and sensors used for instance in the processing industry, with only a few exceptions (e.g. the air-mass meter), the accuracy demands made on vehicle sensors are relatively modest. Generally, the permissible deviations are > 1 % of the measuring-range final value. This applies in particular when considering the unavoidable effects of aging. The permissible deviations are normally achieved by the application of complex techniques to compensate for manufacturing tolerances, and to balance the effective compensation measures used against interference. Particularly since the above-mentioned requirements have for the most part become achievable, continually more demanding and sophisticated systems are imposing higher and higher demands in this sector.

Development trends

Initially, a tightening up of the tolerances in manufacture, and refinement of the calibration and compensation techniques help in this regard. An important step forward here is the hybrid or monolithic integration of the sensor and signal electronics directly at the measuring point, even embracing complex digital circuits such as analog/digital converters and microcomputers (Fig. 30).

Such microsystems are also known as "intelligent sensors". They take full advantage of the sensor's inherent accuracy, and offer the following features:
- Reduction of the load on the control unit
- Uniform, flexible, bus-compatible interface
- Multiple application of sensors
- Use of smaller measuring effects and of radio frequency measuring effects (local amplification and demodulation)
- Correction of sensor deviations at the measuring point, and mutual calibration and compensation of sensor and electronics, is simplified and improved by

Fig. 28
Source:
Prof. Guckel,
University of Michigan
in Madison, USA

storing the individual correction information in the PROM

While simultaneously detecting and digitizing interfering factors, intelligent sensors can calculate the required measured variable practically without error by applying the mathematical sensor model (Fig. 31). Here, the item-specific model parameters are defined in a preceding process which is equivalent to the calibration as previously performed, and stored in a PROM integrated within the sensor. In this manner, it is possible to considerably improve the sensor's static and dy-

namic characteristics (evaluation of the differential equation which defines the dynamic performance).

Local electronic circuitry necessitates the use of multi-sensor structures which use a number of identical sensors, or a number of different sensors, to register a variety of highly complex facts and locally reduce these to their basic information content, as necessary. This applies in particular to imaging sensors which in future will play an ever increasing role in detecting the situation inside and outside the vehicle.

Fig. 30
SE Sensors
SA Signal conditioning
A/D Analog-Digital
 converter
ECU Control unit
MC Microcontroller
BS Bus interface

30 Sensor integration levels

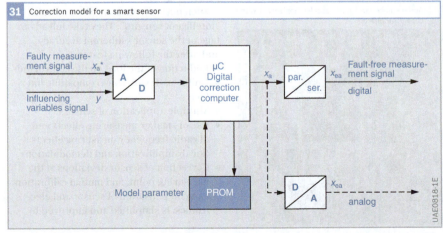

31 Correction model for a smart sensor

With a number of pressure sensors inte-
grated into the smallest of spaces, it is pos-
sible to not only increase the reliability
of measurement, but also to reduce the,
generally irregular, aging drift by applying
mean-value generation. Individual failures
(outliers) can also be detected and elimi-
nated. Sensors of this type are also known
as soft sensors. If the individual sensor
elements are designed for differing mea-
suring ranges – and at the same time fea-
ture high overload capabilities (e.g. capaci-
tive) – such a sensor can be used to con-
siderably extend the high-acccuracy
measuring range. Sensor structures of this
type were developed and tested years ago,
but have not been converted into
marketable products.

Overview of the physical effects for sensors

The sensors are not classified here into a
systematic structure by measuring effect,
but rather by measured variable. For this
reason, the most important physical ef-
fects or measuring concepts exploitable
for measuring purposes will here only be
reproduced in the form of a rough over-
view, which cannot be complete. Overlaps
between the various categories are un-
avoidable. The table specifies principally
the actual electrical effects; mechanical or
fluid prior stages, such as strain springs
(force), membranes (pressure), spring-
mass systems (acceleration), vibration sys-
tems (tuning forks) or even fixed and rotat-
ing turbine blades (flow) are not included.

4 Physical effects for sensors

Physical effect	Example	As at: Series production[1]/ Development[2]
Resistive effects (dependency of electrical resistance):		
Influence of temperature on metallic and semiconductor materials	NTC and thin-film resistors for the measure-ment of air and engine temperature	S
Length or angular proportionality of resistors (potentiometer sensors)	Accelerator-pedal and throttle-valve sensor, fuel tank level	S
(In plane) tension or pressure dependency (piezoresistive): strain-gage resistors	High-pressure sensors (e.g. common rail, ABS): metal diaphragm, low pressure sensors (silicon diaphragm),	S S
Vertical pressure dependency (out of plane)	force sensor	D
Magnetic field dependency (magnetoresistive): semiconductor (magnetoresistors), AMR[3] thin metal layers (e.g. NiFe, also in barber's pole form), GMR[4] sensors (nanolayers)	Speed or delivery angle measurement in diesel distributor injection pumps	S/D
Light dependency: semiconductor photoresistors	Rain sensor, dirt sensor for headlight cleaning, automatic headlights, automatic low-beams	S
Inductive effects (effects of Faraday's law)		
Induction voltage sensors (alternator): movement in the magnetic field	Wheel, camshaft, engine speed, needle lift (injection nozzle)	S
Wiegand effect	Speed of rotation	D
Variation in inductance as a result of the positional change of a ferromagnetic coil core	Solenoid armature sensor	D
Variation in inductance as a result of field limiting conductive elements (eddy current)	Semi-differential short-circuiting ring sensor (diesel pump load sensor)	S

Table 4
[1] Series production, at RB or competitors, possible also already phased out
[2] Development possibly also concluded and in reserve
[3] AMR = Anisotrop Magneto Resistive
[4] GMR = Giant Magneto Resistive

Physical effect	Example	As at: Series production[1]/ Development[2]
Variation of the degree of transformer coupling (by electrical or magnetic conductive elements)	Full-differential short-circuiting ring sensor	D
Variation in inductance or of the degree of transformer coupling (by magneto-elastic conductive elements)	Load bolt (Hitchtronik), braking force	S D
Saturation core sensors (e.g. Foerster probe)	Compass sensor	S
Capacitive effects (influence)		
Capacitance change as a result in change in plate distance and degree of cover	Micromechanical acceleration sensors, e.g. for airbag, ESP, MM2 yaw-rate sensor, pressure sensor	S S D
Capacitance change as a result in a change of the relative dielectric constant	Oil grade, humidity sensors	S S
Capacitance change as a result of change in the electrolyte level with a dielectric medium	Fuel tank level	D
Charge generating effects		
Piezoelectric effect (quartz, piezoceramics)	Knock sensor, airbag sensor, DRS1 yaw-rate sensor	S S
Pyroelectric effect	IR sensor (dynamic)	D
Photoelectric charge generation	CCD and CMOS imaging sensor (also IR range)	D
Voltage generating, galvanic effects		
Hall effect (out-of-plane sensitivity, semiconductor material)	Hall switch (ignition), wheel and engine speed, acceleration sensor (ABS, 2g), front passenger weight detection (iBolt™), ARS1,2 (accelerator pedal etc.)	S
Pseudo-Hall effect (in-plane sensitivity, metal thin film)	LWS2 and LSW4 steering-angle sensors	S
Electrolytic diffusion probes (doped Zr oxide ceramic)	Lambda oxygen sensors	S
Thermocouple, thermopile	IR sensor (Bolometer)	D
Photoelectric and fiber-optic effects		
Photoelectric cells, photodiodes, phototransistors (also in IR range)	Rain sensor, dirt sensor for headlight cleaning, automatic headlights, automatic low beams	S
Medium-dependent absorption	Soot particles, humidity	D
Extrinsic and intrinsic fiber-optic effects: influence on intensity, interference (influencing phases), influencing polarization; e.g. microbending effect	Finger protection (windows, sliding sunroof), pedal force, collision	D
Thermal effects		
Resistor cooling as a function of the flow velocity, of the medium, of the density or fill level of a medium	HFM air-mass sensors, analysis, concentration, fill level (fuel tank)	S – – D
Wave propagation effects		
Sound waves: propagation time effects (echo sounding), superposition with medium speed, Doppler effect (moving source/receiver)	Parking-aid assistant, volume flow rate, speed over the ground	S D S
Light waves: total reflection on boundary surfaces, optical resonators (color analysis)	Rain sensor, fiber-optic finger protection, fill level (analog, limit value),	S D D
Sagnac effect Propagation time	yaw rate: fiber and laser gyroscopes, lidar (light wave radar)	D D
Electromagnetic radiation: Doppler, FMCW, propagation time radar	ACC distance sensor	S

[1] Series production, at RB or competitors, possible also already phased out

[2] Development possibly also concluded and in reserve

[3] AMR = Anisotrop Magneto Resistive

[4] GMR = Giant Magneto Resistive

Overview and selection of sensor technologies

The various sensor technologies for exploiting the measuring effects described are naturally closely bound up with the measuring principle. They will first be listed here in a rough overview:

- Wound or photolithographically produced induction coils (with electrical or magnetic conductors), eddy current and short-circuiting ring sensors
- Flux gate probes (Metglas etc.) for magnetic field measurement
- Pulse jump sensors (Wiegand)
- Wire-wound (inductance-free) resistors
- Foil resistors (laminated onto plastic substrates)
- Sintered ceramic resistors
- Thin- and thick-film techniques (especially resistors and capacitances)

- Semiconductor technology (monocrystalline or polycrystalline resistors, depletion layers, charge-storage cells, etc.), electronics for signal conditioning: Si (bipolar, CMOS, BICMOS, EEPROM), GaAs
- Micromechanics (silicon and other materials, e.g. quartz, metal (LIGA technology) etc.)
- Piezoceramics
- Piezofoil
- Isolating ceramics as a spring material (e.g. as a pressure-sensor diaphragm)
- Ceramic solid electrolytes (e.g. as Lambda oxygen sensor)
- Quartz and other piezoelectric crystals
- Glass or plastic optical fibers or plates

The matrix shown in Figure 32 indicates which sensor technologies should be taken for various requirements. Thus, for instance, should the requirements be for small size, high reliability and high accuracy, it is appropriate to take a thin-film technology, as long as this is available for the measured variable desired.

32 Selection of suitable sensor technologies

Sensor measuring principles

There is a great number of sensors at work in motor vehicles. They act as the sensory organs of the vehicle and convert input variables into electrical signals. These signals are used in control and regulation functions by the control units in the engine-management, safety and comfort and convenience systems. Various measuring concepts are applied, depending on the task.

1	Travel/angular positions as direct measured variables	
Measured variable		**Measuring range**
Throttle-valve position in the gasoline engine		90°
Accelerator/brake-pedal position		30°
Seat, headlamp, rear-view mirror position		
Control-rack travel and position for diesel in-line fuel-injection pump		21 mm
Angular setting of the injected fuel-quantity actuator on the diesel distributor-type injection pump		60°
Fill level in the fuel tank		20 to 50 cm
Travel of clutch servo unit		50 mm
Distance: vehicle – vehicle or vehicle – obstacle		150 m
Steering (wheel) angle		±2·360° (±2 revolutions)
Tilt angle		15°
Angle of direction of travel		360°
GPS (Global Positioning System)		360° geographical latitude/longitude, geographical altitude
Near-range distance (US parking-aid assistant)		1.5 m
Near-range radar (precrash)		10 m
External video		40 m
Long-range and near-range infrared viewer		100 m

Table 1

2	Travel/angular positions as indirect measured variables	
Measured variable		**Measuring range**
Spring compression travel (headlamp range, vehicle tilt)		25 cm
Torsion angle (torque)		1 to 4°
Deflection of a sensor plate (volume flow)		30 to 90°
Deflection of a spring-mass system (acceleration)		1 to 500 µm
Deflection of pressure sensor diaphragm		1 to 20 µm
Deflection of force measuring spring (passenger weight)		10 to 500 µm

Table 2

Position sensors

Characteristics

Position sensors record the widest variety of one or multi-dimensional travel and angular positions (translational and rotary variables) in a wide variety of areas. This includes gaps, distances, displacements, fill levels and even minor expansions – anything that can be measured in meters and angles.

In this sector of applications, activities have long since been directed at changing over to proximity or non-contacting sensor principles. Such sensors are wear-free and thus have a longer service life as well as being more reliable. The costs involved though often force vehicle manufacturers to retain contacting sensor principles which still perform sufficiently well for many measuring tasks.

Position variables are "extensive measured variables", the measured variable or the measuring range of which has a substantial influence on the size of the sensor (by comparison, pressure and temperature, for instance are "intensive measured variables"). Sensors for extensive variables of medium and larger ranges (e.g. including flow and force sensors) are thus initially less suited to miniaturization on the one hand and hence for cost-effective mass production. As this is much less true for angular-position sensors than for travel sensors (the angle value does not depend on the radius or leg length), the former are clearly preferred in the vehicle.

In the case of extensive variables with a large measuring range, however, the extensive variable can often be first converted into an intensive variable, which can then be measured using microsensors. Thus flow measurement can be converted to flow density measurement (partial flow measurement, e.g. 1:100) where the flow tube becomes an essential component part of the sensor without which the sensor ultimately cannot be tested and calibrated.

In the case of travel and distance sensors covering large measuring ranges (approximately 0.1 to 150 m), the measured variable is converted into a pulse or phase runtime by means of wave-propagation sensors (sound and electromagnetic waves), which can then be easily measured electronically with relatively small transceiver devices. In the case of sound waves, however, the propagation medium (e.g. air) and the time standard required (e.g. oscillating crystal) then become an essential component part of the measuring device.

Measured variables: overview

There are a large number of applications in which position represents the actual measured variable. This is shown in overview by Table 1. In other cases, the measured position or angle represents a different measured variable (Table 2).

In practice, "incremental sensor systems" are also often referred to as angular-position sensors, even when they are used for measuring rotational speed. These sensors are in reality not angular-position sensors. Since the increments (steps with which a given variable increases) which have to be measured with these sensors in order to measure the deflection angle must be counted with the correct preceding sign (in other words, added), these sensors are in reality not angular-position sensors. Due to the danger of the counter being falsified as a consequence of spurious pulses, such angular-position measuring systems are only in limited use. Fixed, directly locatable reference marks only provide very little help in this dilemma. Another disadvantage of such angular-position measuring systems is the fact that the absolute position is lost when the operating voltage is switched off. It is no use storing the final position in a non-volatile memory here, since most angular positions can also change mechanically when power has been removed (e.g. steering angle).

Although it is true that the crankshaft angle is measured incrementally, this measurement is recalibrated after each revolution by a reference mark detectable by the sensor, in case a fault should have occurred. This also takes account of the fact that the position of the crankshaft when the engine is started is not known; it must first rotate almost one revolution under the starter so that the reference mark is passed at least once. The crankshaft's motion is also very monotonous, i.e. it is always pretty much smooth (without sudden changes) and is always in one direction (that is forwards, and never backwards). You can thus be sure that no more than one revolution will ever be incorrectly detected.

The disadvantages of an incremental angle measurement cannot be absorbed if the engine starts immediately without being cranked, as is also the case with the steering angle. An absolute angle sensor for a 360° range is required here.

Potentiometer sensors

The wiper potentiometer, usually made as an angular-position sensor (Fig. 1), uses the correspondence between the length of a wire or film resistor (Cermet or conductive plastic) and its resistance value for measuring purposes. At present, this is still the lowest-priced travel/angular-position sensor. The shape of the characteristic curve can be influenced by shaping the measuring track (or just sections of it).

1 Principle of the wiper potentiometer

Fig. 1
1 Wiper
2 Resistance track
3 Contact conductor
 track
I_A Wiper current
U_0 Supply voltage
U_A Measurement
 voltage
R Resistance
φ_{max} Maximum angle
 of rotation
φ Measured angle

The standard wiper connection is furnished by a second contact track consisting of the same material mounted on a low-resistance substrate. The supply voltage is usually applied to the measurement track through low-resistance series resistors R_V (these can also be used for calibration of zero point and progression-rate adjustment) to protect against overloading.

Wear and falsification of measured values can be kept to a minimum by keeping the electrical loading of the pick-off as low as possible ($I_A < 1$ mA) and by dust-proof and waterproof encapsulation. One of the prerequisites for low wear is the optimal friction pairing between the wiper and the conductor track. To this end, wipers can be of "spoon" or "scraper-shape" design, and one or more can be mounted. Brush-shaped wipers are also in use.

A whole range of clear advantages are faced by a considerable number of serious disadvantages:

Advantages of the potentiometer-type sensors
- Low costs
- Simple design
- Very broad measuring effect (measuring range ≈ supply voltage)
- No electronic circuitry required
- High level of interference immunity
- Broad temperature range (up to 250 °C)
- High precision (better than 1 % of measuring range)
- Wide measuring range (almost 360° is possible)
- No difficulty with redundant design
- Calibration possible (laser ablation etc.)
- Flexible characteristic curve (variable conductor track width)
- Flexible assembly (on curved as well as on flat surfaces)
- Numerous manufacturers
- Rapid prototyping

Disadvantages of potentiometer-type sensors
- Mechanical wear from abrasion

- Measuring errors due to abraded particles
- Problematic when operated in fluids
- Variable contact resistance between wiper and measurement track
- Strong acceleration or vibration can result in wiper lift-off
- Testing is costly
- Limited possibilities for miniaturization
- Background noise

Applications
Examples of potentiometer-type sensors:
- Accelerator-pedal sensor or accelerator-pedal module for capturing the torque request for the engine-management system
- Fuel-level sensor (Fig. 2)
- Sensor plate potentiometer (KE-Jetronic and L-Jetronic) for detecting the air mass taken in by the engine
- Throttle-valve position sensor for capturing the position of the throttle valve in a gasoline engine (Figs. 3 to 5)

2 Potentiometric fuel-level sensor (design)

UMK1790Y

Fig. 2
1 Electrical connections
2 Wiper spring
3 Contact rivet
4 Resistor board
5 Bearing pin
6 Twin contact
7 Float lever
8 Float
9 Fuel-tank floor

3 Throttle-valve sensor (design)

4 Throttle valve sensor with two characteristic curves

5 Throttle-valve sensor (circuit)

Magnetically inductive-type sensors

Of all the sensors using proximity and non-contacting principles for position measurement, the magnetic sensors have proved to be the most rugged and most insensitive to interference. This applies in particular to those principles relying on alternating current, in other words magnetically inductive principles. Compared with a micromechanical sensor though, the coil configuration needed here requires far more space. This means therefore that there is no favorable possibility of redundant (parallel measurement) design. Furthermore, coil contacting is less favorable from the costs and reliability viewpoint.

Although there are a multitude of different principles in use for this form of sensor, only two have come to the forefront for motor vehicle applications. Regarding their operating concepts, these are very similar to each other. However, they are currently no longer in use in their macromechanical version by Bosch for motor vehicle sensors in new developments.

Eddy-current sensors

When an electrically conductive flat or curved disk (for instance in aluminum or copper) approaches a coil (usually nonferrous) (Fig. 6) to which high-frequency alternating current has been applied, it has

6 HF damping and eddy-current principle

Fig. 3
1 Throttle-valve shaft
2 Resistance track 1
3 Resistance track 2
4 Wiper arm with wipers
5 Electrical connection

Fig. 4
A Internal stop
1 Characteristic curve for high resolution in the angular range 0° to 23°
2 Characteristic curve for angular range 15° to 88°
U_A Measurement voltage
U_V Operating voltage
α_W Effective measured angle

Fig. 5
1 Throttle valve
2 Throttle-valve sensor
U_A Measurement voltages
U_V Operating voltage
R_1, R_2 Resistance tracks 1 and 2
R_3, R_2 Calibration resistors
R_5, R_6 Protective resistors

Fig. 6
1 Spoiler
2 Eddy currents
3 Air-core inductor
4 Variable-damping oscillator
5 Demodulator
s Measured travel
$A(s)$ Oscillator voltage
$U_A(s)$ Output voltage

an effect upon the coil's equivalent resistance and its inductance. This is the result of the eddy currents generated in the disk (otherwise known as a spoiler) due to the increasing magnetic coupling. The disk's position represents the measured travel s.

The generally high operating frequency (MHz range, a low-frequency feed would take up too much current because of the low coil inductance) requires a direct assignment of the electronics to the sensor or a shielded supply lead, which is generally not supportable. In order to convert the measuring effect into an electrical output voltage, either the damping effect (equivalent resistance) or field-displacement (inductance) principles can be applied. In the first (damping effect) case, a variable-amplitude oscillator can be used, and in the second a variable-frequency oscillator or a constant frequency supplied inductive voltage divider (differential configuration).

There are many ways to adapt the eddy-current principle to the measuring assignment. It is suited both to the determination of large travel distances or angles (there has already been a practically production-ready development, for the throttle valve and accelerator pedal, for instance) and also for very small variables (e.g. torque sensors). Because of the self-compensating properties, this principle generally only shows a very small temperature sensitivity over a wide temperature range. As eddy-current sensors can also, in principle, be manufactured micromechanically, their future use cannot be completely written off given their very advantageous properties. This sensor principle is, however, widely used in manufacturing and quality measuring technology, for the precise detection of the smallest displacements/distances or for the measurement of layer thicknesses in the μm range.

Short-circuiting ring sensors

In contrast to eddy-current sensors, the coil of a short-circuiting ring sensor has a straight or curved U or E-shaped soft magnetic, (usually) laminated iron core (Fig. 7). The coil, or short-circuiting ring ("spoiler"), is of highly conductive material (copper or aluminum), and is located around one (or all) of the core limbs. Because of their iron core, sensors of this type have a far greater inductance than eddy-current sensors and also a far higher measuring effect because of the guidance of the magnetic flux being very concentrated by the iron core. This means that they operate well at low frequencies, and do not necessarily require their signal electronics locally, that is directly on the sensor itself. The iron core also protects the measuring space between the core limbs very well against external interference fields.

The alternating field generated in and around the iron core by the coil current I cannot pass through the short-circuiting ring, since the eddy currents in the short-circuiting ring reduce it practically to zero.

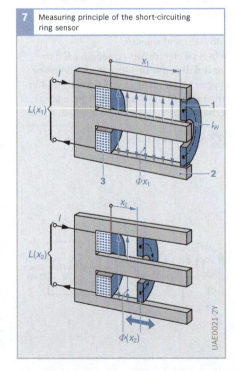

7 Measuring principle of the short-circuiting ring sensor

UAE0021-2Y

Fig. 7
Imaging for two
different measured
travels
1 Short-circuiting
 ring
2 Soft-magnetic core
3 Coil
I Current
I_W Eddy current
$L(s)$ Inductance and
$\Phi(s)$ Magnetic flux in
 measured travel s

Thus the eddy currents in the short-circuiting ring, acting like a "magnetic insulator", limit the extent of the magnetic flux Φ to the space between the coil and the short-circuiting ring (hence its other name of shading ring) and thus make it dependent on the position x of the short-circuiting ring $\Phi = \Phi(x)$.

Inductance is defined:

(1) $L = \Phi/I$

This is why the position x of the short-circuiting ring also directly influences the inductance L of the field coil. The relationship $L = L(x)$ shows a good linear fit over a wide domain. Virtually the entire length of the sensor can be utilized for measurement purposes. A mechanically tight guidance of the short-circuiting ring is not necessary here.

The mass of the short-circuiting ring to be moved is very low. Shaping the gap between coil and limb has an effect upon the shape of the characteristic curve: reducing the gap towards the end of the measuring range further improves the already good linearity. Operation is generally in the 5 to 50 kHz range, depending on material and structural shape. This sensor can also be used in very severe operating conditions, for instance on diesel fuel-injection pumps (Figs. 8, 9).

The cores are laminated (e.g. NiFe plate, 0.2 mm thick) because of the alternating current operation. To achieve the necessary mechanical stability to keep the plates stable under acceleration, they are not only bonded in the normal fashion, but also interlocked. To this end, the laminations are given a knobbly pattern in selected places which can be used to connect the laminations together before soft annealing (just like LEGO bricks), without interfering too much in the electrical insulation from lamination to lamination required.

The half-differential short-circuiting ring sensor with moving measuring ring and fixed reference short-circuiting ring is very accurate; it is evaluated as an inductive voltage divider (analysis of the inductances L_1/L_2, or $(L_1 - L_2)/(L_1 + L_2)$ or even as a frequency determining element of an oscillating circuit to generate a frequency analog signal (very interference-proof, easy to digitize).

Applications

Examples for short-circuiting ring sensors:
- Rack-travel sensors for measuring the control-rack setting on in-line diesel injection pumps (Fig. 8)
- Angular-position sensors in the injected-fuel quantity actuators of diesel distributor-type injection pumps

| 8 | Half-differential short-circuiting ring sensor |

| 9 | Half-differential short-circuiting ring angular-position sensor |

Fig. 8
Design of the rack-travel sensor (RWG) for diesel in-line injection pumps
1 Soft-magnetic core
2 Reference coil (L_2)
3 Reference short-circuiting ring
4 Control rack
5 Measuring coil (L_1)
6 Measuring short-circuiting ring
s Control-rack travel

Fig. 9
1 Measuring coil
2 Measuring short-circuiting ring
3 Soft-magnetic core
4 Control-collar shaft
5 Reference coil
6 Reference short-circuiting ring
φ Measured angle
φ_{max} Adjustment-angle range for the control-collar shaft

Figure 9 shows the structure of a half-differential short-circuiting ring sensor for diesel distributor-type injection pumps. A sensor coil and a reference coil are attached to each leg of the core. As the alternating current transmitted by the control unit permeates the coils, it creates alternating magnetic fields. The reference short-circuiting ring is fixed in position, while the measuring short-circuiting ring is attached to the control-collar shaft (adjustment angle φ). As the measuring short-circuiting ring moves along, the magnetic flux changes and the voltage across the coil changes, since the control unit maintains the current constant (load-independent current). An evaluation circuit forms the ratio of the output voltage U_A to the measurement inductance L_1 to the reference voltage U_{Ref} to the reference inductance L_2. It is proportional to the excursion of the measuring short-circuiting ring and can be evaluated by the control unit.

Sensors with rotatable alternating fields

With alternating current of the angular frequency ω fed coils or coil-like structures (such as meandering conductor track structures) two or more pole alternating fields structures may be arranged either in a circle or even linearly. These pole structures with fixed pole pitch can be displaced in relation to a generally fixed set of receiver coils having the same pole pitch, by the movement of the system to be measured, either rotary or translatorial. Here the amplitudes of the receiver signals are changing with the movement in a sinusoidal pattern. If the receiver coils are displaced in relation to one another by a specific portion of the pole pitch T (e.g. $T/4$ or $T/3$), the sinus pattern is phase-shifted by a corresponding angle (e.g. by 90° or 120°). The voltages arise as follows, for instance:

(1) $\qquad u_1 = U \cdot \sin\varphi \cdot \sin\omega t$
(2) $\qquad u_2 = U \cdot \sin(\varphi - 90°) \cdot \sin\omega t$
$\qquad\qquad = U \cdot \cos\varphi \cdot \sin\omega t$

or even:

(3) $\qquad u_1 = U \cdot \sin\varphi$
(4) $\qquad u_1 = U \cdot \sin(\varphi - 120°) \cdot \sin\omega t$
(5) $\qquad u_1 = U \cdot \sin(\varphi - 240°) \cdot \sin\omega t$

After rectification, the angle of rotation α can be calculated with great precision from these voltages. The sensors preferably made as angle sensors identified in classical measurement technology as synchro, resolver or even inductosyn methods function in this way.

In a simple bipolar arrangement, the electrical phase angle also corresponds directly to the mechanical angle of rotation α. In an arrangement of n pole pairs, the mechanical full angle $\varphi = 2\pi$ is converted into a phase displacement of $\alpha = n \cdot 2\pi$, so that the phase displacement α corresponds only to an angle of rotation of only φ/n, which considerably enhances the resolution of the measuring system. However, the unambiguity of the measurement signal must be restored by additional means, such as a simple angle sensor, in larger measuring ranges.

Hella sensor

The angle sensor developed by Hella does not strictly correspond to any of the methods described above, but is most similar to the inductosyn method. Figure 10 shows the sensor with a 6-count pole structure (n = 6), for instance, which electrically seen converts an angle of rotation of $\varphi = 60°$ into a phase displacement of the signal amplitudes of $\alpha = 360°$. All the conductor track structures required are applied to multilayer PCBs, at least in the case of the fixed part (stator). The rotor may also be made as a stamped part, if necessary, either freestanding or mounted on a plastic carrier (hot stamping).

A circular conductor track loop is made on the stator. This induces an eddy current at an operating frequency of 20 MHz in a closed meandering loop of approximately the same outside diameter on the rotor.

This eddy current naturally generates a secondary magnetic field, just like the exciter loop, which superimposes itself on the excitation field in the sense that it attempts to eliminate it. If the track on the rotor were only a circular conductor track congruent with the stator loop instead of the meandering track, this would simply extinguish the primary field to a great extent. The meandering structure, however, creates a resulting multi-pole field which can be rotated with the rotor and the total flux of which is also naturally almost zero.

This multi-pole alternating field is sensed by concentric receiver coils or meanders of practically identical form located on the stator. These are offset within a pole pitch (of e.g. 60°) by 1/3 each, i.e. electrically in their signal amplitude by 120° each (Fig. 10b). The receiver coils however extend over all n pole pairs (series circuit) and use the sum of all pole fields.

According to Figure 10c, the receiver coils are connected in a star connection. Their signals are sent to an ASIC to determine the electrical phase angle α or mechanical angle of rotation φ, which applies the necessary (sign-correct) rectification, selection and ratio formation. A version ASIC 1 receives the digital control signals necessary for this from an adjacent microcontroller. A different version ASIC 2 is, however, also capable of operating the sensor completely independently (on a stand-alone basis). The ASICS also permit an end-of-line calibration of the mechanical and electrical tolerances in production. It is also possible, for the case of applications with elevated safety requirements, to construct a redundant system with two galvanically isolated signal paths and two ASICs. The sensor principle can also very advantageously be applied in a "cut-open" form as a travel sensor.

The principal benefits of this sensor, in addition to the very favorable production costs, are as follows:
- Contact-free and wear-and-tear resistant measuring principle
- Complete temperature independence (up to 150 °C)
- High accuracy (up to approximately ±0.09° for a measuring range of 360°)
- Flexibility (may be adapted to any angular range up to 360°)
- Possibility of a redundant design
- High EMC
- Use of standard materials exclusively, without the need for ferromagnetic components
- Flat design

The sensor can therefore be used at very many points on the motor vehicle.

10 Hella sensor

a

$i(t)$ U_3 U_2 U_1

Stator Rotor

b

U_1

60°

U_2

U_3

Angle α ⟶

c

Stator Rotor

SAE1016-1E

Fig. 10
a Schematic design
b Circuit
c Output signals

Magnetostatic sensors

Magnetostatic sensors measure a DC magnetic field. In contrast to the magnetically inductive (coil) sensors they are far more suitable for miniaturization and can be manufactured at reasonable cost using microsystem technology. As DC magnetic fields can reach through housing walls of plastic, and also those of non-ferromagnetic metals, with no problems, magnetostatic sensors have the advantage that the sensitive, generally fixed part, can be well encapsulated and protected against the rotating part, generally a permanent magnet or soft-magnetic conductive element – and also against the environment. Above all, galvanomagnetic effects (Hall and Gaussian effects) are used, as well as anisotropic magnetoresistive (AMR and GMR) effects.

Galvanomagnetic effect

The Hall effect is normally evaluated using thin semiconductor wafers. If a current-carrying wafer of this type is exposed perpendicularly to magnetic induction B, the charge carriers are deflected by the Lorentz force normal to the field and the current I by angle φ from their otherwise straight path (Fig. 11). This means that it is possible to pick-off a voltage U_H transverse to the current direction between two opposing points on the edge of the wafer proportional to the field B and to the current I (Hall effect).

(2) $U_H = R_H \cdot I \cdot B / d$
where R_H = Hall coefficient,
d = wafer thickness

At the same time, the series impedance of the wafer increases independently of the field direction following an approximately parabolic characteristic curve (Gaussian effect, magnetoresistor).

The coefficient determinant for the measurement sensitivity of the wafer R_H is only comparatively small for silicon. As the wafer thickness d can be made extremely thin, however, using diffusion techniques, the Hall voltage amounts to a technically usable magnitude. When silicon is used as the basic material, a signal-conditioning circuit can be integrated on the wafer. Sensors using these principles are very cheap to produce. Silicon is by a long way not the best semiconductor material for Hall sensors in respect of measurement sensitivity and temperature sensitivity. III-V semiconductors such as gallium arsenide or indium antimonide possess better characteristics.

Hall switch

In the most simple case, the Hall voltage is applied to an electronic threshold circuit (Schmitt trigger) built into the sensor which then outputs a digital signal. If the magnetic induction B applied at the sensor is below a given minimum threshold level, the Schmitt trigger's output value corresponds to a logical "0" (release status); if it exceeds a given upper threshold the output value corresponds to a logical "1" (operate status). Since this behavior is guaranteed across the complete operating-temperature range and for all sensors of a given type, the two threshold values are relatively far apart (approx. 50 mT). A con-

11 Galvanomagnetic effect

siderable induction jump ΔB is thus neces-
sary to operate the Hall switch.

Sensors manufactured in bipolar tech-
nology of this kind were used in Hall vane
switches (Fig. 12), for example, as were fit-
ted in the housing for the camshaft-driven
ignition distributor. This Hall vane switch
contains the sensor and a permanent mag-
net and soft magnetic conductive ele-
ments. The magnetic circuit is U-shaped or
forked in such a way that a trigger wheel
vane in soft magnetic material can travel
through the open end, alternately screen-
ing or exposing the magnet, thereby tog-
gling the Hall sensor between the operate
and release states.

A further application is the digital steer-
ing-angle sensor LWS3 (see "Hall angular-
position sensor up to 360°").

Hall sensors of this type may be very
cost-effective, but are really only good for
a switch operation and are too imprecise
for capturing analog values.

*Hall sensors using the spinning-current
principle*

A disadvantage of a simple silicon Hall
sensor is its simultaneous sensitivity to the
unavoidable mechanical strain (piezoelec-
tric effect) resulting from the packaging,
leading to unfavorable temperature sensi-
tivity of the offset. Applying the spinning
current principle (Fig. 13), associated with
a transition to CMOS technology, has over-
come this disadvantage. Although the
piezoelectric effect still occurs here, it is
compensated for by the signal being aver-
aged over time, as it occurs with a different
sign in the case of very rapid electronically
controlled interchange (rotation) of the
electrodes. If it is wished to save the ex-
pense of the complex electronics for
switching the electrodes, it is possible also
to integrate several Hall sensors (two, four
or eight) with correspondingly different
alignment of the current paths in close
proximity and to add their signals together
in order to form an average. It is only in
this way that the Hall ICs were also suit-
able for analog sensor applications.

12 Hall vane switch

13 Hall sensor using the spinning-current principle

Fig. 12
a Magnetic flux:
 unhindered
b Magnetic flux:
 short-circuited
1 Vane width b
2 Soft-magnetic
 conductive element
3 Hall IC
4 Air gap
U_0 Supply voltage
U_s Sensor voltage

Fig. 13
a Rotary phase φ_1
b Rotary phase
 $\varphi_2 = \varphi_1 + 45°$
1 Semiconductor
 wafer
2 Active electrode
3 Passive electrode
I Supply current
U_H Hall voltage

These measures though did not result in a reduction of the considerable effects of temperature on the sensor's measurement sensitivity.

Such Hall ICs are suitable above all for the measurement of small travel distances, in which they register the fluctuating field strength of a permanent magnet as it approaches (e.g. the iBolt force sensor detects the passenger's weight to optimize the triggering of the airbag). Similarly good results could, in the past, only be achieved by using single Hall elements, e.g. in III-V compounds with a hybrid amplifier connected downstream (e.g. Hall acceleration sensor).

Differential Hall sensors
The double Hall sensor (differential Hall sensor, Fig. 14) has two complete Hall systems mounted at a specified distance from each other on one chip. The associated electronics evaluate the difference between the two Hall voltages. The advantage of such sensors lies in the fact that their output signal is for the most part independent of the absolute value of the magnetic-field strength, and as differential sensors they only register the magnetic induction's change in space, in other words the field gradient (thus the common name of gradient sensor).

Since the polarity of their output signal is independent of the air gap between rotor and sensor, these sensors are mostly used for rotational-speed measurement. If a simple Hall sensor only is used to sense a gear wheel, this will not be able to distinguish whether the magnetic flux has modified as a result of the gear wheel turning further or by a change in distance (e.g. vibrations, mounting tolerances). Significant scanning errors occur, as the signal must be passed to a threshold detector. The situation with differential sensors is fundamentally different. Here only the difference in the signals from two Hall sensors located at a suitable distance from one another around the circumference is evaluated. If the signal difference is positive, for example, it is possible to freely vary the distance between the gear wheel and sensor; the difference will remain positive, even if the absolute value may become smaller. The sign can only change if the rotor turns further. A threshold detector connected afterwards thus has no problem in distinguishing between a change in distance and rotation.

Usually, in order to achieve as high an output signal as possible, the distance between the two Hall sensors each located on the edge of the (elongated) chip is then selected to correspond to about half an increment (half the rotor tooth interval).

Fig. 14
a Design
b Field-strength
 distribution (1.5
 times increment
 spacing)
c Signal curve for
 air-gap widths *L*
1 Ring gear
2 Differential Hall IC
3 Homogenizing
 wafer (soft iron)
4 Permanent magnet

14 Differential Hall sensor

L = 2 mm

3 mm

Signal curve ↑

Angle of rotation φ →

UAE0862-2E

The signal maximum is very wide and covers a broad range of variation of the increment spacing. More pronounced variations in the spacing necessitate a highly complex redesign of the sensor.

A gradient sensor must be precisely aligned to the rotor's direction of rotation.

Hall angular-position sensors in the range up to approximately 180°
Using a rotatable magnetic ring ("movable magnet"), together with a number of fixed soft-magnetic conductive elements, a linear output signal can be generated for a larger angular range without conversion being necessary (Fig. 15). In this case, the bipolar field of the magnet ring is passed through a Hall sensor arranged between semicircular conductive elements. The effective magnetic flux through the Hall sensor is dependent on the angle of rotation φ.

This principle is used on accelerator-pedal sensors.

The Type ARS1 Hall angular-position sensor with a measuring range of approx. 90° (Fig. 16) is derived from the basic "movable magnet" principle. The magnetic flux from a practically semicircular permanent-magnet disk is returned to the magnets through a pole shoe, two additional conductive elements each of which contains a Hall sensor in its magnetic path, and the shaft which is also ferromagnetic. Depending upon the angular setting, the flux is led to a greater or lesser degree through the two conductive elements, in the magnetic path of which a Hall sensor is also situated. Using this principle, it is possible to achieve a practically linear characteristic in the measuring range.

The Type ARS2 is a simplified version which does without soft magnetic conductive elements (Fig. 17). In this version, the magnet moves around the Hall sensor in a circular arc. Only a relatively small section of the resulting sinusoidal characteristic curve features good linearity. If the Hall

sensor is located slightly outside the center of the circular arc, the characteristic curve increasingly deviates from the sinusoidal, It now features a short measuring range of just 90°, and a longer measuring range of a little over 180° with good linear-

15 Analog Hall angular-position sensor (movable magnet)

Fig. 15
Linear characteristic curve for angles up to 180°
a Position a
b Position b
c Output signal
1 Magnetic yoke
2 Stator (1, 2 soft iron
3 Rotor (permanent magnet)
4 Air gap
5 Hall sensor
φ Angle of rotation

16 ASR1 Hall angular-position sensor (movable magnet)

Flux density B

Angle of rotation φ

Fig. 16
Linear characteristic curve for angles up to approximately 90°
a Design
b Characteristic curve with working range A
1 Rotor disk (permanent-magnetic)
2 Pole shoe
3 Conductive element
4 Air gap
5 Hall sensor
6 Shaft (soft magnetic)

ity. A great disadvantage though is the low level of shielding against external fields, as well as the remaining dependence on the geometric tolerances of the magnetic circuit, and the intensity fluctuations of the magnetic flux in the permanent magnet as a function of temperature and age.

Angular-position sensors in the range up to approximately 360°
Figure 18 shows an analog Hall angular-position sensor with a measuring range of up to 360°. As shown, a permanent magnet rotates in front of two Hall sensors arranged at right angles to each other. The magnet should be adequate in size in order not to be too dependent on the tolerance of the magnet positioning. The two Hall sensors, fitted as closely together as possible, should here be positioned in the homogeneously directed portion of the permanent-magnet's stray field, which also represents the angular location φ of the permanent magnet. They are perpendicular to one another and parallel with the permanent magnet's rotary axis. When the field-strength vector B rotates past the sensors, therefore, they register its x and y components:

(3) $U_{H1} = U_x = B \cdot \sin\varphi$
(4) $U_{H2} = U_y = B \cdot \cos\varphi$

The angle φ can be calculated using the trigonometric relationship $\varphi = \arctan(U_{H1}/U_{H2})$ in an associated commercially available evaluator chip, through which the sensor signal is generally digitized.

A Hall sensor arrangement of this nature can also, in principle, be integrated vertically with VHD (Vertical Hall Devices) in the manner illustrated so that the plane of the sensor chip is arranged perpendicular to the axis of rotation and the sensor has an in-plane sensitivity, unlike a normal, planar Hall sensor (Fig. 19). Monolithic integration also guarantees the high level of precision as required for the 90° arrangement and for the requisite compact form of

Fig. 18
a Constructed from discrete Hall ICs
b Constructed from planar integrated Hall ICs
1 Signal electronics
2 Camshaft
3 Control magnet
B Induction
I Current
U Voltage
U_A Output voltage
φ Angle of rotation

Fig. 17
Linear characteristic curve for angles over 180°
a Principle of operation
b Characteristic curve
1 Hall IC positioned in the mid-point of the circular path
2 Hall IC located outside the mid-point (linearization)
3 Magnet

17 ASR2 Hall angular-position sensor (movable magnet)

18 Analog Hall sensor for 360° angles

the two Hall systems. Sentron (Melexis), of Switzerland, is working on sensors of this type and will be launching them on the market in the near future.

Angular-position sensors of this type were considered for the implementation of starter motor-less instant start systems, for instance, where the absolute rotary position of the camshaft over a 360° range must be measured.

There is, however, another trick which can be exploited to (additionally) give a normal, planar Hall sensor an in-plane sensitivity. In the case of the method used by Melexis-Sentron (e.g. its 2SA-10 model) a total of four Hall sensors are mounted on the same chip in a tight circle offset at 90° to one another to measure the field components B_x and B_y. To be accurate, in this case, each sensor consists of a Hall element pair rotated at 90° for its part for the reasons described above. Once the chip has been manufactured, a circular disk of soft magnetic material approximately 200 µm in diameter is fixed on its surface as shown in Figure 20 in such a way that the Hall sensors are located exactly beneath the edge of the disk.

Because of its high relative permeability, this disk acts as a flow concentrator (IMC, Integrated Magnetic Concentrator) and compels all field lines to pass into its surface vertically. This compels the field lines running horizontally (in-plane) without the concentrator into a vertical direction (out-of-plane) at the position of the Hall sensors and thus they can modulate the Hall elements. As diagonally opposing elements each see fields in opposite directions, the difference between the two Hall voltages running in opposite directions is evaluated to capture a field component. At the same time, any vertical components B_z present have their effect eliminated.

The additional evaluation of the summary signals of diagonally opposing elements can also be used to capture the vertical field component B_z; since these are not in the slightest influenced by the ferromagnetic disk and the elements have also not lost their normal out-of-plane sensitivity

Fig. 19
The currents directed into the interior of the chip I are deflected by the Lorentz forces because of the flux density aligned in parallel with the chip surface B in such a way that a Hall voltage U_H can be picked off between the surface electrodes A and A'.

Fig. 20
IMCIntegrated Magnetic Concentrator

The Hall elements are mounted on the chip in pairs at an angle of 90° to one another directly below the edge of a small ferromagnetic disk. Thanks to their presence, the flux density parallel to the chip surface (induction) B_{\parallel} is converted into a flux density B_{\perp} that can be sensed by the Hall sensors.

20 Set of 4 x 2 planar Hall sensors with IMC

B_x B_y

B_{\parallel} B_{\perp} B_{\perp} B_{\parallel}

1 2 3

SAE1020-1Y

19 Vertical hall device (VHD) in cross-section

$I/2$ A I A' $I/2$

U_H

Lorentz force B ⊗

SAE1019-1E

by it being fitted. In this way, an arrangement of this type can be used to capture the induction vector *B* acting at the sensor location in all its three components. The electronic circuitry required to evaluate the signals including a μC-based digital signal processor (DSP) used to calculate the arctan function and the means required for an end-of-line calibration of the sensor (EEPROM) can cost-effectively also be integrated on the same chip as the sensor.

In order to measure angles up to 360°, it is sufficient to rotate a permanent magnet, preferably round in form, magnetized in parallel with the chip over the sensor chip with its integrated signal processing (Fig. 21). As the angle of rotation is formed using the arctan function from the ratio of the two sine wave and cosine wave sensor signals, the strength of the magnetic field, and thus also the aging of the magnets, its temperature sensitivity and its distance from the sensor surface are largely irrelevant. Only the rotary position of its magnetization is detected.

The maximum intrinsic deviation of the sensor is stated by the manufacturer as + 2° for a measuring range of 360°. This can, however, be significantly improved by the user on calibration of the sensor. The accuracy of the digital output is then 10 bits at a resolution of 12 bits. Because of the processing time required by the signal processor, the smallest sampling rate for a timing frequency of 20 MHz is 200 μs. A pulse width modulated signal output can be operated at a maximum of 1 kHz. The sensor needs 15 ms for initialization. A typical value for the measuring field strength is approximately 40 mT. The concentrator plate becomes saturated at flux densities in excess of 0.7 T.

The sensor may also be programmed for any measurement range < 360° to good effect, with additional functional security being generated by out-of-range monitoring (application as an angular-position sensor in the accelerator-pedal module FPM2.3, for instance).

Angles up to 360° may also be captured with the simplest Hall ICs (Hall switches), as are also used for the measurement of speeds of rotation, (application in LWS1 type steering-angle sensors). Here, in order to obtain an n-bit resolution, n Hall switches are arranged on a single track in a circle at equal distances from each other (Fig. 22). Depending upon its position, a rotatable soft-magnetic code disk blocks the magnetic field of the individual permanent magnets located above each Hall switch, or opens it when it rotates further so that one after another the Hall switches generate n different code words (serial code). It is useful to apply the Gray code (cyclic binary code) to prevent errors in transitional states of the code disk. The Gray code is designed such that the code words for two adjacent positions – unlike the case with a natural binary code, for instance – differ in only 1 bit, and so any incorrect indication on the transition from

Fig. 21
Angle measurement at the end of a rotating shaft with quadruple Hall sensor manufactured by Melexis containing an integrated flux concentrator (IMC) to deflect the field.
a Measuring system
b Primary output signals

21 Angular position measurement over 360° with quadruple Hall sensor

a

b

2

1

S N

Hall voltage →

0° 180° 360° 540°

Angle of rotation α

—— sin α

---- cos α

UAE0997-2E

one position to the next can never amount to more than one angular step.

For the practical implementation of a steering-wheel-angle sensor, the code disk is connected to the steering shaft for instance, and the sensor's non-moving parts to the vehicle body. One difficulty with this sensor is that the soft magnetic code disk must be mounted floating on the steering column for reasons associated with assembly tolerances, which leads to a complex and space-hungry mounting there using a flexible plastic sleeve. Because of the air gap tolerance between the upper disk and the lower disk, the code disk is mounted in a wiping guide. A corresponding optoelectronic solution using photoelectric light barriers is less satisfactory, as it is not possible to construct an adequate encapsulation against the susceptibility of the sensor elements to dirt.

In order to measure a number of complete rotations, an additional 3-bit configuration can be used in which the code disk is rotated by a reduction gear. The resolution of such configurations is mostly no better than 2.5°.

Magnetoresistive sensors

In addition to the transversally directed Hall effect, there is also a longitudinal resistance effect, also known as the Gaussian effect, that occurs on semiconductor wafers. Elements that exploit this effect are known as magnetoresistors which are manufactured from a III-V-semiconductor (crystalline indium antimonide (InSb)). In contrast with the Hall sensors, the optimum wafer shape for a magnetoresistor tends to be shorter and squatter, and represents a very low electrical resistance. In order to arrive at technically applicable values in the $k\Omega$ range, it is therefore necessary to connect a large number of these wafers in series. This problem is solved elegantly by adding microscopically fine, highly conductive nickel-antimonide needles to the semiconductor crystal. These are located transverse to the current direction. A further measure is to apply meander techniques to the semiconductor resistor (Figs. 23 and 24).

The dependence of the resistance on the magnetic induction B follows a square-law function up to inductances of approx. 0.3 T, and above this point it is increasingly linear. There is no upper limit to the control range, and dynamic response can be regarded as practically free from lag, as is the case for the Hall sensor.

Fig. 22
Angular position measurement up to 360° with a circular, equidistant arrangement of simple Hall switches
1 Housing with permanent magnets
2 Code disk
3 Printed-circuit board with Hall switches

22 Digital Hall angular-position sensor with n Hall switches

UFL0029-2Y

23 Magnetoresistive differential sensor

φ

S
N

R_2 R_1

U_0

$U_A(\varphi)$

UAE0022Y-1

Fig. 23
Magnetic activation of a magnetoresistive differential sensor for gear wheel scanning (incremental angular position measurement, speed of rotation sensing)
1 Magnetoresistors R_1, R_2
2 Soft-magnetic substrate
3 Permanent magnet
4 Gear wheel
U_0 Supply voltage
U_A Output voltage at angle of rotation φ

Since their temperature sensitivity has a pronounced effect upon their resistance value (approximately 50 % reduction over 100 K), they are usually supplied only in the dual-configuration form in voltage-divider circuits (differential magnetoresistors). For the particular application, each of the two resistor sections must then be magnetically triggered (as far as possible with opposite polarities). Notwithstanding the high temperature coefficients of the individual resistors, the voltage-divider circuit guarantees good stability of the working-point (that point at which both resistor sections have the same value).

In order to achieve good measurement sensitivity, it is best to operate the magnetoresistors at a magnetic working point between 0.1 to 0.3 T. Generally, the required magnetic initial-tension is supplied by a small permanent magnet, the effects of which can be increased by using a small magnetic return plate.

The magnetoresistor's main advantage is its high signal level which is usually in the volts range. This means that amplification is unnecessary, as well as the local electronic circuitry and the associated protec-

tive measures which would otherwise be needed. Furthermore, in their role as passive, resistive components they are highly insensitive to electromagnetic interference and, as a result of their high bias voltage, practically immune against external magnetic fields.

Because of its pronounced temperature sensitivity, the magnetoresistor is used almost exclusively in incremental angular-position and rev counters, or in binary limit-value sensors (with switching characteristic).

Anisotropic magnetoresistive sensors (AMR)

Thin, only around 30 to 50 nm thick NiFe alloy coatings show an electromagnetically anisotropic behavior. That means that their electrical resistance changes under the influence of a magnetic field. Resistance structures of this type are thus known as anisotropic magnetoresistive (abbreviated to AMR) elements or sensors. The metal alloy generally used is also known as permalloy.

Technology and designs
Even without an external control fields, a small spontaneous magnetization M_S will arise in the longitudinal direction of a conductor track on a long slim resistor strip, as shown in Figure 25a, (shape anisotropy). In order to give this magnetization a clearly defined direction – theoretically, it could be in the other direction – AMR sensors are often provided with weak bias magnets, as shown. In this state, the longitudinal resistance achieves its greatest value R_\parallel. If the magnetization vector is rotated under the effect of an additional external field H_y through angle ϑ, the longitudinal resistance declines gradually until it reaches its minimum value R_\perp at $\vartheta = 0$. Here, the resistance depends only on the angle ϑ which is enclosed by the resulting magnetization M_S and the current I, and it has an approximately cosine wave pattern as a function of ϑ:

24 Microscopic sections of a magnetoresistor

a

b

UAE0864-1Y

Fig. 24
a Microscopic
 section (cut-out)
b On ferrite substrate
 in Super-8 film
 carrier

(5) $R = R_0 \cdot (1 + \beta \cdot \cos^2 \vartheta)$
 where $R_\| = R_0 \cdot (1 + \beta)$; $R_\perp = R_0$

The coefficient β here identifies the maximum possible variation in resistance. It amounts to around 3 %. If the external field is much stronger than the spontaneous generated magnetization, and this is usually the case when permanent magnets are used, the effective angle ϑ is almost completely a function of the direction of the external field. The field strength is now irrelevant, and in other words the sensor is now operating in a "saturated state".

Highly-conductive short-circuiting strips (for instance of gold) on the AMR film force the current to flow at an angle of below 45° to the spontaneous magnetization (longitudinal direction) without the application of an external field. As a result of this "trick", the sensor curve shifts by 45° compared to that of the simple resistor. This results in the so-called "Barber Pole" sensor. This means, therefore, that even with the external field strength $H_y = 0$, the curve is at the point of maximum measurement sensitivity (reversal point).

The striping of two resistors in opposite directions causes them to change their resistances in opposite directions under the effects of the same field. In other words,

one of the resistances increases while the other drops.

Being thin-film sensors, anisotropic magnetoresistive sensors also have the advantage that they can be calibrated to a setpoint value (e.g. zero point) by laser trimming, for instance. In principle, the oxidized silicon wafers which serve as the substrate material can also incorporate the electronic circuitry for signal conditioning. At present, cost considerations dictate that sensor chips and electronic chips are for the most part manufactured separately, and then mounted and packaged in a common leadframe, for instance. The controlling magnetic field B is usually generated by magnets moving side-to-side or rotating over the sensor.

In addition to the simple, two-pole AMR elements, there are also pseudo Hall sensors in practically square NiFe thin layer structures, that have four connections, just like normal Hall sensors. Two for the current path and two transverse to this for pick-off a (pseudo) Hall voltage (Fig. 26). Unlike a normal Hall sensor, the pseudo Hall sensor, however, has its sensitivity to magnetic fields in the plane of the layer, and not perpendicular to it. In addition, it does not have a proportional characteristic curve, but a sinusoidal one with very a

25 AMR basic principle, barber pole structure

high match to a sinus wave, which does not depend in the least on the strength of the control field and the temperature. The output voltage vanishes for a field parallel to the current path, to describe a sinus half-period on rotation up to an angle $\varphi = 90°$. The sinus voltage obtained in this way combined with the amplitude results in:

(6) $U_H = u_H \cdot \sin 2\varphi$

If the external control field is rotated once through $\varphi = 360°$, the output voltage follows two full sinus periods. The amplitude u_H is, however, very dependent on the temperature and the air gap between the sensor and the control magnet; it declines with increasing temperature and widening air gap.

The measurement sensitivity of the pseudo Hall element can be considerably increased (without excessively falsifying the sinusoidal shape) by "hollowing out" the element from the inside so that only the "frame" remains (Fig. 26b). This modification converts the pseudo Hall sensor to a full bridge consisting of four AMR resistors from its geometrical form (Fig. 26c). Even when the bridge resistors are meander-shaped, provided a given minimum conductor width is not fallen below, this still has negligible effect of the signal's sinusoidal shape.

Simple AMR angular-position sensor for ranges < 30°
Barber pole AMR angular-position sensors with limited precision and a restricted measuring range (max. ±15°) rely on the unbalance of a magnetoresistive voltage divider (differential sensor) comprising an elongated (possibly meander-shaped) permalloy resistor with highly conductive gold cross strips (Fig. 27). Although in the case of such sensors the zero point is largely independent of the distance between the magnet and the sensor, the gradient of the characteristic curve, which for its part also depends on the temperature (TK around -3·10⁻³/K), on the other hand, is not. A sensor of this kind has already been considered as a cost-effective alternative for the measurement of pedal positions.

26 Pseudo Hall sensor

27 Magnetoresistive angular-position sensor (barber pole)

Simple AMR travel sensor for the mm range
Similarly, the simplest travel sensors may
be built for a measuring range of typically
just a few millimeters using individual
differential barber pole sensors (Fig. 28).
However, to achieve good accuracy a con-
stant distance (air gap) is required be-
tween sensor and moving magnets, which
embodies the position s to be measured.
The temperature sensitivity of the charac-
teristic curve gradient can be very easily
compensated for by means of an additional
metallic thin-film sensor on the same car-
rier substrate, which will have around the
same temperature coefficient, but with the
inverse sign (e.g. Pt, Ti, Ni).

High-precision multisensor (POMUX) for
travel ranges > 10 mm
There is an intelligent multisensor for po-
sition measurement built up with simple
barber poles also known under the com-
mercial name of POMUX (position multi-
plexer). It exploits the fact that the symme-
try point or zero point of the characteristic
curve is absolutely independent of the air
gap and the temperature. Its use in a motor
vehicle has often been considered (e.g.
shock absorber stroke measurement), as it
has excellent characteristics, in particular,
extreme accuracy. It has also already been
illustrated in a circular form as an angular-

position sensor for scanning a gravity pen-
dulum tuned to a low frequency, which
may be used to determine the orientation
of the vehicle (tilt sensor).

Angular-position sensor for 180°
Pseudo Hall type magnetoresistive angu-
lar-position sensors utilize the practically
100 % sinusoidal signal that is picked off
from the output terminals of the four-pole,
sensor structure, whereby two complete
periods of the electrical output signal cor-
respond to a mechanical rotation of the
magnets through 360°. Using a second ele-
ment, which has been turned through 45°,
a cosine signal is also generated (Fig. 30).
By applying the arc-tangent function for
instance, the relationship between the
two signal voltages can be applied to
determine the measured angle using a
microcontroller or an ASIC. This applies
throughout a range of 180°, and is for the
most part independent of temperature
changes and magnetic-field fluctuations
(distance, aging).
 A further prerequisite for the high accu-
racy of this sensor principle is that the
field at both bridges is at least in the same
direction (above a given magnitude, field
strength is irrelevant). This can only be
guaranteed when both bridges are directly
above one another. A design was drawn up

28 Displacement measurement with single differential barber pole

SAE1062E

in which the two bridges, which are at 45° to each other, were interwoven so that they can be regarded as being at the same point and quasi "on top of one another" (Fig. 29).

Angular-position sensor for 360°

One serious drawback with AMR angular-position sensors is their natural limitation to an unambiguously detectable angular range of 0° to 180°. This restriction may, however, be overcome by modulation of the direction of the magnetic flux by means of an alternating auxiliary magnetic field (Fig. 31). By contrast with the COS and SIN sensor signals which have a 180° periodicity, these modulation signals have a 360° periodicity. In this way it is possible, simply by evaluating the sign of the modulation signals, to distinguish between

Fig. 29
"Nested" design of two AMR bridges offset from each other by 45°

Fig. 30
Measuring range up to 180°
a Measuring principle
b Output signals
1 Thin NiFe film (AMR sensor)
2 Rotatable permanent magnet with control induction B
I_V Supply current
U_{H1}, U_{H2} Measurement voltages
φ Angle of rotation

Fig. 31
a Bridge circuit
b Output signals from the Wheatstone bridges
c Influence of the auxiliary field on the signals
d Output signal from the evaluation circuit
B_H Control induction
U_{DD}, U_{SS} Supply voltages
U_{SIN}, U_{COS} Measurement voltages
φ Angle of rotation

29 Nested AMR bridges

30 Magnetoresistive angular-position sensor (pseudo Hall)

$U_{H1} = U_0 \cdot \sin 2\varphi; \quad U_{H2} = U_0 \cdot \cos 2\varphi$

Angle of rotation φ →

31 AMR360 angular-position sensor with field direction modulation

$$\alpha = \frac{1}{2} \cdot \arctan\left(\frac{U_{sin}}{U_{cos}}\right)$$

$\alpha_{0°-180°} : \Delta U_{cos} < 0$

$\alpha_{180°-360°} : \Delta U_{cos} > 0$

ranges and thus to double the unambiguous range to 0° to 360°. Because the auxiliary field for the two sensor bridges points in different directions, the two modulation signals are phase-shifted. It is sufficient, to distinguish ranges, merely to determine the sign of the largest modulation signal in respect of absolute value. This means that distinguishing the ranges is not critical and is still easily possible, even in the case of modulation signals suffering heavy interference.

The magnetic auxiliary field is generated by a planar coil integrated into the AMR sensor chip (Fig. 32). The auxiliary field in the COS and SIN full bridge points in directions 45° apart in accordance with the wiring arrangement.

The sensor can only be mounted on the end of a rotating shaft. Thus, for instance, it may be mounted on the end of the steering spindle for measuring the steering-wheel angle in systems that do not require determination of multiple rotations of the steering spindle (e.g. steering-angle sensor LWS4).

Angular-position sensor in the range over 360° (multiple revolutions)
A dual-configuration of pseudo-Hall angular-position sensors (180°) can be used to measure multiple rotations of a rotating component, a steering spindle, for instance. The two associated permanent magnets are rotated via a step-up gear train (Fig. 33). As the two driving smaller gear wheels, which carry the control magnets, differ by one tooth (number of teeth m and $m+1$), their mutual phase difference (difference of the angle of rotation: $\Psi - \Theta$) is an unambiguous measure of the absolute angle position φ of the steering spindle. The system is designed such that this phase difference does not exceed 360° in a total of four revolutions of the steering spindle and so the unambiguity of the measurement is ensured. Each individual sensor also offers an indeterminate fine resolution of the angle of rotation. Using such a configuration, it is possible for instance to determine the complete steering-wheel range with a resolution of better than 1° (this principle is applied in the LWS3 steering-angle sensor).

Giant magnetoresistive sensors (GMR)
GMR sensor technology was developed a few years ago and is now finding its first applications in sensing angles and speeds of rotation in the automotive sector.

32 | Layout of the AMR360 angular-position sensor

UAE1074Y

33 | Arrangement for angular measurement in excess of 360°

UAE0704-2Y

Fig. 33
1 Steering axle
2 Ring gear
3 Gear wheel with m teeth
4 Gear wheel with m + 1 teeth
5 Magnets
6 AMR sensor elements
φ Angle of rotation of the steering column
Ψ, Θ Angle of rotation of the small gear wheels

The essential advantages of GMR sensors over AMR sensors are the natural 360° range of unambiguity in angle sensing and the higher sensitivity to magnetic fields in speed of rotation sensing.

Unlike AMR sensors, GMR sensors consist not of one magnetic function layer, but rather of a complex system of layers. A distinction is essentially made between two systems: on the one hand the GMR multilayers with typically some 20 sequences of alternating soft magnetic and nonmagnetic individual layers (Fig. 34a). On the other hand there are the GMR spin valves, comprising antiferromagnetic, ferromagnetic and nonmagnetic functional layers. The individual layer thickness are in the 1 to 5 nm range in both systems, and so embrace just a few atomic layers.

The electrical resistance of a GMR layer stack depends on the angle between the magnetizations of adjacent ferromagnetic individual layers. It is at its maximum when they are aligned antiparallel and at their minimum when in parallel alignment (Fig. 34b).

GMR multilayers
The relative change in resistance (GMR effect) is 20 to 30 % in GMR multilayers and thus is ten times greater than the AMR effect. A hard magnetic layer, which will provide a suitable bias field, is added to the GMR multilayer stack to place the working point in a sensitive region of the characteristic curve. The sensitivity of a sensor structure of this nature is about 0.8 %/mT which makes it greater than that of AMR barber pole structures.

GMR spin valves
Unlike the GMR multilayers, GMR spin valves (Fig. 35) are also suitable for angle sensing. In the latter, the reference magnetization necessary for angle sensing is generated, fixing (pinning) the direction of magnetization of one of the ferromagnetic layers (FM1) through interaction with an adjacent antiferromagnetic (AFM) layer. This is hence also known as the "pinned layer". On the other hand, the magnetization of the second ferromagnetic layer (FM2), which is largely magnetically de-

Fig. 34
a Design
b Change in
 resistance as a
 function of the
 magnetic field

FM Soft magnetic
 single layer
NM Nonmagnetic
 single layer

Fig. 35
a Design
b Change in
 resistance as a
 function of the
 magnetization
 angle

FM1 Antiferromagnetic
 single layer
FM2 Ferromagnetic
 single layer
NM Nonmagnetic single
 layer

34 GMR multilayer stack with characteristic curve

a

FM
NM
FM
NM
FM
NM
FM

■ magnetic
□ nonmagnetic
⇨ magnetization

b

Resistance change (%)

UAE1070E

Magnetic field (mT)

35 GMR spin valve layer stack with characteristic curve

a

M_1
M_2
FM2
NM
FM1
AFM

b

Resistance change (%)

Magnetization angle Θ

UAE1071E

coupled by a nonmagnetic interlayer, is freely rotatable using the external magnetic field. This is accordingly known as the "free layer".

The resistance varies with a cosine-wave dependency on the angle φ between the external field direction and the reference direction. The stability of the reference magnetization against the action of the external field is decisive for the accuracy of the angular measurement. This stability is significantly enhanced by the use of an additional synthetic antiferromagnet (SAF) (Fig. 36). This consists of two ferromagnetic layers which are strongly antiferromagnetically coupled through a very thin nonmagnetic interlayer and, because of their vanishing magnetic net moment, no longer offer a face for the external field to attack. The magnetization of one of these two layers is pinned by the adjacent natural antiferromagnet (AFM). Both the interaction of the pinned layer (PL) with the natural antiferromagnet (AFM) and the mutual interaction of the ferromagnetic layers (PL, RL, FL) are determinant for the magnetic behavior of the layer system. Furthermore, the behavior is also formed by the shape and crystal anisotropy of the individual magnetic layers, as is the case with the AMR sensor structures.

Like an AMR angular-position sensor, a GMR angular-position sensor consists of two full bridges, of which one supplies a cosine-wave signal and the other a sine signal depending on the external field direction (Fig. 37). The different magnetic reference directions M_R required to achieve this are generated by local heating of the individual bridge resistors and cooling in the presence of a magnetic field of suitable orientation. The field direction can be unambiguously determined over the full angular range from 0° to 360° by an arctan function.

36 GMR spin valve with synthetic antiferromagnets

FL
NML
SAF { RL
NML
PL
AF

UAE1072Y

37 GMR angular-position sensor with two full bridges

a

├200 μm┤

b

UAE1073Y

Fig. 37
a Layout
b Circuit
M_R Magnetic reference
 magnetization

Fig. 36
FL Free layer
NML Interlayer
RL Reference layer
PL Pinned layer
AF Antiferromagnet

Wave-propagation sensors

In recent years, the development of sensors has concentrated increasingly on sensors that acquire information from the immediate and more distant surroundings of the vehicle. This refers to the distance to other vehicles or traffic and to obstacles (all-round view from the vehicle, Fig. 38), either for safety reasons or to support vehicle handling.

Ultrasound propagation time methods are suitable for vehicle-to-vehicle distance monitoring in the near range (0.5 to 5 m), with optical propagation time or triangulation methods using short-wave infrared light (lidar) for the medium range up to 50 m, and electromagnetic radar for the long range (up to 150 m).

Ultrasound sensors

According to the echo-sounding process (Fig. 39), the sensors transmit ultrasound pulses with a frequency of around 40 kHz, and record the time taken for the echo pulses to arrive having been reflected back from an obstacle. The distance l from the send/receive head to the closest obstacle is determined from the propagation time t of the first echo pulse to arrive and the velocity of sound c thus:

(8) $\qquad l = 0.5 \cdot t \cdot c$

In practice, the geometrical distance a between an obstacle and the front of the vehicle is determined using the triangulation method from the results measured (distance b and c) from two ultrasound heads, mounted at a distance d from one another (Fig. 40):

Long-range radar sensors (76 GHz)

The vehicle-speed controller fitted is equipped with a long-range radar sensor (adaptive cruise control, ACC) for the automatic detection of vehicles preceding in the same lane or even of obstacles in the path of the vehicle which could eventually necessitate application of the brakes. The working frequency is 76 GHz (wavelength $\lambda \approx 3.8$ mm) permits the relatively compact construction as needed for automotive applications. A Gunn oscillator (Gunn diode in the cavity resonator) feeds in parallel four adjacently arranged patch antennas, for example, which also serve to receive the reflected signals. Referred to the vehicle axis, a plastic Fresnel lens in front concentrates the transmit and receive beam horizontally and vertically.

The signal reflected on the vehicle in front is delayed in accordance with the propagation time. The distance a to the vehicle in front and its relative speed Δv

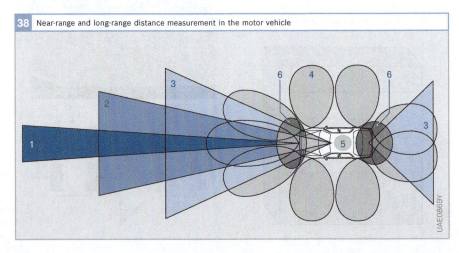

38 Near-range and long-range distance measurement in the motor vehicle

can be determined from the signal propagation time and/or the phase displacement between the signal transmitted and the echo received. The antenna transmit/receive characteristics are aligned in four different directions due to the antennas

being off-set from the center. The angle of the vehicle in front can be determined from this (see "Adaptive cruise control" chapter).

GPS position and distance traveled measurement

GPS is the global positioning system using satellites, originally created by the USA for military purposes, but later increasingly released for civil uses. The system was completed in 1993 with a total of 24 satellites (21 operating, three in reserve) orbiting at an altitude of 20,183 km. More than 24 satellites have since become available. They are distributed over six circular orbital planes taking 12 hours to complete an orbit and so at least four (but generally up to eight) satellites are always visible above the horizon from any point on the earth's surface. They transmit continuous (digital) signals on a carrier frequency of 1.57542 GHz. The signals contain the following message, in simple terms:

• Satellite identification code
• Satellite position
• Message transmit time (including date)

The satellites each have two cesium and two rubidium clocks which vary by less than 20 to 30 ns on board for the high-pre-

39 Principle of distance measurement using ultrasound

40 Calculation of the distance (triangulation)

$$a = \sqrt{c^2 \frac{(d^2 + c^2 - b^2)^2}{4\,d^2}}$$

41 Position determination by GPS

Fig. 40
a Distance between the fender and the obstacle
b Distance sensor 1 to obstacle
c Distance sensor 2 to obstacle
d Distance sensor 1 to sensor 1
1 Transmit/receive sensor
2 Receive sensor
3 Obstacle

Fig. 41
Representation in the plane (two-dimensional):
If the position of the satellites is known, with the propagation times t_1 and t_2 measured, the possible reception locations are on two circles around the satellites which intersect at points A and B. The point A located on the earth's surface is the location of the receiver.

cision determination of the transmission time. The signal propagation times to be evaluated lie typically in the region of approximately 70 ms. The satellites identify themselves with a pseudo random code (PRN number), which is 1,023 bits long and is continuously repeated after every 1 ms. It is imposed on the carrier as a phase modulation.

A user can take this information and determine his own geographical position in three dimensions with a GPS receiver, consisting of a GPS antenna, signal receiver, precision clock and microprocessor. Theoretically, the signals from three satellites are sufficient for this; since, however, the less complex clocks in mobile receivers (oscillating crystals) can differ from the satellite clocks somewhat, a fourth satellite must be used in order to eliminate the errors generated by the receiver clock. The distance to the visible satellites is derived from the signal propagation times measured.

The position is determined – as shown in Figure 40 in only two dimensions of space – then using a trilateration method, according to which there is precisely one point in space which satisfies the three dimensional conditions. In addition, there is also the knowledge that the motor vehicle, unlike an airplane, must be located on the earth's surface. Measurement is thus the more accurate the more satellites can be received and the better, i.e. the more uniformly the satellites are distributed across the sky. Since complete release for civil use (May 2000), the accuracies obtainable are about 3 to 5 m in a plane and around 10 to 20 m in the determination of altitude. With the refinement of GPS to DGPS (Differential GPS, mobile stations have contact with a fixed reference station via long wave signals) this variance can be significantly further reduced. This makes resolutions down to less than 1 m possible today.

These accuracies are the more astonishing when the manifold sources of error in this very complex measurement is taken into consideration. For example, simply the fact that at the very high velocity of approximately 12,000 km/h at which the satellites move, relativistic effects already play a not insignificant role. Time passes more slowly at high velocities than it does on earth. Time passes more quickly in weaker gravitational fields as are found in the orbital planes. The high velocity causes a dilation of time of 7.2 µs per day, the lower gravitation even has an effect that is around six times greater, although opposing.

If the position data determined in a close sequence is stored in a mobile GPS receiver, such as in a motor vehicle, it is easy to derive the orbit and the speed of the vehicle. If the data is sent via a radio module to a fixed central location, the location of the vehicle at any time will then be known.

Speed and rpm sensors

Measured variables

Speed and rpm sensors measure the number of revolutions or the distance traveled per unit of time. When automotive applications are concerned, these are in both cases measured variables which occur between two components or with respect to the road surface or another vehicle. However, it is also necessary to measure the absolute rotational speed in space or about the vehicle axes (yaw rate). For instance, for the electronic stability program, the yaw rate about the vehicle's vertical (or yaw) axis must be determined by sensing.

Measuring principles

Conventional sensors used for rotational-speed measurement are based on pronounced measuring effects (e.g. inductive). They are therefore for the most part electrically passive. That is, they are usually not provided with any form of local/on-site electronics. With the newer sensors, however, measurement is based on less-pronounced measuring effects (for instance, the Hall effect), and these sensors thus need local, integrated electronics for signal conditioning. They belong in the broadest sense to the category of "intelligent sensors" (which are also often referred to as "active" sensors).

In fact, the sensors used for measuring absolute rotating speed (yaw rate) need highly complex electronic circuitry directly at the sensor since the measuring effects used here are not only particularly small, but also require complex signal conditioning.

Incremental rotational-speed measurement takes advantage of a wide variety of different physical effects (some of which can be applied in sensors at a very reasonably price). Optical and capacitive sensors, though, are highly unsuitable for the rough operating conditions encountered in the motor vehicle. Here, magnetic-effect sensors are used almost exclusively.

Relative rpm and speed measurement

Examples of relative rotational speed are as follows:
- Crankshaft and camshaft rotational speed
- Wheel speeds (for ABS/TCS/ESP)
- Rotational speed of the diesel fuel-injection pump

Here, measurement usually makes use of an incremental pick-up system comprised of a gear wheel and a speed sensor.

In the detection of relative yaw rate, depending upon the number and size of the scanned rotational rotor markings, a distinction is made between the following types of sensor (Fig. 1):
- Increment sensor with closely spaced rotational markings. Up to a certain point, this form of sensor permits instantaneous speed to be measured at points on the circumference, or the registration of very fine angular divisions.
- Segment sensor, with only a small number of scanned rotational segments (for instance, equivalent to the number of engine cylinders).
- Simple speed sensor, with only a single scanned marking per revolution, so that only the average rotational speed can be registered.

1 Determination of the rotational speed (rotor forms)

a b c

SAE0780-2Y

Fig. 1
a Incremental sensor
b Segment sensor
c Speed sensor

Sensor shapes

The following shapes of sensor are presently in use (Fig. 2) for speed and rpm sensors:
- Rod sensor
- Fork-shaped sensor
- (Internal and external) toroidal sensor

The preferred form is the rod sensor, in which the rotor teeth approach the sensor and which is also the easiest to install. The fork-shaped sensor is permissible in some cases, and is also in service in the field. This sensor must be roughly aligned to the rotor when installed. The sensor type in which the sensor surrounds the rotor shaft in the form of a ring is practically no longer used. From time to time, an inner-ring sensor is used which is introduced into the end of a hollow shaft featuring an incremental internal structure.

Unfortunately, although it is the most widely used sensor type, the rod-shaped sensor features the lowest measurement sensitivity and is problematical when air gaps become excessive. The fork-shaped sensor on the other hand is practically impervious to axial and radial play. Regarding toroidal sensors, the most widely used form combines a large measurement signal with pronounced insensitivity to geometric tolerances.

In many respects, the traditional inductive-type sensors are highly unsatisfactory. Their output signal features an rpm-dependent amplitude, and for this reason they are unsuitable for measuring low rotational speeds. In addition, they only permit comparatively low-level air-gap tolerances and are usually unable to differentiate between air-gap fluctuations and genuine rotational-speed pulses. And at least the sensor tip must resist high temperatures when installed near hot components (e.g. the brakes). These disadvantages are behind the additional features aimed at with new, innovative sensors:
- Static detection (that is, at zero speed, or at extremely low cranking or wheel speeds)
- Larger air gap (non-aligned mounting with air gaps > 0)
- Small size
- Independence from air-gap fluctuations
- Temperature stability (≤ 200 °C)
- Detection of the sense of direction (optional for navigation)
- Reference-mark identification (ignition)

Magnetostatic sensors (Hall, magnetoresistors, AMR) are highly suitable for complying with the first two demands. And, as a rule, they also permit compliance with the second and third stipulations.

Figure 3 shows three fundamentally suitable sensor shapes, which generally are insensitive to air-gap fluctuations. Here, one must differentiate between sensors which sense radially and those which sense tangentially. This means that independent of the air gap, magnetostatic sensors are always able to differentiate between the north and south poles of a magnetically active pole wheel or rotor ring.

In the case of magnetically passive rotors, the sign of the output signal is then no longer independent of the air gap when they register the tangential-field strength (here though, the fact that the air gap is

2 Sensor shapes

SAE0778-2Y

Fig. 2
a Fork shape
 (vane principle)
b Rod shape
 (proximity
 principle)
d_L Air gap

often enlarged due to the sensor is a disadvantage).

Radially measuring differential-field or gradient sensors are often used. These always register only the gradients of the radial-field components, the signs of which do not change with the air gap but only with the angle of rotation.

Rotors

The rotor is of decisive importance when measuring rotational speed. It is usually provided by the vehicle manufacturer, while the sensor itself comes from a component supplier. Previously, magnetically passive rotors have been used almost exclusively. These are made from magnetically soft materials (usually iron), and are less expensive than magnetically hard rotors (also known as pole wheels). Apart from that, since they are not magnetized they are easier to handle, and there is no danger of mutual demagnetization (for instance during storage). Unfortunately, this form of rotor is the most difficult to scan, particularly in combination with rod sensors.

3 Sensors insensitive to air-gap fluctuations

As a rule, presuming the same increment width and output signal, the pole wheel's intrinsic magnetism (a pole wheel is defined as a magnetically active rotor) permits a considerably larger air gap.

Normally, passive rotors are in the form of gear wheels. In many cases, these are already present on the engine (for instance the starter ring gear). Otherwise they must be installed in a specific position in order to generate the required signal (as needed for instance for ABS). In the latter case, both planar-toothed and axial-scan versions are in use.

An easily identifiable reference mark is required for each revolution when pick-off the crankshaft speed or position (for instance, at the starter ring gear). This reference mark must allow a fault-free and optimum timing of ignition and injection. The reference mark can be in the form of a completely (or partially) removed tooth. Due to the fact that a tooth space "takes more time", the reference mark is immediately identifiable, particularly since the engine speed can only vary gradually and never abruptly.

In addition to gear wheels, stamped perforated disks or wave-shaped metal rings have been introduced in the meantime as low-priced rotors (ABS).

It was the integration of ABS sensors in the vehicle's wheel bearing which led to the introduction of pole wheels some of which also assume the role of a sealing ring (plastic-bound magnetic powder). Small, for the most part encapsulated, tachometer sensors connected through a short flexible shaft with one of the vehicle's wheels, also use pole wheels (with only very few poles) for generating a speed signal. These are usually picked off by means of integrated Hall sensors.

Fig. 3
a Radial-field sensor
 with pole wheel
b Tangential sensor
c Differential sensor
 with gear wheel

Inductive-type sensors

Inductive-type sensors were available on the market as coil versions long before the first suitable microstructure sensor versions (e.g. using the Hall principle) became available. Such inductive-type sensors use Faraday's law when measuring rotational speed. In other words, they generate a voltage U_A at their two-pole output which is proportional to the change (referred to time) of a magnetic flux Φ (w = number of turns).

$$U_A = U_{ind} = w \cdot d\Phi/dt$$

The magnetic flux Φ is also a function of the rotational position x and the air gap d_L:

where
$\Phi = \Phi(x, d_L)$ and $d_L = const$
the following applies:
$$U_A = U_{ind} = w \cdot \partial\Phi/\partial x \cdot dx/dt$$

where dx/dt represents the (rotational) speed to be measured.

The inductive-type sensors' weak point is underlined though by the equation: if it is impossible to keep the air gap d_L constant (due to flutter or other forms of mechanical play), the air-gap fluctuations induce the same change of flux as does a fluctuation of speed. This effect can cause the generation of voltage pulses which are either impossible, or at least difficult, to separate from the genuine engine-speed signals. Since the flux varies exponentially along with the change in air gap, and the air-gap fluctuations are often of the high-frequency type (e.g. brake flutter), these unwanted pulses can feature a high voltage amplitude.

Inductive-type sensors are therefore always of the dynamic type, and being as their output signal tends to fall to zero in such cases, they are in principle also unsuitable for the detection of extremely low speeds (quasi-static or static). The only exception here are the coil sensors which are powered by a carrier frequency and which are based on the eddy-current or damping principle. These though are hardly ever used in automotive applications.

Basically, inductive speed sensors are comprised of three important magnetic component parts (Fig. 4):
- Fixed coil
- Soft-iron component
- Permanent-magnet component

The change in flux needed for the generation of the output voltage results from the motion or rotation of the hard or soft-iron magnetic component. The inductive-type sensors are preferably composed of a rod magnet (1) with a soft-magnetic pole pin (2) carrying the 2-connection induction coil (3). When a ferromagnetic gear wheel (5), or some form of similar rotor, rotates in front of this sensor (pick-up) the changing magnetic flux (which varies as a function of time) induces a practically sinusoidal voltage.

For scanning very fine tooth structures, the end of the pole pin is sometimes pointed and acts as a flux-concentrating element. In other words it is shaped like a pole "blade" which usually projects through the metal or plastic housing and is adapted to the increment structure regarding shape and direction.

Fig. 4

1 Rod magnet
2 Soft-magnetic pole pin
3 Induction coil
4 Air gap d_L
5 Ferromagnetic gearwheel (or rotor or pulse wheel)
6 Rotational or reference mark
λ Tooth interval

4 Inductive speed sensor

The rotor can be provided with one or more rotational markings (6). Figure 5 shows the flux curve and the voltage induced by a single rotational or reference mark (slot, cam, or pole pin).

Normally, the steep passage through zero which takes place at the mid-point of maximum flux is utilized for the electronic

detection of such a rotational or reference mark. According to Faraday's Law, in all phases the signal's amplitude is proportional to rotational speed.

In order to ensure adequate, interference-free evaluation in the control unit, the spacing between the peaks of a double pulse (or of a periodic voltage pulse) U_{ss}

5 Flux and voltage curve for an inductive-type sensor

Fig. 5

a Single magnetic marking per revolution
b Periodic increment structure (e.g. gear wheel, pole wheel)
c Evaluation circuit
1 Cam
2 Slot
3 Switching point
4 Steep passage through zero evaluated
5 Priming edge
6 Switching flank
7 Shifting point
8 Pole pin
9 Sensor
$\gamma/2$ Phase shift
L Inductance
U Voltage
Φ Magnetic flux

should be at least 30 mV. The major disadvantage of the inductive-type sensors is the fact that at high speeds/rotational speeds their output voltages can reach levels far in excess of 100 V which are difficult to process electronically.

If Zener diodes are used to clip the high voltage peaks, the resulting changes in the sensor's load impedance lead to considerable phase-angle errors. With camshaft and crankshaft sensors this can have highly undesirable results with regard to the ignition where the correctness of phase relationship must be better than approx. 0.2°.

Normally, the prepulse generated by the magnetic return field can be ignored at low speeds. With some magnetically passive or active rotational markings though, at high speeds the prepulse voltage can increase to such an extent that it exceeds the threshold value of the downstream threshold discriminator and can cause an even greater error (Fig. 5a). For this reason, the threshold values of the control unit input circuit are adapted dynamically to the speed in question.

Provided that the air gap is not too narrow, a uniform tooth structure results in the practically sinusoidal voltage curve shown in Figure 5b. The speed is reflected in the periodic interval between the voltage's zero transition points, while the amplitude is also proportional to the rotating speed.

The signal amplitude is highly dependent (exponentially) upon air gap and tooth size. As is the case with all magnetic increment processes, up to air gaps of d_L, teeth can be efficiently detected as from half or 1/3 of a tooth interval I.

$$d_L \leq \lambda/(2...3)$$

The conventional ABS gear wheels and crankshaft applications cover air gaps of up to 0.8 and 1.5 mm respectively. The reference mark needed for the ignition and injection results from leaving out one or two teeth or by closing a tooth space.

The reference point is detected when the distance between the passages through zero changes abruptly and causes a far higher signal voltage (corresponding to an apparently larger tooth) which has a negative effect upon the previous and upon the subsequent incremental voltage – this can under certain circumstances be undesirable.

Advantages
- Low manufacturing costs
- High immunity to interference: low static internal resistance (dynamically higher)
- No local electronics (electrically passive), that need to be protected
- No problems with DC voltage drift (dynamic measuring principle)
- Broad temperature range (limited primarily by the sealing compound characteristics)

Disadvantages
- Conventional coil technology imposes limits on size reduction
- The output signal is rpm-dependent, unsuitable for quasi-static movements
- Sensitive to air-gap fluctuations

Examples of application
- Inductive engine-speed sensor (crankshaft-speed sensor)
- Inductive wheel-speed sensor
- Inductive camshaft sensor (transistorized ignition with induction-type pulse generator TZ-I)
- Needle-motion sensor (diesel-fuel injection)

Magnetostatic sensors

Quasi-static speed measurement is best performed using magnetostatic sensors. Their output signal is independent of rotational speed and depends solely on field strength, so that at high rotational speeds their low signal voltages are easier to handle electronically. Furthermore, they not only have the advantage of imposing practically no limits with regard to miniaturization, but also signal amplification and/or signal preprocessing can be integrated locally. The fact that they are very small means that multiple systems such as differential configurations, or arrangements with integrated recognition of direction, present no difficulties.

On the other hand, such active sensors have a serious disadvantage, and that is the fact that their operating-temperature range is for the most part defined by the relevant Si (silicon) evaluation electronics which as a rule cannot withstand such high temperatures as the sensor element itself.

For some time now, active sensors have been available with the option of a two-pole current output, so that the two-core connection can no longer be regarded as a specific advantage of the inductive-type coil sensor.

Hall vane switches

If Hall Si sensors are to be used for incremental rpm measurement, pronounced manufacturing scatter together with the effects of temperature mean that they must be provided with an adequate induction jump of typically 40 to 50 mT in order to ensure reliable high-speed switching. With conventional Hall sensors and acceptable air gaps, this was only possible with the sensor in the form of a "Hall vane switch" (for instance when used as ignition-triggering sensors in the ignition distributor for earlier electronic ignition systems). The sensor and the electronic circuits for supply and signal evaluation are integrated on the sensor chip.

This "Hall IC" (using bipolar technology, for sustained temperatures $\leq 150\,°C$, and for direct connection into the vehicle's onboard electrical supply), is located in a practically closed magnetic circuit comprising permanent magnet and pole pieces (Fig. 6). A soft-magnetic trigger wheel (driven, for instance, by the camshaft) rotates through the remaining air gap. With a trigger-wheel vane in the air gap, the magnetic flux is short-circuited past the sensor element. On the other hand though, when a trigger-wheel opening passes through the air gap the flux is unhindered on its way to the sensor. This principle ensures that the sensor also performs perfectly when the trigger-wheel varies in how far it penetrates into the air gap, or in case the air-gap shifts radially, that is vertically to the direction of rotation.

Since Hall vane switches of this type feature limited rotational resolution they are mainly used as segment sensors. If the vane slot is too narrow, it is practically impossible for the magnetic field to pass through, with the result that the required induction jump is not generated.

6 Hall vane switch

Fig. 6
a Magnetic flux: unhindered
b Magnetic flux: short-circuited
1 Vane width b
2 Soft-magnetic conductive element
3 Hall IC
4 Air gap
U_0 Supply voltage
U_S Sensor voltage

UAE0639-2Y

Simple Hall rod sensors
In contrast to the Hall vane switch, the working-point magnetization depends too much upon the width of the air gap, and the induction excursion is too small for reliable switching with this configuration. Simple rod-shaped Hall sensors provided with a working-point magnet are therefore unsuitable for static or quasi-static scanning of a magnetically passive rotor (Fig. 7a, gear wheel). The switching point of a downstream threshold comparator (Schmitt trigger) would have to be continually adapted to the varying working point. Such applications are only feasible when DC coupling is dispensed with and therefore also static signal evaluation. The coupling capacitors required for such configurations (also known as $\Delta\Phi$ sensors) involve high costs though and lead to reduced operational reliability.

However, simple Hall sensors are highly suitable for scanning a magnetically active rotor (Fig. 7b, pole wheel). The sensor is triggered by the pole wheel with varying polarity only in the vicinity of the magnetic zero point. Although the magnetic control excursion reduces along with increasing air gap, the position of the working point

($B = 0$) remains unchanged. Considering that the working point on new types of Hall sensors is to a great extent thermally stable, the switching points of the downstream threshold comparator can be set relatively close together. This means that relatively wide air gaps become possible. Air-gap fluctuations in this configuration cannot cause missing pulses as they do not lead to pole change. The pole change only characterizes the continuing measurement motion (rotation).

Gradient sensors
When it comes to scanning passive rotors, gradient sensors (Fig. 8) designed on the basis of differential Hall sensors or differential magnetoresistive sensors, are far better than simple Hall sensors. They are provided with a permanent magnet. This pole face of this magnet facing the gear wheel is homogenized by a thin ferromagnetic wafer (2). Two galvanomagnetic elements (generic term for Hall sensors and magnetoresistors) are located on each element's sensor tip, at a distance of roughly one half a tooth. Thus, one of the elements is always opposite a tooth space when the other is opposite a tooth. The sensor mea-

Fig. 7
a　Passive rotor
b　Active rotor
1　Incremental rotor
2　Simple Hall sensor
3　Permanent magnet
4　Pole wheel
5　Housing

Fig. 8
1　Magnetoresistors
　　R_1, R_2 or Hall
　　elements H_1, H_2
2　Ferromagnetic wafer
　　(soft-magnetic
　　substrate)
3　Permanent magnet
4　Gear wheel
U_0　Supply voltage
$U_A(\varphi)$　Measuring
　　voltage for an angle
　　of rotation φ

7　Hall rod sensors

8　Gradient sensor for gear wheel scanning

sures the difference in field strength be-
tween two neighboring points on the cir-
cumference of the rotor. The output signal
corresponds roughly to the field strength
derived as a function of the angle at the
circumference, and its sign is therefore in-
dependent of the air gap. Being as they do
not alter the gradient signal's sign, air-gap
fluctuations does not cause missing pulses.

For signal evaluation, it is a simple mat-
ter to connect the two magnetoresistors
as a voltage divider which is supplied by
a constant voltage and whose (usually un-
loaded) output signal is registered by the
control unit. At room temperatures and
with the customary air gap, this signal is in
the volts range, and even at high tempera-
tures it is suitable for transmission to the
control unit without any form of preampli-
fication.

Provided appropriate circuitry is used,
measuring the loaded output current of
the magnetoresister divider instead of its
open-circuit voltage permits the sensor's
pronounced temperature measurement
sensitivity to be compensated for to a great
extent.

In the case of a gradient sensor based on
the Hall effect, the current paths of both
Hall elements can be connected in parallel,
and their opposite-polarity output voltages
in series, so that their differential voltage
can be picked off directly and inputted to
the downstream amplification and evalua-
tion stages.

Tangential sensors

Tangential sensors differ from gradient
sensors by their reaction to variations in
polarity and intensity in the components
of a magnetic field located tangentially to
the periphery of the rotor. Using AMR
thin-film techniques, tangential sensors
are available as barber-pole or permalloy
resistor types in full-bridge or half-bridge
circuits (Fig. 9). In contrast to gradient
sensors, they need not be matched to the
particular tooth pitch of the rotor and can
in fact be designed to sense practically at a

given point. Local amplification is neces-
sary, even though their measuring effect is
approximately 1 to 2 orders of magnitude
larger than that of the silicon Hall sensors.

In the case of the crankshaft speed sen-
sor integrated in the bearing (composite
seal with sensor), the AMR thin-film sen-
sor is mounted together with an evaluation
IC on a common leadframe.

For the purposes of space saving and
temperature protection, the evaluation IC
is bent at an angle of 90° and also located
further away from the sensor tip.

Giant magnetoresistive (GMR) elements

The giant magneto resistance effect (GMR
effect) was first demonstrated on multilay-
ers consisting alternately of thin soft mag-
netic and nonmagnetic layers (thickness of
a few nanometers). An antiparallel align-
ment of the magnetization of adjacent fer-
romagnetic layers leads to the maximum
electrical resistance, while parallel align-
ment shows a significantly lower resistance.
The relative change in resistance for appli-
cation-relevant layer structures thus var-
ies in the range 20 to 30 %. This represents
an increase over the AMR effect by a factor
of around 10.

9 AMR speed sensor in the form of a
tangential-field sensor

UAE0643-2Y

Fig. 9
1 Gear wheel (Fe)
2 Permanent magnet
3 Sensor
B Control field
 strength with
 tangential
 components B_t and
 radial component
 B_r (B' off position,
 $B_t = 0$),
R_1, R_2 permalloy
 thin-film resistors
 (AMR)
φ Angle of rotation
U_0 Supply voltage
U_A Measurement
 voltage

The simplest layer system that shows a GMR effect is a spin valve (see also "Position sensors" section). It consists of two soft magnetic layers, separated by a nonmagnetic layer. While the magnetization of the one soft magnetic layer is fixed by an antiferromagnet, the magnetization of the second soft magnetic layer can ideally follow the external field without interference (free layer). The resistance is then minimal if the magnetizations of the two layers are aligned in parallel and is increased by approximately 5 % if the magnetizations are aligned antiparallel. The interaction of the reference magnetization with the external field may be minimized with the help of a synthetic antiferromagnet (SAF), comprising two ferromagnetic layers, that are strongly antiferromagnetically coupled by means of very thin nonmagnetic interlayer (Fig. 10). This significantly increases the robustness against interference fields.

Figure 11 shows a typical characteristic curve for a spin valve. The two characteristic states for the two opposing directions of the external magnetic field can clearly be distinguished. The spin valve is sensitive to field strength changes in the transitional range between high and low resistance.

A bridge circuit of four GMR resistance elements, designed as a gradiometer is suited for the determination of rotational speed (Fig. 12). The resistors are interconnected here in such a way that only one magnetic field difference at the two bridge locations results in a signal. The effects of homogenous interference fields can be suppressed in this way. As the reference direction for all resistors is identical, local heating in the magnetic field, as is necessary to write differing reference magnetization directions in an angular-position sensor, is not required.

11 Characteristic curve for a GMR spin valve

Fig. 10
1 Free layer
2 Interlayer
3 Reference layer
4 Pinned layer
5 Antiferromagnet
SAF Synthetic
 antiferromagnet

Fig. 12
a Electrical wiring
b Layout of a
 gradiometer sensor

10 Layer stack for a GMR spin valve

12 Bridge circuit for the GMR sensor

The maximum signal amplitude for a gradi-ometer sensor then results when the direction of the field at the two bridge locations is precisely 180° phase-shifted. This is the case on a multipole wheel, for instance, if the distance between the resistance elements corresponds to the distance between poles in the multipole ring (Fig. 13).

Examples of application
- Hall sensor (transistorized ignition TZ-H)
- Hall phase sensor (camshaft)
- Transmission Hall sensor (RS50, RS51)
- Active Hall speed sensor
- Active AMR speed sensor
- Magnetoresistive sensor (for diesel radial-piston distributor pumps)
- Active GMR speed sensor

Absolute rotating-speed measurement
Measuring principle of the oscillation gyroscope
Basic principles
Mechanical gyroscopes (gyros for short) utilize inertial forces in precisely measuring angular movements in space independent of the reference system. Notwithstanding their pronounced measuring effect, rotating gyroscopes as well as optical sensors based on the interferometric Sagnac effect (laser and fiber gyroscopes),

are out of the question for automotive applications due to the high costs involved.

On the other hand, the slightly less-severe requirements on precision for applications in the motor vehicle can be complied with by gyroscopes manufactured using fine-mechanical and micromechanical processes. Instead of a rotational movement, these units use an equivalent, elastic, oscillatory movement to generate the measuring effect. In principle, these sensors are similar to mechanical gyroscopes. They exploit the Coriolis accelerations which are generated when rotary motions are coupled with an oscillatory component (velocity v) (Fig. 14). These sensors are known as oscillation gyroscopes and are used in the electronic stability program (ESP) for the detection of the vehicle breaking away, in navigation systems for the detection of the direction of travel and in cameras to stabilize the image against wobbling. They also comply adequately with all automotive stipulations such as freedom from maintenance, long service life, switch-on time constant etc., not to forget the question of costs.

Oscillation gyroscopes measure the yaw rate around their "sensitive" axis. The yaw rate Ω_z around the vehicle's vertical axis (yaw axis) is measured in

13 Speed of rotation measurement with GMR sensor

14 Relationships in the rotary motion

Fig. 13
Field lines in the space surrounding a multipole ring
The two squares symbolize the two bridge locations
The arrows indicate the direction of the field lines at these points

Fig. 14
Interrelationship between yaw rate, oscillation, and Coriolis acceleration at a lumped mass m.
Ω_z Yaw rate
v_y Speed of the oscillation
$a_{Coriolis}$ Coriolis acceleration
m Lumped mass

the electronic stability program (ESP) and in the navigation system. Systems for triggering rollover protection systems need data on the yaw rates Ω_x and Ω_y around the vehicle's pitch and roll axes.

Measurement of the Coriolis acceleration
Oscillation gyroscopes measure the Coriolis acceleration in the following way: as the distance between a mass and the axis of rotation increases, the mass must cover a greater distance in the time of one revolution because of the greater radius. Circumference increases with the radius. The mass must, therefore, acquire a higher tangential speed and must be accelerated to achieve this. This acceleration (Coriolis acceleration) is measured by suspending the mass on springs and measuring the bend in the suspension. If the distance of the mass, now rotating at a higher tangential speed, from the axis of rotation is now reduced, the mass must now become slower, as it needs only cover a shorter distance for each revolution. The braking deceleration required is also measured via the bend of the suspension.

In the oscillation gyroscopes, the distance of one (or more) masses from the axis of rotation is periodically increased and then reduced again by an oscillation excitation (Fig. 14, movement in the y direction). This means that the mass must be accelerated and then braked again in the same cycle (movement in the x direction). The forces necessary for this depend on the amplitude of the exciting oscillation and the current yaw rate. If the exciting oscillation is kept constant, the yaw rate Ω can be determined from the acceleration forces.

The Coriolis force acts in accordance with a known vector law perpendicular to the rotary motion and speed of the mass in motion (Fig. 14). This gives the Coriolis acceleration as:

(1) $a_{Coriolis} = a_x = 2 \cdot v_y \cdot x \cdot \Omega_z$

Whereby, the velocity v_y changes sinusoidally in accordance with the oscillatory movement:

(2) $v_y = v_y \cdot \sin\omega t$

This means that, at a constant yaw rate Ω_z, a sinusoidal Coriolis acceleration $a_{Coriolis}$ of the same frequency is measured. The amplitude is then:

(3) $a_{Coriolis} = 2 \cdot v_y \cdot \Omega_z$

The acceleration a_y which is also applied to the mass m in the oscillatory direction is usually several orders of magnitude higher than the useful Coriolis acceleration.

(4) $a_y = dv_y/dt = \omega \cdot v_y \cdot \cos\omega t$

As the Coriolis acceleration has the same frequency as the exciting function, the useful signal (the yaw rate) can be obtained by multiplication of the exciting signal and the Coriolis signal with subsequent averaging (phase-correct rectification). Interference signals of other frequencies are filtered out (principle of the lock-in amplifier). Averaging (with a low-pass filter) frees the output signal from the excitation frequency. An output voltage is generated which is proportional to the yaw rate:

(5) $U_A = const \cdot a_{Coriolis} = const' \cdot \Omega$

Examples of application
- Piezoelectric yaw-rate sensors
- Micromechanical yaw-rate sensors

Acceleration sensors

Measured variables

Acceleration sensors are suitable
• For knock control in gasoline engines
• For triggering restraint systems (e.g. air-
 bag and seat-belt pretensioners)
• For detecting the accelerations of the
 vehicle for the antilock brake system
 (ABS) or the electronic stability program
 (ESP) or
• For the evaluation of body acceleration
 for use by the chassis and suspension
 control systems

The measured variable is acceleration a,
often stated as a multiple of the accelera-
tion due to gravity g_n (1 g ≈ 9.81 m/s²) (see
Table 1 for typical values in motor vehicles).

1 Measuring range of acceleration sensors	
Applications	**Measuring range**
Knock control	40 g
Passenger protection	
– Airbag, seat-belt pretensioner	35 to 100 g
– Side impact, front sensing	100 to 400 g
– Roll-over detection	3 to 7 g
ESP, HHC, ABS	0.8 to 1.8 g
Chassis and suspension control	
– Design	1 g
– Axle/damper	10 to 20 g
Car alarm	1 g

Measuring principles

Acceleration sensors measure the force F
executed on an inert mass m by an acceler-
ation a:

$$(1) \qquad F = m \cdot a$$

Here, just as in force measurement, sys-
tems are available to measure displace-
ment as well as mechanical strain.

Displacement or travel-measuring systems

In the case of travel-measuring systems
(Fig. 1), a mass m - the seismic mass - is
elastically attached to the body, the accel-
eration a of which it is intended to mea-
sure.

Constant acceleration
In the event of a constant acceleration, the
acceleration force is in equilibrium with
the return force applied to the spring with
spring constant c deflected by x:

$$(2) \qquad F = m \cdot a = c \cdot x$$

The system's measurement sensitivity S is
therefore:

$$(3) \qquad S = x/a = m/c$$

This indicates that a large mass together
with low spring stiffness result in high
measurement sensitivity.

Table 1
g Acceleration
 due to gravity
 1 g ≈ 9.81 m/s²
HHC Hill hold control
ABS Antilock brake
 system
ESP Electronic stability
 program

Fig. 1
Schematic:
a Excursion-
 measuring
b Closed-loop
 position controlled
a Measured
 acceleration
x System excursion
F_M Measuring force
 (inertial force on
 the mass m)
F_K Compensating
 force
I_A Output current
U_A Output voltage

| 1 Displacement or travel-measuring acceleration sensors |

Variable acceleration
In the dynamic case, a damping force and an inertia force are to be taken into consideration in addition to the spring force. The essential damping force is proportional to the velocity \dot{x} and is described using the damping coefficient p. The inertia force is proportional to the acceleration \ddot{x}. The resulting equation (4) describes a resonant system:

(4) $F = m \cdot a = c \cdot x + p \cdot \dot{x} + m \cdot \ddot{x}$

Assuming negligible damping ($p \approx 0$), this system has a resonant frequency of:

(5) $\omega_0 = \sqrt{\dfrac{c}{m}}$

This means that in accordance with equation (3), the measuring sensitivity S is directly linked to the resonant frequency ω_0 in the following manner:

(6) $\omega_0^2 \cdot S = 1$

In other words, sensitivity can be expected to be reduced by factor 1/4 when the resonant frequency is doubled. It is only clearly below their resonant frequency that such spring-mass systems display adequate proportionality between measured variable and deflection.

The resulting excursion is converted into an electrical signal by a suitable measuring technique (cf. Table 2).

Position control
Travel measuring systems allow the application of the compensation principle, in which the system deflection due to acceleration is controlled out by an equivalent return force (Fig. 1b). The sensor element is now a component part of a closed control loop. The return force or the variable generating it (e.g. current or voltage) acts as a measure of the acceleration. The system operates through closed-loop control very close to the zero point of the excursion and achieves a high degree of linearity.
 Closed loop position control produces a greater measuring range, which is only limited by the return force and a higher cutoff frequency than similar, non-motion-controlled systems (Fig. 1a).

Damping
To prevent an excessive resonance, which would cause interference, pure deflection systems demand defined damping that is independent of the temperature. If the damping coefficient p is related to the other parameters in equation (4), this results in Lehr's damping factor D:

(7) $D = \dfrac{p}{2 \cdot c} \cdot \omega_0 = \dfrac{p}{2 \cdot \sqrt{c \cdot m}}$

2	Measuring principles		
Electrical pick-off	**Principle**	**Technical implementation**	**Examples of application**
Piezoresistive	Stress measurement via the change in resistance as a result of elongation caused by acceleration on the spring-mass system	Silicon volume microme-chanics, thin-layer systems	Laboratory applications, previously also crash sensors
Piezoelectric	Charge shifts in the crystal as a result of the force applied (spring-mass system)	Ceramic material, PZT, quartz, PVDF	Knock sensors, previously also crash sensors
Capacitive	Excursion of the spring-mass system is determined through capacitance measurement of a plate capacitor	Silicon volume microme-chanics, surface microme-chanics	Used throughout the motor vehicle
Thermal	Lateral excursion of a heated gas zone and detection of the asymmetry with respect to the heated zone	Volume micromechanics	Rollover sensing

Table 2

This dimensionless variable permits a simple description and comparison of differing resonant systems. Natural vibration and increase in resonance sharpness are largely determined by this damping factor. Values of $D = 0.5$ to 0.7 are preferred in practice (Fig. 2).

In the case of position-controlled systems, damping is implemented and adjusted in the closed control loop.

Physical implementation
The majority of the acceleration sensors currently in use in the motor vehicle are manufactured in silicon surface microme-

chanics. Because of the capacitive evaluation, the measurement signal is determined only by the geometry and stable material parameters and is barely influenced by variables such as temperature. The very small measuring capacities of these sensors, however, requires that evaluation electronics are provided in the immediate vicinity.

The advantage of surface micromechanics is the opportunity it provides to control by position or to deflect the systems by applying electrostatic forces (to the measuring electrodes or to an additionally fitted electrode set). The latter, being a genuine electro-mechanical self test, is an effective means of checking the entire signal path.

Systems for measuring mechanical stress
Applications
Piezoelectric sensors are only used in the motor vehicle in acceleration and yaw-rate sensors.

Longitudinal effect principle of action
Under the action of a force F, piezoelectric materials generate a charge Q on their surfaces to which electrodes are attached (Fig. 3). This charge is proportional to the mechanical stress generated by the force F.

2 Resonance curves for the transfer function G

a — Value G
$D = 0.1$
$D = 0.3$
$D = 0.5$
$D = \sqrt{2}/2$
$D = 1$
$D = 1.5$
$D = 2$
$D = 3$

b — Angle G
$D = 0.1$
$D = 0.3$
$D = 0.5$
$D = \sqrt{2}/2$
$D = 1$
$D = 1.5$
$D = 2$
$D = 3$

Standardized angular frequency

SAE0810-2E

3 Types of piezoelectric effects

a F Q Q F
b F F Q
c F Q Q F

SAE0812-1Y

Fig. 2
a Amplitude resonance curve
b Phase resonance curve of the complex transfer function $G(i \cdot \Omega) = [x(i \cdot \Omega)]/[a(i \cdot \Omega)]$

$x(i \cdot \Omega)$
 Deflection amplitude
$a(i \cdot \Omega)$
 Acceleration excitation amplitude
$\Omega = \omega/\omega_0$
 Standardized circular frequency
D Damping

Fig. 3
a Longitudinal effect
b Transverse effect
c Tangential force
F Force
Q Charge

Materials for piezoelectric components can be broadly classified into monocrystals, such as quartz, and ceramics.

Piezoelectrically active ceramics are produced by sintering finely ground ferroelectric materials. These ceramics then consist of very many small crystals randomly oriented in space. These are then polarized in production by the application of a high electric field strength. This largely aligns the randomly oriented polarizations of the microcrystals.

Depolarization, with the consequent loss of the piezoelectric properties, can be caused in piezoceramics by temperatures above the Curie temperature, by intensive mechanical stresses or by polarization with opposite field strengths.

The charges generated are discharged through the external resistance of the measuring circuit or through the piezo sensor's internal resistance. Such sensors cannot measure statically, and are only used where dynamic measurement is needed. The typical limit frequencies of this high pass response, however, lie above 1 Hz, depending on the application.

Transverse effect and shear effect
In addition to the longitudinal piezoelectric effect (Fig. 3a), there are also transverse and shear effects (Figs. 3b and 3c).

In practice, these effects occur together. The relationship between the charge generated and the mechanical stress acting can thus be formulated as a tensor equation.

The transverse effect is used for example in a bimorph. These are composed of two oppositely polarized piezoceramics used for the measurement of bending stresses (Fig. 4). When the two-layer ceramic composite bends, one half is stretched ($\varepsilon > 0$) and the other compressed ($\varepsilon < 0$). The opposed-polarity of the ceramic plates means that the resulting part voltages U_1 and U_2 now add to form a total voltage U which can be picked off on the two outside metal layers.

Electrical signal evaluation
Piezoelectric sensor elements may be modeled as a voltage supply with a capacitive internal resistance. The internal resistance is infinitely high for ideal elements. Signal processing must be rated with corresponding care.

There are two options for matching the impedance to subsequent systems:
• Electrometer amplifier (Fig. 5a): in this case, the voltage applied across the electrodes is detected and amplified by an amplifier with extremely high impedance input. Parasitic capacitances

4 Piezoelectric bimorph

5 Pick-off on piezoelectric sensors

however contribute to the transmission function.

- Charge amplifier (Fig. 5b): here the charge is stored on a second capacitor. Parasitic capacitances have no influence.

In both cases, resistors parallel to the sensor element must be taken into consideration for the transfer function.

Thermal acceleration sensors

Thermal acceleration sensors generate a "bubble of heated gas" over a heater element. The tight heated gas space has a lighter density than the surrounding, cooler gas. In the event of a lateral acceleration, the low density area of gas is displaced within the surrounding, cooler gas. The ensuing asymmetry is detected by thermocouples or resistors as a bridge circuit. The bridge voltage represents the acceleration signal.

Packaging

A packaging which is tailor made to suit each individual application plays a decisive role for the quality of the sensor. Acceleration sensors detect the measured variable without a movable connection to the outside world, it is no problem to encapsulate them hermetically. A rigid mechanical coupling to the body to be measured must, however, be achieved, as additional elastic or loose elements could lead to falsification of the measurement. On the other hand, this fixed coupling must not lead to thermal expansion which may occur at the structure being transferred to the sensor so that measurements are falsified.

Examples of application

- Piezoelectric acceleration sensors (bimorphous spring elements, longitudinal elements such as knock sensors)
- Micromechanical acceleration sensors, made in bulk micromechanics and surface micromechanics

6 Principle of thermal acceleration sensors

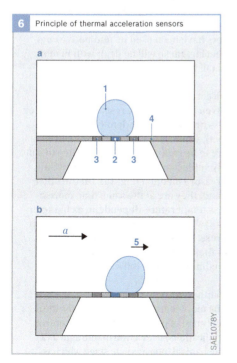

SAE1078Y

Pressure sensors

Measured variables

Pressure measurement takes place either directly, by way of diaphragm deformation, or using a force sensor for the following applications in the motor vehicle (examples):

- Intake-manifold or boost pressure (1 to 5 bar) for gasoline injection
- Brake pressure (10 bar) on electropneumatic brakes
- Air-spring pressure (16 bar) on pneumatic-suspension vehicles
- Tire pressure (5 bar absolute) for tire-pressure monitoring
- Hydraulic reservoir pressure (approximately 200 bar) for ABS and power-assisted steering
- Shock-absorber pressure (approximately 200 bar) for chassis and suspension control
- Coolant pressure (35 bar) for air-conditioning systems
- Modulation pressure (35 bar) for automatic transmissions
- Brake pressure in master cylinder and wheel-brake cylinder (200 bar), and automatic yaw-moment compensation on the electronically-controlled brake
- Overpressure/underpressure of the tank atmosphere (0.5 bar)

- Combustion-chamber pressure (100 bar, dynamic) for detection of misfiring and knock detection
- Element pressure on the diesel fuel-injection pump (1,000 bar, dynamic) for electronic diesel control
- Fuel pressure on the diesel common rail (up to 2,000 bar)
- Fuel pressure on the gasoline direct injection system (up to 200 bar)

Measuring principles

Pressure as a measured variable is a non-directional force acting in all directions which occurs in gases and liquids. It is propagated in liquids, and also very well in gel-like substances and soft sealing compounds. There are static and dynamic measuring sensors for the measurement of these pressures.

The dynamic pressure sensors include for instance all microphones which, since they are insensitive to static pressures, are used to measure pressure pulsations in gaseous and/or liquid media. Since up to now, practically only static pressure sensors have been used in automotive engineering, these will be dealt with in more detail here.

Direct pressure measurement

Given that all known resistors are more or less pressure-dependent (volumetric effect), when very high pressures ($>10^4$ bar) are to be measured it would suffice to simply subject an electrical resistor to the pressure medium (Fig. 1a). On the other hand, they are at the same time more or less temperature-dependent, a characteristic which is usually very difficult to suppress. Furthermore, the sealed lead-out of their connections from the pressure medium presents difficulties.

Diaphragm-type sensors

The most common method used for the measurement of pressure (also in automotive applications) uses a thin diaphragm as a mechanical intermediate stage which is

1 Pressure measurement

UAE0295-1Y

Fig. 1

a Direct measurement, pressure-dependent resistor (3)

b Measurement using a force sensor (1)

c Measurement through diaphragm deformation via strain gage (2)

d Capacitive measurement using the deformation of a diaphragm cell (4)

exposed on one side to the pressure to be measured and which deflects to a greater or lesser degree as a function of the pressure. Within a very wide range, its diameter and thickness can be adapted to the particular pressure range. Low-pressure measuring ranges lead to relatively large diaphragms which can easily deform in the range 1 to 0.1 mm. Higher pressures though demand thicker, small-diameter diaphragms which generally only deform by a few μm.

In case (capacitive) pick-offs for spacing or distance measurements are also required, voltage-measuring methods dominate in the medium-pressure to high-pressure ranges. Here, practically only strain gage techniques are used.

Capacitive pick-off
In contrast to their application in inertia sensors (see acceleration/yaw-rate sensors), capacitive pressure sensors are still only rarely encountered even though they could possibly provide similar advantages (particularly with respect to their accuracy). This is more than likely the result of one important difference compared to the other sensors referred to above: pressure sensors need a direct contact with the monitored medium. The dielectric characteristics of which practically always affect the calibration of such capacitive pressure sensors. This means that the calibration would then not only be dependent upon the medium in question, but would also be impossible without it (that is, in the "dry" state). Clear separation of the sensor from the monitored medium has up to now only been achieved at the cost of considerable technical outlay.

Strain gage pick-off
The elongations of the diaphragm occurring on the deflection of a diaphragm-type sensor are detected using strain-gage technology (strain-gage strips or strain-gage resistors). Strain-gage resistors are affixed to the diaphragm (e.g. vapor-deposited). Their electrical resistance

varies under the influence of mechanical stresses. The resistors are connected together to form a Wheatstone bridge. Its voltage is a measure for the pressure.

Table 1 presents a systematic overview of the proven pressure-measurement techniques which to a great extent have already been used in automotive applications. The list is arranged according to the type of diaphragm material and the applied strain gage technology. The combinations have been marked to indicate either those which are already actually in production and which will be described as actual examples in the "sensor designs" chapter (x), or those which are already being closely considered for production (marked in blue):

1 Strain gage pick-off and diaphragm material			
Strain gage pick-off	**Diaphragm material**		
	Ceramics	**Metal (steel)**	**Silicon**
Foils [1] (glued)			
Thick-film			
Metal thin-film		x	
Silicon thin-film		x	
Diffusion resistors			x

Table 1
[1] Low suitability for large-scale production, (x) current examples
▬ Suitability considered

With regard to the particular measuring effect's magnitude and type, the strain-gage techniques listed above have widely varying characteristics. The gage factor (K) defines the magnitude of the measuring effect of strain-gage resistors. It gives the relative change in such a resistor's resistance R referred to the relative change in its length l (equation 1):

$$(1) \quad K = \frac{\Delta R/R}{\Delta l/l} = 1 + 2 \cdot v + \frac{d\rho/\rho}{\varepsilon}$$

Here, the symbol ε (elongation) is often used for the relationship $\Delta l/l$, and is used in multiples of 10^{-6} (ppm) as "micron" or "micro strain".

ν is the material's Poisson ratio, ρ representing its electric conductivity. ν characterizes the reduction in cross-section of the material on elongation and, in the idealized case of a volume being kept constant, amounts to $\nu = 0.5$ (in reality $\nu = 0.3$ to 0.4).

Whereas the conductivity term in equation 1 is of hardly any importance in the case of metallic resistors, with regard to Si resistors it plays a dominant role.

One refers to a longitudinal K factor when the resistor is expanded in the current direction and to a transverse K factor when it is expanded transverse to the current direction (Fig. 2). Table 2 provides an overview of typical values for the most important gage factors.

"Creep" (slight mechanical give under the effects of long-term unidirectional loading) is a highly-feared phenomenon which, when it occurs at all, is only encountered on glued foil strain gage. The other strain gage techniques all apply non-glued techniques and are not affected by this phenomena.

To be precise, a diaphragm's deformation depends upon the difference in the pressure applied to its top and bottom sides. This means that there are four different basic types of pressure sensor (Table 3):
- Absolute pressure
- Reference pressure
- Barometric pressure and
- Differential pressure

Transfer to a force sensor

Instead of directly using the force taken up by their diaphragm, a number of sensors transfer it to a force sensor whose measurement range can remain constant due to the fact that the purely mechanical diaphragm has already performed the adaptation to the pressure-measuring range. Perfect linkage from measuring diaphragm to force sensor (for instance by a tappet) must be ensured though.

Examples of application
- Thick-film pressure sensors
- Micromechanical pressure sensors
- Si combustion-chamber pressure sensors
- Metal-diaphragm high-pressure sensors

2	Gage factors (K) for different materials		
Material	**Gage factors (K)**		
	longitudinal	**transverse**	
Foil strain gage	1.6 to 2.0	≈ 0	
Thick-film	12 to 15	12 to 15	
Metal thin film	1.4 to 2.0	−0.5 to 0	
Si thin film	25 to 40	−25 to −40	
Si-monocrystalline	100 to 150	−100 to −150	

3	Basic sensor types for pressure measurement		
Pressure on diaphragm bottom side p_u	**Pressure on upper side of diaphragm p_o**		
	Measured pressure	**Ambient pressure**	**Vacuum**
Measured pressure	Differential pressure	Reference pressure	Absolute pressure
Ambient pressure	Reference pressure	–	Barometric pressure
Vacuum	Absolute pressure	Barometric pressure	–

2	Gage factor (K), physical quantities

a F

$K_\text{l} = \dfrac{\Delta R/R}{\varepsilon_\text{l}}$

$\varepsilon_\text{l} = \dfrac{\Delta l}{l}$

b

$K_\text{t} = \dfrac{\Delta R/R}{\varepsilon_\text{t}}$

$\varepsilon_\text{t} = \dfrac{\Delta w}{w}$

UAE0819Y

Table 2

Table 3

Fig. 2
a Longitudinal
b Transverse
F Force
I Current
R Resistance
l Length
w Width
ε Elongation
K Gage factor

Force and torque sensors

Measured variables

The following list underlines the wide variety of applications for force and torque sensors in automotive engineering:

- In the commercial-vehicle sector, coupling force between the tractor vehicle and its trailer or semitrailer for the closed-loop controlled application of the brakes, whereby neither push nor pull forces are active at the drawbar
- Damping force for use in electronic chassis and suspension control
- Axle load for electronically controlled braking-force distribution on commercial vehicles
- Pedal force on electronically-controlled brake systems
- Braking force on electrically actuated, electronically-controlled brake systems
- Drive and brake torque
- Steering and steering servo torque
- Finger protection on power windows and electrically operated sliding sunroofs
- Wheel forces
- Weight of vehicle occupants (for occupant-protection systems)

In many cases, initial developments failed to lead to the expected results since generally the costs involved in achieving the stipulated accuracy were excessive for the systems in which the sensors were to be installed. Contrary to expectations, it proved impossible to force down the costs for the production of good torque sensors below those for pressure and acceleration sensors. In fact the torque sensors cost more. Matters are aggravated, and this applies particularly to torque sensors, when the measured variable has to be transferred using non-contact methods from a rotating shaft (e.g. steering spindle or drive shaft) to a sensor mounted on the chassis. Collector rings are not acceptable for the motor vehicle.

Since any form of measurement of only part of the force or torque is very problematical and can easily lead to false results, force and torque sensors must be directly connected into the power flux (in other words, the complete measured variable must pass through them). Sensors that measure force are extensive sensors, in other words, their size depends directly on the measuring range.

Although, in line with automotive-industry demands, there are compact force and torque sensors on the market, these only measure accurately enough when the forces are introduced to the sensor in a precisely defined manner, a stipulation that normally can only be complied with under laboratory conditions. The tolerances and misalignment normally encountered in practice generally dictate the connection of homogenization elements which in turn then lead to the sensors becoming too large.

If force and torque-transfer components must be cut in order for previously testable sensors to be fitted, this generally results in an interface problem. This can only be solved by close cooperation between the sensor supplier and the various suppliers of the parts (or vehicle manufacturer) which must be cut in order to install the sensors. Up to now, this problem has not been encountered with other sensor types, at least not with this severity and with such wide-ranging implications.

Even if the force and torque-transfer components do not have to be cut, and mechanical elements are used as measuring springs which only need modifying for installation of the sensor elements, very precise alignment is still necessary.

Even though hardly any series production force and travel sensors are currently available from Bosch for the motor vehicle, there follows a brief overview of the most important measuring principles and proposed developments that have made it to the shortlist.

Measuring principles

Basically speaking, when considering force and torque measurement, a difference must be made between static and dynamic measuring principles, and between measuring principles based on displacement and mechanical strain. Past requirements were predominantly for static sensors, and in the case of force sensors, generally non-resilient strain-measuring principles were preferred. "Soft", resilient sensor systems are only accepted for torque sensors, particularly in the current example of steering-wheel torque sensing, which can also be implemented using angle-measurement pick-offs. This is possible, particularly because this characteristic proved to be tolerable in earlier hydraulic systems which were not equipped with sensors. The use of magnetically active coil systems previously dominated both sectors, force and torque sensing, in automotive and industrial uses.

More recently, however, displacement measuring magnetostatic sensors (Hall sensors) are being used here, too, to allow the application of micro-structured elements manufactured in series production (e.g. measurement of the passenger's weight) and, despite the familiar problems with dirt and the complexity of their installation, optoelectronic pick-offs are also being considered (e.g. for electronic power-assisted steering systems), which simultaneously permit torque measurement and high-resolution angle of rotation measurement.

Strain measuring force sensors
Magnetoelastic principle
Under the influence of a magnetic field, ferromagnetic materials change their length in the direction of the field (magnetostriction effect). The length can either increase (positive magnetostriction) or reduce (negative magnetostriction) here depending on the material, for the same field direction. The inversion of this effect, the change in the magnetic characteristics under the effect of tensile and compressive stresses, or elongation and compression, is known as the magnetoelastic effect. This effect is expressed in an anisotropic (directional) behavior of relative magnetic permeability μ_r (relationship between magnetic induction B and magnetic field strength H). If this still has the same value in all directions (isotropic) in a force-free situation, it takes on a somewhat different value (μ_{rI}) in the direction of the force under the influence of a force F introduced than that transverse to it (μ_{rq}) (Fig. 1). The effect is not only observed on crystalline or polycrystalline materials, but also in amorphous substances.

In fact, the permeability change in the direction of force is a true reflection of the sign of the force. Even though practically all ferromagnetic materials demonstrate this effect, it can be optimized by using a specific alloy composition. Unfortunately though, the materials which have good linearity, low hysteresis, and low temperature sensitivity, are not identical with the ones which have a high measuring effect. Whereas the maximum measuring effects observed up to the present are about 30 % (referred to the fundamental isotropic values), and can be utilized without electronics being necessary, the effect displayed by materials which have been optimized from

1 Magnetoelastic anisotropic effect

Fig. 1
a Magnetoelastic measurement structure
b Measuring effect
F Force
μ_r Relative magnetic permeability
μ_{rq} Transverse to the direction of force
μ_{rI} In the direction of force

the point of view of measurement techniques is only in an area of a few percent and needs electronic amplification.

The advantages of the magnetoelastic effect are to be found in its broad temperature range and the fact that technical applications are possible up to temperatures as high as approximately 300 °C. On the other hand, it represents a marked volume effect. This means that the coils used for detection not only register a local variation of permeability (as with a strain gage, for instance) caused by the application of force, but more or less the effects throughout the whole of the coil's cross section. This makes the sensor somewhat less sensitive to the possibility of force being applied asymmetrically.

Since the changes in permeability as a function of applied force are practically always registered with the help of alternating fields, the penetration depth of these fields, which is highly frequency-dependent, must be observed:

Only those mechanical forces can contribute to the measuring effect which are present in the measuring field's effective penetration depth. In order to be able to put the measuring effect to maximum use, the magnetically active air gap should be

kept as small as possible. Often, this means that the magnetically active measuring circuit is closed with ferromagnetic material, even when this is not included in the power flux.

Figure 2 shows the two most important possibilities of evaluating the magneto-elastic effect: If a coil is arranged on the measurement structure so that its direction of field coincides with the direction of applied force, the change in inductance L can be picked-off and applied directly. Independent of the magnitude of the applied force, the excitation field strength H and the induction B always have the same direction (Fig. 2a).

If the field strength H of the supply coil is not axially parallel to the applied force, the effect of the latter not only changes the magnetic induction B, but also its direction (due to the anisotropy of the permeability, Fig. 2b). Assuming that with no force applied the directions of H and B are superposed one upon the other in the normal manner, these assume increasingly different directions when force is applied and increased. This can be exploited particularly advantageously in the variation of the magnetic coupling of two coils affixed at 90° to one another (Kreuzduktor) (Fig. 3).

2 Influence of the magnetoelastic effect

a

b

UAE0836-1Y

3 Magnetoelastic tensile-force/compressive-force sensor for measuring brake torque (design specimen)

NAE0838Y

Fig. 2
a With direction of force parallel to the direction of field
b For different directions of field strength H and force F
B Induction
α Enclosed angle

Strain gage principle (piezoresistive)
Strain-gage measuring resistors (strain-gage strips) represent the most widespread and probably the most reliable and precise method for measuring force and torque (Fig. 4). Their principle is based on the fact that in the zone of the elastic-member material to which Hooke's Law applies there is a proportional relationship between the mechanical strain σ in the member, caused by the introduction of force, and the resulting elongation ε. In this case, in accordance with Hooke's law:

(1) $\varepsilon = \Delta l/l = \sigma/E$

whereby the proportionality constant E is the modulus of *elasticity*. Since it does not directly measure the strain resulting from the applied force, but rather the - locally - resulting elongation, the strain-gage method can be regarded as an indirect measuring method. For instance, if the modulus of *elasticity* decreases by 3 % above 100 K, a figure which is normal for metals, then the force indicated by the strain-gage method is 3 % too high.

Strain-gage resistors in the form of film resistors - by contrast with cemented-on strain gages - are so closely bonded to the surface of the selected elastic member that they follow its surface elongation perfectly. The change in resistance resulting from the resistor's elongation is defined by the particular gage factor K of the resistor in question (refer to "pressure sensors"):

(2) $\Delta R/R = K \cdot \varepsilon$

The K factor generally does not exceed a value of 2 for metal-film resistors, but is generally somewhat below this in practice. Strain gages are designed such that they (in conjunction with a specific elongating material (substrate) and their thermal elongation) have no inherent temperature sensitivity where possible ($T_{CR} \approx 0$). Any residual temperature sensitivity is usually eliminated by depositing the resistors on the elastic member in the form of a half or full bridge. Since temperature effects result in same-direction changes on the strain gage this results in no output signal.

The auxiliary bridge resistors can be (but need not be) located within the elastic member's elongation zone. They can also be fitted as purely compensation resistors (Fig. 4c). It must be noted that often the K factor itself also has a temperature coefficient (TC_K). Usually, this decreases along

Fig. 4
a Rod-shaped
b Toroidal-shaped
c Electronic
 evaluation
F Force
$R_{l,q}$ Metal-film resistors,
 lengthwise,
 crosswise
R Auxiliary bridge
 resistors
U_0 Supply voltage
U_A Output voltage

4 Strain-gage force sensors

with increasing temperature, which means that in favorable cases it can compensate for the signal increase caused by the modulus of *elasticity*. Apart from this, signal reductions caused by the K factor are usually compensated for by a bridge supply voltage U_0 which increases accordingly.

In addition to the longitudinal K factor K_l described above, there is also the transverse K factor K_t, which must then be applied if the direction of strain is transverse to the current flow. It has an inverse sign (negative) and is no greater than 0.5 in value.

Disadvantages/limitations:
Notwithstanding their high levels of accuracy and reliability, since the elongations and therefore also the changes in resistance (at least in the case of metal film resistors) are only in the percentage range of the original state, the strain-gage sensors only generate output signals in the mV range so that in general local amplification is required. A further disadvantage of small strain-gage sensors is the fact that they measure precisely the strain

(and only this strain) at the point at which it is applied. This, of course, no longer holds true in cases in which the strain-gage structure is distributed across the elastic member's complete surface. This necessitates extremely precise and reproducible application of the force to be measured if measuring errors due to uneven force introduction are to be avoided.

Application:
As a rule, in order to carry out force measurements, very small strain-gage resistors must be attached to larger force-carrying parts or elastic members. The traditional method of applying the strain-gage resistors to the elastic member by means of a foil backing (as applied in such devices as high-precision scales), is not inexpensive enough for large-batch "low-cost" production. First attempts are therefore being made to apply low-cost, large batch film/layer techniques by depositing the strain-gage resistors on small metal wafers which are then pressed into, or welded onto the elastic element.

Orthogonally printed resistors
Practically all electrical film resistors change their resistance not only under the influence of laterally applied deformation strain but also when compression is applied vertical to the film plane (orthogonal). Here, the so-called "conductive plastic", commonly used in potentiometers, features very high sensitivity. "Cermet" and "carbon layers" are also highly sensitive (Fig. 5). Up to a certain limit, the resistance of the above materials decreases along with increasing compression. The values that can be achieved without permanent resistance change are similar to those which apply for lateral elongation. In both cases, the limit is a function of substrate strength and not of the resistance material. Of course, sensors of this type are almost only suitable for loading by compression but not by lateral strain.

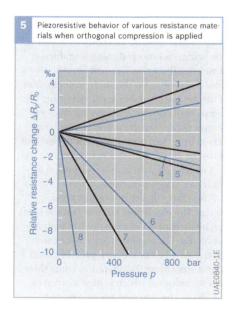

5 Piezoresistive behavior of various resistance materials when orthogonal compression is applied

Relative resistance change $\Delta R_p/R_0$

UAE0840-1E

Fig. 5
1 84.5 Ag 15.5 Mn
2 Manganin
3 Cu
4 Au
5 Ag
6 Carbon film/layer
7 Cermet
8 Conductive plastic

Travel measuring force sensors

The question of whether an integrated travel-measuring weight sensor still has a little additional elasticity or not is not important when it is built into seats for motor vehicle occupants, which are anyway suspended.

This principle is used in the iBolt™ force sensor (see sensor types). Four of these sensors are built into the frame of the front passenger seat for measuring the weight of the passenger. This makes it possible to modify the triggering of the airbag.

Examples of application for force sensors

- Magnetoelastic load measuring bolts (agricultural tractor with plow force control)
- iBolt™ force sensor for sensing passenger weight

Torque sensors

A fundamental distinction is made in torque measurement, too, between methods using angle or strain measurement. In contrast to strain-measurement methods (strain-gage resistors, magnetoelastic), angle-measurement methods (e.g. eddy

current) require a certain length l of the torsion shaft via which the torsion angle (approximately 0.4 to 4°) can be picked-off. The mechanical stress σ proportional to the torque is aligned at an angle of less than 45° to the shaft axis (Fig. 6).

The principles described below are all suited for the non-contacting (proximity) transfer of measured values, even from rotating shafts. In the case of steering-torque measurement, it is also desirable for the employed measuring system to measure the steering angle very accurately (through a full rotation of 360°). This is in the sense of modular integration and is to be implemented with only slight modification.

Strain-measuring sensors
Even though magnetoelastic sensors which enclose the shaft are available on the market, these involve very high costs. Since it is often impossible to optimize the shaft material with regard to its magnetoelastic properties, a search is being made for methods to coat the measuring shaft with a magnetoelastic layer. Such a coating, which has to exhibit good measuring qualities, has still not been found.

This is why the strain gage principle has become predominant (Fig. 7): a strain-gage bridge measures the mechanical stress. The bridge is powered using transformer principles (by a rectifier and control-electronics circuitry on the shaft, independent of the air gap). Further local electronic components on the shaft permit the measurement signal to be amplified and converted to an AC form which is independent of the air gap (for instance, frequency-analog) which can then also be outputted using transformer principles.

When large-batch quantities are involved, the electronic circuitry on the shaft can be integrated in a single chip without problems. The strain-gage resistor can be inexpensively formed on a prefabricated round steel plate (for instance, using thin-film techniques) which is then welded to the shaft. The two ring transformers re-

6 Torque measurement: basic principle

Fig. 10
1 Torsion rod
Φ Torsion angle
σ Torsional stress
M Torque
r Radius
l Rod length

quired for power and signal transmission can be manufactured at an extremely low cost from a soft-magnetic molding compound. High precision levels can be achieved with such a configuration in spite of reasonable manufacturing costs.

Angle-measuring (torsion-measuring) sensors
Sensors for measuring angular difference
It is a relatively easy matter to determine the torsion angle when two mutually-independent incremental speed sensors, or an absolute-measuring (analog or digital) non-contacting angular-movement pick-off, are provided at each end of the section of the shaft subject to torsion ($L \approx 5$ to 10 cm) (Fig. 8). The difference in their outputs $\varphi_2 - \varphi_1$ is a measure for the torsion angle:

(3) $M = \text{const} \cdot L \cdot (\varphi_2 - \varphi_1)$
where L = length of the
 section subject
 to torsion

Up to now, since adequate accuracy demands extremely precise bearings, together with the necessity to provide correspondingly accurate angular or incremental subdivision around the complete periphery, this method has been regarded as too complicated. Nevertheless work is forcing ahead with solving this problem (magnetically or optically), because such a system would incorporate two distinct advantages:
• Possibility of simultaneous measurement of the angle of rotation with the same system
• Possibility of measurement without far-reaching modifications to the torsion shaft being necessary, so that essentially the sensor could be in the form of a plug-in sensor providing an efficient interface for a supplier component

The angle may be measured with high precision using the vernier principle, for example: this is done by marking the circumference of the shaft with a sinusoidal track, of which an integer number of N cycles fits in one circumference.

| **7** | Strain-gage torque sensor with non-contacting (proximity) transformer pick-off |

| **8** | Determining torque by measuring angular difference |

Fig. 7
1 Torque indicator
σ Torsional stress
M Torque
U_0 Supply voltage
R_1 to R_4 Strain-gage
 measuring resistors

Fig. 8
1, 2 Angle/speed
 sensors
3 Angle markings
l Torsion-
 measurement
 section
M Torque to be
 measured
$\varphi_{1,2}$ Angle signals

By contrast with a pure incremental track, which would permit only a resolution of an N-th of the circumference, this sinusoidal signal structure, however, theoretically permits as fine a resolution per circumference as is required by the application of the arcsin function within a single sine cycle. This fine resolution is, however, only useful if it is always known in which of the N similar cycles one is located.

The unambiguity can easily be created if a second track with a little less spatial frequency in which only N-1 sine cycles fit on the circumference is applied around the circumference. Although the two signals do not supply an unambiguous signal for an angle-of-rotation measurement, it is, however, possible to draw firm conclusions from the phase difference Δφ between the two signals, in which sine cycle one is currently located. The pitch difference between the two signals varies over the entire circumference by precisely 360°. Figure 10 shows an example with N = 10. If the phase difference between the two signals is, for instance, in the range between 36° and 72°, then one is clearly located in the second sine cycle. If the value is between 216° and 252°, one is located in the 6th cycle. The precise measuring angle φ is then formed by adding another angle, in the first case of 360° and in the second case of 216° to the measured, but not unambiguous fine signal φ.

In practice however, the arcsin function is not used to resolve the individual sine cycles finely, as it would be necessary to guarantee a constant and normalized signal amplitude for this. It is rather the case that a second, cosine-wave track, phase-shifted against the first by precisely 90°, is applied next to the sine-wave marking. Because of the spatial vicinity, it may be assumed that both tracks are sensed with the same amplitude u, so that the angle φ within the sine cycle can be determined by the arctan function from the two individual signals u_1 and u_2 independently of u:

$$(4) \qquad \varphi = \arctan (u_1/u_2)$$
$$\text{where } u_1 = u \cdot \sin \varphi \text{ and}$$
$$u_2 = u \cdot \cos \varphi$$

9 Optoelectronic angular difference steering torque sensor

UAE1029-1Y

Fig. 9

Optical pick-off of the angle marking tracks

1 Steering shaft with torsion rod
2 Code disks with bar code
3 Steering gear housing
4 Lens
5 Optical waveguides
6 Optical ASIC
7 LED
8 Sensor module
9 PCB

10 Angle measurement using the vernier principle

UAE1030-1E

Thus a total of four tracks are required for the determination of a precise and unambiguous angle of rotation φ. The steering angle and TAS (Torque Angle Sensor) steering-angle sensor illustrated in Figure 9 thus requires a total of eight tracks, which are read optoelectronically, to measure two angles φ_1 and φ_2. Its angular precision over 360° is 1°, the angular resolution is 0.0055°, the resolution of the difference angle $\Delta\varphi$ is 0.0044° with a measuring range of + 9°.

Eddy-current sensors
Two slotted sleeves of electrically conducting aluminum are attached to each end of a sufficiently long section of the measuring shaft. These are inserted one inside the other (Fig. 11). Each sleeve is provided with two rows of slots so that when the shaft is twisted, it becomes increasingly visible through one row of slots, and is hidden more and more by the other row. This leads to increasing, or decreasing, damping of the two high-frequency coils (approx. 1 MHz) situated above each row of slots so that coil inductance also varies accordingly. The slotted sleeves

must be precisely manufactured and assembled in order to achieve the stipulated accuracy. The associated electronics are appropriately accommodated very near to the coils.

Although this sensor principle was developed at Bosch for motor vehicle applications, it was only used in electric power tools (torque sensing in industrial power screwdrivers). A license-holder in Japan has further developed this, very cost-effective, sensor principle for motor vehicle applications to product maturity.

Examples of application for torque sensors
- Strain gage steering torque sensor (developed up to A prototype for electrical power-assisted steering)
- Angle-measuring eddy-current torque sensors (electric power tools, third-party product in Japan)
- Angle-measuring optoelectronic steering angle sensor (electrical power-assisted steering)

11 Eddy-current torque sensor

UAE0710-2Y

1 Slotted sleeves
2 Air gap
3 High-frequency coils
M Torque

Flowmeters

Measured variables

The purpose of flow measurement in the motor vehicle is the detection of the intake air flow rate. This air flow rate must be known precisely so that the engine-management system – both in diesel and in gasoline engines – can set a defined air-fuel mixture. This value can be determined by a flowmeter. The sensors which are used for measuring air flow rate or gas flows in general are also referred to as "anemometers".

As such, the often-used term "air quantity" is incorrect because it does not stipulate whether volume or mass is concerned. Since the chemical processes involved in fuel combustion are clearly based on mass relationships, the object of the measurement is the mass of intake air. On gasoline engines, the air-mass flow rate is the most important load parameter. In diesel engines, the exhaust-gas recirculation rate is regulated using the air-mass flow.

Depending upon engine power, the average (over time) maximum air-mass flow rate to be measured is between 400 and 1,200 kg/h. Due to the low air requirements at engine idle in modern gasoline engines, the ratio of minimum to maximum

flow is 1:50 to 1:100. Because of the higher idle air demand in diesel engines, these ratios must be assumed to be 1:20 to 1:40. The severe exhaust gas and fuel-consumption requirements dictate accuracies of 2 to 3 % of the measured value. Referred to the measuring range, this can easily correspond to a measuring accuracy of $2 \cdot 10^{-4}$, which is unusually high for a motor vehicle.

The air though, is not drawn in continuously by the engine, but rather in time with the opening of the intake valves. Particularly with the throttle valve wide open (WOT) in gasoline engines, this leads to considerable pulsation of the air-mass flow, also at the measuring point which is always in the intake tract between air filter and throttle valve, or between air filter and turbocharger (Fig. 1). Intake-manifold resonance leads to the pulsation in the manifold sometimes being so pronounced that brief return flows can occur. This applies in particular to 4-cylinder engines in which there is no overlap of the air-intake phases. An accurate flowmeter must be capable of registering these return flows with the correct direction.

On a 4-cylinder engine, the pulsations are generated at twice the crankshaft speed. This means that they can easily be in the range 50 to 100 Hz. With a flowmeter featuring a linear characteristic curve and a narrower frequency bandwidth than above, it would suffice for it to follow the mean value of these rapidly fluctuating air flows. The mean value is in any case positive, so that the meter need not necessarily detect the correct sign.

Practically all of the air-mass meters actually in use feature a characteristic curve which is far from linear, so that the measurement signal must be linearized electronically before it can be evaluated. Averaging before linearization can lead to significant dynamic errors (mean-value errors). This fact means that such air-mass meters must be sufficiently rapid in following the pulsations which mostly have a

1 Pulsating air-mass flow Q_{LM} in the intake tract

1 crankshaft rotation

Q_{LM}

Air-mass flow Q_{LM} — kg/h

300

200

100

0 5 10 15 20 ms

Time t

UAE0802-1E

Fig. 1
At full load with
engine speed
n = 3,000 rpm,
intake-manifold
pressure
p_s = 0.96 bar,
mean air-flow rate
Q_{LMm} = 157.3 kg/h

pronounced non-sinusoidal characteristic, and therefore also have a considerable harmonic content. This necessitates a bandwidth of about 1,000 Hz. Apart from this considerable bandwidth, the air-mass meters must also have a high switch-on time constant in order for them to be able to measure correctly during the engine starting phase.

As is the case in all flowmeters, those used in the motor vehicle are calibrated for flow through a pipe with a symmetrical flow profile. This means that a flow with a velocity vector v at practically any point of the flow cross-section of area A depends only on the radius to the central axis, and not on the angle at the circumference. The flow profile (laminar or turbulent, Fig. 2) is related to the Reynolds number R_e:

$R_e = v \cdot D / \eta$
where
D typical cross-section dimension and
η kinematic ductility of the medium.

The flow profiles illustrated by way of example only establish themselves after the flow has been uniformly formed for some time. In the motor vehicle, a profile lying

between the laminar and turbulent profiles illustrated will be established for the case of laminar flow – caused by the short inflow. The laminar or turbulent characteristic of the flow profile depends on whether R_e is below or above a value of approximately 1,200.

A further important influencing factor is the degree of turbulence in the flow, which is determined by the aerodynamic design of the inflow to the flowmeter. If the laminar-turbulent transition is within the measuring range, marked irregularity of the characteristic curve can be expected at this point. As far as motor vehicle applications go, a purely turbulent flow (square profile: $v = const_r$) cannot be assumed at the smallest air masses (gasoline engine idling). A careful design of the inflow, to achieve square profiles even in the event of laminar flow, is then necessary.

Assuming a homogeneous density ρ, and assuming a square profile, the flow is simple to calculate as follows:

$Q_V = v \cdot A$ Volume flow rate
$Q_M = \rho \cdot v \cdot A$ Mass flow rate

2 Flow profiles

Fig. 2
1 Laminar flow profile
2 Turbulent flow profile
A Cross-sectional area of the tube
Q Flow
R Tube radius
r Distance from the tube center
$v(r)$ Flow profile

UAE0803-1Y

Whereas in measurement techniques, long, straight, advance and overshoot sections of constant cross-section are stipulated in order to guarantee a symmetrical profile, such conditions cannot be complied with in a vehicle's cramped under-hood installation space. If pronounced asymmetries occur, the flowmeter must be calibrated as a function of the actual installation conditions.

Measuring principles

Up to now, of the practically unlimited variety of flowmeters on the market, only those which operate according to the impact-pressure principle have come to the forefront for air-quantity measurement in the vehicle. This principle still depends upon mechanically moving parts, and in principle correction measures are still needed to compensate for density fluctuations.

Today, true air-mass meters applying thermal methods (hot-wire or hot-film air flowmeters) are used which can follow sudden flow changes without mechanically moving parts.

Variable orifice plates (sensor plates)

The calculation of the pressure drop across fixed orifice plates is based on two physical laws:

Continuity equation:

$$\rho \cdot v_1 \cdot A_1 = \rho \cdot v_2 \cdot A_2 = const$$

Bernoulli's equation:

$$p_1 + \frac{1}{2} \cdot \rho \cdot v_1{}^2 = p_2 + \frac{1}{2} \cdot \rho \cdot v_2{}^2 = const$$

These laws are to be applied for two different measuring cross-sections A_1 and A_2 (Fig. 3). Assuming constant density $\rho = \rho_1 = \rho_2$, this results in the pressure drop:

$$\Delta p = Q_V{}^2 \cdot \rho \cdot \left(\frac{1}{A_2{}^2} - \frac{1}{A_1{}^2}\right)$$

This pressure drop can be measured either directly with a differential-pressure sensor, or by means of the force acting against a so-called sensor plate. It should be noted here that impact-pressure gages of this type measure a flow value that corresponds neither to the volume flow rate nor to the mass flow rate, but to a geometrical mean value of the two:

$$Q_{St} = const \cdot \sqrt{\rho} \cdot v = const \cdot \sqrt{Q_V \cdot Q_M}$$

Fig. 3
a Ring orifice
b Sensor plate
1 Orifice plate
A_S Disk cross-section
$A_{1,2}$ Measuring cross-section
$p_{1,2}$ Measured pressures
Δp Pressure drop
Q_{LM} Air-mass flow

Fig. 4
1 Sensor plate
2 Air-temperature sensor
3 To control unit
4 Potentiometer
5 Damping chamber
6 Compensation flap
Q_L Intake-air flow

3 Impact-pressure flowmeter

4 Impact-pressure air-mass meter

Because of the square relationship to the flow rate, fixed orifice plates permit only a variation in measured variable of 1:10, as otherwise the accuracy of the differential pressure recording is no longer adequate and significant pressure losses would result which would lead to power losses in the engine. When larger ranges are to be covered, several orifice plates must be used, or such versions which automatically adapt themselves to the measuring range by opening up a larger flow cross-section A_2 in line with the impact pressure.

With such variable, moving sensor plates it is an easy matter to increase the variation to 1:100. Here, the increasing air flow causes the sensor plate to be deflected (usually against a constant counterforce) into an area whose cross-section is specifically shaped so that the resulting deflection/angle relationship complies with the desired characteristic. In other words, linear for K-Jetronic and non-linear for L-Jetronic. The sensor plate's (Fig. 4) setting is then a measure for the air flow which is in relationship to the impact pressure defined above.

The limit frequency for such sensors is typically approx. 10 Hz. Such sensor plates though are unable to follow the high pulsation frequencies which often occur. From the point of view of the pulsation, they can be regarded as fixed orifice plates with a square-law characteristic curve. Under certain load conditions this leads to considerable mean-value errors which can only be compensated for roughly by the use of suitable software.

If the density ρ of the intake air changes because of temperature fluctuations or the altitude, the measurement signal here will only change by $\sqrt{\rho}$. An air-temperature sensor and a barometric pressure sensor must also be used to determine the density fluctuation fully.

Hot-wire/hot-film anemometers

When current I_H flows through a thin wire with electrical resistance R, its temperature increases. If at the same time a medium with density ρ flows across it at velocity v, a balance is set up between the electrical power input P_{el} and the power P_V drawn off by the air flow, whereby

$$P_{el} = I_H^2 \cdot R = P_V = c_1 \cdot \lambda \cdot \Delta\vartheta$$

Here, the power drawn off by the air flow is proportional to the temperature difference $\Delta\vartheta$ established and the coefficient of thermal conductivity λ. The following applies in close approximation:

$$\lambda = \sqrt{\rho \cdot v} + c_2 = \sqrt{Q_{LM}} + c_2$$

Although λ is primarily a function of the mass flow Q_{LM}, with the static medium at standstill ($v = 0$), a certain heat loss takes place (convection) represented by the additive constant c_2. This results in the familiar interrelationship between the heating current I_H and the mass flow Q_{LM}.

$$I_H = c_1 \cdot \sqrt{(\sqrt{Q_{LM}} + c_2)} \cdot \sqrt{\frac{\Delta\vartheta}{R}}$$

With the supply of a constant heat $I_H^2 \cdot R$, a temperature rise $\Delta\vartheta$ decreasing reciprocally with the root of the air-mass flow Q_{LM} would establish itself. If, on the other hand, the heating current I_H is controlled such that a constant temperature increase (for instance, $\Delta\vartheta = 100$ K) is maintained even when the flow rate increases, this will lead to a heating current which increases approximately with the fourth root of the mass flow, and at the same time serves as a measure for the mass flow.

The advantage of such a control loop lies in the fact that the electrical heating resistor always remains at the same temperature so that its heat content need not be changed by means of time-wasting heat transfer. In fact, with a 70 µm platinum wire for instance, it is possible to achieve

time constants in the 1 ms range for changes in air-flow rate. In cases where closed-loop control is not used the time constants would be 40 to 100 times higher.

If the heater temperature were to be maintained constant simply by keeping its (temperature-dependent) resistance constant, with constant mass flow and higher medium temperature, this would result in a current drop and therefore a false measurement. In practice, this error is avoided by using a bridge circuit containing a second high-ohm compensation resistor R_K of the same type (e.g. platinum). Here, the heating resistor is kept at a constant over-temperature $\Delta\vartheta$ compared to the medium (Fig. 5) by a control loop. In case of a sudden jump in the medium temperature, the sensor reacts with a long time constant since in this case the heat content of the hot wire must be changed. The heating current generates a control voltage U_M proportional to the air-mass flow on a precision resistor (measuring resistor R_M).

The heating resistors in the first air-mass meters (hot-wire air-mass meters) used for automotive applications were of very fine platinum wire. This wire was stretched in trapezoidal form across the flow cross-section so that it was able to

average out any asymmetries in the flow profile (Fig. 6). Adequate service lives only became possible when the platinum wire was stabilized by alloy additives so that its resistance no longer changed due to deposits and cracks on its surface. This meant though that the deposits on the hot wire had to be burned-off automatically (approximately 1,000 °C) following every operating phase.

Notwithstanding a number of functional advantages, this sensor concept was far too costly. A thick-film version (HFM2 hot-film air-mass meter) was able to combine all the resistors concerned with the measurement on a single ceramic substrate. Due to the ceramic substrate's considerable thermal capacity, it was difficult not to exceed the maximum permissible switching constants. Furthermore, a complicated saw cut had to be made to reduce the undesirable heat coupling between heating and compensation resistors. On the other hand though, this version permitted the burn-off process to be dispensed with since the special flow conditions no longer led to unwanted deposits.

In contrast to both its predecessor types, a further silicon-based micromechanical version (HFM5 and HFM6 hot-

Fig. 5
Q_{LM} Air-mass flow
U_M Measurement
 voltage
R_H Hot-wire resistor
R_K Compensation
 resistor
R_M Measuring resistor
$R_{1,2}$ Calibration
 resistors

Fig. 6
1 Temperature
 compensation
 resistor R_K
2 Sensor ring with
 hot wire R_H
3 Precision
 measuring
 resistor (R_M)
Q_M Air-mass flow

5 Hot-wire air-mass meter (circuit)

6 Hot-wire air-mass meter (components)

film air-mass meters) fulfilled practically all expectations. In particular, this version is able to measure in both flow directions with the correct sign (Fig. 7). This means that the brief return flows that occur as a result of pulsation no longer lead to measuring errors (Fig. 8).

To this end, in addition to the heater control circuit used in the previous versions, a temperature sensor is located on each side of the heating resistor, in other words upstream and downstream. This principle is similar to the Thomas process often encountered in the literature. When there is no flow ($Q_{ML} = 0$), each of these sensors indicates the same temperature. When flow starts though, since the upstream sensor is cooled by the medium, the higher the flow the higher the temperature difference between the two resistors. The output signal derived from the temperature difference has a similar characteristic curve to the anemometers used up to now, whereby its sign is a clear indication of the flow direction.

Due to its small size, the micromechanical flowmeter is only a partial-flowmeter. In other words, it is no longer in any way able to average-out any non-homogeneity in the flow velocity as a function of the flow cross-sectional area. Rather,

this flowmeter must ensure that the partial flow it measures represents the same fraction of the total flow throughout the whole measuring range. This precondition can be achieved by means of a flow rectifier (grill) built into the measuring pipe.

Examples of application
- HFM5 and HFM6 hot-film air-mass meters for gasoline and diesel engine applications

7 Micromechanical hot-film air-mass meter with air-quantity measurement in both directions

Signal voltage U / Q_R / Air-mass flow Q_{LM}

UAE0807-1E

8 Determination of the pulsating air-mass flow in a 4-cylinder engine

Air-mass flow Q_{LM} (kg/h) / Time t (ms)

UAE0808-1E

Fig. 7
Q_R Return flow

Fig. 8
At full load and engine speed
n = 900 rpm
1 Hot-wire air-mass meter
2 Hot-film air-mass meter

Gas sensors and concentration sensors

Measured variables

The concentration of a given material or medium defines the mass or volume percent of a given material in another given material or in a mixture or combination of other materials. With a concentration sensor (also known as a concentration probe, the important thing is that in the ideal case it is sensitive to only one medium, while at the same time practically "ignoring" all other mediums. Of course, in practice, every concentration sensor has its own cross sensitivity to other mediums even though, as is often the case, "temperature" and "pressure" are maintained constant.

In the vehicle, the following parameters must be measured:

- Oxygen content in the exhaust gas (closed-loop combustion control, catalytic-converter monitoring)
- Carbon-monoxide and nitrogen-oxide content, as well as air humidity inside the vehicle (air quality, misting of vehicle windows)
- Humidity in the compressed-air brake system (air-drier monitoring)
- Dampness of the outside air (black ice warning)
- Concentration of soot in diesel-engine exhaust gas. A still unsolved problem. In contrast to the above-mentioned gas concentrations, this is a particle concentration. The difficulties inherent in the measuring assignment are further aggravated by the possibility of the sensor being blocked by particles so that it no longer functions.

The introduction of the fuel cell as an automotive drive means that further gas sensors will have to be developed, for instance for the detection of hydrogen.

Measuring principles

Measured mediums occur in gaseous, liquid, or solid state, so that in the course of time countless measuring methods have been developed. For automotive applications, until now only the gas-analysis area, and in particular the measurement of gaseous humidity, has been of any interest. Table 1 presents an overview of the processes applied in general measurement techniques.

Gas measurement in general

Gas sensors are usually in direct unprotected contact with the monitored medium (in other words with foreign matter) so that the danger of irreversible damage exists. This form of damage is referred to as sensor "contamination". For instance, the lead that may be contained in fuel or the exhaust gas can make the electrolytic oxygen concentration sensors (Lambda oxygen sensors) unusable.

Moisture measurement

In addition to the outstanding significance of the Lambda oxygen sensor in dealing with exhaust gases, moisture measurement also plays an important role.

In the broader sense, moisture indicates the moisture content of gaseous, liquid, or solid substances. In the narrower sense, we are dealing here with the gaseous-water (water vapor) content in gaseous media – above all in the air.

1	Gas-analysis processes (without particular attention being paid to the moisture-measurement process) (X) = for automotive applications		
Physical process	**Physical-chemical process**		**Chemical process**
Thermal conductivity	Heat of reaction		Selective absorption
Magnetic processes	Heat of absorption		Selective absorption with prior chemical conversion
Radiation absorption	Characteristic color reaction		
Gas chromatography	Electrolytic conductivity	X	
Radioactive processes	Electrochemical processes	X	

Table 1

When a damp gas is cooled in an isobaric process, it reaches its saturation point at a specific temperature (known as the dew point τ).

A number of important definitions are given below in connection with humidity measurement (refer also to Figure 1):

m_w Mass of water
m_s Mass of water in the
 saturation state
m_{tr} Mass of dry gas

M_w Mole mass of water
M_{tr} Mean mole mass of the
 dry gas

p Total pressure of the
 gas mixture
p_w Partial pressure of the
 water vapor
p_s Saturated vapor pressure
 (vapor pressure of the water
 at mixture temperature)

Absolute humidity:

$$X = \frac{m_w}{m_{tr}} = \frac{M_w}{M_{tr}} \cdot \frac{p_w}{p-p_w} \text{ (in \%)}$$

$$f_a = \frac{m_w}{V_{tr}} \qquad\qquad \text{(volume-related)}$$

Relative humidity:

$$\Phi = \frac{p_w}{p_s} \text{ (in \%)}$$

For low-cost applications (for instance in the vehicle), resistive and capacitive sensors are used almost exclusively. They are provided with hygroscopic layers which can store water as a function of the relative humidity (and release it again), and thus trigger a usually drastic change in a resistor's value or in the value of a planar capacitor.

On capacitive humidity sensors, a hygroscopic insulating layer (e.g. Al_2O_3, or a polymer plastic), which can possibly also

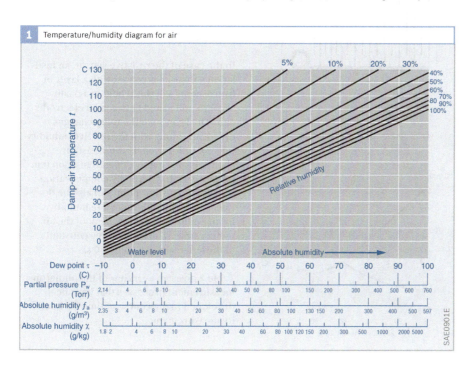

1 Temperature/humidity diagram for air

2 Humidity-measurement procedures. X = technologically important				
Processes	**Serial no.**		**Measuring device**	**Measuring method**
Saturation processes	1	X	Dew point hygrometer	Direct methods
	2	X	LiCl dew point hygrometer	(measurement of absolute
Evaporation processes	3	X	Psychrometer	humidity)
Absorption processes	4		Volume hygrometer	
	5	X	Electrolysis hygrometer	
	6		Condensate volume hygrometer	
Energetic processes	7	X	Infrared hygrometer	
	8		Microwave hygrometer	
	9		Electrical-discharge hygrometer	
	10		Diffusion hygrometer	
Hygroscopic processes	11	X	Electrical conductive-foil	Indirect methods
	12		hygrometer	(measurement of relative
	13	X	Capacitor hygrometer	humidity)
	14	X	Hair hygrometer	
	15		Bimetal strip hygrometer	
	16		Color hygrometer	
	17		Quartz hygrometer	
			Gravimetric hygrometer	

Table 2

2 Capacitive sensor wafer with comb electrodes

SAE0903Y

Fig. 2
1 Plastic wafer
2 Comb-like,
 embossed gold-foil
 electrodes

3 Resistive and capacitive humidity sensors (typical characteristic curve)

Resistance R / Capacity C vs. Relative humidity ϕ

SAE0902E

be the mounting wafer, serves as the dielectric of a capacitor. Either one of the electrodes is permeable to water, or the electrodes have a comb-like structure (Fig. 2). With increasing humidity, the dielectric absorbs more water and the sensor's capacitance increases considerably (relative dielectric constant of water $\varepsilon_{rW} \approx 81$, Fig. 3).

In the case of the resistive sensor, an insulation substrate is coated with a layer of hygroscopic salt (LiCi) held in a paste binder and located between an electrode pair. The layer's conductivity changes drastically along with the relative humidity (Fig. 3). Unfortunately this resistance change is also highly dependent upon temperature so that normally compensation is needed. The dew point can also be defined, and with it the absolute humidity, by an additional measurement of the air temperature. The typical time constants of these sensors are about 30 s.

Table 2 gives an overview of the numerous measurement processes that have been developed in the course of time for humidity measurement.

▶ The piezoelectric effect

In 1880 Pierre Curie and his brother Jacques discovered a phenomenon that is still very little known today, but is present in the everyday lives of millions of people: the piezoelectric effect. For example, it keeps the hands of a quartz clock operating in time.

Certain crystals (e.g. quartz and turmaline) are piezoelectric: Electric charges are induced on the crystal surface by exerting a compression or elongation force along certain crystal axes. This electrical polarization arises by shifting positive and negative ions in the crystal relative to each other by exerting force (see Fig., Item b). The shifted centers of charge gravity within the crystal balance each other out automatically, but an electric field forms between the end faces of the crystal. Compressing and elongating the crystal create inverse field directions.

On the other hand, if an electrical voltage is applied to the end faces of the crystal, the effect reverses (inverse piezoelectric effect): The positive ions in the electric field migrate toward the negative electrode, and negative ions toward the positive electrode. The crystal then contracts or expands depending on the direction of the electric field strength (see Fig., Item c).

The following applies to piezoelectric field strength E_p:

$$E_p = \delta \, \Delta x/x$$

$\Delta x/x$: relative compression or elongation
δ: piezoelectric coefficient, numeric values 10^9 V/cm to 10^{11} V/cm

The change in length Δx when a voltage U is applied results from the following:
$U / \delta = \Delta x$ (Example of quartz: deformation of approximately 10^{-9} cm at $U = 10$ V)

The piezoelectric effect is not only used in quartz clocks and piezo-inline injectors, it has many other industrial applications, either as a direct or inverse effect:

Piezoelectric sensors are used for knock control in gasoline engines. For example, they detect high-frequency engine vibrations which is a feature of combustion knock. Conversion of mechanical vibration to an electric voltage is also used in the crystal audio pickup of a record player or crystal microphones. In the piezoelectric igniter (e.g. in a cigarette lighter), mechanical pressure produces the voltage needed to generate a spark.

On the other hand, if an alternating voltage is applied to a piezoelectrical crystal, it vibrates mechanically at the same frequency as the alternating voltage. Oscillating crystals of this type are used as stabilizers in electrical oscillating circuits or as piezoelectric acoustic sources to generate ultrasound, for instance.

When used in clocks, the oscillating quartz is excited by an alternating voltage whose frequency is the same as the quartz's natural frequency. This is how an extremely time-constant resonant frequency is generated. In a calibrated quartz, it deviates by only approx. 1/1,000 second per year.

Principle of the piezoelectric effect
(illustrated on the example of a unit cell)

a Quartz crystal SiO_2

b Piezoelectric effect: When the crystal is compressed, negative O^{2-} ions shift upward, positive Si^{4+} ions shift downward: electric charges are induced at the crystal surface.

c Inverse piezoelectric effect: When an electrical voltage is applied, O^{2-} ions shift upward, Si^{4+} ions shift downward: the crystal contracts.

a Si^{4+} O^{2-} b c

SAN0170Y

Temperature sensors

Measured variables

Temperature is defined as a nondirectional quantity which characterizes the energy state of a given medium, and which can be a function of time and location:

$$T = T(x, y, z, t) \tag{1}$$

where: x, y, z are the spatial coordinates, t is time, and T is measured according to the Celsius or Kelvin scale.

Generally speaking, with monitored media which are in gaseous or liquid form, measurements can be taken at any point. In the case of solid bodies, measurement is usually restricted to the body's surface. With the most commonly used temperature sensors, in order for it to assume the medium's temperature as precisely as possible, the sensor must be directly in contact with the monitored medium (direct-contact thermometer). In special cases though, proximity or non-contacting temperature sensors are in use which measure the medium's temperature by means of its (infrared) thermal radiation (radiation thermometer = Pyrometer, thermal camera).

Generally speaking, a temperature sensor should reflect this dependency correctly, in other words, it should reflect as accurately as possible the local distribution of the temperature and its change as a function of time.

In special cases, possibly for functional reasons, this stipulation can be relaxed somewhat. The call for high local resolution and for high-speed response both demand that the sensor should be as small as possible, since it should not falsify the temperature readings by itself absorbing heat. In other words it should have a low thermal capacity.

In order to ensure that the temperature assumed by the sensor remains independent of the usually very different temperature of its mounting, it should be thermally well insulated from it. Since this also falsifies the measurement, the heat developed by the majority of sensors in the active state should be kept to a minimum (e.g. < 1 mW).

The temperature sensor's dynamic response is specified by the time constant τ. This defines the time taken by the sensor to reach 63 %, 90 %, or 99 % of its final reading when subjected to a jump in temperature. This time depends not only upon the sensor's thermal capacity, but also to a great extent upon the heat-transfer coefficient between the sensor and the monitored medium. The higher it is, the faster the sensor reaches its final reading. Naturally, this figure is far higher for liquid mediums than for gaseous mediums. It must also be noted that the heat-transfer coefficient is highly influenced by any flow velocity v in the medium. It increases with approximately \sqrt{v}. In other words, the time constant of a temperature sensor should always be specified with reference to a specific flow velocity of a defined medium.

1	Temperature measuring points in the motor vehicle
Measuring point	**Temperature range °C**
Intake/charge air	−40 to 170
External environment	−40 to 60
Interior	−20 to 80
Exhaust air/heating system	−20 to 60
Evaporator (air-conditioning system)	−10 to 50
Cooling water	−40 to 130
Engine oil	−40 to 170
Battery	−40 to 100
Fuel	−40 to 120
Tire air	−40 to 120
Exhaust gas	100 to 1,000
Brake caliper	−40 to 2,000

Table 1

Temperature measurement in the vehicle makes use almost exclusively of the temperature-dependence of electrical resistance material with positive (PTC) or negative (NTC) temperature coefficient. The direct-contact thermometers apply this phenomena. For the most part, the conversion of the temperature change to an analog voltage takes place by adding a second resistor to form a voltage divider (which also has a linearizing effect). Recently though, non-contacting (pyrometric) temperature sensing has been considered for occupant-protection (passenger-position monitoring for airbag triggering), and for comfort and convenience (air conditioning in accordance with skin-temperature measurement, prevention of windscreen mistup). It needed the advent of microsystem technology for this (pyrometric) method to become feasible from the costs viewpoint. Table 1 presents a listing of the temperatures which have to be measured in the vehicle.

It is not only the highly differing measuring ranges that demand a variety of different sensor concepts and technologies, but the different accuracies and dynamic response requirements also lead to very different sensor shapes. At many locations, temperature is also measured in order that it can be compensated in cases where temperature variations trigger faults or act as an undesirable influencing variable.

Measuring principles for direct-contact sensors

The fact that practically all physical processes are temperature-dependent means that there are almost just as many methods for making temperature measurements. The preferable methods though are those in which the temperature effect is very distinctive and dominant and as far as possible features a linear characteristic curve. Furthermore, the measuring elements should be suitable for inexpensive mass-production, whereby they should be adequately reproducible and non-aging. Taking these considerations into account, the following sensor techniques have come to the forefront, some of which are also applied in automotive technology:

Resistive sensors

In the form of 2-pole elements, temperature-dependent electrical resistors are particularly suitable for temperature measurement, no matter whether in wire-wound, sintered ceramic, foil, thin-film, thick-film, or monocrystalline form. Normally, in order to generate a voltage-analog signal they are combined with a fixed resistor R_V to form a voltage divider, or load-independent current is applied (Fig. 1). The voltage-divider circuit changes the original sensor characteristic $R(T)$ to a slightly different characteristic $U(T)$:

$$U(T) = U_0 \cdot \frac{R(T)}{R(T) + R_V} \qquad (2)$$

On the other hand, the application of a supply current I_0 permits the resistor curve to be reproduced exactly:

$$U(T) = I_0 \cdot R(T) \qquad (3)$$

The measurement sensitivity is reduced more or less depending upon the construction of the voltage-divider circuit. Notwithstanding this fact, in the case of resistance characteristics which display a

1 Methods for resistance/voltage conversion

NAE0820Y

Fig. 1
I_0 Current supply
U_0 Supply voltage
R_V Temperature-independent series resistor
$R(T)$ Temperature-dependent measuring resistor
$U_A(T)$ Output voltage

slightly progressive curve it does have a linearization influence (an effect which is usually very welcome). Very often, in this connection, the auxiliary resistor is so dimensioned that it equals the measuring resistor at a given reference temperature T_0 (e.g. 20 °C):

$$R_V \approx R\,(T_0) \tag{4}$$

If the precision is inadequate, a resistance sensor can be calibrated to the setpoint value both with regard to its resistance (referred to a reference temperature), and its temperature coefficient (TC), by means of an adjustable parallel resistor R_P and a serial resistor R_S (Fig. 2). Of course, when fixed resistors are added the TC is reduced and the characteristic changes somewhat.

Sintered-ceramic NTC resistors
As a result of their pronounced measuring effects and inexpensive manufacture, the most common semiconductor resistors in use are based on heavy-metal oxides and

oxidized mixed crystals. These are sintered in bead form or disc form (Fig. 3) and have a polycrystalline structure. Because of their very steep temperature curve, they are also known as NTC resistors or thermistors. To a good approximation, and by applying the exponential law, their characteristic curve can be defined as follows:

$$R(T) = R_0 \cdot e^{\,B \cdot \left(\frac{1}{T} - \frac{1}{T_0}\right)} \tag{5}$$

Where $R_0 = R(T_0)$,
$B = 2,000$ to $5,000$ K = const,
T absolute temperature

Here, the characteristic-curve gradient (TC) or the percentage resistance change as a function of temperature, is highly dependent upon the working point. This means that it can only be defined at given points:

$$TC = -B/T^2 \tag{6}$$

Fig. 2
a Resistance sensor
b Characteristic
 curves
1 Auxiliary contacts
2 Bridge
R_{Ni} Nickel-film resistor
$R_{tot}(T)$ Total resistance
 referred to
 temperature T
R_P Adjustable
 parallel resistor
R_S Adjustable
 series resistor

Fig. 3
a Pearl form
b Disk form
c Characteristic
 curve with limits
 of variation

2 Calibration of a resistor to its setpoint value

3 NTC resistors (examples)

It reduces considerably along with increasing temperature, and the resistance value itself frequently varies by 4 to 5 powers of ten, e.g. typically from a few 100 kΩ to a few 10 Ω. This pronounced temperature-dependence means that applications are restricted to a "window" of about 200 K. This though can be selected in the range from –40 to approximately 850 °C. Tighter tolerances of up to ± 0.5 K at a given reference point are complied with by using a selection process, or possibly even by grinding in oil, a method which of course has an affect upon costs. The aging stability of these sensors has been vastly improved compared to earlier versions which means that it is quite possible for the very close tolerances to apply throughout the sensor's useful life.

PTC thin-film/thick-film metallic resistors
The thin-film metallic resistors placed with two additional temperature-neutral calibration resistors on a common substrate wafers have particularly high accuracy values. They can be tightly toleranced in respect of their characteristic curve and can be manufactured to be stable over the long term and are additionally trimmed by laser cuts. The use of layer technology makes it possible to adapt the substrate (ceramic, glass, plastic foil) and the upper layers (plastic molding or paint, sealed foil, glass and ceramic coatings) to the respective application, and thus provide protection against the monitored medium. Although metallic layers are less sensitive to thermal variations than the ceramic-oxide semiconductor sensors, both linearity and reproducibility are better: The following applies for the computational definition of these sensors:

$$R(T) = R_0 (1 + \alpha \cdot \Delta T + \beta \cdot \Delta T^2 + ..) \qquad (7)$$

where $\Delta T = T - T_0$ and
$T_0 = 20$ °C (reference temperature),
α linear temperature coefficient (TC),
β square temperature coefficient.

Even though the β coefficient is generally very small for metals, it cannot be ignored completely. This is why the measurement sensitivity of such sensors is usually characterized by means of a mean TC, the "TC 100". The TC 100 corresponds to the mean curve gradient between 0 °C and 100 °C (Table 2 and Figure 4).

$$\text{Where } TC\ 100 = \frac{R(100\,^{\circ}\text{C}) - R(0\,^{\circ}\text{C})}{R(0\,^{\circ}\text{C}) \cdot 100K} \qquad (8)$$

2 Temperature coefficient TC 100

Sensor material	TC 100 10^{-3}/K	Characteristic curve	Measuring range
Nickel (Ni)	5.1	slightly progressive	–60 to 320
Copper (Cu)	4.1	slightly progressive	–50 to 200
Platinum (Pt)	3.5	slightly degressive	–220 to 850

Table 2

4 Definition of the mean temperature coefficient
TC 100 = α_{100}

UAE0823E

Although platinum (Pt) resistors have the lowest TC, they are not only the most precise resistive temperature sensors, but can also boast the best aging stability. They are available on the market under the designation "PT 100" or "PT 1000" (100 Ω or 1,000 Ω nominal resistance at a reference temperature of 20 °C) in a variety of different tolerance classes (up to 0.1 °C) (Fig. 5). At temperatures up to about 1,000 °C, thick-film platinum sensors are suitable whose platinum layer has been stabilized by special additives.

Thick-film resistors (PTC/NTC)
Thick-film pastes with both high resistance (low surface-area requirement) and positive and negative temperature coefficients are generally employed as temperature sensors for compensation purposes. They have a non-linear characteristic curve (which though is not as "bent" as the curve of the solid NTC resistors), and are suitable for laser-trimming. The measuring effect is improved by forming voltage-divider circuits from NTC and PTC materials.

Monocrystalline silicon semiconductor resistors (PTC)
Basically speaking, with temperature sensors of monocrystalline semiconductor materials such as silicon (Si), it is possible to incorporate further active and passive circuit elements on the sensor chip (initial signal conditioning at the measuring point). Due to the closer tolerances involved, their production uses the "spreading resistance" principle (Fig. 6a). The current flows through the measuring resistor and through a surface-point contact before arriving at the silicon bulk material. It then proceeds, widely distributed, to a counter-

6 Spreading-resistance principle (dual-hole version)

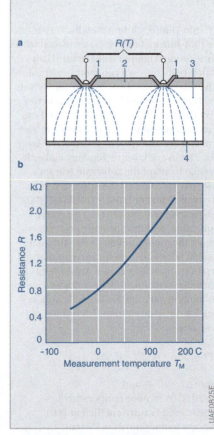

Fig. 6
a Design
b Characteristic curve
1 Contacts
2 Passivation (Nitride, oxide)
3 Si substrate
4 Counter-electrode without connection
$R(T)$ Temperature-dependent resistor

5 Pt resistor (tolerance diagram)

electrode covering the base of the sensor chip. In addition to the high reproducibility material constants, the high current density behind the contact point (high accuracy thanks to photolithographic production) mainly defines the sensor's resistance. In order to make the sensor highly independent of polarity, these sensors are usually series-connected in pairs in opposite alignments (dual-hole version, Fig. 6). The base electrode can be in the form of a temperature contact (no electrical function).

Measurement sensitivity is practically double that of a Pt resistor (TC = $7.73 \cdot 10^{-3}$/K). The progressive bend of the temperature curve is more pronounced than on a metallic sensor. The measuring range is limited to approx. +150 °C by the material's intrinsic conductivity (Fig. 6b). There are special versions (Fig. 7) available for uses up to 300 °C.

Thermocouples

Thermocouples are used in particular for measuring ranges ≥ 1,000 °C. They rely on the Seebeck effect, according to which there is a voltage across the ends of a metallic conductor when these are at different temperatures T_1 and T_2. This "thermoelectric voltage" U_{th} depends (independently of the development of this) exclusively on the temperature difference ΔT at the ends of the conductors (Fig. 8). It is expressed by the equation:

$$U_{th} = c\,(T_2 - T_1) = c\,\Delta T, \tag{9}$$

where the proportionality constant is material-specific and termed the Seebeck coefficient.

Since the instrument leads used to measure this voltage across the metallic conductor must themselves be equipped with terminals (for instance made of copper),

7 Spreading resistance sensor (unipolar version for temperatures up to 300 °C)

A

1
2
3
1

n^+

Detail A

NAE0826Y

Fig. 7
1 Metal wire
2 Glass
3 Si crystal

8 Seebeck effect

U_{th}

T_2 > T_1

+ −

1 2

NAE0827Y

Fig. 8
1 Heat source
2 Metallic conductor
+ High,
− low thermal velocity
 of the electrons
T_2 High temperature
T_1 Low temperature
U_{th} Thermoelectric
 voltage

3 Thermoelectric voltage U_{th} of a number of metals

Material	Thermoelectric voltage U_{th} mV/100 °C
Constantan	−3.40
Nickel	−1.90
Paladium	−0.28
Platinum	0.00
Copper	+0.75
Manganite	+0.60
Iron	+1.88
Silicon	+44.80

Table 3

these are also subject to the same temperature difference, so that unfortunately only the difference between the metallic conductor and the connecting cables is measured. Thermoelectric voltages are always listed based on Platinum as the reference material (Table 3).

In order that the generated voltages are as high as possible, a number of material pairs have established themselves (Fig. 9, e.g. iron/constantan etc.). It is important that the "limbs" of such a thermocouple are joined at the end to which the heat is applied in such a manner that the joint is electrically conductive (by means of twisting, welding, soldering, etc., Fig. 10).

Thermocouples themselves are usually short, and the extensions up to the point where the signal is captured can be made with compensating cables which use the same material pair as in the thermocouple itself. It is important that both free ends

of the thermoelement configuration are at the same (reference) temperature, otherwise the temperature difference at the free ends will also be included in the measurement. Thermocouples, therefore, always measure only the temperature difference to a given reference point. If the measuring point's absolute temperature is to be measured, other devices (such as resistive sensors) must be used to also measure the temperature at the reference point.

The characteristic curve of the thermoelectric voltage against temperature is usually not as linear as that given in equation (9). The sensor signals are usually small, and ICs are used for their amplification and for their linearization. In order to increase the measurement voltage, it is common practice to connect a number of identical thermocouples in series. These have their "hot" junctions at the temperature to be measured, and their "cold" junctions at the reference temperature (Fig. 11, thermopile).

9 Common thermocouples (characteristic curves)

11 Thermocouples connected in series

10 Thermocouple measuring set-up

Although thermocouples are robust (for instance, high-level EMC due to low internal resistance), they are not particularly accurate as a measuring device. Their deviation can easily be in the 5° to 15 °C range, and they are not outstanding regarding their resistance to aging, which means that individual calibration does not result in a permanent improvement of their accuracy.

Of course, thermocouples can be manufactured using both thin-film and thick-film techniques. Metallic films stacked one on top of the other provide for excellent thermal contact, and extremely small thermocouples can be produced by applying microsystem technology. Thermocouples are particularly suitable for use in thermopiles comprised for instance from 50 to 100 individual thermocouples. They are used in non-contacting radiation thermometers (pyrometers).

Semiconductor depletion layers

Presuming a constant current, the forward voltage of semiconductor depletion layers (Fig. 12) such as those in diodes and in the basis-emitter path of a transistor, demonstrate very good linearity as a function of temperature:

$$U_F(T) = \frac{k \cdot T}{q} \cdot \ln\left(\frac{I_F}{I_{sat}} + 1\right) \tag{10}$$

Where:
$I_{sat} = I_{sat}\,(T)$ and I_F = constant,
$q = 1.6 \cdot 10^{-19}$ C (elementary charge),
$k = 1.88 \cdot 10^{-23}$ JK^{-1} (Boltzmann's constant),
T absolute temperature.

It is advantageous here when the sensor directly outputs a voltage which is a function of the temperature. The two-pole sensor is of course dependent upon polarity. Whereas, for all sensors, the forward voltage decreases by almost exactly 2 mV/°C, the absolute voltage at each depletion layer differs considerably from sensor to

sensor, and may necessitate additional calibration elements in order for a precise measurement to be made. Above all, it is the temperature-dependent saturation current I_{sat} which is responsible for the negative temperature coefficient (NTC). This current increases dramatically with rising temperature. Here, the silicon's intrinsic conductivity limits this sensor's application to temperature ranges <150 °C.

Sometimes, emitter-coupled transistor pairs are used in a similar manner for temperature measurement. With this form of temperature measurement, the ratios of the collector currents to each other represent a very good reproducible measure for the temperature. Usually, an integrated supplementary circuit is used for the "on-chip" conversion to an analog output voltage.

Zener diodes operated in the reverse direction can also be used as highly practical temperature sensors. Their voltage changes are highly dependent upon the Zener voltage. Here, the option exists of various levels of voltage reduction at Zener voltages <4.7 V, and voltage increases at Zener voltages >4.7 V.

Such sensors are often used for temperature compensation on the chip itself.

12 Semiconductor depletion layers

a b

Fig. 12
a Diode
b Transistor
B Base
C Collector
E Emitter
I_F Conducting-state current
U_F Forward voltage
U_{BE} Voltage between base and emitter

Measuring principles for non-contacting temperature measurement

The radiation emitted by a body is used for the non-contact measurement (pyrometry) of its temperature. This radiation is for the most part in the infrared (IR) range (wavelength: 5 to 20 µm). Strictly speaking, the product of the radiated power and the emission coefficient of the body is measured. The latter is a function of the material, but for materials which are technically of interest (including glass) it is usually around 1, although for reflective and IR-permeable materials (e.g. air, silicon) it is far less than 1.

The measuring point is reproduced on a radiation-sensitive element, which heats up slightly with respect to its environment (typically 0.01 to 0.001 °C). The element's temperature is a measure for the tempera-

ture of the body being measured. A given temperature difference at the object often corresponds to only 1/1,000th of this difference at the measuring point. Nevertheless, the object's temperature can be determined with an accuracy of 0.5 °C.

Bolometer

The Bolometer is a highly sensitive resistance temperature sensor for measuring minute temperature increases (Fig. 13). A further sensor is needed for measuring the temperature of the sensor housing. If this device is to operate efficiently across a wide temperature range, it is necessary though for both these sensors to feature an extremely high degree of synchronism. The Bolometer housing is therefore usually thermostatically controlled (and well-insulated to the housing) so that the primary detecting element (sensor) always operates at the same temperature.

Thermopile sensor

When a very extensive temperature range is concerned, it is more practical for the temperature difference generated by the radiation from the object to be measured using thermocouples. In order to increase the measuring effect, a number of thermocouples are connected in series (thermopile). Such a thermopile sensor (Fig. 14) is inexpensive to manufacture micromechanically. All its "hot" junctions are located on a thermally well insulated thin diaphragm,

13 Prototype of a bolometric sensor array for automotive applications

14 Pyrometric sensor produced using micromechanical techniques, with thermopile pick-off.

and all its "cold" junctions are in contact with the thicker chip rim (heat sink). Typically, the sensor's response time is approx. 20 ms. Using such a so-called "single-pixel sensor", it is an easy matter to determine the windshield's surface temperature so that measures can be taken to prevent misting should the temperature fall below the dew point.

Single-point sensors, imaging sensors

If a number of pixels are combined on a single chip (e.g. 4 x 4) to form an array, this provides the basis for a rough form of image analysis (Fig. 16). However, there should not be too much insensitive surface area between the pixels and the pixels themselves must be thermally well insulated against each other. As all the pixels can optionally respond electrically, the chip has a large number of terminals. On a TO5 housing, for example, the ASIC for signal preamplification and series connection of the signal must be located directly adjacent to the sensor chip. Usually, in the case of thermopile sensors, this ASIC also includes a reference-temperature sensor which measures the pixel's absolute temperature. This permits object temperatures to be measured with an accuracy of approx. ±0.5 K.

In order to reproduce a scene thermally on the sensor array, the array requires an IR optical imaging unit. The very inexpen-

sive curved mirror is usually ruled out due to it needing too much room. Glass lenses are impermeable for IR light, and plastic lenses can only be used for operating temperatures of up to approx. 85 °C max. On the other hand, Si lenses are highly suitable for thermal radiation and up to diameters of approx. 4 mm micromechanical techniques can be used to inexpensively manufacture them in the form of a Fresnel or refraction lens. Fitted in the cover of a TO5 housing, these then also serve to protect the sensor against direct damage (Fig. 15). Even though filling the housing with an inert gas improves the crosstalk between the individual pixels, it also negatively affects their response time.

16 Micromechanical thermopile array

UAE0714-1Y

Fig. 16
1 Silicon chip
2 Pixel
3, 4 Pixel connections

15 Thermal image analysis

UAE0834Y

Fig. 15
a IR imaging sensor
1 Silicon IR lens
2 TO5 housing
3 Terminal posts
4 Sensor chip
5 Evaluation ASIC
α Viewing angle

b Simple IR camera
1 Electronics
2 Lens system
3 Camera's field
 of view
4 Si IR lens
5 Sensor array

Imaging sensors (video)

In particular, imaging sensors are beginning to gain a hold in the motor vehicle using visible light or infrared light. They can be used for passenger-compartment monitoring, but are principally aimed at observation outside the vehicle.

All of these sensors have one objective in view, and that is the simulation of the superior capabilities of the human eye and its mental recognition capabilities (of course, only to a very modest degree at first). These were introduced in large numbers to industrial measurements some time ago – in particular on robot handling equipment. The costs of imaging sensors, and the associated very high-performance processors needed for the interpretation of a scene, are already of interest for applications in the automotive sector.

In contrast to the human eye, common imaging sensors are sensitive in the near IR range (wavelength approximately 1 μm). With appropriate non-visible IR illumination, therefore, all imaginable applications in the vehicle become feasible, including nighttime operation. For applications in the visible light range, the imaging sensors generally have an IR filter in front of them to prevent false recordings due to color.

In future, imaging sensors will be able to play a highly variegated role for the observation of the vehicle's interior (seating position, forward shift in case of a crash, size of seat occupants etc.), and of the vehicle's surroundings (lane guidance, collision prevention, parking and reversing aids, road-sign recognition). A night vision device, that takes the IR image recorded for the road and displays it on a screen in the dashboard, thus offering enhanced driving safety in poor visibility (darkness, fog, etc.) is already in series production.

Imaging sensors are a special case of multisensor structures comprising light-sensitive elements (photodiodes), arranged as pixels in a row or matrix form and receiving the light through a conventional imaging optical system. In the case of the silicon imaging sensors principally available (CCD, Charge-Coupled Devices, Figs. 1 and 2), free charge carriers are generated proportionally to the intensity and exposure time by the light arriving through a transparent electrode; they are then collected in a "potential pot" (Si-SiO$_2$ boundary layer). Further electrodes move these charges into an opaque zone which are further transported in analog shift registers (bucket chain principle) line-by-line into an output register which is read serially at a high clock rate.

1 CCD principle

Fig. 1
1 Photodiode
2 Light
3 Storage electrode
4 Shift gate
5 Transfer electrode
6 Optical masking

While CCD sensors can only be used in motor vehicles to a very limited degree because of their limited brightness/darkness dynamics (50 dB), their read time and their temperature range (< 50 °C), newer, smart, imaging sensors based on CMOS technology appear to be fully suitable for use in the motor vehicle. The logarithmic brightness/signal curve possible here, corresponds to the human eye and has a dynamic response of 120 dB. This makes aperture control, for instance, unnecessary and supplies a constant contrast resolution across the entire brightness range. These sensors permit a digitally controlled, random access to the individual pixels in a matrix sensor (Fig. 3) with simultaneous higher sensitivity (higher read rate).

The first steps in preprocessing the signals on the imaging-sensor chip have already been implemented (intelligent imaging sensor). A resolution of 1,280 × 1,024 pixels with typical individual dimensions of 6 μm × 6 μm is already possible in sensors currently on the market.

3 CMOS imaging sensors with random pixel access

Fig. 3
Pixel processors at each pixel in the active pixel matrix ensure logarithmic signal compression. Reading is random access and non-destructive to the signal, transmission is loss-free.

2 CCD imaging-sensor structure

Fig. 2
1 Column clock pulse A_1/A_2
2 Photosensors
3 CCD array
4 Line clock pulse B_1/B_2
5 Output register
6 Video output

Sensor types

Engine-speed sensors

Application

Engine-speed sensors are used in engine-management systems for

- Measuring the engine speed and
- Determining the crankshaft position (position of the pistons)

The engine speed is calculated from the interval between the speed sensor's signals.

Inductive speed sensors

Design and operating principle

The sensor is mounted directly opposite a ferromagnetic pulse wheel (Fig. 1, Item 7) from which it is separated by a narrow air gap. It has a soft-iron core (pole pin, 4), which is enclosed by a winding (5). The pole pin is also connected to a permanent magnet (1), and a magnetic field extends through the pole pin and into the pulse wheel. The level of the magnetic flux through the coil depends on whether the sensor is opposite a pulse-wheel tooth or space. Whereas the magnet's leakage flux is concentrated by a tooth, and leads to an increase in the magnetic flux through the coil, it is attenuated by a space. When the pulse wheel rotates, these changes in magnetic flux induce a sinusoidal output voltage in the coil which is proportional to the rate of change of the flux and thus the

engine speed (Fig. 2). The amplitude of the alternating voltage increases strongly along with increasing pulse-wheel speed (several mV to >100 V). At least about 30 rpm are needed to generate an adequate amplitude.

The number of teeth on the pulse wheel depends on the particular application. In Motronic systems, a 60-pitch pulse wheel is normally used, although 2 teeth are omitted (Fig. 1, Item 7) so that the pulse wheel has 60 - 2 = 58 teeth. The space where the missing teeth would be situated is allocated to a defined crankshaft position and serves as a reference mark for synchronizing the control unit.

The geometries of the pulse-wheel teeth and the pole pin must be matched to each other. An evaluation circuit in the control unit converts the sinusoidal voltage, which is characterized by strongly varying amplitudes, into a constant-amplitude square-wave voltage for evaluation in the control unit microcontroller.

Active speed sensors

Active speed sensors operate according to the magnetostatic principle. The amplitude of the output signal is not dependent on the rotational speed. This makes it possible for very low speeds to be sensed (quasistatic speed sensing).

Fig. 1

1 Permanent magnet
2 Sensor housing
3 Crankcase
4 Pole pin
5 Winding
6 Air gap
7 Pulse wheel with
 reference mark

Fig. 2

1 Tooth
2 Tooth space
3 Reference mark

1 Inductive engine-speed sensor (design)

2 Signal from an inductive engine-speed sensor

Output voltage

Time

Differential Hall sensor

A voltage U_H proportional to the magnetic field (Hall voltage) can be picked off horizontally to the current direction at a current-carrying plate which is permeated vertically by a magnetic induction B. In a differential Hall sensor, the magnetic field is generated by a permanent magnet (Fig. 3, Item 1). Two Hall sensor elements (2 and 3) are situated between the magnet and the pulse wheel (4). The magnetic flux by which these are permeated depends on whether the speed sensor is opposite a tooth or a space. Determining the difference between the signals from two sensors reduces the magnetic interference signals and achieves and improved signal/noise ratio. The edges of the sensor signal can be processed without digitization directly in the control unit.

Multipole wheels are used instead of the ferromagnetic pulse wheel. Here, a magnetizable plastic is attached to a nonmagnetic metallic carrier and alternately magnetized. These north and south poles adopt the function formerly performed by the teeth of the pulse wheel.

AMR sensors

The electrical resistance of magnetoresistive material (AMR, Anisotrop Magneto Resistive) is anisotropic, i.e. it depends on the direction of the magnetic field to which it is exposed. This property is utilized in an AMR sensor. The sensor is located between a magnet and a pulse wheel. The field lines change direction when the pulse wheel rotates (Fig. 4). This generates a sinusoidal voltage, which is amplified in an evaluation circuit in the sensor and converted into a square-wave signal.

GMR sensors

The use of GMR technology (Giant Magneto Resistive) is a further development of active speed sensors. Because of the higher sensitivity by comparison with the AMR sensors, larger air gaps are possible, which makes applications in difficult circumstances possible. The higher sensitivity also results in less background noise from the signal edges.

All two-wire current interfaces previously used for Hall speed sensors may also be used on GMR sensors.

3 Principle of differential Hall sensor

4 Principle of speed sensing with an AMR sensor

Time

Hall phase sensors

Application

The engine's camshaft rotates at half the crankshaft speed. Taking a given piston on its way to TDC, the camshaft's rotational position is an indication as to whether the piston is in the compression or exhaust stroke. The phase sensor on the camshaft provides the control unit with this information. This is required, for example, for ignition systems with single-spark ignition coils and for sequential fuel injection (SEFI).

Design and operating principle

Hall rod sensors

Hall rod sensors (Fig. 1a) utilize the Hall effect: a rotor of ferromagnetic material (7, pulse wheel with teeth, segments or aperture plate) rotates along with the camshaft. The Hall IC (6) is located between the pulse wheel and a permanent magnet (5), which generates a magnetic field strength perpendicular to the Hall element.

If one of the pulse-wheel teeth (Z) now passes the current-carrying sensor element (semiconductor wafer), it changes the magnetic-field strength perpendicular to the Hall element. This results in a voltage signal (Hall voltage) which is independent of the relative speed between sensor and pulse wheel. The evaluation electronics integrated in the sensor's Hall IC condition the signal and output it in the form of a square-wave signal (Fig. 1b).

1 Hall rod sensor (design)

Fig. 1
a Positioning of sensor and single-track pulse wheel
b Output signal characteristic U_A

1 Electrical connection (plug)
2 Sensor housing
3 Crankcase
4 Sealing ring
5 Permanent magnet
6 Hall IC
7 Pulse wheel with tooth/segment (Z) and space (L)
a Air gap
φ Angle of rotation

Fig. 2
TIM = Twist Intensive Mounting (i.e. the sensor may be twisted around its axis any number of times, without losing accuracy. Important for minimizing type diversity).
TPO = True Power On (i.e. the sensor detects directly on switching on whether it is located over a tooth or a space. Important for short synchronization times between crankshaft and camshaft signals).

2 Generations of camshaft sensors

	PG-1	PG-3-3	PG-3-5	PG-3-8
TIM	no	no	yes	yes
TPO	no	yes	no	yes
Accuracy	low	medium	medium	high

Technological progress

PG-1 PG-3-3 PG-3-5 PG-3-8

Speed sensors for transmission control

Application

Transmission speed sensors determine the shaft speeds in AT, ASG, DKG and CVT transmissions. These are the turbine and output shaft speeds on AT transmissions with a hydrodynamic torque converter, the speed of the primary and secondary pulleys in CVT transmissions and the speeds of the two input shafts and the driven shaft of double-clutch transmissions. The speed of the power take-up element is also detected for high dynamic demands on power take-up control.

Direction of rotation detection can also be necessary in high-end transmissions in order to optimize clutch management and for rollback prevention.

Both stand-alone sensors and sensors built into electronics modules which project into the transmission from the outside or are mounted internally are used.

Requirements

Transmission speed sensors are exposed to very severe operating stresses because of

- Extreme ambient temperatures between -40 and +150 °C
- An aggressive operating environment caused by the transmission oil, also known as ATF (contains special additives for transmissions and has a low condensate content)
- High mechanical stress with vibrational accelerations up to 30 g
- Abraded metallic materials and a build-up of particles in the transmission

The package of the electronics used in the sensor is hence subject to high demands as a consequence of these severe stresses. A service life of more than 15 years in transmission oil has been achieved by using a suitable oil-resistant package.

Because of the very compact design of the transmissions, the mechanical customer interface cannot generally be covered by standard geometries. This means that custom sensor types are needed for each transmission model. Module-integrated types differ in respect of insertion length, direction of recording and mounting flange (Fig. 1). A further variation found in stand-alone sensors is the location of the mounting socket and the design of the connector.

Hall ASICs (Application Specific Integrated Circuit) with varying complexities of the evaluation algorithms are used to cover the complete range of the functional requirements (Fig. 2).

If a ferromagnetic trigger wheel or a trigger area (toothed, stamped or embossed) is available on the rotating transmission component, the magnetic field required to operate the Hall sensor is generated by a back-bias magnet.

1 Sensor types

a

↓1

b

c

Fig. 1
a Bottom read
b Side read
c Slant read

1 Direction of detection

SAE1081Y

It is mounted in the sensor immediately behind the ASIC.

Compact transmissions increasingly require speed measurement over greater distances (magnetic air gap) by rotating, nonmagnetic components or through a housing wall. Multipole wheels (magnetized rings) are used for these applications, the back-bias magnet is omitted from the sensor.

Design

The Hall ASICs used in the transmission speed sensors are mounted in a retainer with or without a back-bias magnet – depending on the magnetic interface – with the electrical contact being made by a welding process, then inserted in a housing, potted with epoxy resin or – in the case of types mounted on the outside of the transmission – have an oil-tight enclosure sprayed onto them (Fig. 3).

The sensor has a two-wire interface, which combines optimum diagnostic options with a minimum number of electrical connections. The two terminals both supply the Hall IC and transmit the signal.

Operating principle

Transmission speed sensors use the differential Hall effect. The difference between the Hall voltages from two Hall plates on the ASIC is determined. The majority of the similar interferences can be compensated for in this way. The difference signal is first amplified on some ASIC types, then converted into a digital signal by trigger algorithms of varying complexities. This constitutes the controlled variable for modulating the output current by means of a power source. A digital signal with two current levels (typically 7 mA for the

3 Hall sensor with two-wire current interface

2 Requirements complexity

low level and 14 mA for the high level) is obtained, the modulation frequency of which corresponds to the frequency of tooth change on the trigger wheel and thus represents the speed of rotation. The sensor signal is evaluated in the electronic control unit by means of a measuring shunt R_M, which converts the sensor current I_S into the signal voltage U_{RM}.

On principle, it does not matter for the operation of a differential Hall ASICs whether the sensor is operated with a steel trigger wheel or on a multipole wheel (Figs. 4a and 4b).

4 Transmission speed sensor operating principle

Some transmission controls include functions that require detection of a standstill. For this application, the sensor must have as great an immunity as possible against vibration-induced variations in the air gap and the torsional vibration of the trigger wheel. This property of the sensor, known as immunity against vibration, can only be implemented to a very limited degree in differential Hall sensors with only two Hall plates, by the use of adaptive trigger thresholds, for example. Two phase-shifted differential signals become available with the use of a third Hall plate. This makes both the detection of the direction of rotation (Figs. 4c to f) and additional function algorithms for increasing immunity against vibration possible.

The typical values for "value" and "high feature" sensors differ in the air gap range achievable (distance between the sensitive area on the sensor and the trigger wheel), the signal frequency range and the additional functionalities implemented (Table 1).

The complexity of transmission type, mounting space restrictions including all the derived marginal design conditions and the functional requirements leads to application-specific solutions in most applications. These are characterized by a combination of ASIC, package design and sensor mechanical and magnetic interface suitable for the system requirements.

Fig. 4
a Configuration with trigger wheel
b Configuration with multipole wheel
c Sensor signal (difference signal between Hall plates R and C)
d Sensor signal (difference signal between Hall plates C and L)
e Output signal for clockwise rotation
f Output signal for counterclockwise rotation

1 Trigger wheel
2 Hall plates L and R (C optional for direction of rotation detection)
3 Permanent magnet (back-bias)
4 Multipole wheel
5 Phase shift dependent on the direction of rotation

1 Typical characteristics

Type	Value	High feature
Maximum air gap		
at trigger wheel	2.5 mm	3.5 mm
at pole wheel	5 mm	7 mm
Signal frequency	0 to 8 kHz	0 to 12 kHz
Direction of rotation detection	No	Yes
Trigger wheel vibration	–	± 1.5°

Table 1

Wheel-speed sensors

Application

Wheel-speed sensors are used to measure the rotational speed of the vehicle wheels (wheel speed). The speed signals are transmitted via cables to the ABS, TCS or ESP control unit of the vehicle which controls the braking force individually at each wheel. This control loop prevents the wheels from locking up (with ABS) or from spinning (with TCS or ESP) so that the vehicle's stability and steerability are maintained.

Navigation systems also use the wheel-speed signals to calculate the distance traveled (e.g. in tunnels or if satellite signals are unavailable).

Design and operating principle

The signals for the wheel-speed sensor are generated by a steel pulse generator that is fixed to the wheel hub (for passive sensors) or by a multipole magnetic pulse generator (for active sensors). This pulse generator has the same rotational speed as the wheel and moves past the sensitive area of the sensor head without touching it. The sensor "reads" without direct contact via an air gap of up to 2 mm (Fig. 2).

The air gap (with strict tolerances) ensures interference-free signal acquisition. Possible interference caused for instance by oscillation patterns in the vicinity of the brakes, vibrations, temperature, moisture, installation conditions at the wheel, etc. is therefore eliminated.

Since 1998 active wheel-speed sensors have been used almost exclusively with new developments instead of passive (inductive) wheel-speed sensors.

Passive (inductive) wheel-speed sensor

A passive (inductive) speed sensor consists of a permanent magnet (Fig. 2, 1) with a soft-magnetic pole pin (3) connected to it, which is inserted into a coil (2) with several thousand windings. This setup generates a constant magnetic field.

The pole pin is installed directly above the pulse wheel (4), a gear wheel attached to the wheel hub. As the pulse wheel turns, the continuously alternating sequence of teeth and spaces induces corresponding fluctuations in the constant magnetic field. This changes the magnetic flux through the pole pin and therefore also through the coil winding. These fluctuations in the magnetic field induce an alternating voltage in the coil suitable for monitoring at the ends of its winding.

The frequency and amplitude of this alternating voltage are proportional to wheel speed (Fig. 3) and when the wheel is not rotating, the induced voltage is zero.

Tooth shape, air gap, rate of voltage rise, and the control unit input sensitivity define the smallest still measurable vehicle speed and thus, for ABS applications, the minimum response sensitivity and switching speed.

1 Passive (inductive) wheel-speed sensors

a

b

SAE0974Y

2 Figure illustrating the principle of the passive wheel-speed sensor

SAE0975Y

Fig. 2
1 Permanent magnet
2 Solenoid coil
3 Pole pin
4 Steel pulse wheel
5 Magnetic field lines

Various pole-pin configurations and installation options are available to adapt the system to the different installation conditions encountered with various wheels. The most common variants are the chisel-type pole pin (Fig. 1a, also called a flat pole pin) and the rhombus-type pole pin (Fig. 1b, also called a lozenge-shaped pole pin). Both pole-pin designs necessitate precise alignment to the pulse wheel during installation.

Active wheel-speed sensor
Sensor elements

Active wheel-speed sensors are used almost exclusively in today's modern brake systems (Fig. 4). These sensors usually consist of a hermetic, plastic-cast silicon IC that sits in the sensor head.

In addition to magnetoresistive ICs (the electrical resistance changes as the magnetic field changes) Bosch now uses Hall sensor elements almost exclusively. These sensors react to the smallest changes in the magnetic field and therefore allow greater air gaps compared to passive wheel-speed sensors.

3 Signal output voltage of passive wheel-speed sensor

SAE0976E

4 Active wheel-speed sensors

SAE0977Y

Fig. 3
a Passive wheel-speed sensor with pulse wheel
b Sensor signal at constant wheel speed
c Sensor signal at increasing wheel speed

Pulse wheels

A multipole ring is used as a pulse wheel for active wheel-speed sensors. The multipole ring consists of alternately magnetized plastic elements that are arranged in the shape of a ring on a nonmagnetic metal carrier (Fig. 6 and Fig. 7a). These north and south poles adopt the function formerly performed by the teeth of the pulse wheel. The sensor IC is exposed to the continuously changing fields generated by these magnets. The magnetic flux through the IC therefore changes continuously as the multipole ring turns.

A steel pulse wheel can also be used instead of the multipole ring. In this case a magnet is mounted on the Hall IC that generates a constant magnetic field (Fig. 7b). As the pulse wheel turns, the continuously alternating sequence of teeth and spaces induces corresponding fluctuations in the constant magnetic field. The measuring principle, signal processing and IC are otherwise identical to the sensor without a magnet.

Characteristics

A typical feature of the active wheel-speed sensor is the integration of a Hall measuring element, signal amplifier and signal conditioning in an IC (Fig. 8). The wheel-speed data is transferred as an impressed current in the form of square-wave pulses (Fig. 9). The frequency of the pulses is proportional to the wheel speed and the speed can be detected until the wheel is practically stationary (0.1 km/h).

The supply voltage is between 4.5 and 20 volts. The square-wave output signal level is 7 mA (low) and 14 mA (high).

Fig. 7
a Hall IC with multipole pulse generator
b Hall IC with steel pulse wheel and magnet in sensor

1 Sensor element
2 Multipole ring
3 Magnet
4 Steel pulse wheel

Fig. 5
1 Wheel hub
2 Ball bearing
3 Multipole ring
4 Wheel-speed sensor

Fig. 6
1 Sensor element
2 Multipole ring with alternating north and south magnetization

5 Explosion diagram with multipole pulse generator

6 Sectional drawing of active wheel-speed sensor

7 Diagram illustrating principle for measuring speed

This type of data transmission using digital signals is less sensitive to interference than the signals from passive inductive-type sensors. The sensor is connected to the control unit by a two-conductor wire.

Compact dimensions combine with low weight to make the active wheel-speed sensor suitable for installation on and even within the vehicle's wheel-bearing assemblies (Fig. 10). Various standard sensor head shapes are suitable for this.

Digital signal conditioning makes it possible to transfer coded additional information using a pulse-width-modulated output signal (Fig. 11):
- Direction of wheel rotation recognition: This is especially significant for the "hill hold control" feature, which relies on selective braking to prevent the vehicle from rolling backwards during hill starts. The direction of rotation recognition is also used in vehicle navigation systems.
- Standstill recognition: This information can also be evaluated by the "hill hold control" function. The information is also used for self-diagnosis.
- Signal quality of the sensor: Information about the signal quality of the sensor can be relayed in the signal. If a fault occurs the driver can be advised that service is required.

8 Block diagram of Hall IC

9 Signal conversion in Hall IC

10 Wheel bearing with wheel-speed sensor

11 Coded information transfer with pulse-width-modulated signals

Fig. 9
a Raw signal
b Output signal

O_{S1} Upper switching threshold
U_{S1} Lower switching threshold

Fig. 10
1 Wheel-speed sensor

Fig. 11
a Speed signal when reversing
b Speed signal when driving forwards
c Signal when vehicle is stationary
d Signal quality of sensor, self-diagnosis

Micromechanical yaw-rate sensors

Applications

In vehicles with electronic stability program (ESP), the rotation of the vehicle about its vertical axis is registered by micromechanical yaw-rate (or yaw-speed) sensors (also known as gyroscopes) and applied for vehicle-dynamics control. This takes place during normal cornering, for instance, but also when the vehicle breaks away or goes into a skid. These sensors are reasonably priced as well as being very compact. They have forced out the conventional high-precision mechanical sensors.

A further area of application is rollover detection in airbag electronic control units for triggering restraints (e.g. sidebags/window airbags, rollover bar) in the event of the vehicle rolling over. MM2 yaw-rate sensors have been specially developed for this area. These are particularly suited for the detection of rotary motion around the longitudinal axis because of their optimized sensing direction. They permit a very compact design and space-saving installation in airbag control units, installed along the longitudinal axis of the vehicle. Packaging the sensor element and the evaluation electronics together in a standard IC housing contributes to the cost reductions in this area.

MM1 micromechanical yaw-rate sensor

A mixed form of technology is applied in order to achieve the high accuracies needed for vehicle-dynamics systems. That is, two somewhat thicker oscillating elements (mass plates) which have been machined from a wafer using bulk micromechanics oscillate in counter-phase to their resonant frequency which is defined by their mass and their coupling-spring stiffness (>2 kHz). On each of these oscillating elements, there is a miniature, surface-type micromechanical capacitive acceleration sensor. When the sensor chip

1 Structure of the MM1 yaw-rate sensor

200 μm

Fig. 1
1 Retaining/guide spring
2 Part of the oscillating element
3 Coriolis acceleration sensor

2 MM1 micromechanical yaw-rate sensor

0 2 mm

Fig. 2
1 Frequency-determining coupling spring
2 Permanent magnet
3 Direction of oscillation
4 Oscillating element
5 Coriolis acceleration sensor
6 Direction of Coriolis acceleration
7 Retaining/guide spring
Ω Yaw rate
v Oscillation velocity
B Permanent-magnet field

rotates about its vertical axis at yaw rate Ω, these register the Coriolis acceleration in the wafer plane vertical to the direction of oscillation (Figs. 1 and 2). These accelerations are proportional to the product of yaw rate and the oscillation velocity which is maintained electronically at a constant value.

To drive the sensor, all that is required is a simple, current-carrying conductor track on each oscillating element. In the permanent-magnet field B vertical to the chip surface, this oscillating element is subjected to an electrodynamic (Lorentz) force. Using a further, simple printed conductor (which saves on chip surface), the same magnetic field is used to directly measure the oscillation velocity by inductive means. The different physical construction of drive system and sensor system serves to avoid undesirable coupling between the two sections. In order to suppress unwanted external acceleration effects (common-mode signal), the opposing sensor signals are subtracted from each other. The external acceleration effects can be measured by applying summation. The high-precision micromechanical construction helps to suppress the effects of high oscillatory acceleration which is several factors of 10 higher than the low-level Coriolis acceleration (cross sensitivity far

below 40 dB). Here, the drive and measurement systems are rigorously decoupled from each other.

MM2 micromechanical yaw-rate sensor
If the silicon yaw-rate sensor is manufactured completely in surface micromechanics (SMM) and, at the same time, the magnetic drive and control system is replaced by an electrostatic system, the decoupling of drive and measuring system does not need to be so thorough. Comb-like structures (Figs. 3 and 4) electrostatically force a centrally mounted rotary oscillator to oscillate. The amplitude of these oscillations is held constant by means of a similar capacitive pick-off. Coriolis forces result at

3 MM2 yaw-rate sensor: structure

50 µm

Fig. 3
1 Comb-like structure
2 Rotary oscillator

4 MM2 surface-micromechanical yaw-rate sensor

0,5 mm

C_{Drv1} C_{Drv2} 1
$C_{Drv\,Det1}$
$C_{Drv\,Det2}$
v
2
C_{Det1} C_{Det2}
Ω
$-F_C$ $+F_C$
3

Fig. 4
1 Comb-like structure
2 Rotary oscillator
3 Measuring axis
C_{Drv} Drive electrodes
C_{Det} Capacitive pick-off
F_C Coriolis force
v Oscillation velocity
Ω $= \Delta C_{Det}$, measured yaw rate

the same time in an out-of-plane tilting movement, the amplitude of which is proportional to the yaw rate Ω, and which is detected capacitively by the electrodes underneath the oscillator. To avoid excessive damping of this movement, the sensor must be operated in a vacuum. Although the chip's small size and the somewhat simpler production process result in considerable cost reductions, this miniaturization is at the expense of reductions in the measuring effect, which in any case is not very pronounced, and therefore of the achievable precision. It also places more severe demands on the electronics. The system's high flexural stability, and mounting in the axis of gravity, serve to mechanically suppress the effects of unwanted external acceleration from the side.

DRS MM3x sensor cluster

A new generation of micromechanical elements are used in the DRS MM3x sensor cluster. These can measure and digitally process angular velocities and accelerations. They create a modular concept for hardware and software based on circuit board technology with many new safety features which lead to a versatile and reliable solution for many applications in the vehicle.

Application

The ESP system, the connection to further chassis comfort systems and the development of advanced driving stability systems called for inertia signals with high requirements, in particular in respect of signal quality and robustness, additional measuring axes and high reliability. This is why Bosch has developed a third generation, the versatile yet inexpensive DRS MM3.x sensor cluster, to meet the requirements of functions such as hill hold control (HHC), automated parking brake (APB),

navigation (Travel Pilot), adaptive cruise control (ACC), roll over mitigation (ROM), electronic active steering (EAS), active suspension control (ASC), Steer-by-Wire.

DRS MM3.7k is the base variant of the MM3 generation for ESP applications. It comprises a yaw-rate sensor and an integrated lateral acceleration module.

Operating principle

The new micromechanical measuring element for measuring yaw rate belongs to the familiar group of vibrating gyroscopes operating according to the Coriolis principle (CVG = Coriolis Vibrating Gyros). It consists of an inverse tuning fork with two linear vibration modes perpendicular to one another, the drive circuit and the evaluator circuit. Drive and evaluation take place electrostatically in a comb-like structure. The Coriolis acceleration is measured electrostatically using electrodes that extend into each other. The measuring element consists of two masses connected by a coupling spring. The resonant frequency the same for the two vibration modes. It is typically around 15 kHz and is thus outside the usual interference spectrum in the vehicle and is hence resistant to interfering vibration. The evaluation circuit (ASIC) and the micromechanical measuring element are contained in a prefabricated housing with 20 terminals (Premold 20).

The acceleration module is comparable in form with the yaw-rate sensor module and consists of a micromechanical measuring element, an electronic evaluating circuit and a housing with 12 terminals (Premold12). The spring-mass structure is deflected in its sensitive axis by external accelerations and is evaluated by means of a differential capacitor in a comb-like form.

Piezoelectric "tuning-fork" yaw-rate sensor

Application

In order that it can use the digital road map stored on the CD-ROM to calculate the distance driven, the computer in the vehicle's navigation system needs information on the vehicle's movements (composite navigation).

When the vehicle is cornering, the navigation system's yaw-rate sensor records the vehicle's rotation about its vertical axis, thus permitting determination of the direction of travel. The vehicle's position can hence be calculated in conjunction with the determination of the distance traveled from the tachometer or wheel sensor signals by "composite positioning".

Initially in navigation systems, the vehicle's direction was determined from the earth's magnetic field by a magnetic field sensor. However, these could easily be influenced by the magnetic interference in the vehicle. As oscillation gyroscopes are based on the measurement of mass forces, they are insensitive to magnetic interference.

Design

The piezoelectric "tuning fork" yaw-rate sensor comprises a steel element shaped like a tuning fork. This incorporates four piezo elements (two above, two below, Fig. 1) and the sensor electronics. The tuning forks are about 15 mm long.

Operating principle

When voltage is applied, the bottom piezo elements start to oscillate and excite the upper section of the "tuning fork", together with its upper piezo elements, which then starts counter-phase oscillation. The frequency is approximately 2 kHz.

Straight-ahead driving

With the vehicle being driven in a straight line there are no Coriolis forces acting on the tuning fork, and since the upper piezo elements always oscillate in counter-phase and are only sensitive perpendicularly to the direction of oscillation, they do not generate a voltage.

Cornering

The rotational movement around the vehicle's vertical axis as it travels around a curve causes the upper portion of the tuning fork to leave the oscillatory plane so that an alternating voltage is generated in the upper piezo elements which is transferred to the navigation computer by an electronic circuit in the sensor housing. The voltage-signal amplitude is a function of both the yaw rate and the oscillation velocity. Its sign depends on the direction (left or right) taken by the curve.

1 Piezoelectric "tuning-fork" yaw-rate sensor

UAE1068Y

Fig. 1
A Sensing section of the oscillating element
B Stimulating section of the oscillating element
1 Oscillating element
2 Acceleration sensor
3 Actuator (piezoelectric element for vibration excitation)
4 Regulator for constant vibration excitation
5 Charge amplifier
6 Multiplication (demodulation)
7 Low pass
8 Vibration excitation
U_A Output voltage (proportional to the yaw rate)
Ω Yaw rate

Micromechanical pressure sensors

Application

Pressure is a non-directional force acting in all directions which occurs in gases and liquids. Micromechanical pressure sensors detect the pressure of various media in the motor vehicle, e.g.:

- Intake-manifold pressure, e.g. for load sensing in engine-management systems
- Boost pressure for boost-pressure control
- Ambient pressure for taking into account air density, e.g. in boost-pressure control
- Oil pressure to take engine load into account in the service display
- Fuel pressure for monitoring the level of fuel-filter contamination

Micromechanical pressure sensors determine the absolute pressure of liquids and gases by measuring the pressure differential in relation to a reference vacuum.

Version with the reference vacuum on the component side

Design

The measuring cell of a micromechanical pressure sensor consists of a silicon chip (Fig. 1a, Item 2), in which a thin diaphragm is micromechanically etched (1). Four strain-gage resistors (R_1, R_2), are diffused into the diaphragm. The electrical resistance in these resistors changes in response to mechanical elongation. The measuring cell is surrounded and sealed on its component side by a cap (Fig. 2, Item 6), which encloses the reference vacuum underneath.

A temperature sensor (Fig. 3, Item 1), the signals of which can be evaluated independently, can also be integrated in the pressure-sensor housing.

Operating principle

The diaphragm of the sensor cell is deflected to varying degrees depending

on the external pressure acting on it (the center of the diaphragm is deflected by 10 to 1,000 µm). The four strain-gage resistors on the diaphragm change their electrical resistance as a function of the mechanical stress resulting from the applied pressure (piezoresistive effect).

The four measuring shunts are arranged on the silicon chip so that when the diaphragm is deformed, the resistance of two of them increases and that of the other two decreases. These measuring shunts are arranged in a Wheatstone bridge circuit (Fig. 1b) and a change in their resistance values leads to a change in the ratio

1 Pressure-sensor measuring element with reference vacuum on the components side

2 Measuring element of pressure sensor with cap and reference vacuum on the components side (design)

Fig. 1
a Sectional drawing
b Bridge circuit

1 Diaphragm
2 Silicon chip
3 Reference vacuum
4 Glass (Pyrex)
p Measured pressure
U_0 Supply voltage
U_M Measurement voltage
R_1 Strain-gage resistor (compressed)
R_2 Strain-gage resistor (elongated)

Fig. 2
1, 3 Electrical terminals with glass-enclosed lead-in
2 Reference vacuum
4 Measuring element (chip) with evaluation electronics
5 Glass base
6 Cap
7 Supply for measured pressure p

of the voltages across them. This leads to a change in the measurement voltage U_M. This as yet unamplified voltage is therefore a measure of the pressure applied to the diaphragm.

The measurement voltage is higher with a bridge circuit than would be the case if an individual resistor were used. The Wheatstone bridge circuit thus permits a higher sensor sensitivity.

The signal-conditioning electronic circuitry is integrated on the chip. Its function is to amplify the bridge voltage, compensate for temperature influences, and linearize the pressure curve. The output voltage is in the range of 0 to 5 V and is supplied via electrical connections to the engine control unit (Fig. 3, Item 5). The control unit uses this output voltage to calculate the pressure.

Version with reference vacuum in a special chamber

Design

A pressure sensor with the reference vacuum in a special chamber (Fig. 4) for use as an intake-manifold or boost-pressure sensor is easier to install than a sensor with the reference vacuum on the component side: A silicon chip (Fig. 5, Item 6) with etched diaphragm and four strain-gage resistors in a bridge circuit is located – like the pressure sensor with cap and reference vacuum on the component side – as a measuring cell on a glass base (3). In contrast to the sensor with the reference vacuum on the component side, there is no passage in the glass base through which the measured pressure can be applied to the sensor element. Instead, pressure is

3 Micromechanical pressure sensor with reference vacuum on the components side (design)

1 cm

UAE0722Y

4 Micromechanical pressure sensor with reference vacuum in a special chamber (design)

1 cm

UMK1645-1Y

5 Measuring element of pressure sensor with reference vacuum in a special chamber (design)

p

UMK1644-1Y

Fig. 3
1 Temperature sensor (NTC)
2 Housing base
3 Manifold wall
4 Sealing rings
5 Electrical connection (plug)
6 Housing cover
7 Measuring cell

Fig. 4
1 Manifold wall
2 Housing
3 Sealing ring
4 Temperature sensor (NTC)
5 Electrical connection (plug)
6 Housing cover
7 Measuring cell

Fig. 5
1 Protective gel
2 Gel frame
3 Glass base
4 Ceramic hybrid
5 Chamber with reference vacuum
6 Measuring cell (chip) with evaluation electronics
7 Bonded connection
p Measured pressure

applied to the silicon chip from the side on which the evaluation electronics is situated. This means that a special gel must be used on this side of the sensor to protect it against environmental influences. The reference vacuum (5) is enclosed in the cavity (special chamber) between the silicon chip and the glass base. The complete measuring element is mounted on a ceramic hybrid (4), which incorporates the soldering surfaces for electrical contacting inside the sensor.

A temperature sensor can also be incorporated in the pressure-sensor housing. It protrudes into the air current, and can therefore respond to temperature changes with a minimum of delay (Fig. 4, Item 4).

Operating principle
The operating principle, and with it the signal conditioning and signal amplification together with the characteristic curve, correspond to that used in the pressure sensor with cap and reference vacuum on the component side. The only difference is that the measuring element's diaphragm is deformed in the opposite direction and therefore the strain-gage resistors are "bent" in the other direction.

6 High-pressure sensor

2 cm

1
2
3
4
5

p

UMK1576-1Y

High-pressure sensors

Application
High-pressure sensors are used in a motor vehicle to measure fuel pressure and brake-fluid pressure, e.g:
• Rail-pressure sensor for gasoline direct injection (pressure up to 200 bar)
• Rail-pressure sensor for common-rail diesel-injection system (pressure up to 2,000 bar)
• Brake-fluid pressure sensor in the hydraulic modulator of the electronic stability program (pressure up to 350 bar)

Design and operating principle
High-pressure sensors operate according to the same principle as micromechanical pressure sensors. The core of the sensor is a steel diaphragm, onto which strain-gage resistors have been vapor-deposited in the form of a bridge circuit (Fig. 6, Item 3). The sensor's measuring range is dependent on the thickness of the diaphragm (thicker diaphragms for higher pressures, thinner diaphragms for lower pressures). As soon as the pressure to be measured is applied to one side of the diaphragm via the pressure connection (4), the strain-gage resistors change their resistance values as a result of the deflection of the diaphragm. The output voltage generated by the bridge circuit is proportional to the applied pressure. This voltage is transmitted via connecting lines (bonding wires) to an evaluation circuit (2) in the sensor. The evaluation circuit amplifies the bridge signal to 0 to 5 V and transmits it to the control unit, which calculates the pressure with the aid of a characteristic curve.

Temperature sensors

Application

Engine-temperature sensor
This is installed in the coolant circuit
(Fig. 1). The engine management uses
its signal when calculating the engine
temperature (measuring range
– 40 to +130 °C).

Air-temperature sensor
This sensor is installed in the air-intake
tract. Together with the signal from the
boost-pressure sensor, its signal is applied
in calculating the intake-air mass. Apart
from this, setpoint values for the various
control loops (e.g. EGR, boost-pressure
control) can be adapted to the air tempera-
ture (measuring range –40 to +120 °C).

Engine-oil temperature sensor
The signal from this sensor is used in cal-
culating the service interval (measuring
range –40 to +170 °C).

Fuel-temperature sensor
Is incorporated in the low-pressure stage
of the diesel fuel circuit. The fuel tempera-
ture is used in calculating precisely the
injected fuel quantity (measuring range
–40 to +120 °C).

Exhaust-gas temperature sensor
This sensor is mounted on the exhaust-gas
system at points which are particularly
critical regarding temperature. It is ap-
plied in the closed-loop control of the
systems used for exhaust-gas treatment.
A platinum measuring shunt is usually
used (measuring range –40 to +1,000 °C).

Design and operating principle
Depending upon the particular application,
a wide variety of temperature sensor types
are available. A temperature-dependent
semiconductor measuring shunt is fitted
inside a housing. This resistor is usually
of the NTC (Negative Temperature Coeffi-
cient, Fig. 2) type. Less often a PTC (Posi-
tive Temperature Coefficient) type is used.
With NTC, there is a sharp drop in resis-
tance when the temperature rises, and
with PTC there is a sharp increase.

The measuring shunt is part of a voltage-
divider circuit to which 5 V is applied.
The voltage measured across the measur-
ing shunt is therefore temperature-depen-
dent. It is supplied through an analog-digi-
tal (A/D) converter and is a measure of the
temperature at the sensor. A characteristic
curve is stored in the control unit which
allocates a specific temperature to every
resistance or output voltage value.

1 Coolant-temperature sensor

1 cm

2 NTC temperature sensor: characteristic curve

Fig. 1
1 Electrical
 connection
2 Housing
3 Sealing ring
4 Thread
5 Measuring shunt
6 Coolant

Accelerator-pedal sensors

In Motronic systems with electronic throttle control (ETC), the accelerator-pedal sensor records the travel or the angular position of the accelerator pedal. For this purpose, potentiometers are used in addition to proximity-type sensors. The accelerator-pedal sensor is incorporated with the accelerator pedal in the accelerator-pedal module. These ready-to-install units make adjustments on the vehicle a thing of the past.

Potentiometric accelerator-pedal sensor
The engine control unit receives the measured value picked off at the potentiometer wiper as a voltage. The control unit uses a stored sensor curve to convert this voltage into the relative pedal travel or the angular position of the accelerator pedal (Fig. 1).

A second (redundant) sensor is incorporated for diagnosis purposes and for use in case of malfunctions. It is a component part of the monitoring system. One sensor version operates with a second potentiometer, which always delivers half the voltage of the first potentiometer at all operating points. Thus, two independent signals are available for fault-detection purposes (Fig. 1). Instead of the second potentiometer, another version uses a low-idle switch, which signals the idle position to the control unit. The status of this switch and the potentiometer voltage must be plausible.

For vehicles with automatic transmissions, a further switch can be incorporated for an electrical kickdown signal. Alternatively, this information can also be derived from the rate of change of the potentiometer voltage. A further possibility is to trigger the kickdown function by means of a defined voltage value of the sensor curve; here, the driver receives feedback on a jump in force in a mechanical kickdown cell. This is the most frequently used solution.

1 Characteristic curve of an accelerator-pedal sensor

2 ARS1 Hall angular-position sensor

3 Principle of ARS2 Hall angular-position sensor

Hall angular-position sensors

Hall sensors are used to measure the movement of the accelerator pedal on a non-contact basis. In the case of the Type ARS1 Hall angular-position sensor, the magnetic flux of a roughly semicircular, permanent-magnetic disk is fed back via a pole shoe, two further conductive elements and the similarly ferromagnetic shaft to the magnet (Fig. 2). Depending upon the angular setting, the flux is led to a greater or lesser degree through the two conductive elements, in the magnetic path of which a Hall sensor is also situated. Using this principle, it is possible to achieve a practically linear characteristic curve in the measuring range of 90°.

The Type ARS2 is a simplified version which does without soft magnetic conductive elements (Fig. 3). In this version, the magnet moves around the Hall sensor in a circular arc. Only a relatively small section of the resulting sinusoidal characteristic curve features good linearity. If the Hall sensor is located slightly outside the center of the circular arc, the characteristic curve increasingly deviates from the sinusoidal, and now features a short measuring range of almost 90°, and a longer measurement range of more than 180° with good linearity. A great disadvantage though is the low level of shielding against external fields, as well as the remaining dependence on the geometric tolerances of the magnetic circuit, and the intensity fluctuations of the magnetic flux in the permanent magnet as a function of temperature and age.

In the case of the FPM2.3 type Hall angular-position sensor, it is not the field strength but rather the direction of the magnetic field which is used to generate the output signal. The field lines are recorded by four radially arranged measuring elements lying in one plane in the x- and y-directions (Fig. 4). The output signals are derived in the ASIC from the raw data (cos and sin signals) using the arctan function. The sensor is positioned between two magnets to generate a homogenous magnetic field. The sensor is therefore insensitive to component tolerances and is temperature-resistant.

As with the accelerator-pedal module with a potentiometric sensor, these proximity-type systems also contain two sensors in order to receive two redundant voltage signals.

4 Measuring principle of FPM2.3

5 Exploded view of FPM2.3 accelerator-pedal module

Fig. 4
a Design
b Principle
c Measurement signals

1 Integrated circuit (IC) with Hall elements
2 Magnet (opposing magnet not shown here)
3 Conductive element
4 Hall elements (for recording x-components of B)
5 Hall elements (for recording y-components of B)

Bₓ Homogenous magnetic field (x-components)
B_y Homogenous magnetic field (y-components)

Fig. 5
1 Pedal
2 Cover
3 Spacer sleeve
4 Sensor block with housing and plug
5 Bearing block
6 Shaft with two magnets and hysteresis elements (round magnets not visible)
7 Kickdown (optional)
8 Two springs
9 Stop damper
10 Thrust member
11 Floor cover

Steering-angle sensors

Application

The electronic stability program (ESP) applies the brakes selectively to the individual wheels in order to keep the vehicle on the desired track selected by the driver. Here, a control unit compares the steering angle and the applied brake pressure with the vehicle's actual rotary motion (around its vertical axis) and its road speed. If necessary, the brakes are applied at individual wheels. These measures serve to keep the float angle (deviation between the vehicle axis and the actual vehicle movement) down to a minimum and, until the physical limits are reached, prevent the vehicle breaking away.

Basically speaking, practically all types of angular-position sensors are suitable for registering the steering angle. Safety considerations, though, dictate that only those types are used which can be easily checked for plausibility, or which in the ideal case automatically check themselves. Potentiometer principles are used, as well as optical code-registration and magnetic principles. Whereas a passenger-car steering wheel turns through ±720° (a total of 4 complete turns), conventional angular-position sensors can only measure maximum 360°. This means that with the majority of the sensors actually used for this purpose it is necessary to continually register and store the data on the steering wheel's actual setting.

Design and operating principle

Steering-angle sensor with AMR element

The LWS3 steering-angle sensor uses AMR (Anisotropic Magnetoresistive Sensors) for its operation. The AMR's electrical resistance changes according to the direction of an external magnetic field. In the LWS3, the information on angle across a range of four complete rotations is provided by measuring the angles of two gear wheels which are rotated by a third gear wheel on the steering shaft. The first two gear

wheels differ by one tooth which means that a definite pair of angular variables is associated with every possible steering-wheel position.

By applying a mathematical algorithm (a computing process which follows a defined step-by-step procedure) referred to here as a modified vernier principle, it is possible to use the above AMR method for calculating the steering angle in a mi-

1 LWS3 AMR steering-angle sensor (principle)

UFL0030-1Y

2 LWS3 AMR steering-angle sensor (view)

UFL0031-1Y

Fig. 1
1 Steering shaft
2 AMR sensor
 elements
3 Gear wheel with
 m teeth
4 Evaluation
 electronics
5 Magnets
6 Gear wheel with
 n > m teeth
7 Gear wheel with
 m + 1 teeth

croprocessor. Here, even the measuring inaccuracy of the two AMR sensors can be compensated for. In addition, a self-check can also be implemented so that a highly plausible measured value can be sent to the control unit.

Figure 1 shows the schematic representation of the LWS3 steering-angle sensor. The two gear wheels, with magnets inserted, can be seen. The sensors are located above them together with the evaluation electronics.

A different design is the LWS4 steering-angle sensor which unambiguously measures a steering angle of 180°. It is mounted on the shaft end of the steering axle (Fig. 3)

Steering-angle sensor with GMR element
The LWS5 is the first steering-angle sensor based on the GMR effect (Giant Magneto Resistive). Its mechanical structure and operating principle have been taken from the LWS3. The LWS3 and LWS5 are mechanically and electrically compatible.

The GMR layers are processed on the planarized surface of the evaluation circuit (vertical integration). The GMR resistor bridge and the evaluation module are connected by means of plated-through holes.

These short connections enhance the sensor's robustness against external interference.

The two GMR elements measure the direction of the field lines of the two magnets. The steering angle is calculated from this in the microprocessor. The sensor element and microprocessor communicate via a digital interface (SPI interface). The calculated steering angle value is placed on the CAN by the microprocessor.

Because of the greater sensitivity, by comparison with the AMR effect, the LWS5 can work with weaker magnets and greater air gaps. This brings considerable cost benefits in materials and design. The 360° angular measuring range of a single GMR element (a 180° measuring range is typical for AMR), allows the use of smaller gear wheels on the LWS5. It thus requires a significantly smaller installation space than the LWS3. In addition, it has a significant level of scaleability, reflected in the measuring range (±90° to ±780°) and in the degree of redundancy. This means that the sensor can meet the specific requirements of the various vehicle manufacturers as closely as possible.

3 LWS4 AMR steering-angle sensor for mounting on the end of the steering axle

UFL0032-1Y

Fig. 3
1 Steering column
2 Steering gear
3 Steering-angle
 sensor
4 Steering rack

Position sensors for transmission control

Application
The position sensor detects the positions of an actuator (e.g. selector lever shaft, selector slide or park lock cylinder) in AT, ASG, DKG and CVT transmissions. The sensors may be fitted stand-alone or built into electronics modules.

Requirements
Sensors built into the transmission must be designed to cope with the challenging operating conditions prevailing there:
- Ambient temperatures between -40 and +150 °C,
- Transmission oil, which is an aggressive ambient medium,
- High mechanical stress up to 30 g
- Abraded metallic materials and a build-up of particles in the transmission.

Resistance to the media and temperature in these applications is provided by an oil-resistant enclosure for the electronics and the use of high temperature PCBs and AVT for the electronics. The electronic components of ASIC, capacitors, resistors and interfaces (bonds, soldering points) must be designed to be sufficiently robust as to be able to withstand the severe mechanical stresses over the service life of the vehicle.

Surface-mounted sensors, on the other hand, are made in the standard packages familiar from engine compartment applications. They must meet the corresponding requirements such as being splash-proof and handling operating temperatures from -40 to 130 °C.

Because of the complex requirements (Table 1) arising from the various transmission topologies, and the requirements on mounting space and function (safety concept, required precision, etc.) different physical measuring principles (Hall, AMR, GMR, eddy-current principles) and shapes (linear and rotary detection) may be used. These are explained in greater detail below using the example of the Hall effect and the eddy current principle.

Linear position determination on the basis of Hall switches
Design
Four digital Hall switches are mounted on a printed-circuit board in such a way that they detect the magnetic orientation of a linearly displaceable, multipolar permanent magnet (Fig. 1). The magnet carriage is coupled with the linearly actuated selector slide (hydraulic slide in the transmission control plate) or the parking lock cylinder.

In addition to the Hall switch, the printed-circuit board carries resistors to represent diagnosis functions and EMC capacitors.

The sensor electronics are protected against the influence of the transmission oil by being potted in a dense, oil-resistant epoxy resin.

1 Requirements on position detection	
Installation location	Inside the transmission On the transmission
Design	Linear Rotational
Measuring principle	Hall AMR GMR Eddy current
Signal acquisition	Digital (Gray code) Analog
Output signal	Digital (2-/4-bit) PWM Analog
System security	Redundancy High availability P/N recognition
Shifting type	M and E shifting

Table 1

Operating principle

In the case of automatic transmission with manual shifting, also known as M-shifting, the position sensor detects the positions of the selector slide, P, R, N, D, 4, 3, 2 and the intermediate ranges and passes these to the transmission control in the form of a 4-bit code (Fig. 2). For safety reasons, the position is encoded in one-step encoding, i.e. two-bit changes are always required before a new position is detected. Single-bit changes caused by malfunctions can be detected as false by the control unit by plausibility considerations. The sequence of these bit changes corresponds to a Gray code.

In some applications, the sensor transmits the controlled variable for starter release, the P/N signal, directly via a processor-independent hardware route. This increases the availability in operating states (e.g. when the vehicle's electrical system voltage is low) in which the electronic control unit would not yet be operational.

In the case of automatic transmission with electronic shifting, also known as E-shifting, the position sensor detects only the positions of the park lock cylinder P_{in} and P_{out} and the intermediate range and passes these to the transmission control unit in the form of a 2-bit code.

Rotational position determination on the basis of eddy currents
Design

A position sensor with rotational design is shown in Figure 3. In this example, a specially shaped rotor with a return coil is mounted to rotate on the selector lever shaft outside the transmission. It represents the interface to the transmission actuator. Redundant transmit and receive coils are mounted on the fixed sensor board with their associated evaluating ASICs. In this solution, the actuator (shaft) passes through the circuit board.

1 Position sensor with Hall switch

a

b

UTS0363Y

2 Coding on the magnetic carriage

a

☐ Magnetic North Pole
☐ Magnetic South Pole

4 bit code

b

P R N D 4 3 2

☐ Switching and intermediate ranges
■ Transition range

SAE1084E

Fig. 1
a Front view
b Rear view

1 Oil-resistant package
2 Electrical connection by terminal plate
3 Printed-circuit board potted Hall elements
4 Carriage with permanent magnet
5 Locating pin

Fig. 2
a Magnetic encoding
b Position ranges

1 Moving carriage
2 Fixed position of the Hall elements

Operating principle

The sensor uses the eddy-current principle. The transmitting coil induces in the return coil an eddy current opposite to its cause, the magnetic field of which induces a voltage in the receiver coils. The geometries of the return coil and the receiver coil are tuned to one another, so that continuously changing positions of the rotor can be detected.

Similarly to the example of linear position detection, the rotational position sensor detects the P, R, N, D, 4, 3 and 2 positions. In this case, however, on the selector lever shaft projection out of the transmission. The positions of the shaft correspond to the positions of the selector slide inside the transmission.

One benefit of analog signal conditioning is that the individual shift and intermediate ranges are assigned to the angle positions detected by the sensor in the software, and so can easily be adapted to transmission design variants.

For safety reasons, the sensor makes two independent, opposing output signals available, with the upper and lower 5 % of the signal stroke being used as a diagnostic range. In the event of a fault being detected inside the sensor, a voltage level in the upper diagnostic range is output. Faults that can occur on the transmission path to the control unit (e.g. open circuit or short-circuit) also lead to voltage levels in the diagnostics range. In addition to this, the electronic control unit tests the signal for plausibility by adding the two signals together. Sensor or transmission errors can be detected by the control unit in this way, and an appropriate emergency running program can be selected, depending on the fault.

Table 2 compares the two examples of position sensor and illustrates the complexity of the topic.

The complex requirements on the actuator's mechanical interface, the mounting space, the environmental influences and the safety concept frequently lead to application-specific sensor system solutions.

3 Position sensor using eddy-current principles

Fig. 3
a Components (without housing)
b Printed-circuit board
c Rotor

1 Selector lever shaft
2 Sensor PCB
3 Redundant transmitter and receiver coil
4 Redundant electronics
5 Rotor with return coil

Table 2

2 Typical characteristics		
Position sensor	**Hall switch**	**Eddy current**
Measuring principle	Hall effect (digital)	Eddy current (analog)
Sensitivity to external magnetic fields	Yes	No
Sensitivity to metallic surroundings	No	Yes
Flexible position scaling	No	Yes (software)
Operating voltage range	4 to 12 V	4.5 to 5.5 V
Sensitivity to the air gap	High	Low

Axle sensors

Applications

The automatic headlamp range control (ALWR) adjusts the vehicle's headlamp range automatically. With the headlamps on low beam, ALWR compensates for the vehicle tilt so that the driver has adequate vision without dazzling oncoming traffic. Whereas, the static ALWR compensates for the vehicle tilt resulting from the vehicle loading, the dynamic version also takes into account the front-end up and down pitching movement of the vehicle caused by braking and acceleration. Here, the axle sensors precisely register the bodywork's tilt angle.

Design and operating principle

Axle sensors (angular-position sensors) are used to measure the vehicle tilt. These are attached to the bodywork at the front and rear of the vehicle. Spring compression is transmitted to the sensors by a pivot lever which is connected through a connecting rod to the particular axle or to the wheel suspension. The vehicle's tilt is then calculated from the difference between the voltages of the front and rear axle sensors.

The axle sensors function according to the Hall effect. A Hall IC is built into the stator (Fig. 1, Item 5) where it is located in a homogenous magnetic field. The magnetic field generates a Hall voltage in the Hall IC proportional to the magnetic-field strength. When the ring magnet (6) is rotated by the shaft (2), the magnetic field through the Hall IC changes accordingly.

When the springs are compressed due to loading and/or acceleration/braking, the connecting rod (Fig. 2, Item 4) transfers the spring compression movement to the pivot lever of the axle sensor so that it is converted into a voltage signal which is proportional to the angle of rotation.

The control unit registers the axle-sensor signals and from them generates the difference between front and rear axle. Using this voltage difference, and taking into account the vehicle's speed, the control unit now calculates the setpoint value for the servomotor settings. With the vehicle driving at constant speed, the dynamic ALWR remains in the status which features a high level of damping whereby the stepping motors are adapted slowly to the vehicle tilt. This prevents bumps, unevenness, or holes in the road causing continuous correction of the headlamp settings. When the vehicle accelerates, or the brakes are applied, the system automatically switches to the dynamic mode, and adjusts the headlamp range within a few milliseconds. The system then switches back to the slower mode.

1 Axle sensor (section)

2 Axle sensor (installed in vehicle)

Fig. 1
1 Pivot lever
2 Shaft
3 Housing
4 Ring-magnet mount
5 Stator with Hall IC
6 Ring magnet

Fig. 2
1 Bodywork mounting
2 Axle sensor with plug-in connection
3 Pivot lever
4 Connecting rod
5 Vehicle axle

Hot-film air-mass meters

Application

To provide precise pilot control of the air-fuel ratio, it is essential for the supplied air mass to be exactly determined in the respective operating status. The hot-film air-mass meter measures some of the actually inducted air-mass flow for this purpose. It takes into account the pulsations and reverse flows caused by the opening and closing of the engine's intake and exhaust valves. Intake-air temperature or air-pressure changes have no effect upon measuring accuracy.

HFM5 design

The housing of the HFM5 hot-film air-mass meter (Fig. 1, Item 5) projects into a measuring tube (Fig. 3, Item 6) which, depending upon the engine's air-mass requirements (370 to 970 kg/h) can have a variety of diameters. The measuring tube normally

contains a flow rectifier, which ensures that the flow in the measuring tube is uniform. The flow rectifier is either a combination of a plastic mesh with straightening action and a wire mesh, or is a wire mesh on its own (Fig. 3, Item 8).The measuring tube is installed in the intake tract downstream from the air filter. Plug-in versions are also available which are installed inside the air filter.

The most important component parts in the sensor are the measuring cell (Fig. 1, Item 4) in the air intake (8) and the integrated evaluation electronics (3).

The sensor measuring cell consists of a semiconductor substrate. The sensitive surface is formed by a diaphragm which has been manufactured in micromechanical processes. This diaphragm incorporates temperature-sensitive resistors. The elements of the evaluation electronics (hybrid circuit) are installed on a ceramic substrate. This principle permits very compact design. The evaluation electronics are connected to the control unit by means of electrical connections (1).

The partial-flow measuring passage (6) is shaped so that the air flows past the measuring cell smoothly (without swirl effects) and back into the measuring tube via the air outlet (7). The length and location of the inlet and outlet of the partial-flow measuring passage have been chosen to provide good sensor performance even in the event of sharply pulsating flows.

Operating principle

The HFM5 hot-film air-mass meter is a thermal sensor which operates according to the following principle: A centrally situated heating resistor on the measuring cell (Fig. 3, Item 3) heats a sensor diaphragm (5) and maintains it at a constant temperature. The temperature drops sharply on each side of this controlled heating zone (4).

The temperature distribution on the diaphragm is registered by two temperature-dependent resistors which are mounted upstream and downstream of the heating

1 HFM5 hot-film air-mass meter (schematic diagram)

Q_M

1 cm

UMK1713-1Y

Fig. 1

1 Electric connections (plug)

2 Measuring-tube or air-filter housing wall

3 Evaluation electronics (hybrid circuit)

4 Sensor measuring cell

5 Sensor housing

6 Partial-flow measuring passage

7 Outlet, partial air flow Q_M

8 Intake, partial air flow Q_M

resistor so as to be symmetrical to it (measuring points M_1, M_2). Without the flow of incoming air, the temperature profile (1) is the same on each side of the heating zone ($T_1 = T_2$).

As soon as air flows over the measuring cell, the uniform temperature profile at the diaphragm changes (2). On the inlet side, the temperature characteristic is steeper since the incoming air flowing past this area cools it off. On the opposite side, the temperature characteristic only changes slightly, because the incoming air flowing past has been heated by the heater element. The change in temperature distribution leads to a temperature difference (ΔT) between the measuring points M_1 and M_2.

The heat dissipated to the air, and therefore the temperature characteristic at the measuring cell is dependent on the air mass flowing past. The temperature difference is (irrespective of the absolute temperature of the air flow past) a measure of the air-flow mass. It is also direction-dependent so that the air-mass meter can record both the amount and the direction of an air-mass flow.

Due to its very thin micromechanical diaphragm, the sensor has a highly dynamic response (< 15 ms), a point which is of particular importance when the incoming air is pulsating heavily.

The evaluation electronics integrated in the sensor convert the resistance differential at the measuring points M_1 and M_2 into an analog voltage signal of between 0 and 5 V. Using the sensor curve (Fig. 2) stored in the engine control unit, the measured voltage is converted into a value representing the air-mass flow (kg/h).

The shape of the characteristic curve is such that the diagnosis facility incorporated in the control unit can detect such malfunctions as an open-circuit line. An additional temperature sensor for evaluation functions can be integrated in the HFM5 hot-film air-mass meter. It is not required for measuring the air mass.

Incorrect air-mass readings will be registered if the sensor diaphragm is contaminated with dust, dirty water or oil. For the purpose of increasing the robustness of the HFM5, a protective device has been

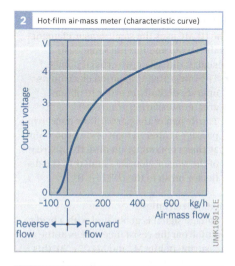

2 Hot-film air-mass meter (characteristic curve)

Reverse ←→ Forward
flow flow

3 Hot-film air-mass meter (measuring principle)

Fig. 3
1 Temperature profile without air flow
2 Temperature profile with air flow
3 Measuring cell
4 Heating zone
5 Sensor diaphragm
6 Measuring tube with air-mass meter
7 Intake-air flow
8 Wire mesh
M_1, M_2 Measuring points
T_1, T_2 Temperature values at measuring points M_1 and M_2
ΔT Temperature difference

developed which, in conjunction with a deflector mesh, keeps dirty water and dust away from the sensor element (HFM5-CI; with C-shaped bypass and inner tube (I), which together with the deflector mesh protects the sensor).

HFM6 hot-film air-mass meter

The HFM6 uses the same sensor element as the HFM5 and has the same basic design. It differs in two crucial points:

- The integrated evaluation electronics operate digitally in order to obtain greater measuring accuracy
- The design of the partial-flow measuring passage is altered to provide protection against contamination directly upstream of the sensor element (similar to the deflector mesh in the HFM5-CI)

Digital electronics

A voltage signal is generated with a bridge circuit from the resistance values at the measuring points M_1 and M_2 (Fig. 3); this voltage signal serves as the measure of the air mass. The signal is converted into digital form for further processing.

The HFM6 also takes into account the temperature of the intake air when determining the air mass. This increases significantly the accuracy of the air-mass measurement.

The intake-air temperature is measured by a temperature-dependent resistor, which is integrated in the closed control loop for monitoring the heating-zone temperature. The voltage drop at this resistor passes through an analog-digital converter to give a digital signal representing the intake-air temperature. The signals for air mass and intake-air temperature are used to address a program map in which the correction values for the air-mass signal are stored.

Improved protection against contamination

The partial-flow measuring passage is divided into two sections in order to provide better protection against contamination (Fig. 4). The passage which passes the sensor element has a sharp edge (1), around which air must flow. Heavy particulates and dirty-water droplets are unable to follow this diversion and are separated from the partial flow. These contaminants exit the sensor through a second passage (5). In this way, significantly fewer dirt particles and droplets reach the sensor element (3) with the result that contamination is reduced and the service life of the air-mass meter is significantly prolonged even when operated with contaminated air.

4 HFM6 with improved contamination protection

Figure 4

1 Diverting edge
2 Partial-flow measuring passage (first duct)
3 Sensor element
4 Air outlet
5 Second passage
6 Particulate and water outlet

Piezoelectric knock sensors

Application
Regarding their principle of functioning, knock sensors are basically vibration sensors and are suitable for detecting structure-borne acoustic oscillations. These occur as "knock", for instance, in gasoline engines when uncontrolled combustion takes place. They are converted by the knock sensor into electrical signals (Fig. 1) and transmitted to the Motronic control unit, which counteracts the engine knock by adjusting the ignition angle.

Design and operating principle
Due to its inertia, a mass (Fig. 2, Item 2) excited by a given oscillation or vibration exerts a compressive force on a toroidal piezoceramic element (1) at the same frequency as the excitation oscillation. These forces effect a charge transfer within the ceramic element. An electrical voltage is generated between the top and bottom of the ceramic element which is picked off via contact washers (5) and processed in the Motronic control unit.

2 Knock sensor (design and mounting)

Fig. 2
1 Piezoceramic element
2 Seismic mass with compressive forces F
3 Housing
4 Bolt
5 Contact washer
6 Electric connection
7 Engine block
V Vibration

Mounting
In four-cylinder engines, one knock sensor is sufficient to record the knock signals for all the cylinders. Engines with more cylinders require two or more knock sensors. The knock-sensor installation point on the engine is selected so that knock can be reliably detected from each cylinder. The sensor is usually bolted to the side of the engine block. It must be possible for the generated signals (structure-borne-noise vibrations) to be introduced without resonance into the knock sensor from the measuring point on the engine block. A fixed bolted connection satisfying the following requirements is required for this purpose:
- The mounting bolts must be tightened to a defined torque
- The contact surface and the bore in the engine block must comply with prespecified quality requirements
- No washers of any type may be used

1 Knock-sensor signals (oscilloscope display)

Without knock

a
b

c

With knock

a
b

c

Fig. 1
a Cylinder-pressure characteristic curve
b Filtered pressure signal
c Knock-sensor signal

SMM acceleration sensors

Application

Surface micromechanical (SMM) acceleration sensors have a variety of acceleration detection roles in the vehicle. These include the detection of acceleration values for a frontal or lateral impact in order to trigger the seat-belt pretensioners, the airbag and the rollover bar in vehicle occupant-protection systems. Surface micromechanical (SMM) acceleration sensors are used in the field of active vehicle safety in ABS, ESP and HHC (Hill Hold Control, low-speed traction control) systems. Further possible applications of acceleration sensors are found in chassis and suspension control (active suspension) and vehicle alarm systems, where a change in the vehicle tilt will trigger an alarm.

Design and operating principle

The surface micromechanical sensor elements are designed for different measuring ranges, depending on the application. These measuring ranges lie between 1 g and 400 g (1 $g \approx$ 9.81 m/s^2). The design of an acceleration sensor for side or frontal impact sensing is described below. The SMM sensor element is fitted together with the evaluation electronics (ASIC) in a first packaging stage (module). Figure 1 shows a module in a SO16 housing. This module on a small PCB, with further circuit components, is pressed into the second packaging stage, a plastic housing. This housing is closed and sealed with a welded plastic lid. This sensor is intended for installation in the side or fender of the vehicle after calibration and testing and sends its acceleration signal via plug and cable to the central airbag control unit.

The functional layers of the sensor element representing a spring-mass system are applied to the surface of the silicon wafer by an additive method (Fig. 3, surface micromechanics).

The mass with its comb-like electrodes (Figs. 2 and 3, Item 1) is connected via spring elements (2) to the anchor points (Fig. 3, Item 4) in the core of the sensor. There are fixed comb-like electrodes (3, 6) on the chip on each side of these movable electrodes. A useful capacitance in the range of 300 fF to 1 pF results from the connection in parallel of the individual capacitances formed by the fixed and moving electrode fingers. The two rows of electrode fingers connected in parallel result in two useful capacitances (C_1 - C_M and C_2 - C_M), which change in opposite directions as the central mass is deflected.

Fig. 1

1 Surface micromechanical acceleration sensors for airbag triggering (example)

Fig. 1

1 First packaging
 stage (module)
 in SO16 housing
2 Evaluator chip
 (ASIC)
3 SMM sensor
 element
4 Populated PCB
5 Second packaging
 stage (device)
6 Plug

SAE1079Y

The acceleration applied to the spring-mass system deflects it by an amount linearly related to the acceleration applied-over the spring return force. An electrical output signal that is linearly dependent on the acceleration is obtained by evaluation of this differential capacitor.

The acceleration signal obtained in the first stage of the evaluation circuit is further conditioned in the ASIC, i.e. it is amplified, filtered and prepared for the output interface.

Analog voltages, pulse-width-modulated signals, SPI protocols or current loop in terfaces are usual as the output interface. Tolerances arising from the manufacture of the sensor element, the evaluation cir cuit and package influences on sensitivity and zero point are eliminated by means of a comparison of programmable memory cells at the end of the production line. A self-test function tests the full mechani cal and electrical signal path. In the course of this self-diagnosis, an electrostatic force deflects the sensor structure, i.e. simulat ing an acceleration in the vehicle, and the response of the measurement signal is compared with a setpoint value.

Up to three acceleration sensors (for ESP and HHC) and one or two yaw-rate sen-sors (for ESP) are incorporated in the ESP sensor cluster. This clustering of sensor modules reduces the number of individual components and signal lines by compari-son with separate sensor devices. It also means that fewer mountings and less in-stallation space are needed in the vehicle.

3 Comb-like structure of the sensor measuring cell

100µm

UAE0800-1Y

Fig. 3
1 Spring-mounted seismic mass with electrode
2 Spring
3 Fixed electrodes
4 Anchor areas

2 Surface micromechanical acceleration sensors with capacitive pick-off

C_1 C_2 C_M

C_2 C_M C_1

UAE0801-2Y

Fig. 2
1 Spring-mounted seismic mass with electrodes
2 Spring
3 Fixed electrodes with capacity C_1
4 Printed Al conductor track
5 Bond pad
6 Fixed electrodes with capacity C_2
7 Silicon oxide
a Acceleration in the sensing direction
C_M Measuring capacity

Micromechanical bulk silicon acceleration sensors

Application

Micromechanical bulk silicon acceleration sensors detect the acceleration signals needed for ABS, ESP and shock absorber adjustment.

This type of sensor is today predominantly used in the low acceleration ranges with high demands on signal to noise ratio ($< 2\,g_n$).

Design and operating principle

Anisotropic and selective etching techniques are used to form the required spring-mass system for the sensor from the solid wafer (bulk or volume micromechanics). Capacitive pick-offs have proven especially effective for the high-precision measurement of this seismic-mass deflection. They require a wafer-thick silicon or glass plate (Fig. 1, Item 4) with counter-electrodes, on each side of the spring-held mass (2). The wafers and their counter-electrodes here also provide overload protection.

This configuration corresponds to a series circuit of two capacitors C_{1-M} and C_{2-M}. Alternating voltages are applied across terminals C_1 and C_2, and their superimpositions picked off between the capacitors at C_M.

In the idle condition, the capacitances C_{1-M} and C_{2-M} are ideally equal. This means that the difference ΔC is equal to zero. If an acceleration a is acting in the measuring direction, the Si center plate (the seismic mass) is caused to deflect. This causes a change in the distance to the upper and lower plates, and with it a capacitance change in the capacitors C_{1-M} and C_{2-M} and thus a difference ΔC not equal to zero. This leads to a change in the electrical signal at C_M which is amplified and filtered in the evaluation electronics.

The air space between the plates allows effective damping with low temperature sensitivity.

1 Bulk silicon acceleration sensor with capacitive pick-off (schematic diagram)

Fig. 1

1 Upper Si plate
2 Center Si plate
 (spring-held or
 seismic mass)
3 Si-oxide
4 Bottom Si plate
5 Glass substrate
a Acceleration in the
 sensing direction
C_M Measuring capacity

Piezoelectric acceleration sensors

Application
Piezoelectric spring elements are used as acceleration sensors in passenger-restraint systems for triggering the seatbelt pretensioners, the airbags, and the rollover bar.

Design and operating principle
The core of this acceleration sensor is a spring element (bending beam). A bonded structure comprising two piezoelectric layers of opposing polarity (bimorphous spring element). An acceleration acting on this element generates a mechanical tensile stress in one layer and a compressive stress in the second layer (Fig. 1).

The metallized surfaces on the top and underneath of the spring element act as electrodes from which the resultant voltage is picked off. This structure is packaged together with the electronics required in a hermetically sealed housing (Fig. 2).

The electronic circuit consists of an impedance converter and a tunable amplifier with predefined filter characteristics. The design means that no static signals can be measured (lower cutoff frequency is typically 1 to 10 Hz).

Piezo-spring elements do not require any additional seismic mass. Their own mass is adequate for an easily evaluated signal.

2 Piezoelectric acceleration sensor (two channel sensor for PCB mounting)

Fig. 2
1 Spring element

1 Spring element in a piezoelectric acceleration sensor

a
$U_A = 0$ 1 $a = 0$

b
$U_A > 0$ 1 $a \neq 0$

Fig. 1
a Not under acceleration
b Under acceleration a
1 Piezoceramic bimorphous spring element
U_A Measurement voltage

iBolt™ force sensor

Applications
NHTSA FMVSS-208 (National Highway Traffic Safety Administration; Federal Motor Vehicle Safety Standards and Regulations 208) came into force in the USA in 2004. These regulations were introduced to prevent or reduce injuries to children traveling in the front passenger seat caused by the airbag triggering. Classifying the front passenger by means of a weight measurement makes it possible to switch the airbag off selectively if a small child is occupying the seat.

The BOSCH iBolt™ (intelligent bolt) sensor was developed to perform this weight classification reliably and robustly. This is done by fitting four iBolts™ in the seat frame of the front-passenger seat (a sensor in each corner of the seat, Fig. 1). A control unit also built into the seat analyses the four analog, electrical weight sig-

1 Fitting four iBolts™ in the seat base frame of the front passenger's seat

Fig. 1
1 Seat base frame
2 BOSCH iBolt™
 sensor

nals and transmits the result of the classification to the airbag control unit.

Design and operating principle
The iBolt™ sensor's operating principle is based on the measurement of the deflection of a bending beam caused by the passenger's weight. The amount of the deflection is detected by measurement of the magnetic field strength in a special Hall sensor/magnet configuration (Fig. 2a).

The iBolt™ is designed such that it is the z component of the passenger's weight that causes the deflection of the bending beam. The vehicle's co-ordinates system here defines the x axis as the direction of travel, the z and y axes being perpendicular and horizontal to this. The configuration of the magnet and the Hall IC in the sensor is selected such that the static magnetic field which penetrates the Hall IC results in an electrical signal linear to the deflection of the bending beam. The special design of the iBolt™ sensor prevents a horizontal deflection of the Hall IC in relation to the magnet, to minimize the influence of transverse forces and moments. In addition, the maximum stress in the bending beam is limited by a mechanical overload stop (Fig. 2b). This protects the iBolt™, especially in the event of overloads in a crash.

Calibration
The linear Hall sensor uses the spinning current measuring principle which allows calibration of the sensitivity, the offset and the temperature sensitivity of measurement. The calibration data is stored in an EEPROM incorporated in the substrate for the Hall sensor.

Linearity of the output signal
A linear output signal has been achieved by means of a special design feature. The force generated by the weight of the front passenger is passed from the upper seat structure through the sleeve into the bending beam (Fig. 2a). The force is then

passed from the bending beam into the lower seat structure. The bending beam was designed as a double bending beam as this has an S-shaped deformation line. The two vertical connection points of the double bending beam remain vertical over the entire deflection range. This guarantees a linear and parallel movement of the Hall IC in relation to the magnet, resulting in a linear output signal (Fig. 3).

Symmetrical measuring range

Tests of the system in vehicle seats have shown that both positive and negative forces can act on the sensors. This has a number of causes: on the one hand, negative forces can act on an individual tension as a result of initial tensions as a consequence of tolerances in the assembly of the seat and of its installation in the vehicle. On the other hand, negative forces arise on individual sensors as a function of the distribution of forces on the sensors

generated by the seat occupant's position, which also depends on the position of the seat back. For this reason, the measuring range of the iBolt™ sensor has been designed such that the forces can be detected in both the positive and negative z directions. This permits an unambiguous determination of the front passenger's weight.

Thanks to its symmetrical measuring range, the iBolt™ sensor detects compression and tensile forces with the same sensitivity and the same tolerances. This means that the same sensors can be used in both vertical mounting orientations for all four points of connection between the upper and lower seat structures.

2 Measuring principle of the BOSCH iBolt™ sensor

a $F_G < 850N$

b $F_G > 850N$

3 Typical output signal from the sensor

Output voltage U_A

Overload range

Overload range

Weight F_G

Fig. 2
a Ratios for weight $F_G < 850$ N (within the measuring range)
b Ratios for weight $F_G > 850$ N (outside the measuring range)
1 Oscillating crank
2 Air gap
3 Sleeve
4 Seat rail
5 Double bending beam
6 Magnet
7 Hall IC

Fig. 3
Output signal as a function of the force applied

Torque sensor

Application

Electromechanical power-assisted steering is increasingly being introduced in small, compact and medium class vehicles. The essential advantages are simple installation and commissioning in the vehicle, energy saving and the suitability of these systems in the vehicle's network of control units for assistance systems enhancing comfort and safety.

Design and operating principle

In order to sense the driver command in electromechanical power-assisted steering, it is necessary to measure the torque applied by the driver. In the sensor currently in series production use for this, a torsion rod is fitted in the steering shaft. The rod is twisted by a defined amount in linear proportion to the torque applied by the driver when the driver turns the steering wheel (Fig. 1). This twisting can, for its part, be measured by suitable means and converted into electrical signals. The measuring range required for a torque sensor to be used in electromechanical power-assisted steering is usually around ± 8 to ± 10 Nm. The maximum angle of rotation is mechanically limited by special components to protect the torsion rod against overload or destruction.

A magnetoresistive sensor is mounted on one side of the torsion rod to allow measurement of the twisting and hence the torque applied. This scans the field of a magnetic multipole wheel mounted on the other side. The number of poles selected for this wheel is such that the sensor issues an unambiguous signal within its maximum measuring range and thus makes a clear statement of the torque applied at any time.

Here the magnetoresitive sensor use supplies two signals over the measuring range which describe a sine and a cosine signal plotted against the angle of rotation of the torsion rod. The angle of rotation and hence of the torque is calculated in an electronic control unit using an arctangent function.

As there is always a fixed assignment of the two signals over the defined measuring range, a deviation from this allows faults in the sensor to be detected and the necessary substitution measures to be introduced.

A coil spring with the necessary number of contacts is used to provide the electrical contacts for the sensor over the rotational range of approximately ± 2 revolutions of the steering wheel. The power is supplied and the measured values are transmitted via this coil spring.

1 Torque sensor

Rain/light sensor

Application of the rain sensor

The rain sensor detects fine raindrops on the windshield, so that the windshield wipers can be triggered automatically. The sensor controls the speed of the wiper system according to the amount of rain measured on the windshield (intermittent, speed 1 and speed 2). Together with the electronically controlled wiper drive, intermittent operation can be varied infinitely. For instance, when masses of rain water flood the windshield when overtaking, the wiper is automatically switched to the highest wiper speed. Various activation strategies are available:

- Permanently active
- Activated via steering column switch
- Activation after steering column switch movement (transients)

This assistance function frees the driver from the other control operations used on conventional wiper systems. Nevertheless, manual control is retained as an additional intervention.

The rain sensor can also be used for other functions such as closing the windows and the sliding sunroof.

Application of the light sensor

A light sensor is also incorporated in the rain sensor, it may be used to control the low beam. This can detect the various situations (e.g. twilight, driving into and out of a tunnel, driving under long bridges) and switch the low beam on or off accordingly. The light sensor can also be used to control all the lighting functions on the vehicle. These functions include adjusting the brightness of the instrument cluster lighting, "coming home leaving home" or selectively switching on the rear lamps.

Design and operating principle

A rain sensor comprises an optical transmission and reception path (Fig. 1). An LED (6) emits light at a determined angle into the windshield (2) which is reflected on the outer boundary layer (glass/play) (total reflection) and then is evaluated on an aligned receiver (photodiode, 4). If there is moisture on the sensitive outer surface, a part of the light is coupled out as a function of the droplet size and number, which attenuates the return signal.

The integrated light sensor generally comprises two or three photodiodes (3, 5) which receive and evaluate the light coming from various directions. Depending on the function of the light sensor, diodes are used that reflect the receptiveness of the human eye (silicon eyes) or have their maximum sensitivity rather in the near infrared range.

Complex filter functions and trigger strategies are stored in the rain/light sensor. The status information derived from this (e.g. wiper speed 1, intermittent wiper, intermittent timing, lights on, lights off) are made available to other control units via a data bus (e.g. CAN, LIN).

1 Rain/light sensor for windshields

Fig. 1
1 Raindrops
2 Windshield
3 Ambient-light sensor
4 Photodiode
5 Light sensor set up for distance
6 LED

UAE0675-3Y

Two-step Lambda oxygen sensors

Application

These sensors are used in gasoline engines equipped with two-step Lambda oxygen sensor. They extend into the exhaust pipe between the engine's exhaust manifold and the catalytic converter, and record the exhaust-gas flow leaving each cylinder. Thanks to the Lambda oxygen sensor being heated, it can be installed further away from the engine so that long periods of full-load engine operation are unproblematical. The LSF4 sensor is also suitable for operation with exhaust-gas control systems featuring a number of sensors (for instance with OBDII).

Two-step Lambda oxygen sensors compare the residual-oxygen content in the exhaust gas with the oxygen content in the reference atmosphere (surrounding air inside the sensor) and indicate whether a rich (λ < 1) or lean air-fuel mixture (λ > 1) is present in the exhaust gas. The sudden jump in the characteristic curve of these sensors permits air-fuel control of $\lambda = 1$ (Fig. 1).

Operating principle

The two-step Lambda oxygen sensors operate in accordance with the principle of the galvanic oxygen-concentration cell with solid-state electrolyte (Nernst principle). The ceramic element is conductive for oxygen ions from a temperature of approximately 350 °C (safe, reliable operation at >350 °C). Due to the abrupt change in the residual-oxygen content on the exhaust-gas side in the range of $\lambda = 1$ (e.g. $9 \cdot 10^{-15}$ % vol. for $\lambda = 0.99$ and 0.2 % vol. for $\lambda = 1.01$), the different oxygen content on both sides of the sensor generates an electrical voltage between the two boundary layers. This means that the oxygen content in the exhaust gas can be used as a measure of the air-fuel ratio. The integrated heater ensures that the sensor

functions even at extremely low exhaust-gas temperatures.

The sensor's output voltage U_S is a function of the oxygen content in the exhaust gas. In the case of a rich air-fuel mixture (λ<1) it reaches 800 to 1,000 mV. For a lean mixture (λ>1), only approximately 100 mV are generated. The transition from the rich to the lean area at U_{reg} is at about 450 to 500 mV.

The temperature of the ceramic element influences its ability to conduct the oxygen ions, and thus the shape of the output-voltage curve as a function of the excess-air factor λ (the values in Figure 1 are therefore temperature-dependent). In addition, the response time for a voltage change when the mixture composition changes is also strongly dependent on temperature.

Whereas response times at ceramic temperatures below 350 °C are in the seconds range, at optimum temperatures of around 600 °C the sensor responds in less than 50 ms. When the engine is started therefore, the Lambda control is switched off until the minimum operating temperature of about 350 °C is reached. During this period, the engine is open-loop-controlled.

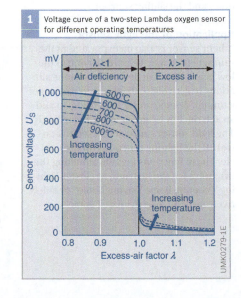

1 Voltage curve of a two-step Lambda oxygen sensor for different operating temperatures

Fig. 1
a Rich mixture
 (air deficiency)
b Lean mixture
 (excess air)

Design
LSH25 finger-type sensor
Sensor ceramic element with protective tube
The solid electrolyte is a ceramic element
which is impermeable to gas. It is a mixed
oxide comprising the elements zirconium
and yttrium in the form of a tube closed at
one end (finger, Fig. 3). The surfaces have
been provided on both sides with elec-
trodes made from a microporous, thin
noble-metal layer.

The ceramic body protrudes into the
exhaust pipe, and the platinum electrode
on its outside surface acts as a catalytic
converter in miniature. Exhaust gas which
reaches this electrode is processed cata-
lytically and brought to a stoichiometrical
balance ($\lambda = 1$). In addition, the side that is
exposed to the exhaust gas has a porous,
ceramic multiple layer (spinel layer) to
protect it against contamination and

3 Configuration of a finger-type Lambda oxygen sensor in the exhaust pipe

Fig. 3
1 Sensor ceramic element
2 Electrodes
3 Contacts
4 Housing contact
5 Exhaust pipe
6 Ceramic protective layer (porous)
7 Exhaust gas
8 Outside air
U_S Sensor voltage

2 LSH25 heated finger-type Lambda oxygen sensor (view and section)

1 cm

Fig. 2
1 Sensor housing
2 Ceramic support tube
3 Connecting cable
4 Protective tube with slots
5 Active sensor ceramic
6 Contact element
7 Protective sleeve
8 Heater element
9 Clamp-type connections for the heater element
10 Disk spring

erosive damage. The ceramic element is also protected against mechanical impact and thermal shocks by a metal tube. A number of slots in the protective tube are specially shaped so that on the one hand they are particularly effective against extensive thermal and chemical stresses, while on the other hand preventing a sharp drop in sensor-ceramic temperature when the exhaust gas is "cool".

The sensor's "open" inner chamber is connected to the surrounding air, which acts as a reference gas (Fig. 3).

Sensor element with heater element and electrical connection
A ceramic support tube (Fig. 2, Item 2) and a disk spring (10) serve to locate, fix, and seal-off the active, finger-shaped sensor ceramic in the sensor housing. A contact element (6) between the support tube and the active sensor ceramic element provides the contact between the inner electrode and the connection cable.

The outer electrode is connected to the sensor housing by the metal sealing ring.

The sensor's complete internal structure is located and held in place by a protective metal sleeve (7) which at the same time acts as the support for the disk spring. The protective sleeve also guards against contamination of the sensor's interior. The connecting cable is crimped to the contact element which protrudes from the sensor, and is protected against moisture and mechanical damage by a temperature-resistant cap.

The tube-type (finger) Lambda oxygen sensor is also equipped with an electrical heater element. This ensures that the ceramic-element temperature remains sufficiently high, even at low engine load and thus low exhaust-gas temperature.

This external heating is so quick that the sensor reaches operating temperature 20 to 30 s after the engine has started and therefore Lambda control can come into operation. Finally, sensor heating provides for an optimal ceramic-element operating temperature above the operating limit of 350 °C and thus ensures low and stable exhaust-gas emissions.

Fig. 4
1 Porous protective
 layer
2 Outer electrode
3 Sensor foil
4 Inner electrode
5 Reference-air-
 channel foil
6 Insulation layer
7 Heater
8 Heater foil
9 Connection
 contacts

Fig. 5
1 Exhaust gas
2 Porous ceramic
 protective layer
3 Measuring cell
 with microporous
 noble-metal layer
4 Reference-air
 channel
5 Heater
U_A Output voltage

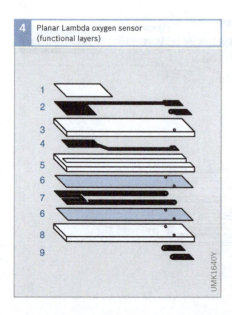

4 Planar Lambda oxygen sensor (functional layers)

5 LSF4 planar Lambda oxygen sensor (schematic diagram)

LSF4 planar Lambda oxygen sensor

Regarding their function, planar Lambda oxygen sensors correspond to the heated finger sensors with their voltage-jump curve at $\lambda = 1$. On the planar sensor though, the solid-state electrolyte is comprised of a number of individual laminated foils stacked one on top of the other (Fig. 4). The sensor is protected against thermal and mechanical influences by a double-walled protective tube.

The planar ceramic element (measuring cell and heater are integrated) is shaped like a long stretched-out wafer with rectangular cross-section.

The surfaces of the measuring cell are provided with a microporous noble-metal layer, which also has a porous ceramic layer on the exhaust-gas side to protect it against the erosive effects of the exhaust-residues. The heater is a wave-shaped element containing noble metal.

It is integrated and insulated in the ceramic wafer and ensures that the sensor heats up quickly even in the event of low power input.

The reference-air channel inside the LSF4 Lambda oxygen sensor – operating as a reference-gas sensor (Figs. 5 and 6) – has access to the ambient air. This enables it to compare the residual oxygen content of the exhaust gas with the oxygen content of the reference gas. In other words, with the air inside the sensor. This means that the sensor voltage also features a characteristic jump in the case of the planar sensor, in the area of stoichiometric air-fuel mixture ($\lambda = 1$) (Fig. 1).

6 LSF4 planar Lambda oxygen sensor (view and section)

1 cm

UMK1606Y

UMK1607Y

Fig. 6
1 Planar measuring cell
2 Double protective tube
3 Sealing ring
4 Seal packing
5 Sensor housing
6 Protective sleeve
7 Contact holder
8 Contact clip
9 PTFE grommet
10 PTFE shaped sleeve
11 Five connection cables
12 Seal

LSU4 planar wide-band Lambda oxygen sensor

Application
As its name implies, the wide-band Lambda oxygen sensor is used across a very extensive range to determine the oxygen concentration in the exhaust gas. The figures provided by the sensor are an indication of the air-fuel ratio in the combustion chamber. The excess-air factor λ is used when defining the air-fuel ratio.

The sensor projects into the exhaust pipe and registers the exhaust-gas mass flow from all cylinders. Wide-band Lambda oxygen sensors make precise measurements not only at the stoichiometric point λ = 1, but also in the lean range (λ>1) and in the rich range (λ<1). In combination with control electronics, these sensors generate an unmistakable, continuous electrical signal in the range from 0.7 < λ < ∞ (air with 21 % O_2) (Fig. 3). This means that the wide-band Lambda oxygen sensor can be used not only in engine-management systems with two-step control (λ=1), but also in control concepts with rich and lean air-fuel mixtures. This type of Lambda oxygen sensor is therefore also suitable for Lambda control with lean-burn concepts on gasoline engines, as well as for diesel engines, gas-powered engines and gas-powered central heaters and water heaters (hence the German designation LSU: Lambda-Sonde-Universal = universal Lambda oxygen sensor).

In a number of systems, several Lambda oxygen sensors are installed for even greater accuracy. Here, for instance, they are fitted upstream and downstream of the catalytic converter as well as in the individual exhaust tracts (cylinder banks).

Design
The LSU4 wide-band Lambda oxygen sensor (Fig. 2) is a planar dual-cell current-limit sensor. It features a measuring cell (Fig. 1) made of zirconium-dioxide ceramic (ZrO_2), and is a combination of a Nernst concentration cell (sensor cell which functions in the same way as a two-step Lambda oxygen sensor) and an oxygen pump cell for transporting the oxygen ions. The oxygen-pump cell (Fig. 1, Item 8) is so arranged with respect to the Nernst concentration cell (7) that there is a 10 to 50 µm diffusion gap (6) between them. Here, there are two porous platinum electrodes: one pump electrode and one Nernst measuring electrode. The diffusion gap is connected to the exhaust gas by way of a gas-access passage (10). A porous diffusion barrier (11) serves to limit the flow of oxygen molecules from the exhaust gas.

Fig. 1
1 Exhaust gas
2 Exhaust pipe
3 Heater
4 Control electronics
5 Reference cell with reference-air channel
6 Diffusion gap
7 Nernst concentration cell with Nernst measuring electrode (on the diffusion-gap side) and reference electrodes (on the reference-cell side)
8 Oxygen-pump cell with pump electrode
9 Porous protective layer
10 Gas-access passage
11 Porous diffusion barrier

I_P Pump current
U_P Pump voltage
U_H Heating voltage
U_{Ref} Reference voltage (450 mV, corresponds to λ = 1)
U_S Sensor voltage

1 Planar wide-band Lambda oxygen sensor (schematic design of the measuring cell and installation in the exhaust pipe)

On the one side, the Nernst concentration cell is connected to the atmosphere by a reference-air channel (5), on the other it is connected to the exhaust gas in the diffusion gap.

The sensor requires control-electronics circuitry to generate the sensor signal and to regulate the sensor temperature.

An integral heater (3) heats up the sensor quickly so that it soon reaches its operating temperature of 650 to 900 °C needed for generating a usable signal. This function decisively reduces the effects that the exhaust-gas temperature has on the sensor signal.

Operating principle

The exhaust gas enters the actual measuring chamber (diffusion gap) of the Nernst concentration cell through the pump cell's small gas-access passage. In order that the excess-air factor λ can be adjusted in the diffusion gap, the Nernst concentration cell compares the gas in the diffusion gap with that in the reference-air channel.

The complete process proceeds as follows: by applying the pump voltage U_P across the pump cell's platinum electrodes, oxygen from the exhaust gas can be pumped through the diffusion barrier and into or out of the diffusion gap. With the help of the Nernst concentration cell, an electronic circuit in the control unit con-

trols the voltage (U_P) across the pump cell in order that the composition of the gas in the diffusion gap remains constant at λ = 1. If the exhaust gas is lean, the pump cell pumps the oxygen to the outside (positive pump current). On the other hand, if the exhaust gas is rich, the oxygen (due to the decomposition of CO_2 and H_2O at the exhaust-gas electrode) is pumped from the surrounding exhaust gas and into the diffusion gap (negative pump current). At λ = 1, no oxygen needs to be transported, and the pump current is zero. The pump current is proportional to the exhaust-gas oxygen concentration and is this a nonlinear measure for the excess-air factor λ (Fig. 3).

3 Pump current I_P of a wide-band Lambda oxygen sensor as a function of the excess-air factor λ of the exhaust gas

2 LSU4 planar wide-band Lambda oxygen sensor (view and section)

1 cm

Fig. 2
1 Measuring cell (combination of Nernst concentration cell and oxygen-pump cell)
2 Double protective tube
3 Sealing ring
4 Seal packing
5 Sensor housing
6 Protective sleeve
7 Contact holder
8 Contact clip
9 PTFE grommet
10 PTFE shaped sleeve
11 Five connecting cables
12 Seal

Electric Actuators

Actuators (final-control elements) form the interface between the electronic signal processor (data processing) and the actual process (mechanical motion). They convert the low-power signals conveying the positioning information into operating signals of an energy level adequate for process control. Signal transducers are combined with amplifier elements to exploit the physical transformation principles governing the interrelationships between various forms of energy (electrical – mechanical – fluid – thermal).

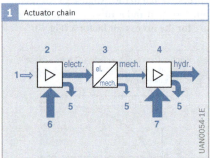

1 Actuator chain

Electromechanical actuators

This type of energy conversion represents one option for classifying electromechanical actuators. The energy emanating from the source is transformed into magnetic or electrical field energy, or converted to thermal energy. The individual force-generation principle is determined by these forms of energy, and is based on either field forces or certain specific material characteristics. Magnetostrictive materials make it possible to design actuators for applications in the micropositioning range. This category also includes piezoelectric actuators, which are built according to a multilayer design similar to ceramic capacitors, and are actuators for high-speed fuel injectors. Thermal actuators depend exclusively on the exploitation of characteristics of specific materials.

Actuators in a motor vehicle are mostly electromagneto-mechanical converters and, by extension, electrical servomotors, translational, and rotational solenoid actuators.

2 Electromechanical converters (system overview)

An exception is the pyrotechnic airbag system. The solenoid actuators can themselves be the servo element, or they can assume a control function by governing a downstream amplifier (e.g. mechanical-hydraulic).

Force generation in the magnetic field
The distinction between the electrodynamic and the electromagnetic actuator principles stems from the manner in which forces are generated in the magnetic field. Common to both principles is the magnetic circuit with soft-magnetic material and the coil for excitation of the magnetic field. A major difference lies in the force which can be extracted from the unit under technically feasible conditions. Under identical conditions, the force produced by application of the electromagnetic principle is greater by a factor of 40. The electrical time constant for this type of actuator is comparable to the mechanical time constants. Both force-generation principles are applied in linear and rotary drive mechanisms.

Electrodynamic principle
The electrodynamic principle is based on the force exerted on moving charges or charged conductors (Fig. 3a, current I) within a magnetic field (Lorentz force).

A field coil or a permanent magnet generates a DC magnetic field (magnetic flux density B). The electrical energy destined for conversion is applied (current I) to the moving armature coil (immersion coil). A high degree of actuator precision is achieved by designing the armature coil with low mass and low inductance. The two energy accumulators (one on the fixed and one on the moving component) produce two active force directions via current-direction reversal in the armature and field coils.

The secondary field produced by the armature current flows in an open magnetic circuit, thereby diminishing the effects of saturation. Approximately speaking, the force (torque) exerted by an electrodynamic actuator over its setting range is proportional to current and independent of travel.

Electromagnetic principle
The electromagnetic principle exploits the mutual attraction displayed by soft iron materials in a magnetic field. The electromagnetic actuator is equipped with only one coil, which generates both the field energy and the energy to be transformed. In accordance with the operating principles, the field coil is equipped with an iron core to provide higher inductance.

3 Electrodynamic and electromagnetic converters

a

or

b

UAE0309-1E

Fig. 3
a Electrodynamic principle:
 Force on moving charge
 $F = B \cdot I \cdot l \cdot \sin \alpha$

b Electromagnetic principle:
 Force between pole faces
 $F = B^2 \cdot A / 2\mu_0$

However, as the force is proportional to the square of the magnetic flux density B, the unit is operative in only a single force-transfer direction. The electromagnetic actuator thus requires a return element (such as a mechanical spring or a magnetic return mechanism).

Dynamic response

The dynamic response of an electromechanical actuator, i.e. the activation and deactivation operations, is defined by the equation of mechanical motion, the differential equation of electrical circuits and Maxwell's equations of dynamics. The current- and position-dependent force follows from Maxwell's equations.

The most basic electrical circuit consists of an inductance with an ohmic resistor. One means of enhancing the dynamic response is through over-excitation at the instant of activation, while deactivation can be accelerated by a Zener diode, for example. In each case, increasing the dynamic response of the electric circuit involves additional expenditure and increased losses in the actuator's triggering electronics.

Field diffusion is a delay effect which is difficult to influence in actuators with high dynamic response. Rapid switching operations are accompanied by high-frequency field fluctuations in the soft-magnetic material of the actuator's magnetic circuit. These fluctuations, in turn, induce eddy currents, which counteract their cause (build-up and decay of the magnetic field). The resultant delay in the build-up or reduction of forces can only be reduced by selecting appropriate materials with low electric conductivity and permeability.

Design

Design selection is essentially determined by operating conditions (i.e. installation space, required force/travel characteristic curve, and dynamic response).

Electromagnetic actuators

A typical form for translational electromagnetic actuators is the switching solenoid (Fig. 4) with a force/travel characteristic curve which falls as a function of the square of positioning travel. The precise shape of the curve (Fig. 5) is determined by the type of working air gap (e.g. conical or solenoid armature).

Rotational electromagnetic actuators are characterized by a defined pole arrangement in stator and rotor. When current is applied to one of the coils, the rotor and stator poles respond with mutual attraction, and in doing so generate a torque.

4 Switching solenoid

5 Switching solenoid (characteristic curves)

Force F →

Travel s →

Fig. 4
1 Armature
2 Coil
3 Magnetic yoke

Fig. 5
1 Solenoid plunger
2 Conical armature
3 Cylindrical
 armature

6 Electromagnetic single-winding rotary actuator

UAE0312-1Y

7 Torque motor

UAE0313-1Y

8 Electromagnetic rotary actuators (characteristics)

Torque $M \longrightarrow$

Control angle $\varphi \longrightarrow$

UAE0314-1E

The single-winding rotary actuator (Fig. 6) incorporates a pair of poles in each of the two main sections, as well as a coil in the stator. Its maximum adjustment range is approx. 45°.

The torque motor (Fig. 7) is a bidirectional electromagnetic rotary actuator featuring a stable operating point and without counterforces. The rotor is maintained in a stable position by the excitation field of the permanent magnet in the stator. The magnetic field generated by one or two stator windings produces torque and provides unilateral compensation for the excitation field. This type of layout is suitable for applications in which substantial torque is required over small control angles. The relationship between the applied current and the torque motor's force is roughly linear (Fig. 8). The torque-motor principle is also employed in translational actuators.

Electrodynamic actuators
In a pot magnet (immersion-coil actuator, Fig. 9) a cylindrical immersion coil (armature coil) is set in motion in a working air gap. The adjustment range is limited by the axial length of the armature coil and by the air gap.
 The short-stroke linear motor (Fig. 10) is an actuator with a virtually round disk coil.

A distinction is made between single-winding and dual-winding rotary actuators (Fig. 11). Both types include a permanent magnet within the rotor and one or two stator windings. The rotor magnet is magnetized at both ends to produce magnetic flux density in the rotary magnet's working air gap, which combines with the armature current to produce a torque. Originating from the position illustrated, the adjustment range is less than ±45°. The setting range of the single-winding rotary actuator also varies according to both the torque requirement and the angle range in which the necessary flux density can be provided.

Fig. 6
1 Coil
2 Stator
3 Armature

Fig. 7
1 Magnets
2 Armature
3 Control coils

Fig. 8
1 Single-winding
 rotary actuator
2 Torque motor

The dual-winding rotary actuator can be described as a combination of two single-winding rotary actuators with a 90° peripheral offset and designed to produce opposed torque flows. A stable operating point is achieved at the zero transition point on the resulting torque curve without auxiliary counterforces.

Application

Electromechanical actuators are direct-action control elements, and without an intermediate ratio-conversion mechanism, they convert the energy of the electrical control signal into a mechanical positioning factor/work. Typical applications include positioning of flaps, sleeves, and valves. The described actuators are final-control elements without internal return mechanisms, i.e. without a stable operating point. They are only capable of carrying out positioning operations from a stable initial position (operating point) when a counterforce is applied (e.g. return spring and electrical control).

A solenoid armature provides a stable static operating point when its force/travel curve is superimposed on the characteristic response of a return spring. A variation of the coil current in the solenoid shifts the operating point. Simple positioning is achieved by controlling the current. However, particular attention must be paid here to the nonlinearity of the force-current characteristic and the positioning system's sensitivity to interference factors (e.g. mechanical friction, pneumatic, and hydraulic forces). The temperature sensitivity of the coil resistance results in positioning errors, making corrective current control necessary. A high-precision positioning system with good dynamic response must incorporate a position sensor and a controller.

Fig. 9
1 Immersion coil
2 Permanent magnet
3 Magnetic yoke

Figure 10
1 Coil
2 Permanent magnet
3 Magnetic yoke

Fig. 11
a Single-winding
 rotary actuator
b Dual-winding
 rotary actuator

1 Coil 1
2 Coil 2
3 Stator
4 Permanent magnet
5 Shaft

9 Electrodynamic immersion-coil actuator

UAE0315-1Y

10 Electrodynamic short-stroke linear motor

UAE0316-1Y

11 Electrodynamic rotary actuator

UAE0317-1Y

12 Electrodynamic rotary actuator (characteristic curves)

a

Torque M

0

b

Torque M

0

$(i_1 \pm i_2)$

1

2

3

−45° 0° 45° 90°

Control angle φ

UAE0318-1E

13 Operating points (A) of a linear solenoid

Force F

A_1

A_2

A_3

1

2

3

4

Travel s

UAE0319-1E

Fluid-mechanical actuators

Hydraulic and pneumatic actuators utilize similar principles for the conversion and regulation of energy. Table 1 shows the differences in characteristics and applications.

In most applications, fluid-mechanical actuator drives are in the form of hydrostatic energy converters. These operate according to the displacement principle, converting the pressure energy of the fluid medium into mechanical work and vice versa (Fig. 14).

In contrast, hydrodynamic transformers operate by converting flow energy (kinetic energy of the moving fluid) into mechanical work (example: hydrodynamic clutch).

Losses during energy conversion stem from leakage and friction. Fluid-thermal losses are caused by flow resistance, in which throttle action transforms the hydraulic energy into heat. A portion of this heat is dissipated into the environment, and some of it is absorbed and carried away by the fluid medium:

$$Q_{heat} = Q_1 \cdot p_1 - Q_2 \cdot p_2$$

With incompressible fluids:

$$Q_{heat} = Q_1 \cdot (p_1 - p_2)$$

Fig. 12
a Single-winding
 rotary actuator
b Dual-winding
 rotary actuator

1 Coil 1
2 Coil 2
3 Coils 1+2

Fig. 13
1, 2, 3 Characteristic
 curves for various
 currents
4 Characteristic curve
 of a return spring

1 Comparison of hydraulic and pneumatic actuators

	Hydraulic actuators	**Pneumatic actuators**
Medium	– Fluid, mainly oil – Supplied from tank, oil sump – Virtually incompressible – Self-lubricating – Viscosity highly temperature-dependent	– Gas, mainly air – Supplied from ambient air – Compressible – Additional lubrication necessary – Viscosity changes practically irrelevant
Pressure section	– To approx. 30 MPa (200 MPa for diesel-fuel injectors)	– To approx. 1 MPa or greater (approx. 0.05 MPa for vacuum actuators)
Line connections	– Supply and return connection (possible leakage connection)	– Pressure connection only, return directly to environment
Applications	– Positioning applications with high load rigidity, demanding requirements for synchronization and positioning precision in closed-loop control system	– Actuators with low power requirement, positioning by mechanical stops, in open control loop

Table 1

The flow develops into turbulence at restrictions. The flow rate of the fluid is then largely independent of viscosity. On the other hand, viscosity does play a role in the case of laminar flow in narrow pipes and apertures.

Fluid-mechanical amplifiers control the conversion of energy from fluid to mechanical state. The regulating mechanism must be designed for control with only a very small proportion of the energy flow required for the ultimate positioning operation.

Switching valves open and close the orifice governing the flow to and from a fluid-mechanical energy converter (Fig. 15). Provided that the control-element opens sufficiently, the throttling losses remain negligible. Pulse-width-modulated opening and closing can be applied to achieve quasi-continuous control of the fluid-mechanical energy conversion process with virtually no losses. In practice, however, pressure fluctuations and mechanical contact between the valve elements result in undesirable vibration and noise.

Electrical machines

Electrical machines are used to convert electrical and mechanical energy. An electric motor converts electrical energy into mechanical energy, and an alternator converts energy in the opposite direction. Electrical machines (Fig. 16) consist of a stationary component (the stator) and a rotating component (the rotor or armature). There are special designs which depart from this configuration, such as linear machines which produce linear motion.

Permanent magnets or several coils (windings) are used to produce magnetic fields in the stator and rotor. This causes torque to develop between the two machine components. Electrical machines have iron stators and rotors in order to control the magnetic fields. As the magnetic fluxes change over time, stators and rotors must consist of stacks of individual laminations which are insulated with respect to one another (to minimize eddy-current losses).

The spatial arrangement of the coils and the type of current used (direct current, alternating current, or three-phase current) permit a number of different electrical machine designs. They differ from one another in the way they operate, and therefore have different applications.

14 Fluid-mechanical actuators

a

b

Fig. 14

a Linear actuator

$F = p_A \cdot A$

$s = Q_A / A$

b Rotary actuator

$M = (p_A \cdot V_{th})/2\pi$

$\varphi = (Q_A / V_{th}) \cdot 2\pi$

15 Energy regulation with 2/2-way switching valves

F_{load}

p_A Q_A

p_1 Q_1

p_2 Q_2

UAH0103-1Y

UAH0104-1E

Direct-current machines

The stator of a direct-current machine contains salient poles which are magnetized by the direct-current excitation windings. In the rotor (here also called the armature), the coils are distributed among slots in the laminated iron core and connected to a commutator. Carbon brushes in the stator housing wipe against the commutator as it rotates, thereby transferring direct current to the armature coils. The rotary motion of the commutator causes a reversal in the direction of current flow in the coils. The different rotational speed vs. torque characteristics result from the method selected for connecting the excitation winding and armature.

Series connection (series characteristics, Fig. 17)
- Clear relationship between speed and load
- High starting torque
- Unacceptably high speed possible at the moment of load reduction, requiring therefore a rigid connection with the load
- Torque reversal (change in direction of rotation) caused by a change in the current direction in the armature or excitation winding
- Application, e.g. as vehicle drive motor or starter for internal-combustion engines

Parallel connection (shunt characteristic, Fig. 18)
- Speed varies only slightly with load
- Change in direction of rotation caused by a change in the current direction in the armature or excitation winding
- Application, e.g. as drive motor for machine tools or direct-current generators

16 Layout of an electrical machine

Fig. 16
1 Stator
2 Rotor
3 Rotor coil
4 Commutator

Stator field generated by permanent magnet or electromagnet

17 Series-wound machine

18 Shunt-wound machine

A shunt characteristic can also be obtained by using a separate power supply for the excitation winding (external excitation) or by using permanent-magnet excitation in the stator.

Applications for permanent-field motors in motor vehicles: starter, windshield wiper, and low-power motors for various drives.

If the motor incorporates both series and shunt excitation windings (compound-wound motor), intermediate levels in the rotational speed/torque characteristic can be obtained.

Application: e.g. large starters.

All direct-current machines are easily capable of speed control over a wide range. If the machine incorporates a static converter which allows adjustment of the armature voltage, the torque and therefore the rotational speed are infinitely variable. The rotational speed can be further increased by reducing the excitation current (field weakening) when the rated armature voltage is reached. A disadvantage of direct-current machines is carbon-brush and commutator wear which makes regular maintenance necessary.

Three-phase machines

A three-phase winding is distributed among the stator slots in a three-phase machine. The three phases of current produce a rotating magnetic field. The speed n_0 (in rpm) of this rotating field is calculated as follows:

$$n_0 = 60 \cdot f / p$$

f = frequency (in Hz)
p = number of pole pairs

Three-phase machines are either synchronous or asynchronous, depending on rotor design.

Asynchronous machines

The laminated rotor contains either a three-phase winding, as in the stator, or a rod winding. The three-phase winding is connected to sliprings which are short-circuited either directly or via series resistors (Fig. 19). In the case of the rod winding, the bars are connected to one another by two short-circuiting rings (squirrel-cage rotor). As long as the rotational speed of the rotor deviates from n_0, the rotating stator field induces current in the rotor windings, thereby generating torque. Deviation of the rotational speed of the rotor n from n_0 is termed slip s:

$$s = (n_0 - n) / n_0$$

Continuous-running operation is only economical in the vicinity of n_0 because losses increase as slip increases (nominal slip $\leq 5\%$). In this range the asynchronous machine has a shunt characteristic. The machine operates as a motor when $n < n_0$, and as an alternator when $n > n_0$. The direction of rotation is changed by reversing two of the phases.

The asynchronous machine is the most frequently used electric motor in the field of drive engineering. With a squirrel-cage rotor it is easy to operate, and requires little maintenance.

19 Three-phase asynchronous motor with squirrel-cage rotor

a b

M 3~ M 3~

UAE0279-1Y

2	Examples of rotating-field speeds		
No. of poles (2 p)	Frequency		
	50 Hz	150 Hz	200 Hz
	Rotating-field speed in rpm		
2	3,000	9,000	12,000
4	1,500	4,500	6,000
6	1,000	3,000	4,000
8	750	2,250	3,000
10	600	1,800	2,400
12	500	1,500	2,000

Synchronous machines

In the rotor (here, also called the pole wheel), the poles are magnetized by direct-current coils. The excitation current is usually transferred via two sliprings to the rotor. The pole wheel can be made of solid steel, because the magnetic flux remains constant over time. Constant torque is generated as long as the rotor rotates at a speed of n_0. At other speeds, the torque fluctuates periodically between a positive and a negative maximum value, and excessively high current is produced. For this reason, a synchronous machine is not self-starting. The synchronous machine also differs from the asynchronous machine in that the reactive power absorption and generation are adjustable.

The synchronous machine is most frequently used as a generator of energy supply in electric power plants. Synchronous motors are used in cases where constant motor speed based on constant line frequency is desired, or where a reactive power demand exists. The automotive three-phase alternator is a special type of synchronous machine.

The rotational speed of all three-phase machines is determined by the stator frequency. Such machines can operate over a wide range of speeds if used in conjunction with static converters which vary the frequency.

Table 2

EC motors

The "electronically commutated direct-current" or "EC" motor is becoming increasingly popular (Fig. 21). It is essentially a permanent-magnet synchronous machine, and is brushless. The EC motor is equipped with a rotor-position sensor, and is connected to the DC power source through its control and power electronics. The electronic transfer circuit switches the current in the stator winding according to rotor position – the magnets which induce the excitation current are attached to the rotor – to provide the interdependence between rotational speed and torque which is normally associated with a separately excited direct-current machine. The respective magnetic functions of the stator and rotor are the opposite of what they would be in a classical direct-current machine.

20	Star-connected three-phase synchronous generator

21	EC motor

UAE0280-1Y

UAE0282-1Y

Fig. 20
Slip-ring rotor with excitation winding

Fig. 21
1 Electrical machine with rotor-position sensor
2 Control and power electronics
3 Input

The EC motor's potential applications are a result of the advantages which this drive principle provides: commutator and carbon brushes are replaced by electronics, dispensing with both brush noise and wear. EC motors are maintenance-free (long service life) and can be constructed to meet high degrees of protection. The electronic control feature makes it easy for drive units with EC motors to incorporate auxiliary functions such as infinitely variable speed governing, direction reversal, soft starts, and antilock protection.

The main areas of automotive application are in the HVAC (Heating/Ventilation/Air-Conditioning) sectors, and for pumps and servo units. In the area of production machinery, EC motors are chiefly employed as precision drive units for feed-control in machine tools. Here the decisive advantages are freedom from maintenance, favorable dynamic properties and consistent torque output with minimal ripple.

Single-phase alternating-current machines

Universal motors

The direct-current series-wound motor can be operated on alternating current if a laminated rather than a solid iron stator is used. It is then called a universal motor.

When operated on alternating current, a torque component at twice the frequency of the current is superposed on the constant torque component.

Single-phase asynchronous motors with squirrel-cage rotor

The simplest design of a single-phase asynchronous motor is a three-phase asynchronous machine in which alternating current is supplied to only two stator phases. Although its operation remains largely the same, the power and the maximum torque are reduced. In addition, the single-phase asynchronous machine is not self-starting.

Machines which are intended only for single-phase operation have only a single-phase main winding in the stator, as well as auxiliary starting circuits. The stator also contains an auxiliary winding connected in parallel with the main winding for this purpose. The necessary phase displacement of the auxiliary winding current can be achieved through increased winding resistance (low breakaway torque) or by means of a capacitor connected in series with the auxiliary winding (somewhat greater breakaway torque).

The auxiliary winding is switched off after the motor starts. The direction of rotation of the motor is changed by reversing the two auxiliary or main winding connections. A motor which has a capacitor in series with the auxiliary winding is called a capacitor motor. Capacitor motors with a starting and running capacitor also operate continuously with capacitor and auxiliary winding. Optimum operation for a specific operating point can be achieved by correct selection of capacitor. An additional capacitor is often used in order to increase the starting torque; this capacitor is then disconnected after the motor starts.

| 22 | Two-value capacitor motor |

Duty-type ratings for electrical machines
(VDE 0530)

S1: Continuous-running duty
Operation under constant load (rated output) of sufficient duration to reach the thermal steady-state condition.

S2: Short-time-duty type
Operation under constant load is so brief that the thermal steady-state condition is not reached. The rest period is so long that the machine is able to cool down to the temperature of the coolant.

 Recommended short-time-duty types: 10, 30, 60 and 90 min.

S3 to S5: Intermittent-periodic duty
Continuous alternating sequence of load and idle periods. The thermal steady-state condition is not reached during the load period or during the cooling period of one duty cycle.
- S3 intermittent-periodic duty without influence of starting on temperature
- S4 intermittent-periodic duty with influence of starting on temperature
- S5 intermittent-periodic duty with influence of starting and braking on temperature

S6: Continuous operation with intermittent loading
Operation with intermittent loading. Continuous alternating sequence of load periods and no-load periods, otherwise as S3.

S7: Uninterrupted duty
Operation with starting and braking.

S8: Uninterrupted duty
Operation with pole-changing.
For S3 and S6, the duty cycle time is 10 mins unless otherwise agreed; and recommended values for relative switch-on duration are 15, 25, 40 and 60 %. For S2, S3 and S6, the operating time or the duty cycle time and the switch-on duration are to be specified after the rating; the duty cycle time is only to be specified if it differs from 10 mins. Example: S2 - 60 mins, S3 - 25 %.

Relative switch-on duration
The relative switch-on duration is the ratio of the loading period, including starting and braking, to the cycle time.

Winding temperature
The mean temperature t_2 of the windings of an electrical machine can be determined by measuring the resistance (R_2) and referring it to an initial resistance R_1 at a temperature t_1:

$$t_2 = \frac{R_2 - R_1}{R_1}\ (\tau + t_1) + t_1$$

where

$$\tau = \frac{1}{\alpha} - 20\text{ K}$$

α = temperature coefficient

Electrohydraulic Actuators

Electrohydraulic actuators form the interface between electrical signal processing (data processing) and the system process (mechanics). They convert the low-power command signals into an actuating force with the increased power required for the process.

Application and Function

The transmission types most commonly in use today (AT, CVT, AST) are equipped with actuators to perform the most varied of functions. Table 1 provides an overview of the most important applications and highlights the link between the transmission functions and the actuator types that can be used.

The actuators are important switching and control elements in an electrohydraulic transmission-control system. They control the oil (fluid) flow and the pressure characteristics in the hydraulic control plate. The following different actuator types are available:
1. On/off solenoid valves (on/off, o/o)
2. Pulse-width modulated solenoid valves (PWM)
3. Pressure regulators (spool-type) (PR-S)
4. Pressure regulators (flat-seat-type) (PR-F)

In most automatic transmissions, these actuators currently serve as pilot-control elements, whose outlet pressure or volumetric flow is boosted in the hydraulic control plate before operating the clutches. In contrast, direct-action control elements without this boosting facility supply the clutches with suitably high pressure and volumetric flow.

Requirements

The installation location (on the transmission inside the oil pan) gives rise to far-reaching requirements and demanding operating conditions for the actuators that are used. Figure 1 provides a summary of these requirements.

As ever more transmissions are now being filled with lifetime oil (i.e., they do not need to have their oil changed) abrasion and dirt particles from breaking-in remain in the oil system throughout the entire operation of the transmission. Even central intake filters and individual filters on the actuators can only retain particles over a specific size. Excessively fine filters would soon clog up.

In addition to these considerations, the working life of transmissions is being increasingly extended: Transmissions in standard passenger cars are now expected to last at least 250,000 km (155,000 miles), and much more for taxis and similar applications.

In contrast to many other electromagnets (e.g. in ABS valves), actuators used in transmission-control applications must be designed over the entire temperature range for 100%

1 Transmission functions and associated actuators				
Automatic transmission (AT)				
Function	Actuator type			
	PWM	PR-F	PR-S	On/Off
• Control main pressure	X	X	X	
• Initiate gear change: 1-2-3-4-5-6		X		X
• Modulate shift pressure	X	X	X	
• Control torque converter lockup clutch	X	X		X
• Reverse-gear lock				X
• Safety functions		X		X
Continuously variable transmission (pulley CVT)				
Function	Actuator type			
	PWM	PR-F	PR-S	On/Off
• Adjust gear ratio		X	X	
• Control belt tension		X	X	
• Control start-up clutch	X	X	X	
• Reverse-gear lock				X
Automated shift transmission (AST)				
Standard electric-motor actuation:				
• Initiate gear change				
• Actuate clutch				
• Safety functions (fail-safe)				

Table 1

ON time, because, for example, they have to maintain the pressure while the gear is held or control a torque converter lockup clutch during driving. This creates a need to limit the power loss and to implement copper windings of suitably generous dimensions.

Vehicle and transmission-specific function characteristics (switching performance, control response, technical characteristic data), electrical and hydraulic interfaces tailored to individual applications, and the drive towards miniaturization and reducing costs are further boundary conditions for developing actuators for transmission-control applications.

PWM switching valves are suitable because of their pulse-width-modulated input signal (constant-frequency current, variable ratio of ON to OFF time) for use as pressure actuators, whose outlet pressure is proportional or inversely proportional to the pulse duty factor. The characteristic curve is referred to as rising or falling.

Pressure-control valves finally are operated with a controlled input current and can likewise be designed with a rising or falling characteristic. These valves are activated by analog means whereas PWM valves are activated digitally.

Design and Operating Concept

Automatic transmissions require *switching valves* for the simple on/off switching operations and/or *proportional valves* for infinitely variable pressure control. Figure 2 shows different ways of converting an input signal (current or voltage) into an output signal (pressure). Basically, actuators can be designed with proportional or inversely proportional performance characteristics.

On/off valves are usually voltage-controlled, i.e., the battery voltage is applied to the copper winding. The hydraulic section of the valve is designed as either a normally closed contact (n.c.) or as a normally open contact (n.o.).

1 Boundary conditions for actuator development

Dirt tolerance (rubbed-off elements, residues)
Fatigue limit ...5000 h (250,000 km)
Media resistance (ATF with additives and water content)
ON time 100%
Temperature load (–40...160°C)
Vibration strength of engine: up to 30 g
Temperature change
Miniaturization (mass and space reduction)
External mounting (splash water, salt spray, industrial climate)

STS0286E

2 Actuators (typical characteristics for switching and proportional valves)

Switching valves — On/Off valve

Input (Current *I*)
Output (Pressure *p*) — n.c. / n.o.
Time *t* →

Proportional-control valve — Pulse-width-modulated (PWM) valve

Current *I* → Time *t* →
Pressure *p* → Duty factor →

Pressure-control valve
Pressure *p* → Current *I* →

STS0285E

Actuator Types

Overview

The following types are examples of actuators commonly used for transmission-control purposes together with their characteristic values and features (overview, Figure 1). The examples are pilot-control actuators which operate within a pressure range of 400 to approx. 1000 kPa and act on a booster element in the transmission's hydraulic control system. Spool plungers in the hydraulic control system boost the pressure and/or the volumetric flow.

The actuators described can be designed in any concrete application at their interfaces in line with the conditions in the transmission, e.g.
- mechanically (mounting),
- geometrically (installation area),
- electrically (contacting), or
- hydraulically (interface to control plate).

The function data must be constructively adapted to the requirements in the transmission, particularly with regard to
- feed pressure and
- dynamics (i.e., reaction speed and control stability).

On/Off Solenoid Valves

On/off solenoid valves (Figure 2) are most commonly used as simple 3/2 valves (3 hydraulic ports/2 positions). They are clearly simpler in design than the 4/3-valves more commonly found in stationary hydraulics, and are therefore inexpensive. Nor do they have the drawbacks of 2/2 valves (such as high leakage or limited flow rate, necessitated by their action in combination with an external restrictor).

2 On/off solenoid valves (view and oil flow)

a

b

F_{mag}

1

p_{clutch}

p_{feed}

F_{feed}

Fig. 2
a View as switching valve with ball seat
b Oil flow in valve
1 Return to tank
F_{mag} Magnetic force
F_{feed} Feed-pressure force
p_{feed} Feed pressure
p_{clutch} Clutch pressure

1 Actuators (overview with examples)

Solenoid valves Hydraulic module Pressure regulators

Switching valves Flat-seat valves

PWM valves Spool valves

On/off solenoid valves are mainly used in simple 3-speed or 4-speed transmissions without overlap control. However, they are being used increasingly less in progressive or complex transmission-control systems (possibly still for safety functions). A pressure regulator controls the gearshift.

In the 3/2 switching valve (n.o.) depicted in Figure 3, the feed pressure p_{feed} generated by the transmission pump is applied before the flange (P) and seals the ball seat. This feature is referred to as self-sealing. Since no pressure is applied at zero current in the working-pressure channel (A), the valve is closed at zero current.

In this state, the working pressure, which in the end supplies the consumer (e.g. a clutch), is directly connected to the return line to the tank (oil pan) so that a pressure that is applied there can reduce or an oil volume contained therein can drain.

When current is applied to the switching-valve winding, the magnetic force generated reduces the working air gap, and the armature moves with its permanently connected push rod towards the ball and opens it. Oil flows to the consumer (from P to A) and builds up the pump pressure there. The return line to the tank is closed off at the same time.

Figure 4 shows the characteristic of the 3/2 switching valve with the pressure switching cycles.

A switching valve of this type offers the following features:
- low costs,
- non-susceptible to dirt and contamination,
- low leakage, and
- simple actuation electronics.

Areas of application for On/off switching valves are:

- **Gear change**
 with multiple use of the same main pressure regulator.

- **Safety functions**
 e.g. hydraulic disabling of the reverse gear during forward travel.

- **Torque converter lockup clutch**
 Cutting in and out (often controlled with a PWM valve or pressure regulator for economy or convenience reasons).

- **Changeover of register pump**
 Selected for two different flow-rate ranges via o/o (mainly for CVT transmissions).

3 3/2 switching valve (sectional view)

A
T
P
p_{feed}

UTS0289E

4 3/2 switching valve (characteristic with switching cycles)

kPa
600
500
Pressure p
400
300
200
100
0

0 10 20 30 40 ms
Time t

UTS0290E

Fig. 3
A Working-pressure channel
P Feed
T Return to tank
p_{feed} Feed pressure

Characteristic data (typical example):
Feed pressure 400...600 kPa
Flow rate > 2.5 $l \cdot$ rpm
Operating voltage 9...16 V
Resistance 12.5 |
Number of switching cycles $2 \cdot 10^6$

PWM Valves

PWM valves (Figure 5) have essentially the same design as switching valves. Since they operate with a frequency of 30...100 Hz, they must be designed for a higher switching speed (dynamics) and higher mechanical load (wear). The latter is particularly applicable if a PWM valve is used for main-pressure control and is operated throughout the life of the vehicle.

These requirements are reflected in the design. The sectional view in Figure 6 shows the incorporation of a rim on the armature, which ensures a high magnetic force and a high hydraulic damping action when closing. The relatively long push rod absorbs pulse forces. An annular seat seals the feed pressure in the zero-current state shown against the working-pressure channel.

Compared with the ball seat, the advantage of this type is that the feed pressure acts only on minimal surface areas. These surfaces are still pressure-compensated to the greatest extent, i.e., the applied feed pressure has an opening and closing effect such that essentially the compression spring ensures reliable closing. Only minimal opening forces have to be applied here, which results in high dynamics (switching speed) and keeps the coil size and inductance low.

Additional requirements in relation to the accuracy of the characteristics place stricter demands on precision during the manufacture of the valve components and during their installation than is the case in a pure on/off switching valve.

5 PWM valve (view and oil flow)

Fig. 5
a Sectional view of PWM valve
b Oil flow in valve
1 Return to tank

F_{mag} Magnetic force
F_{feed} Feed-pressure force
p_{feed} Feed pressure
p_{clutch} Clutch pressure

6 3/2 PWM valve with rising characteristic (sectional view)

Fig. 6
1 Feed from pump
2 Return to tank
p_{work} Working pressure

7 3/2 PWM valve (with rising characteristic)

In all, PWM valves exhibit the following features:
- low costs,
- non-susceptible to dirt and contamination,
- free from hysteresis,
- low leakage, and
- simple actuation electronics.

However, drawbacks are
- pressure pulsation and
- dependence of the characteristic on the feed pressure.

Areas of application for the PWM valves with rising characteristic or high flow rate described in the following are:
- control of the torque converter lockup clutch,
- clutch control, and
- main-pressure control.

3/2 PWM Valve with Rising Characteristic

The pressure-compensated tubular design of the closing element of this PWM valve (Figure 6) results in low inertial forces. Even this circumstance meets the demand for fast switching times, low noise emission, and long service life. The pronounced linearity of the characteristic (Figure 7) in a broad characteristic and temperature range is an important advantage for use in the vehicle.

The ends of the characteristic demonstrate slight discontinuity. This is caused by the transition from the switching to the holding operating state (when closed or open). This inaccuracy is tolerable in these closely limited areas.

3/2 PWM Valve with High Flow Rate

The PWM valve with high flow rate has essentially the same design as the standard PWM valve described above, but provides larger opening cross-sections with a greater diameter and a longer opening stroke of the closing element (comparison in Table 1). This design calls for a larger copper winding with a higher magnetic force (Figures 8 and 9).

1	Technical data of PWM valves in comparison		
PWM valve type		**Rising characteristic**	**With high flow rate**
Feed pressure	kPa	300...800	400...1200
Flow rate (at 550 kPa)	l·rpm	> 1.5	> 3.9
Timing frequency	Hz	40...50	40...50
Coil resistance	Ω	10	10
Dimensions			
Diameter	mm	25	30
Free length	mm	30	42

Table 1

8	3/2 PWM valve with high flow rate (sectional view)

9	3/2 PWM valve with high flow rate (characteristic)

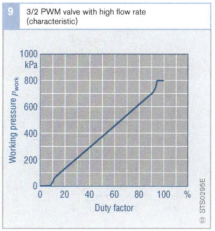

Fig. 8
1 Feed from pump
2 Return to tank
p_{work} Working pressure

Pressure Regulator

Two principles are used in the analog valves for pressure control (Figure 10):

- The *spool-type* pressure regulator opens a metering notch at the feed port and simultaneously closes a metering notch to the tank return line (two-notch regulator). The position of the regulator plunger is derived from the equilibrium of forces, depending on the stamped magnetic force, the regulated pressure, and the spring force.
- The *flat-seat* pressure regulator operates as an adjustable pressure relief valve (single-notch regulator).

With both principles, the regulated pressure intervenes directly in the equilibrium of forces, which is why complete closed control loops are provided. The following text explains the flat seat and spool principles by way of examples.

Spool-Type Pressure Regulator PR-S

The spool-type pressure regulator (Figures 11 to 13) operates as a two-notch regulator. The regulated pressure is tapped between inlet and outlet metering notches, and is produced as a function of their opening ratio. Maximum pressure is obtained with the inlet opened and the outlet closed, while zero pressure is obtained under reversed conditions. In between these settings, this pressure regulator changes its plunger position in the equilibrium between force of pressure, spring force, and magnetic force in proportion to the current passing through the coil and adjusts the control pressure accordingly.

The recirculation of the regulated pressure to the end face of the plunger via an oilway in the transmission control plate closes the closed control loop (external recirculation). The regulated pressure can also be included by means of a stepped plunger or other measures as a resulting force in the equilibrium of forces and thus in the closed control loop (internal recirculation).

Fig. 10
a Spool type
b Flat-seat type
1 Feed from pump
2 Return to tank
F_{spring} Spring force
F_{hyd} Hydraulic-pressure force
F_{mag} Magnetic force
p_{reg} Regulated pressure to clutch
p_{feed} Feed pressure from pump

Fig. 11
a Cutaway view of spool-type pressure regulator
b Oil flow in valve
1 Feed from pump
2 Return to tank
p_{reg} Regulated pressure to clutch
p_{feed} Feed pressure from pump

10 Pressure regulator (basic layout)

11 D30 spool-type pressure regulator (view and oil flow)

The advantages of the spool-type pressure regulator are:
- high accuracy,
- non-susceptible to influencing quantities,
- low temperature sensitivity,
- non-susceptible to system leakage,
- low leakage, and
- zero pressure can be achieved.

The disadvantages of the spool-type pressure regulator are:
- expensive production of the precision components, and
- precision electronics required.

By combining a frictionless mounting (diaphragm spring) with a Teflon-coated plain bearing, this pressure regulator ensures minimum hysteresis and optimum accuracy. The presence of a spool makes the regulator largely resistant to the influences of system leakage, fluctuations in feed pressure, or temperature changes.

The spool-type pressure regulator shown in Figures 11 and 12, with its typical technical data listed in Table 2, exhibits none of the drawbacks of the flat-seat pressure regulator discussed below, but is more expensive than the latter due to its higher-quality individual parts (expensive flange, precision-machined plunger).

As with the flat-seat pressure regulator, there are versions available with rising and falling characteristics (Figure 13).

Flat-Seat Pressure Regulator PR-F

Like its spool-type counterpart, the flat-seat pressure regulator is also a proportional valve. The magnetic force is proportional to the current passing through the coil. The hydraulic pressure acts via the sensing surface on the equilibrium of forces. The pressure regulator can also be fitted with a compression spring to be able to achieve a defined initial state. The regulated pressure is established by a pressure drop caused by oil flowing back to the tank through a varying cross-section.

2	Technical data of spool-type pressure regulator PR-S (typical)		
Feed pressure	kPa	700...1600	
Regulated pressure	kPa	typical 600...0	
Current range	mA	typical 0...1000	
Activation frequency	Hz	≤ 600	
Dimensions			
Diameter	mm	32	
Free length	mm	42	

Table 2

12 D30 spool-type pressure regulator (sectional view)

P_{work}

Fig. 12
1 Feed from pump
2 Return from tank
P_{work} Working pressure

13 Spool-type pressure regulator (falling characteristic)

The flat-seat-type pressure regulator is therefore essentially an adjustable pressure limiter with a hydraulic control function.

The essential features of the flat-seat pressure regulator are:
- high accuracy,
- inexpensive,
- non-susceptible to influencing quantities,

- non-susceptible to dirt and contamination,
- high leakage,
- residual pressure present (dependent on temperature), and
- expensive electronics.

Flat-seat pressure regulator, falling characteristic

Figures 14 to 16 show the example of a D30 flat-seat pressure regulator with falling characteristic.

The pressure regulator operates in conjunction with a restrictor (diameter 0.8... 1.0 mm) which is either arranged externally in the hydraulic-control system or integrated directly in the pressure regulator. The advantage of the latter type of restrictor lies in the fact that the pressure-regulator characteristic (of the variable restrictor at the flat seat) can be more

precisely matched with the upstream permanent restrictor. Suitable measures can be used even to compensate temperature sensitivity to a certain extent.

This pressure regulator described by way of example has been specially optimized such that the pressure/current characteristic exhibits low hysteresis and a narrow tolerance range. These properties are achieved by using high-quality magnetic-circuit materials and modern manufacturing procedures.

Since the ratio of the hydraulic resistance of both restrictors cannot be indiscriminately small, the pressure/current characteristic of a typical flat-seat pressure regulator exhibits a residual pressure which increases as the temperature decreases. The hydraulic-control system of the transmission must take this cir-

14 D30 flat-seat pressure regulator, falling characteristic (view and oil flow)

a

b

15 D30 flat-seat pressure regulator, falling characteristic (sectional view)

P_{work}

16 Flat-seat pressure regulator (falling characteristic)

Regulated pressure p_{work}

Current I

Fig. 14
a Cutaway view of flat-seat pressure regulator
b Oil flow in valve
1 Feed from pump
2 Restrictor
3 Clutch
4 Return to tank

Fig. 15
1 Feed from pump
2 Return to tank
p_{work} Working pressure

cumstance into account; accordingly, the effective range for control does not begin at 0 kPa, but rather at a suitably higher pressure.

Another drawback of the flat-seat pressure regulator manifests itself particularly when several of these pressure regulators are used in a transmission: The system dictates that there is a permanent return oil flow through the opened pressure regulator to the oil pan, which results in energy losses and under certain circumstances may require the use of a transmission-oil pump with increased volumetric flow. These drawbacks can be avoided by increasing the level of sophistication in the actuator or in the hydraulic-control system. However, this cancels out to some extent the main benefits of the flat-seat pressure regulator (simple design and low costs). The options offered by the closed-end function are discussed in the following.

Flat-seat pressure regulator, rising characteristic (miniature version)
Thanks to the use of high-precision plastic technology, the D20 flat-seat pressure regulator (Figures 17 to 19) has undergone a significant reduction in both size and cost.

With a diameter of little more than 20 mm, it achieves a high level of characteristic precision while taking up the minimal amount of space.

17 D20 flat-seat pressure regulator, rising characteristic (cutaway view)

UTS0314Y

18 D20 flat-seat pressure regulator, rising characteristic (sectional view)

1

P_{work}

2

UTS0304E

Fig. 18
1 Feed from pump
2 Return to tank
P_{work} Working pressure

3	Technical data of flat-seat pressure regulators in comparison		

Pressure-regulator type		D30 falling characteristic	D20 rising characteristic
Feed pressure	kPa	500...800	500...800
Regulated pressure typical	kPa	40...540	40...540
Current range typical	mA	150...770	150...770
Activation frequency	Hz	600...1000	
Chopper frequency	Hz		600...1000
Dimensions Diameter Free length	mm mm	30 33	23 42

19 D20 flat-seat pressure regulator (rising characteristic)

Regulated pressure P_{work}

600 kPa

400

200

0

0 0.2 0.4 0.6 A 0.8

Current I

STS0305E

Table 3

Deciding on the Pressure-Regulator Type

The decision as to whether a flat-seat pressure regulator, a spool-type pressure regulator, or even a PWM valve is used in the transmission depends on a whole range of considerations. Some technical criteria have already been discussed in the descriptions of the individual types.

These criteria are again set out and compared in Table 4. The figures in this table representing accuracy are not absolute, but only relative to each other. They are only approximate values and the conditions may differ completely according to the vehicle type used, for example depending on constructive details such as

- variability of the equilibrium of forces,
- variability of the magnetic force,
- use of special tools in the magnetic circuit,
- gap between feed pressure and maximum control pressure,

and system-conditioned criteria such as
- type of control,
- damping properties of the follow-up system,

- installation position (horizontal, vertical),
- installation location (above or below oil level or varying),
- adjustability of the closed control loop (end of line programming), and
- dirt concentration and dirt composition.

However, selection criteria such as the following are not to be underestimated:
- The user's own experience with a particular type (familiarity level, risk).
- In-house user traditions which can also lead to a wealth of single-sided experiences.
- Existing control-system concepts, for which the reworking expense in changing the regulator type is deemed to be too great.
- A cost analysis is conducted partly from a standpoint of reducing outsourcing costs. An analysis of the overall costs including, for example, the expenditure in internal production on control-plate machining is difficult for the user (existing equipment …).

4 Criteria for actuator selection

Criterion	Spool-type pressure regulator PR-S	Flat-seat pressure regulator PR-F	3/2 PWM valve
Sophistication in hydraulic system	non-susceptible to fluctuations in feed pressure	Constant feed pressure Feed restrictor	Constant feed pressure Damping
Accuracy: Comparison value (manufacturing tolerance)	\approx7% (recirculation) ±5...±25 kPa (dependent on characteristic range)	\approx11% ±5...±30 kPa (dependent on characteristic range)	\approx13% (control) ±20 kPa (constant)
Influence of feed pressure at $p_{feed} = 800 \pm 50$ kPa	$p_C = 400 \pm 0.2$ kPa	$\Delta p_C \approx 0.2 \cdot \Delta p_{feed}$	$\Delta p_C \sim \Delta p_{feed}$
Leakage	typ. 0.3 $l \cdot$ rpm	0.3...1.0 (...0) $l \cdot$ rpm	0...0.5...0 $l \cdot$ rpm (without elasticity)
Noise	–	–	If necessary damping required
Costs	High	Medium	Low

Table 4

Simulations in Development

Requirements
The development times (time to market) for new transmission generations are getting ever shorter. Testing-compatible actuators must be available with function data specific to the transmissions virtually immediately after the start of any project. The strict demands on quality and reliability necessitate comprehensive tests and trials with prototypes. The previously standard iterative method using recursions and modifications will in future not be practicable from time and cost standpoints.

In future, function prognoses and analyses of conspicuous features in the product-creation process using virtual prototypes will increasingly take place at an earlier stage and will at least support experimental development.

Computer-aided design of magnetic and hydraulic circuits thus forms the basis for simulation of the static and dynamic properties of the actuators in the transmission system. This simulation can be used to optimize functions and analyze properties under limit conditions.

Function Simulation
1D simulation enables, for example, the properties of pressure-control valves to be simulated in their system environment. These include pressure/current characteristics, temperature sensitivity, dynamics, etc., under different boundary conditions or geometric variables.

Function simulation of the actuator in the (sub)system is performed for example for the variables:
- characteristic,
- dynamics,
- temperature influence,
- influence of feed-pressure fluctuations, and
- worst-case studies, manufacturing tolerances.

Flow Simulation
The Fire tool used by Bosch to calculate hydraulic losses and flow forces enables, for example, the geometries of the hydraulic section of a pressure-control valve to be optimized is such a way that flow influences on the pressure/current characteristic are minimized under all operating conditions and flow cross-sections are optimized.

By way of example, Figure 1 shows the simulated flow characteristic and the pressure distribution in a flat-seat pressure regulator.

1 Simulation of flow characteristic and pressure distribution in a PR-F pressure regulator

b

0...21.7 m·s⁻¹

a

c

55.5...350 kPa

STS0306E

Fig. 1
a Pressure regulator (hydraulic section)
b Flow characteristic
c Pressure distribution

Magnetic-Circuit Calculation

A finite-element program (such as MAXWELL 2D or Edison) is used to design the magnetic circuit and thus the pressure/current characteristic of a pressure-control valve. But it can also be used to utilize to optimum effect the available space (size reduction) and the material of a magnetic circuit (magnetic-force increase, Figure 2) or to adapt the force/travel characteristic to the requirement.

The magnetic force is simulated, for example, for:
● dimensioning,
● the layout for the magnetic-force characteristic,
● eddy-current losses, and
● manufacturing tolerances.

The following improvements have been implemented as the result of the simulation (example in Figure 2):
● position and shape of the working air gap,
● optimized cross-section for a stronger magnetic flux, and
● reconfiguration of the working air gap and parasitic air gap (geometry, position).

Basic Principles of 1D Simulation

Motion Equation with Force Balance
The motion equation with force balance is as follows:

$$F_{tot} = \Sigma(F_M + F_p + F_{Fl} + F_{Fr} + F_D + F_S) = 0$$

With the variables
F_M magnetic force,
F_p pressure force,
F_{Fl} flow force,
F_{Fr} frictional force,
F_D damping force, and
F_S spring force

or according to Figure 3:

$$F_{tot} = m\ddot{x} + d\dot{x} + cx$$

With the variables
m moving mass,
d damping,
\ddot{x} acceleration,
\dot{x} velocity,
x travel,
c spring rate.

| 2 | Optimization of magnetic force with MAXWELL 2D (example) |

a b

$F_M = 5.8$ N $F_M = 10.2$ N

STS0307E

| 3 | "Force balance" model |

STS0308E

Fig. 2
a Basic model
b Optimization
F_M Magnetic force

Pressure Calculation in a Chamber

Pressure calculation in a chamber is governed by:

$$\frac{dp}{dt} = \frac{\beta_{oil}}{V_{oil}} \cdot Q$$

Where:
β_{oil} compression module of pressurized fluid,
V_{oil} chamber volume,
Q cumulative volumetric flow.

The cumulative volumetric flow flowing into the chamber, which is responsible for the pressure build-up, is calculated from (Figure 4):

$$Q = (Q_E - Q_A)$$

Where:
Q cumulative volumetric flow,
Q_A outlet volumetric flow,
Q_E inlet volumetric flow.

Flow-Rate Calculation at a Restrictor

The volumetric flow at a restrictor is calculated from (Figure 5):

$$Q = \alpha_f \cdot A_o \cdot \sqrt{\frac{2 \cdot \Delta p}{\varrho}} \; ;$$

where

$$\Delta p = p_1 - p_3$$

And where:
α_f flow-rate coefficient,
A_o cross-sectional area of a restrictor,
ϱ density of medium,
Δp pressure differential,
p_1 pressure at point 1,
p_3 pressure at point 3.

4 "Chamber pressure" model

5 "Restrictor flow" model

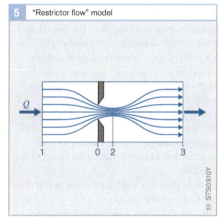

Electronic Transmission Control

In complicated traffic situations, unfamiliar surroundings, or poor weather conditions (e.g., heavy rain, snow, or fog), manual gear changing can distract car drivers to such an extent as to create situations that are difficult to control. This also applies to the annoying, incessant process of engaging and disengaging the clutch when driving in stop-and-go traffic. Automatic transmissions with electronic control assist drivers in these and other traffic situations so that they can concentrate fully on the road conditions and what is happening around them.

Drivetrain Management

As the number of electronic systems in the vehicle increases, so too does the complexity of the overall network of the various ECUs. Controlling such networked structures requires hierarchical order concepts, such as, for example, the "Cartronic" system from Bosch. Coordinated drivetrain management is integrated as a substructure in the "Cartronic" system. It facilitates optimally matched management of the engine and transmission in the various vehicle operating conditions.

The engine is operated as often as possible in the fuel-saving ranges of its program map. If the driver adopts a sporty driving style, however, the high, less economical speed ranges are increasingly utilized. Such a situation-dependent mode of operation presupposes on the one hand that the driver command is recognized and on the other hand that its implementation is left to the electronic drivetrain-management system and a higher-level driving strategy. When the driver presses the accelerator pedal, the system interprets this action as an acceleration request. From this request, the

drivetrain-management system calculates the conversion into torque and engine speed and implements them. In order for such a strategy to be implemented, it is essential for the system to be equipped with an electrically actuated throttle valve (drive by wire).

Figure 1 shows the organizational structure of drivetrain management as part of the overall vehicle structure. The vehicle coordinator forwards the requested propulsion movement to the drivetrain coordinator while taking into account the power requirements of other vehicle subsystems (e.g., body or electrical-system electronics). The drivetrain coordinator distributes the power demand to the engine, converter, and transmission. In the process, the various coordinators may also have to solve conflicts of interest that arise in accordance with defined priority criteria. A whole range of different external influencing factors (such as environment, traffic situation, vehicle operating status, and driver type) plays a role here.

The Cartronic concept is based on an object-oriented software structure with physical interfaces, e.g., torque as an interface parameter of drivetrain management.

1 Monitoring architecture of monitored drivetrain management

Market Trends

The statutory requirements relating to fuel consumption and exhaust-gas emissions will play a significant role in transmission development over the next few years. There follows a brief comparison of the requirements of the main markets for this purpose.

ACEA, JAMA and KAMA

The ACEA (Association des Constructeurs Européens d'Automobiles, i.e. Association of European Automobile Manufacturers) has agreed to reduce the corporate average in CO_2 emissions in the period from 2002 to 2008 from 170 mg CO_2 to 140 mg CO_2 (Figure 1).

The Japanese and Korean manufacturers' associations (JAMA and KAMA) have adopted the same limits for the year 2009. In order for this target to be achieved, the next few years will see an increased acceptance of transmission types such as the 6-speed transmission CVT (continuously variable transmission) and AST (automated shift transmission).

CAFE Requirements

In contrast to Europe, the USA, the most important market for automatic transmissions, has seen no change in the CAFE fuel-consumption requirements (corporate average fuel efficiency) since 1990 (Figure 2). All attempts to bring about a tightening of these requirements have proven unsuccessful.

1 CO_2-emission requirements

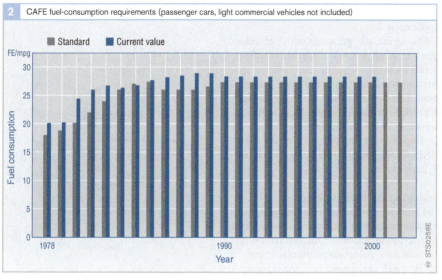

2 CAFE fuel-consumption requirements (passenger cars, light commercial vehicles not included)

Control of Automated Shift Transmission AST

Requirements

Current market developments reveal a marked trend towards an increase in in-car safety and operating comfort and convenience. This is accompanied by an increase in vehicle mass and in the final analysis an increase in fuel consumption. The emission guidelines laid down by legislators (140 g/km CO_2 by 2008) only intensify the situation.

The automated shift transmission (AST) combines the advantages of a manually shifted transmission with the functions of an automatic transmission. The automated version of the classic manual transmission is characterized by its high efficiency. Slip losses do not occur as in conventional torque-converter transmissions.

Specific fuel consumption in the automatic mode is below the low level of the manual transmission.

AST development is founded on the knowledge and findings gained in connection with electric-motor clutch management (ECM).

Electric-Motor Clutch Management (ECM)

Application

Following initial experiences with hydraulic clutch management, users are now concentrating their efforts on the use of electric motors as clutch actuators in the small-car segment. This new approach allows for savings to be made on costs and weight, as well as providing a higher degree of integration. Corresponding ECM systems are used in the Mercedes A-Class, the Fiat Seicento, and the Hyundai Atoz.

Design and Operating Concept

The most important step in minimizing costs was switching from a hydraulic actuator to an electric-motor actuator. This step removed the need for a pump, an accumulator, and valves as well as the need for a travel sensor in the clutch-release system. Instead the clutch-travel sensor is integrated in the electric-motor actuator.

The small electric motor offers a low power density in comparison with a hydraulic pump and accumulator. The electric-motor actuator is therefore only suitable if it can achieve sufficiently short declutching times for rapid gearshifts.

Shifting without Torque Correction

In the conventional system without torque correction (Figure 1), the clutch torque is far in excess of the engine torque. The reason behind this is that the dry clutch, which has to transfer at least the engine torque under all extreme conditions, normally offers a reserve of 50 to 150%. The engine torque drops when the driver wishes to change gear and at the same time takes his foot off the accelerator pedal. Operating the shift lever initiates the intention to change gear, and the clutch must now be moved from the fully closed to the fully open position. This defines the declutching time.

If the declutching time is too long, the clutch will still transfer torque while the next gear is being synchronized, a process which may result in rattling or damage to the transmission.

1 Shifting without torque correction

Shifting with Torque Correction

Shifting with Torque Correction
The most technically sophisticated solution for avoiding transmission damage is to combine a reduced-force clutch with torque correction.

Figure 2 shows the shifting operation with torque correction, in which the clutch torque is only marginally above the engine torque. When the driver takes his foot off the accelerator pedal to change gear, the clutch torque drops as well as the engine torque. When the intention to change gear is initiated, the clutch is thus already almost open and the remaining declutching operation follows very rapidly.

Figure 3 features a schematic representation of electric-motor clutch management (ECM) as a partial-automation solution, and the automated shift transmission (AST) as the complete automation of the manual transmission, both as add-on systems.

Electric-Motor Automated Shift Transmission AST

Application
Today the AST is used primarily in the lower torque classes (e.g., VW Lupo, MCC Smart, Opel Corsa Easytronic, see also chapter entitled "Transmission Types"), where, in comparison to the fully automatic transmission, the cost benefit makes up for the downside of the interruption in tractive force.

Design and Operating Concept
The electric-motor AST features the automated clutch operation of the ECM system. With an additional electric-motor actuator for the transmission, the driver is able to change gear without there being a mechanical connection between the selector lever and the transmission (shift by wire).

In the case of the AST, all modifications to the transmission are to be avoided. This will enable the transmission manufacturer to mount either manual transmissions or ASTs on the production line. As the hardware for this system (e.g., for the Opel Corsa Easytronic), Bosch supplies the electric motors for clutch engagement, shifting, and selection (see chapter entitled "Transmission, Automated Shift Transmission"), and the ECU. Automated and cost-effective mass production is made possible by the use of standard components in all AST applications.

2 Shifting with torque correction

3 ECM and AST as add-on systems

Fig. 2
M_C Clutch torque
M_E Engine torque
t_D Declutching time
S_1 Signal for gearshift command

Fig. 3
a ECM
b AST

1 Available signals
2 Clutch actuator with integrated ECM ECU
3 Gear recognition
4 Shift-intention recognition on shift lever
5 Clutch actuator with integrated AST ECU
6 Transmission actuator
7 Selector lever

Software Sharing

The vehicle manufacturer (OEM), the supplier and if necessary a system integrator share the AST software tasks. The operating system, signal conditioning, and the hardware-specific routines for activating the actuators are provided by Bosch. Bosch's extensive knowledge and experience in the field of automatic transmissions is also applied in establishing the AST target gears. This includes, among others, driver recognition, uphill/downhill recognition, cornering recognition, and other adaptive functions (see also chapter entitled "Adaptive Transmission Control, ATC").

The tasks of activating the transmission and coordinating the gearshift sequence (clutch, engine, transmission) are the responsibility of the OEM or the system integrator.

This also applies to clutch control, significant parts of which can be taken over from the ECM system. Each vehicle manufacturer brings its marque-specific philosophy regarding shifting time, shifting points, and shifting sequences to bear.

Shifting Operation and Interruption of Tractive Force

The basic problem associated with the AST is the interruption of tractive force. This is represented in Figure 4 by the "hollow" of the vehicle acceleration between the two shifted gears. In terms of what is required of the actuators, these phases can be divided into two blocks:

- phases which have an effect on the vehicle acceleration,
- phases which represent pure response times.

In the phases which have an effect on the vehicle acceleration, it transpires that a throttle action is needed because excessively quick changes in vehicle acceleration are felt to be unpleasant. The optimum interaction of engine, clutch, and transmission intervention results in the best possible performance.

The synchromesh can be supported for example by double-declutching. In the response times, however, the maximum speed of the actuators is demanded. It is important here that the synchromesh does not experience too hard a shock after the gear has been disengaged and the following rapid phase.

4 Phases of AST shifting operation

Fig. 4
1 Current gear
2 Next gear

ΔM_1 Torque reduction
ΔM_2 Torque increase
t_0 Tractive-force
 interruption
t_1 Shifting operation
t_2 Acceleration
t_3 Disengage and
 select gear
t_4 Synchronization
t_5 Shift through gear

Figure 5 shows a comparison of the shifting times that can be achieved with a hydraulic system and an electric-motor system and the shifting time necessary for a comfortable shifting operation. The bar lengths equate to the times required for the individual phases and the same shading schemes are used.

When the capacity of the actuators is exploited to maximum effect, the electric-motor system only demonstrates a time disadvantage compared with its hydraulic counterpart in the clutch-operation phase. This could be reduced in particular in the torque-reduction phase by an intelligent con-

trol strategy such as torque correction or by the interaction of the engine and the clutch.

It is important to highlight here that there is hardly any difference between the two systems in the phases for response time, gear disengagement, and gear engagement. The response times are not extended practically, even in the case of comfortable shifting. However, the phases relevant to acceleration must be two to four times as long as in the extreme case, both for the hydraulic and the electric-motor actuator systems.

5 Comparison of achievable shifting times

6 Automated shift transmission (AST) in a diesel vehicle (example: system diagram)

Fig. 6
1 Engine electronics
 (EDC)
2 Transmission
 electronics
3 Transmission
 actuator
4 Diesel engine
5 Dry interrupting
 clutch
6 Clutch servo unit
7 Intarder electronics
8 Display
9 Driving switch
 (selector lever)
10 ABS/TCS
11 Transmission
12 Air supply

——— Electrics
•••• Pneumatics
——— CAN communication

Control of Automatic Transmissions

Requirements

The control system for an automatic transmission must fulfill the following essential requirements or functions:

- always shifting the correct gear or setting the correct gear ratio as a function of assorted influencing variables,
- executing the shifting operation through adapted pressure characteristics as comfortably as possible,
- implementing additional manual interventions on the part of the driver,
- detecting maloperations, e.g. by preventing non-permitted gearshifts,
- providing ATF oil for cooling, lubrication and for the converter.

Current control systems are exclusively electrohydraulic in nature.

Hydraulic Control

The main function of the hydraulic-control system (Figures 1 and 2) is to regulate, boost, and distribute hydraulic pressures and volumetric flows. This includes generating the clutch pressures, supplying the converter, and providing the lubricating pressure. The housings of the hydraulic-control system are made from diecast aluminum and contain several precision-machined slide valves and electro-hydraulic actuators.

2 Main control with hydraulic valves

1 Exploded view of a hydraulic-control system (example: GM HYDRA-MATIC 4L60-E)

Electrohydraulic Control

Due to their extensive range of functions, all modern automatic transmissions with four to six gears and continuously variable transmissions are exclusively controlled by electrohydraulic means. In contrast to earlier, purely hydraulic control systems with mechanical regulators, the clutches are activated individually by pressure regulators, which facilitate precise modulation and regulated overlapping gearshifts (without a one-way clutch).

Clutch Control

Clutch control is always performed with either pilot-controlled or directly controlled pressure.

Pilot Control

With pilot control, the required pressure and throughflow for rapid clutch filling are provided via a slide valve in the control housing. Pressure regulation is effected by pilot pressure acting on the sensing surface of the slide valve. An actuator generates this pilot pressure (Figure 3).

This results in greater degrees of freedom in the packaging and the use of standardized actuators, high dynamics, and small electromagnets.

Direct Control

With direct control, the required pressure and throughflow for rapid clutch filling are provided directly by the actuator (Figure 4).

This results in a compact clutch-control system with reduced hydraulic sophistication.

Shifting-Sequence Control

Conventional Shifting-Sequence control
The following two shifting scenarios are examples of conventional control of a simple 4-speed automatic transmission with one-way clutches (Figure 5).

Upshift Under Load
Unlike in a manual transmission, throttling upshifts in an automatic transmission take place without an interruption of tractive force. The graphic in Figure 6 shows the time curve of the characteristic variables during an upshift into the direct gear (ratio 1). Shifting begins at time t_0: The clutch is filled with fluid and the friction elements are pressed against each other. The clutch transfers a torque from time t_1 onwards. As the clutch torque increases, so the torque supported at the one-way clutch decreases. The one-way clutch is released at time t_2. Now the engine speed begins to change. The clutch torque increases to t_3. The clutch slips up to t_4, after

3 Pilot control (pictorial diagram)

STS0266Y

4 Direct control (pictorial diagram)

STS0267Y

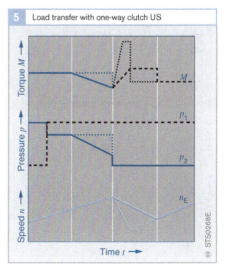

5 Load transfer with one-way clutch US

STS0268E

Fig. 3
1 Supply to actuator
2 Oil pan
3 Supply to slide valve
4 Actuator
5 Slide valve in control housing
6 Clutch

Fig. 4
1 Supply to actuator
2 Actuator
3 Clutch

Fig. 5
p_1 Pressure, cutting-in clutch
p_2 Pressure, cutting-out clutch
n_E Engine speed
M Torque

which it sticks. After the end of the shifting operation, the clutch pressure is controlled upward to a level of safety.

The speed difference between the engine and the transmission output speed remaining after t_4 is caused by the converter, which always operates with slip when locked up.

The progression of the output torque in the phase $t_1...t_4$ determines the gear-shift sophistication (ease of shifting). To ensure good shift quality, the clutch pressure must be set so that the output torque is between the level at $t < t_1$ and the level at $t > t_4$. The torque jump at t_4 should also be as low as possible.

The load on the friction elements is determined by the clutch torque and the slip time $(t_4 - t_1)$. It is clear here that controlling the shifting sequence always involves a compromise.

Downshift Under Load

In contrast to upshifts, downshifts take place with an interruption of tractive force. Figure 7 shows the time progression.

At t_0 shifting begins with the clutch being drained. Engine torque is no longer transferred from t_1 onwards and the engine revs up. At t_2 the synchronization speed of the new gear is reached and the one-way clutch is engaged; up to t_3 the converter slip is set according to the engine torque. The shifting operation is completed at t_3. The gear-shift sophistication (ease of shifting) is determined by the torque drop in the phase $t_0...t_1$ and depends quite significantly on the torque increase between t_2 and t_2.

All the shifting scenarios that occur are primarily controlled by the electronic system; the hydraulic system is left above all with the function of clutch power control.

In all newer transmissions (5-speed and 6-speed types), one-way clutches are replaced by regular clutches for weight reasons. However, during the shifting operations, they require overlap control

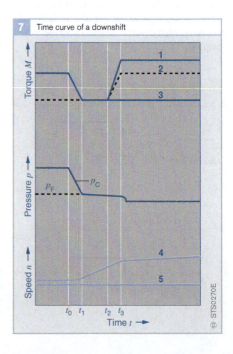

Fig. 6
p_C Clutch pressure
p_F Filling pressure
p_S Safety pressure
t_0 Start of shift
t_1 Start of torque transfer, clutch torque rises, one-way clutch torque drops
t_2 One-way clutch released
t_3 Clutch slips, clutch torque remains constant
t_4 Clutch sticks, clutch torque decreases, converter operates with slip

1 Output
2 One-way clutch
3 Clutch
4 Engine
5 Transmission output

Fig. 7
p_C Clutch pressure
p_F Filling pressure
t_0 Start of shift, clutch drains
t_1 End of torque transfer, engine revs up
t_2 Synchronization speed of new gear reached, one-way clutch engaged, converter operates with slip
t_3 Shifting operation completed

1 Output
2 One-way clutch
3 Clutch
4 Engine
5 Transmission output

6 Pressure, torque and speed curve during an upshift

7 Time curve of a downshift

for the clutches (Figure 8). This means that while clutch 1 opens for gear x, clutch 2 must close for gear y. Since this type of control is very elaborate and time-critical, it is necessary to provide considerably higher computing power in the ECU than for the simple shifting sequences with one-way clutch shifts (see also chapter entitled "ECUs").

The most important features of overlap control are:
- low mechanical complexity,
- minimal space requirement,
- multiple use for different gear steps possible,
- high control precision for load transfer required,
- high software complexity for torque control,
- in event of incorrect control: excessive speed (engine races) or onset of a braking torque (extreme case: transmission blocking).

Adaptive Pressure Control
The function of adaptive pressure control is to achieve a consistently good shift quality over the entire working life of the transmission and the accompanying changes in the friction coefficients at the clutch surfaces. It also compensates for any potential deviation of the calculated torque or the torque transferred by the engine which can occur on account of changes to the engine or manufacturing tolerances.

In this case, an important role is played by pressure adaptation with the aid of the shifting times applied by the manufacturer. To this end, the applied shifting times are compared with the real shifting times that occur. If the measurements are repeatedly outside a pre-specified tolerance range, the pressure parameters pertaining to the shifting operation are incrementally adapted. A distinction is made here between the fill time and the slip time of the clutch.

8 Overlap control US

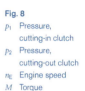

Fig. 8
p_1 Pressure, cutting-in clutch
p_2 Pressure, cutting-out clutch
n_E Engine speed
M Torque

9 Fill time

Fill-Time Measurement
The fill time t_{fill} (Figure 9) is the time from the start of the gearshift t_{shift} to the start of synchronization (a drop in speed is identified during the upshift [US]):

$$t_{fill} = t_{vertex} - t_{shift}$$

Slip-Time Measurement
The slip time t_{slip} (Figure 10) of the clutch is the time from recognition of the speed vertex (start of synchronization) to complete synchronization of the speed in the new gear.

10 Slip time

$t_{slip} = t_{sync} - t_{vertex}$

The speed thresholds used for measuring the slip time t_{slip} (Figure 10) are calculated in advance of the start of shifting, where the following relationship applies to upshifts: Start of fill-time measurement = start of shifting

Vertex: A decrease in the turbine speed n_{tu} by at least n_{tu} (vertex) revolutions is detected.
$$n_{tu}(t-1) - n_{tu}(t) > n_{tu} \text{ (diff)}$$

Synchronization speed: An increase in the turbine speed n_{tu} by at least n_{tu} (sync) revolutions is detected.
$$n_{tu}(t) - n_{tu}(t-1) > \Delta n_{tu} \text{ (sync)}$$

Pressure Correction
Pressure adaptation is only permitted within specific limits on account of operational reliability. The typical adaptation width lies in the range of ±10% of the modulation pressure calculated for the shift. The correction values are also still distinguished according to speed bands.

 The adaptation values are stored in a non-volatile memory so that the optimum modulation pressure can be reapplied when the vehicle is restarted. The overall pattern of pressure adaptation can also be evaluated as a sign of changes in the transmission.

Shifting-Point Selection
Conventional Shifting-Point Selection
In the majority of automatic transmissions currently available, the driving program is selected using a selector switch or a button. The following driving programs are generally available in this respect:
- Economy (very economical),
- Sport, or
- Winter.

The individual programs differ in the position of the shifting points in relation to the position of the accelerator pedal and the driving speed. The Economy and Sport shifting maps of a 5-speed transmission are used here as examples (Figure 11).

If the current driving speed or the accelerator-pedal position corresponding to the driver command (accelerator-pedal value) intersects the shift curve, a gearshift is triggered. A requested gearshift can be either canceled or converted into a double shift within a specific period of time (which depends on the hydraulic system of the automatic transmission)

 For example, the driver is driving in fifth gear on an interstate highway and would like to overtake. To do so, he presses the accelerator pedal to the floor, whereupon a downshift is requested.

11 US and DS characteristics in Economy mode (XE) and Sport mode (XS)

Fig. 11
1 Upshift
XE Economy mode
XS Sport mode

When the accelerator pedal is firmly pressed to the floor, the 4-3 DS shift curve is intersected directly after the 5-4 DS shift curve, and a 5-3 double downshift is performed instead of a sequential downshift. Special shifting points for kickdown (forced downshift) allow the maximum possible engine power to be utilized at this point.

Adaptive Transmission Control (ATC)

All newer transmission-control systems have – instead of active driving-program selection by the driver – software which enables the driver to adapt to the special ambient conditions while driving. This includes first and foremost driver-type recognition and driving-situation recognition. Examples which are currently in use are adaptive transmission control (ATC) from BMW and the dynamic shift program (DSP) from Audi.

Driver-Type Recognition

A driver type can be identified by means of an evaluation of the actions he or she performs. This includes:
- kickdown operation,
- brake operation, and
- restriction via selector lever.

For example, the kickdown evaluator counts the number of times the driver engages kickdown during a presettable period of time. If the counter exceeds a specific threshold, the driver-type recognition facility selects the next, more sporty driving program. It automatically switches back to a more economical driving program once this time has elapsed.

Driving-Situation Recognition

For driving-situation recognition, different transmission-control input variables are linked to conclusions about the present driving condition. The following situations can generally be recognized:
- uphill driving,
- cornering,
- winter operation, and
- ASC operation.

Uphill Driving

Recognition of uphill driving by comparing the current acceleration with the requested acceleration by way of the engine torque, results in upshifts and downshifts at higher engine speeds and thus prevents gearshift hunting.

Cornering

This facility uses the difference in wheel speeds to calculate whether the vehicle is in a curve or bend. With active cornering recognition, requested shifts are delayed or prohibited in order to increase vehicle stability.

Winter Recognition

Winter operation is recognized on the basis of slip detection from analysis of the wheel speeds. This serves primarily to
- prevent the wheels from spinning and
- select a higher gear during starting so that less torque is transferred to the drive wheels, thereby preventing premature wheel spin.

ASC Operation

If the system detects while driving that the ASC ECU (anti-slipping control or traction control system, TCS) is in control mode, requested gearshifts are suppressed in order to support the ASC function.

Engine Intervention

Application

A precisely controlled time characteristic of engine torque during the shifting operations of an automatic transmission offers the possibility of optimizing transmission control with regard to gear-shift sophistication (convenience), clutch service life, and transferrable power. The engine management system implements the torque command (reduction) of the transmission control by retarding the moment of ignition.

The theoretical principles, processes, and measurement results are presented using the example of engine intervention in ignition.

Symbols and Abbreviations

C	Spring stiffness of drivetrain
i	Gear ratio
J	Mass moment of inertia
k	Constant
M	Engine torque
n	Rotational speed
q	Specific lost work
Q	Lost work
t	Time
W	Running resistance
x	Spatial coordinate
δ	Temperature
ω	Angular velocity
Φ	Angle of rotation
φ	Angle of rotation, linearized

Indices

o	Output
v	Vehicle
limit	Permitted limit value
C	Clutch (friction element)
kin	Kinetic share
E	Engine (transmission input)
red	Reduced value
s	Slip time
com	Share of combustion energy (engine torque)
•	Reference variable
1	Clutch drive side
2	Clutch output side

Requirements

The ever-increasing demand for more economical fuel consumption in motor vehicles dictates to a significant degree the development objectives in the field of automatic transmissions as well. In addition to measures for improving the efficiency of the transmission itself (such as, for instance, the torque converter lockup clutch), these objectives include introducing transmissions with more gears. However, additional gear steps inevitably call for increased shift frequency. This in turn results in increased demands placed on gear-shift sophistication (convenience) and the load capacity of the friction elements.

Engine intervention takes into account both requirements and institutes an additional degree of freedom for controlling an automatic transmission. "Engine intervention" covers all those measures which allow the engine torque generated by the combustion process during the shifting operation in the transmission to be specifically influenced and in particular reduced. Engine intervention can be used in both upshifts and downshifts.

The primary aim of engine intervention in upshifts is to reduce the lost energy that occurs in the friction elements during the shifting operation. This is done by reducing the engine torque during synchronization without interrupting the tractive force. The margin acquired in this process can be used to:

- Increase the service life by shortening the slip time (if all other operating parameters in the transmission, such as clutch pressure and number of plates, remain unchanged).
- Improve the convenience by reducing the clutch torque, brought about by lowering the clutch pressure during the slip phase.
- Transmit higher power, provided the mechanical strength of the transmission permits this; in most cases, however, the power loss in the clutches is the limiting factor.

Naturally, it is also possible to adopt a sensible combination of these measures within the framework of the specified margin.

The aim of engine intervention in downshifts is to reduce the jolt which occurs when the one-way clutch or a friction element engages at the end of the synchronization processes. This results in
- improved convenience and
- supported and improved synchronization in transmissions without one-way clutches.

Interventions in the Mechanical Shifting Sequence

The following explanations illustrate which possibilities present themselves for intervention in the mechanical shifting sequence. The individual phases of upshifts and downshifts are described in the section entitled "Shifting-Sequence Control".

Upshifts
Engine intervention is discussed using the example of an upshift from the direct gear ($i = 1$) to overdrive ($i < 1$). The following simplifications serve to illustrate the physical relationships more clearly:
- The influence of the torque converter is disregarded.
- There is no overlap of friction elements, i.e., only *one* friction element participates in the gearshift.
- The engine torque remains constant during the gearshift, thereby providing linear speed characteristics.
- The vehicle speed during the gearshift is taken to be constant.
- The heating of the friction linings by briefly successive shifting operations is disregarded.

Upshifts take place without an interruption of the tractive force. Synchronization of engine and transmission takes place via a friction element in slipping-intervention operation. The following relative speed ensues between the drive and output side of the clutch:

$$\Delta\omega = \omega_1 - \omega_2 \tag{1}$$

In relation to the lost energy which must be absorbed or forwarded by the friction elements during the shifting operation, the following equation applies:

$$Q = \int_0^{t_s} M_C(t) \cdot \Delta\omega(t) \cdot dt \tag{2}$$

Furthermore, the angular-momentum principle applies to the drive and output sides of the clutch. For the rotational masses of the drive side:

$$\Delta\omega = \omega_O \cdot \frac{1-i}{i} + \frac{M_E - M_C}{J_O} \cdot t \tag{3}$$

Under the above-mentioned preconditions, this produces:

$$\Delta\omega = |\omega_E - \omega_O| = \omega_E\,(t=0) - \omega_O + \omega_E \cdot t$$

or

$$Q = M_C \left[\omega_O \cdot \frac{1-i}{i} \cdot t_s + \frac{M_E - M_C}{J_E} \cdot \frac{t_s^2}{2} \right]$$

From (1), (2), and (3), this produces for a time-constant clutch torque the lost energy as a function of the shifting-sequence parameters.

$$J_1 \cdot \dot\omega_1 = M_E - M_C$$

The slip time itself is dependent on the clutch and engine parameters, where

$$t_s = \frac{\omega_O}{|\dot\omega_E|} \cdot \frac{1-i}{i} = \omega_O \cdot \frac{1-i}{i} \cdot \frac{J_E}{|M_E - M_C|} \tag{4}$$

This produces the lost energy to be absorbed by the friction element

$$Q = \frac{1}{2} \cdot \frac{M_C \cdot J_E}{M_E - M_C} \cdot \omega_O^2 \left(\frac{1-i}{i}\right)^2 \tag{5}$$

i.e., the lost energy is dependent only on the clutch and engine torques, the driving speed, and the gear ratios.

When the clutch torque determined by (4) is applied in (5), this produces the lost energy as

the sum of a share of the kinetic energy which is released when the rotational masses are braked to the synchronization speed and a share of the engine combustion energy:

$$Q = Q_{kin} + Q_{com} = J_E \cdot \frac{\omega_0^2}{2} \cdot \left(\frac{1-i}{i}\right)^2 + M_E \cdot t_S \cdot \frac{\omega_0}{2} \cdot \frac{1-i}{i} \quad (6)$$

Both these shares are roughly of the same order of magnitude. At speeds of $n = 3000$ rpm and typical values for the gear step and the engine-drag torque ($i = 0.8$, $J_E = 0.3$ kg \cdot m^2, $M_E = 100$ Nm, $t_S = 500$ ms), this produces:

$$Q_{com}/Q_{kin} \approx 1...4$$

This clearly shows the possibilities of engine intervention for reducing the power loss in the friction elements.

A further significant aspect is derived from (6): Only the share of lost energy stemming from the combustion energy is dependent on the slip time t_S. The decisive factor is the product of the engine torque and the slip time. This means, however, that the slip time can be extended accordingly when the engine torque is reduced without an increase in the total lost energy. In actual fact, the wear of the friction elements even decreases with constant total lost energy when the slip time is extended. The temperature of the friction linings corresponds to the load on the friction elements.

Figure 12a shows the lost energy absorbed by the friction element as a function of the engine torque and the slip time. The maximum permitted lost energy Q_{limit} and the engine torque to be transferred during this gearshift determine the maximum slip time, for instance in accordance with point S. The maximum permitted energy Q_{limit} corresponds in accordance with (5) to the clutch torque determined by the slip time M_{Climit} (point 1 in Figure 12b).

To reduce the lost energy, the clutch torque in relation to point S would have to be increased and thereby the slip time shortened. However, this would lead in equal measure to

12 Lost energy Q_{com} (a) and clutch torque M_C (b) of a clutch as a function of slip time t_S and engine torque M_E

a reduction in gear-shift sophistication (convenience). A reduction of the clutch pressure is not permitted in this case, as otherwise Q_{limit} will be exceeded.

It is now easy to tell from Figure 12 which possibilities are offered by engine intervention. It is taken that the engine torque to be transferred $M_E = 100$ Nm can be reduced during the slip phase to an average of 50%. When first the case of constant clutch torque (shift quality) is considered, reducing the engine torque to 50 Nm results in a shortening of the slip time from 400 ms to 245 ms (point 1 → point 2) with a simultaneous reduction in lost energy to 61% (point 3). If, on the other hand, the slip time is kept constant, the

clutch torque can be reduced from 179 Nm to 128.5 Nm (point 1 → point 4) with a reduction in lost energy to 72% (point 5).

The maximum sensible slip time is then obtained if the minimum clutch torque M_{Cmin} during the gearshift does not drop below the value after the end of the shift. On the one hand, a smaller engine torque as a result of engine intervention would result in a deterioration in convenience; on the other hand, the clutch torque for safety reasons should at any rate be so large that the non-reduced engine torque can be transferred by the friction element after the end of the shift.

In this example, it is taken that the torque to be transferred at least by the clutch is 100 Nm in accordance with the engine torque (direct gear). This means that the slip time can be stretched from 400 ms to max. 625 ms (point 6), again with a simultaneous reduction in lost energy to 88% (point 7).

Finally, it can be gleaned from Figure 12 that even an engine torque of 200 Nm, which without intervention would require a maximum slip time of 200 ms with a minimum clutch torque of 360 Nm (point 8), can be traced back to the example with a torque of 100 Nm (point 1).

The results of this analysis are on the safe side in this respect because extending the slip time with constant lost energy results in reduced friction-lining temperature and thus protects the friction linings. Table 1 lists the numerical values for these examples.

Downshifts
In contrast to upshifts, downshifts take place in the throttling mode with an interruption of load. The engine is decoupled from the drivetrain and runs as a result of the torque it has generated up to the synchronization speed. Only after the one-way clutch or the friction element has engaged is the frictional connection re-established. The torque ratios when the synchronization speed is reached

1	Numerical values for text examples and Figure 12				
M_E	M_{redE}	M_C	M_E	t_0	Q/Q_{100}
Nm	Nm	Nm	Nm	ms	%
100	100	179	400	3740	100
100	50	179	245	2285	61
100	50	128.5	400	2693	72
100	50	100	628	3290	88
200	200	360	200	3740	100
200	100	179	400	3740	100

Table 1

determine the gear-shift sophistication (convenience) to a substantial extent.

For a better understanding of the characteristic relationships, damping in the drivetrain is disregarded in the following analysis. It also applies on the assumption that the overall vehicle dynamics can be reduced to engine mass, input-shaft rigidity, and vehicle inertia.

In the case of all the moments of inertia relating to the transmission output, the engine and the vehicle are governed by the following:

$$J_E \cdot \ddot{\Phi}_E = M_E, \quad J_V \cdot \ddot{\Phi}_V = -W \tag{9}$$

At the point when the one-way clutch engages, the drivetrain resembles a torsion damper (Figure 13, next page), and the motion equations are as follows:

$$J_E \cdot \ddot{\Phi}_E = c(\dot{\Phi}_V - \dot{\Phi}_E) + M_E \tag{10a}$$

$$J_V \cdot \ddot{\Phi}_V = c \cdot (\dot{\Phi}_V - \dot{\Phi}_E) - W \tag{10b}$$

Since in this case it is not the absolute angle of rotation but rather only the deviations from the basic rotation (i.e., the rotation of the input shaft) which are significant, these two equations can combined. If the driving speed v for the short time segments to be considered is taken as constant, this produces with

$$\Phi_E = \Phi_{E0} + \varphi_E, \quad \Phi_V = \Phi_{V0} + \varphi_F,$$
$$\dot{\Phi}_{E0} = \dot{\Phi}_{V0} = v = \text{const.}$$

and

$$\psi = \varphi_V - \varphi_E,$$

the motion equation

$$\ddot{\psi} + c \cdot \left(\frac{1}{J_E} + \frac{1}{J_V} \right) \cdot \psi = -M_E \qquad (11)$$

The natural frequency ω_0 for this system as follows

$$\omega_0 = \sqrt{c \cdot \left(\frac{1}{J_E} + \frac{1}{J_V} \right)}$$

The general solution produces the acceleration acting on the driver:

$$\psi = A \cdot (\sin \omega_0 \cdot t) + B \cdot \cos (\omega_0 \cdot t) \qquad (12)$$

(9) produces as the end condition of the engine run-up phase:

$$\ddot{\psi}_E = \frac{M_E}{J_E} \text{ for } t < t_0 \qquad (13)$$

The energy to be applied here simply to accelerate the engine is transformed when the one-way clutch engages (at time t_0) abruptly into a torque, which causes the input shaft to rotate:

$$\varphi_{E_0} = \frac{M}{c}$$

13 Diagram of drivetrain during load interruption and establishment of frictional connection

Fig. 13
a Load interruption
b Frictional connection

J_V Mass moment of inertia of vehicle drivetrain
J_E Mass moment of inertia of engine
M_E Engine torque
Φ_V Angle of rotation of vehicle drivetrain
Φ_E Angle of rotation of engine
W Running resistance

(12) produces the relative acceleration

$$\psi = \omega_0^2 \cdot \frac{M}{c} \cos (\omega_0 \cdot t)$$

and (10b) the vehicle acceleration

$$\ddot{\Phi}_V = \frac{M}{J_V} \cdot [1 + \cos (\omega_0 \cdot t)] \qquad (14)$$

This means that, at time t_0 when the one-way clutch engages, an acceleration jump takes place, namely

$$\text{from } \ddot{\Phi}_V = 0 \text{ for } t < t_0 \qquad (15)$$
$$\text{to } \ddot{\Phi}_V = 2 \cdot \frac{M}{J_V} \text{ for } t = t_0$$

followed by a drivetrain vibration damped in the real vehicle.

Similar conditions are present in a gearshift from friction element to friction element (overlapping gearshift), only this involves the additional problem of cutting in the friction element of the new gear exactly when the synchronization speed is reached. As the damping effects of the torque converter and the rest of the drivetrain have been disregarded in this analysis, the possibility which engine intervention offers is all the more clear:

According to (13), the initial acceleration acting on the driver at time t_0 is directly proportional to the engine torque and thus the engine acceleration during the run-up phase. With precisely timed control of the engine torque in the time segment $t \leq t_0$ to $t \gg t_0$, it is possible to create an almost continuous transition from the range of tractive-force interruption to the range of tractive-force transfer.

Implementation takes the form of a marked reduction in engine torque at time t_0 followed by renewed control-up in accordance with a time function. The convenience can be varied within broad limits with this control-up.

There is an equally clear possibility in gearshifts without a one-way clutch of influencing the engine acceleration by control-

ling the engine torque in the time range $t \leq t_0$ and thereby reducing the time demands on the cut-in precision of the friction element at the synchronization point.

Sequence Control

The process of reducing the engine torque is essentially very simple. However, effective control requires precise coordination as the entire process only lasts approximately 500 ms.

Pure time control of engine intervention is not practicable because different variables which determine the sequence (such as clutch fill times, plate friction coefficients, and similar) fluctuate within broad limits depending on the temperature and the service life.

As engine intervention is directly linked to the shifting sequence, a speed sequence control system suggests itself. The characteristic variable that characterizes the shifting sequence exactly is the transmission input sequence. The engine speed is also suited with limitations to transmissions with hydrodynamic converters as the controlled variable. This is therefore important because it requires a separate sensor to record the transmission input speed, and not every transmission is fitted with this sensor for cost reasons. In the interests of clarity, the following text describes the control system with the transmission input speed as the characteristic variable and reference is made, where necessary, to the limitations or changes when the engine speed is used.

Upshifts

The time curve of the characteristic variables for an upshift is depicted in Figure 14.

The ratio of the old gear is retained up to the overrunning point t_2; only then does only a slipping clutch intervene. For this reason, the engine torque cannot be reduced before the overrunning point is reached, otherwise this would entail an intensified dip in the output torque in the phase $t_1 \dots t_2$.

The overrunning point is identified through continual monitoring of the transmission input speed in the time phase after t_0. For this purpose, the maximum speed is calculated in the time phase $t_0 \dots t_3$ as well as the speed gradient.

The overrunning point is identified in the event of a reduction in the gradient by more than a pre-specified threshold value, and engine-torque control begins with control-down to a pre-specified value in accordance with a pre-specified time function.

In order to determine the speed n_3 at the start of control-up, the synchronization speed $n_4 = n_1/i$ in the new gear is calculated from the maximum speed n_1 at the overrunning point and the ratio jump i of the gear change to be carried out. A speed-dependent share Δn is added to this synchronization speed in order to obtain a derivative action for control-up. When the speed $n_3 = n_4 + \Delta n$ is reached, torque control-up begins in accordance with a pre-specified time function. The end of the gearshift is identified as soon as the value of the non-corrected torque is achieved.

14 Sequence control, upshifts with engine intervention

⊕ STS0277E

Fig. 14

1 Control-down phase
2 Control-up phase

a Acceleration
n_T Transmission input speed
n_E Engine speed
M_E Engine torque
S Shift signal

For upshifts in the upper load range (greater than half load), the engine speed is used as the controlled variable instead of the transmission input speed because here the shifting points are at such great engine speeds that the converter operates in the clutch range and is thus subject to roughly constant slip.

Gearshifts at part load on the other hand take place in the converter range. This means that the slip can change very substantially during a gearshift. Here the engine speed is no longer suitable for determining the synchronization speed. In this case, a superimposed form of time control, which terminates engine intervention after a pre-specified time, is suitable for the part-load range $t_3...t_4$.

Downshifts
The time curve of the characteristic variables for a downshift is depicted in Figure 15. Precise determination and recording of the synchronization speed are crucial to engine intervention in the case of downshifts because

- retarding the ignition angle too early extends the engine revving phase and thus the time of tractive-force interruption and
- engine intervention after the one-way clutch has engaged does not bring about any improvement in convenience, but rather a deterioration as this causes a dip in torque for the duration of the engine intervention.

The synchronization speed is calculated via the gear step from the transmission input speed at the start of the gearshift. The engine torque is abruptly reduced approximately 200 rpm before the synchronization speed is reached until this speed is reached or slightly exceeded. Then the engine torque is slowly controlled up again.

The synchronization speed cannot be calculated directly by means of the engine speed as a result of the slip at the hydrodynamic torque converter. A consideration of the converter program map with the requisite accuracy requires too much computation effort in the microcontroller.

It is possible however to calculate the synchronization speed from the transmission output speed by multiplying it by the corresponding gear step. It is now possible to identify by means of the engine speed when the synchronization point is reached, since the speed difference between the engine and the turbine is approximately zero when the engine revs up freely (interruption of tractive force) to the synchronization point.

The different possibilities of torque reduction are now discussed in the following text.

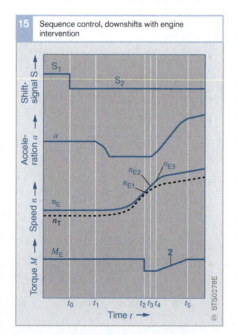

15 Sequence control, downshifts with engine intervention

Fig. 15
2 Control-up phase
a Acceleration
n_T Transmission input
 speed
n_E Engine speed
M_E Engine torque
S Shift signal

Shifting the Ignition Angle

The oldest version of engine intervention involves intervention through shifting of the ignition angle. This form of intervention offers the following advantages:
- continuous regulation of the engine torque within broad limits,
- short response time, and
- availability in all vehicles with gasoline-engines.

Figure 16 shows in schematic form the dependence of the engine torque on the ignition angle for different load conditions and engine speeds. It is clear from this figure that adjusting a pre-specified engine torque generally requires an ignition map as a function of engine load and engine speed.

The response time τ between the initiation of engine intervention and the start of reduction in engine torque is specified by the ignition angle, therefore

$$\tau \approx \cdot \frac{1}{(z/2) \cdot n_E}$$

where z is the number of cylinders and n_E the engine speed. In the effective speed range $n \geq 2000$ rpm, the maximum delay for a 6-cylinder engine is 10 ms for initiation and 30 ms for a complete reduction in engine torque.

Engine-Torque Specification

In appropriately equipped vehicles with their CAN network of all the ECUs in the drivetrain (Figure 17), torque reduction is performed on the basis of a torque interface between engine management (TI-Motronic) and electronic transmission control (ETC). The torque reductions of the ABS and TCS ECUs must also be taken into consideration.

Figure 18 shows how a current transmission-control system calculates the desired engine torque intervention (ETI_Etc).
 The next torque intervention is determined as a function of the available torque (actual torque). The torque M is the engine torque of the engine-management system without intervention by transmission control.

16 Engine torque as a function of ignition angle

Engine torque M_E

M_{F1} — α_{F1} **a**

M_{F2} — 25 °KW — α_{F2}

M_{P1} — **b** α_{P1}

M_{P2} — 20 °KW — α_{P2}

50 40 30 20 10 0 10 °KW

Ignition angle α ← | → advance retard

STS0281E

17 ECUs in the drivetrain

TI-Motronic ETC

High-Speed CAN

ABS TCS

STS0282E

Fig. 16
a Full load (index F)
α_{F1}, M_{F1} Ignition angle or engine torque without engine intervention
α_{F2}, M_{F2} Reduced ignition angle or engine torque with engine intervention
b Part load (index P)
α_{P1}, M_{P1} Ignition angle or engine torque without engine intervention
α_{P2}, M_{P2} Reduced ignition angle or engine torque with engine intervention

18 Calculation of engine-torque intervention (ETI) in current transmission control

Torque without transmission intervention

Limitation of torque gradient

Parameter of gradient limitation

Calculation of torque intervention from program maps and functions

Engine torque M

Gradient limitation

ETI_Dyn_Lim

Parameter of shift status

Calculate reduced torque

No intervention

S_Phase

Status check

TI_State

Torque-request calculation

Torque limitation

Shift-sequence phase

Torque-intervention status

Determination of torque intervention as a function of phase and status of shift sequence

Limitation of torque intervention

Desired torque intervention

UTS0279E

Key to Figure 18:
ETI_Etc = f(ETI_Dyn_Lim, TI_State)

where

ETI_ETC: Engine-torque request of transmission control

ETI_Dyn_Lim: Dynamically limited engine torque (reduction of torque change in the engine to ensure convenience)

TI_State: Current status of torque intervention

Torque Converter Lockup Clutch
Application and Operating Concept

The hydrodynamic converter is (necessitated by its operating principle) subject to a level of slip which is required particularly for convenience reasons during startup and in certain driving situations to increase torque. Since this slip also involves a simultaneous loss of power, the torque converter lockup clutch (TCLC) was developed with this in mind (see also section entitled "Torque Converter").

It is only advisable to lock up the converter from a specific speed because at low speeds the irregular rotation of the engine would cause uncomfortable vibrations in the drivetrain. The controlled torque converter lockup clutch (CTCC) was developed in order to be able to utilize these ranges for lockup as well.

Controlled Torque Converter Lockup Clutch

The controlled torque converter lockup clutch (CTCC) sets a very low level of slip (40...50 rpm) and thus an almost stationary state. In this way, it keeps unwanted vibrations away from the drivetrain. The converter lockup clutch therefore has three different states:

- open,
- controlled, and
- closed.

These states are defined by means of characteristic curves, which are plotted like shift curves for each gear against throttle-valve opening and driving speed (Figure 19). As is the case with shift curves, fuel consumption and tractive force are crucial criteria for the torque converter lockup clutch.

In slipping controlled operation, the speed differential between the converter impeller and turbine is constantly set to a low value. A closed control loop constantly compares the speed differential with a pre-specified set-point value and corrects the pressure continually. Special functions perform changeovers between the individual states and provide for a comfortable shift performance.

19 Characteristic (pictorial diagram) of controlled torque converter lockup clutch (CTCC)

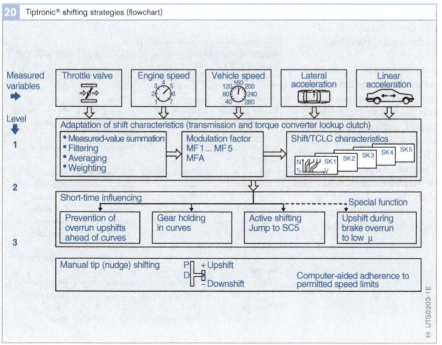

20 Tiptronic® shifting strategies (flowchart)

Control of Continuously Variable Transmission

Requirements

Continuously variable transmissions that operate according to the wrap principle have a whole variety of different equipment specifications (Table 1). The following equipment packages are widely used in compact and mid-range class vehicles:

- When the master/slave concept is used, the primary pulley (transmission input side) has double the surface of the secondary pulley (transmission output side). The pressure in the primary chamber can thus always be below the secondary pressure.
- The converter with torque converter lockup clutch as power take-up element offers very good starting convenience and facilitates a good starting response through torque increase so that the large ratio span of the CVT is completely beneficial to the overdrive range.
- Two wet clutches for the forward and reverse gears.
- Variable-capacity pump.
- Convenient fail-safe and limp-home strategies.

In the event of a control-electronics failure, the fail-safe and limp-home requirements partially determine the hydraulic concept. Engine overspeeding and associated high slip at the driven wheels must be avoided at all costs. An adjustment in the overdrive direction would satisfy this requirement, but starting from a standing stop would no longer be possible.

Open and Closed-Loop Control Functions

The aforementioned transmission equipment requires the following open and closed-loop control functions:

- contact-pressure control,
- ratio control,
- driving program,
- clutch activation,
- activation for converter and torque converter lockup clutch,
- pump activation,
- reverse-gear lock, and
- deactivation of limp-home function.

Contact-Pressure Control

The belt contact pressure is adjusted in accordance with the current load situation with the aid of the measured secondary pressure. To achieve a high level of efficiency, the secondary pressure is reduced to such an extent that the current engine torque can still be transferred to a specific degree of safety and reliability without the belt slipping.

1	Variations of CVT based on wrap principle			
Assembly, function	**Variants**			
Wrap element	Band	Chain	Belt	
Variator principle	Master/slave	Partner principle	Partner principle	
Converter:	Fitted	Not fitted	Not fitted	
– Converter clutch	Yes	No	No	
– Slip duration	Brief	Permanent	Permanent	
Clutch:				
– Type	Friction surfaces	Magnetic-powder	Magnetic-powder	
– Pressures	Low	High	High	
– Slip duration	Brief	Permanent	Permanent	
Pump adjustment	Constant	2-stage	Continuous	
Limp home	Not possible	Limited (loss of convenience)	Unlimited (increased fuel consumption)	
Vehicle:				
– Class	Small to mid-range class	Compact class	Mid-range and Luxury class	
– Engine size	$<3\,l$	$<2\,l$	$>2\,l$	
– Type of drive	Front transversal	Front longitudinal	Rear	

Table 1

Ratio Control

The gear ratio can be changed by means of the primary pulley. The enclosed fluid volume determines the axial position of the moving part of the primary pulley and thus the radius on which the belt circulates on the pulley. The primary pressure adjusts itself in response to the secondary pressure.

The requirements of driveability determine the necessary adjustment speed. For example, in the case of kickdown, it is necessary to switch from overdrive to low within 1.5 s. On the other hand, the pump delivery limits the adjustment speed.

Driving Program

A driving program ascertains the desired gear ratio. In addition to different program maps for normal operation, in which there is the option of choosing between economical and sporty operation (see also section entitled "Adaptive Transmission Control, ATC"), it is also possible to implement special functions such as kickdown, downhill driving, etc.

It is also possible to simulate range transmissions, where any intermediate variations between copying a manually shifted transmission and an automatic transmission are realizable (see also chapter entitled "Transmission for Motor Vehicles").

Clutch Activation

The interrupting clutch between engine and drivetrain is designed as a function of the position of the selector lever (P-R-N-D), the engine speed, and the engine load.

Activation of Converter and Torque Converter Lockup Clutch

In order to achieve the greatest possible efficiency, it is essential for the converter to be locked up as early as possible. Depending on the power requirement, the torque increase is used up to different speeds for acceleration.

Pump Activation

A variable-capacity pump must be used to ensure high transmission efficiency. This pump enables the delivery flow to be limited at high speeds.

Suitable suction-throttled pumps which operate without additional activation have been in development for years, but have failed to make a breakthrough as yet. An initial step towards a "variable-capacity pump" has been taken in the form of a two-stage version, in which the more favorable delivery flow can be selected as a function of the current demand.

Further concepts are feasible with continuously variable pumps, in which secondary-pressure control and pump adjustment can be combined.

Reverse-Gear Lock

Engagement of the reverse gear is disabled during forward driving at speeds above a limit to be defined (e.g., 7 km/h).

Deactivation of Limp-Home Function

Limp-home is an emergency function which is shut down during normal control operation.

The fail-safe function remains permanently activated so that engine overrevving is avoided even in the event of a partial failure or late identification of partial failures.

ECUs for Electronic Transmission Control

Application

In implementing electronic transmission control, engineers can choose to locate the ECUs in different positions in the vehicle. There are, for example, separate, combined, mounted, or integrated ECUs (Figure 1).

The way in which these ECUs are distributed in the vehicle is essentially determined by
- the ratio of vehicles with automatic transmissions to vehicles with manually shifted transmissions and
- the demands made by the transmission on the control system (performance of the microcontroller used).

In Europe, the market is still dictated by the separate printed-circuit-board ECUs ME and

TC (Figure 1a). In the USA, on the other hand, it is predominantly the combined drivetrain ECU (MEG) that is used for 4-speed and 5-speed automatic transmissions (Figure 1b), the reason for this being that automatic transmissions dominate the market in America with a market share of over 85 %.

The newer 6-speed or 7-speed transmission types and the increasing demands on engine-management systems (emission-control legislation, CARB requirements) have now started a trend in the USA moving away from combined ECUs in favor of separate ECUs. This trend is consolidated still further by the latest generation of 6-speed transmissions. These transmissions are already equipped with electronic modules with integrated electronic circuitry.

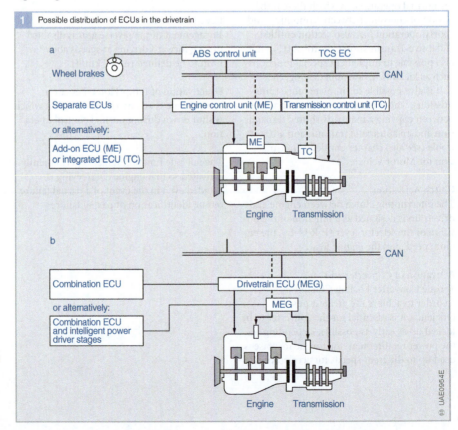

1 Possible distribution of ECUs in the drivetrain

Fig. 1
a Layout with separate ME and TC printed circuit board ECUs
b Layout with combined drivetrain ECU

Design and Operating Concept

The various ECUs and their technical and functional details will be discussed below.

Printed Circuit Board ECUs

The most widely used ECUs currently are printed circuit board (PCB) units.

Figure 2 shows an ECU with a 32-bit microcontroller (Motorola 683xx) for a ZF 5-speed transmission. This transmission has been in mass production at BMW for some years now. The figure depicts the essential layout and the data flow of the ECU in a block diagram. The ECU itself can be roughly divided into three sections:

1. Input side

The input side comprises the power supply (terminals 15 and 30), signal acquisition, and the communication interface.

The input signals include the signals for engine speed, turbine speed, output speed, and wheel speeds. The transmission control unit usually receives the engine-speed and wheel-speed signals via the CAN interface from the acquiring ECUs (engine and ABS control units).

The ECU acquires the transmission-fluid temperature as an analog input signal because the properties of the fluid have a significant effect on shift quality, especially when the engine is cold. The ECU acquires the position of the selector lever in the form of a digital signal. The following information can also be obtained and evaluated via the CAN interface:

- accelerator-pedal position (driver command),
- kickdown switch,
- engine temperature, and
- engine torque.

2. Computer Core

The computer core comprises microcontroller, flash, RAM, EEPROM, analog-digital converter, and CAN bus system.

3. Output Side

The output side features the driver stages for the on/off valve, ASICs, current control (CG205), and low-level signal driver stages.

2 Printed circuit board ECU for electronic transmission control (block diagram)

Drivetrain ECUs

MEG drivetrain ECUs (MEG = engine ETC/EGAS transmission) are based on the standard printed circuit board ECUs for engine and transmission, and are commonly used in the USA. As the block diagram in Figure 3 shows, the main advantage of this ECU lies in the fact that specific electronic components only have to be fitted once, thereby reducing costs.

An MG ECU (MG = engine/transmission) was the first example of electronic transmission control to be mass-produced. This ECU was developed back in 1983 for BMW for use with a ZF 4HP22 automatic transmission (Figure 4).

4 BMW drivetrain ECU from 1983

UAE0946Y

3 MEG drivetrain ECU for electronic transmission control (block diagram)

☐ Not necessary in MEG 7 integrated ECU

UAE0952E

5 MG7.9 combined drivetrain ECU

Figure 5 shows the current configuration of a drivetrain ECU for the block diagram in Figure 3.

Since then, the demands on ECU computing power and memory capacity have changed dramatically (see Figure 6 and Table 1).

As the figures in Table 1 demonstrate, these demands are constantly rising and there is no end in sight to their development.

1	Development of computer performance		
Year	Computer	Memory	RAM
Year	Computer	Memory	RAM
1983	Cosmac	8 k ROM	128 byte
1988	80515	32 k ROM	256 byte
1992	80517	64 k ROM	256 byte
1996	80509	128 k flash	2 k
199x	C167	256 k flash	4 k
1996	683xx	256 k flash	8 k
2001	MPC555	448 k flash	28 k
2003	MPC555	1 MB flash	28 k
2005	?	1.5 MB flash	66 k

Table 1

6 Increasing demands on ECU computing power and memory capacity

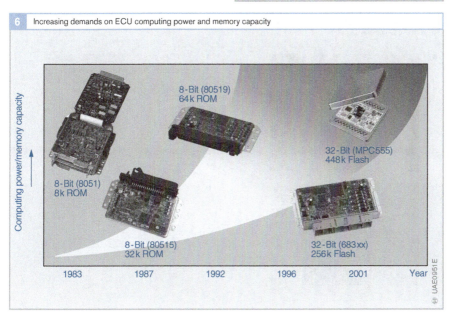

Microhybrid ECUs

The introduction of new transmissions (such as the ZF 6HP26) has seen a transformation in the type of ECU from the PCB to the microhybrid. This development has been influenced by the changing demands, mainly because of the environmental conditions under which the ECU is used (Table 2).

The microhybrid ECU basically contains the same circuitry as the PCB, but now with unencapsulated semiconductor components being used, i.e. as "bare" silicon chips. Electrical contacting is provided by wire bonding (in the PCB ECU with soldering).

Passive components are electrically contacted by means of conductive bonding agents.

In contrast to the currently mass-produced circuits with LTCC (low-temperature cofired ceramics), finer layout structures are used in the new systems with 32-bit processors. This relates in particular to the via density and the bondland size.

The previous bondland grid of 450 µm would require four bond rows and at least three wiring layers to route the computer core. With the via grid of 260 µm used, two bond rows are sufficient, and this even with a reduced space requirement and only two wiring layers.

Figure 7a shows the ABS computer bond zones (44 bonds) on an LTCC standard substrate compared with the 32-bit controller in the diesel-control system (240 bonds) on LTCC-fine-line in Figure 7b.

In addition, Figure 8 shows a comparison of the wiring density of the inner layers (Figure 8a) and the reverse side of the hybrid with the integrated resistors (Figure 8b).

2	Technical limits for PCB and microhybrid	
Type	**PCB**	**Microhybrid**
Installation location	Interior or engine compartment	In transmission
Temperature	–40...+85/+105°C	–40...+140°C
Vibration	...5 g	...30 g
Protection	IP 40 / IP 69	IP6K9K in ATF

7 Bondzone microcontroller LTCC compared with LTCC-fine-line

a

b

UTS0318Y

Fig. 7
a On LTCC standard substrate
b On LTCC-fine-line substrate

Fig. 8
a Inner layers
b Reverse side with resistors

8 Hybrid wiring densities

a

b

UTS0319Y

The following significant measures were taken to bring about an improvement in the process for microhybrid ECUs:
- use of finer punch needles,
- finer screens,
- adaptation of the pastes used, and
- tolerance optimization through adapted process management.

This compression of the layout makes it possible to produce the circuit for transmission control on an area measuring 2 x 1.2≤. In other words, a substrate with the working format of 8 x 6≤ alone can process 20 circuits in parallel.

For optimum cooling of ICs subject to high heat loss, thermal vias with a diameter of 300 µm are filled in parallel to the function vias. This increases the thermal conductivity of the substrate from approx. 3 W/mK to effectively 20 W/mK.

Figure 9 shows the complete microhybrid in its housing. The following procedures are used for the assembly process:
- All the components are bonded with conductive bonding agent.
- Bonding is carried out using a 32 µm gold wire and a 200 µm aluminum wire.
- The hybrid is bonded to the steel plate with heat-conducting bonding agent.
- The connection to the glass bushing is made by a 200 µm aluminum bond.
- The housing is hermetically sealed tight.

9	Microhybrid ECU in steel housing

UAE0948Y

ASIC Chips

In addition to computer and memory chips, application-specific integrated circuits make up a significant proportion of the electronic components in the ECUs.

Different functions have been combined into ASIC chips for the purpose of reducing costs and standardizing the electronic design of transmission-control systems. These ASICs are available in encapsulated and unencapsulated form and used for both microhybrid and PCB ECUs. Transmission control features various ASICs which are currently in volume production; the three ASICs used in the microhybrid ECU are discussed in the following.

Current-Regulator ASIC CG205

The current-regulator ASIC CG205 with integrated shunt was developed for high-precision pressure control in the transmission. It achieves a control precision of 1% over the entire temperature range.

Figure 10a shows the encapsulated ASIC, as is used in PCB ECUs.

There is also the option of adjusting the current range and the PWM output frequency with the aid of an external circuit.

Watchdog ASIC CG 120

Due to the fact that the 6-speed transmission with integrated electronic module no longer has a mechanical connection between the selector lever and the transmission, the control system requires specific safety mechanisms. This function is performed by the ASIC CG120 (Figure 10b), which monitors the function of the microcontroller (see also the chapter entitled "Diagnostic functions"). The ASIC CG120 performs the following functions:

- power supply with 5 or 3.3 V,
- sensor supply,
- watchdog,
- serial interface,
- CAN interface,
- ISO 9141 interface, and
- programmable via SPI interface.

I/O ASIC CG115

In order to achieve the high level of integration in the microhybrid, it is necessary to combine as many functions as possible in one ASIC. Achieving this with individual components would therefore take up too much space on the substrate.

The following functions are integrated in the I/O ASIC CG115 (Figure 10c):

- voltage monitoring,
- inputs and outputs for digital signal transmission,
- 2 inputs for inductive Hall-sensor signals,
- 8-channel analog multiplexer,
- serial interface, and
- programmable via SPI interface.

10 ASIC chips

a CG 205

b CG 120

c CG 115

UAE0955Y

Fig. 10
a Current-regulator ASIC CG205
b Watchdog ASIC CG120
c I/O ASIC CG115

Thermo-Management

Dissipating the heat losses generated in the ECUs represents a prime consideration in the design of mechatronic modules, particularly in those cases where "hotspots" cause a highly uneven distribution of the heat losses. Figure 1 shows the heat-dissipation model in an ECU to the point of a heat sink which is situated in the valve housing in this example. This is an LTCC microhybrid in a welded steel housing which is mounted on the aluminum housing of the hydraulic main control stage.

Effective heat management of the IC necessitates a close contact between the chips and the housing. Materials with high thermal conductivity are suitable for use in this respect.

As other tests on the various substrates for high-temperature applications have shown, these materials demonstrate very different thermal-conductivity properties. LTCC glass ceramic shows itself to be less effective than aluminum-oxide ceramic (Al_2O_3) by a factor of almost 10. However, this drawback is compensated by thermal vias in the microhybrid to such an extent that LTCC technology demonstrates equally good levels of thermal conductivity as aluminum-oxide technology.

Figure 2 shows an area of thermal vias. The production process creates these heat-dissipating thermal vias (heat spreaders) in parallel to the electrical connections.

Essentially, the terms "power loss", "depletion-layer/junction temperature", and "heat dissipation" define the limits for a microhybrid ECU.

The power loss P_l can be described in simplified form for stationary operation as follows:

1 Heat-dissipation model in the housing of a microhybrid ECU

Fig. 1
1 Bonding agent
2 Substrate
3 Si chip
4 Thermal vias
5 Al valve housing
6 ATF
7 Steel housing base

2 Cross-section through a 4-layer microhybrid with thermal vias

1 mm

$$P_l = (T_j - T_a) / R_{jth}$$

where

T_j Junction temperature
R_{jth} Thermal internal resistance
T_a Ambient temperature

The thermal resistance R_{jth} (auxiliary quantity) is dependent on the geometric parameters and the specific thermal conductivity of the material and is determined from measurements. The maximum permitted junction temperature T_j determines the maximum permitted power loss P_{lmax}, where T_j is dependent on the material (for silicon $T_{jmax} = 150...200°C$). Current specifications laid down by microcontroller manufacturers establish an upper limit of $T_{jmax} = 150°C$. Since T_j is also dependent on the design, a

3 Comparison of thermal resistance of aluminum-oxide and glass ceramics (LTCC)

a

b

c

Thermal resistance R_{th}

25 K⁄W
20
15
10
5

0 5 10 15 20 25% 30

Thermal vias, surface

UTS0322E

Fig. 3
a Aluminum-oxide
 ceramic
R_{th} = 12 K/W (CS 200
 on aluminum)
b LTCC
R_{th} = 10...11 K/W (CS
 200 in ABS 5.3)
c Comparison

1 Chip
2 Al_2O_3
3 LTCC (thermal vias)
4 Aluminum

consideration of design rules is important for optimization of the junction temperature, e.g.:
● Circuit sections subjected to thermal load should not be designed in accordance with the otherwise applicable minimum criteria, rather affected driver-stage areas (transistors/pn junctions), for example, should be geometrically enlarged.
● "Hotspots" should not be positioned in the corners of the ICs, this enables the IC substrate material to act in all directions as a heat spreader.

The design of the ASICs for transmission control described in the "ASIC Chips" section is suitable for maximum junction temperatures of T_{jmax} = 175°C.

In relation to the complete microhybrid system, designers are working towards optimizing the level of sophistication between increasing IC size, using thermal vias, and mounting on special substrates such as DBC or PCs.

Explanation of abbreviations:
DBC (**d**irect **b**onded **c**opper): copper-coated ceramic
and
PCs (**p**ower **c**hips): chips soldered onto copper plates.

Figure 3 shows a comparison of the thermal resistance of aluminum oxide (Figure 3a) and glass ceramic, i.e., LTCC (Figure 3b), and also specifically in chart form (Figure 3c).

Processes and Tools Used in ECU Development

Simulation Tools

Simulation of individual components and of the complete system is playing an increasingly important role in improving and stepping up the development process. The advantages of mathematical modeling over actual physical models (prototypes) are as follows:
- frequent reproducibility as desired,
- deeper understanding of system behavior,
- individual parameter analyses possible,
- lower costs,
- less time required for model modifications, and
- flexible application in all technical fields.

The following text deals briefly with some tools for simulation that are used.

Figure 1 shows by way of example the thermo-simulation of a microhybrid ECU.

It clearly shows the high temperature of the hotspots in the electronic circuitry which in this case occurs at a voltage regulator. This simulation is used to optimize the positioning of the thermal vias and the distribution of the components before the real ECU is built.

Circuit Simulation with SABER

Before a circuit is built with real components, it is possible to test its function using the SABER® simulation tool. Many component manufacturers already offer the data on their products in a SABER library for this purpose (Figure 2).

These data can be used to test the circuit with regard to its robustness, thermal properties, worst-case performance, and EMC behavior and thereby facilitate any necessary circuit optimization at a very early stage in its development.

This approach can reduce the number of redesigns.

1 Thermo-simulation of a microhybrid layout

UTS0323Y

2 Simulation with SABER

UTS0324Y

Modules for Transmission Control

Modules are compact function and constructional units that enable different standardized components to be integrated with the minimum component and space requirements and simplified interfaces. Hydraulic, electronic, and electrohydraulic modules are available to suit the level of integration required.

Application

Mechatronic Modules

Since the number of sensors and actuators deployed in the automatic transmission is increasing while the amount of space available to accommodate them is diminishing, the drive towards ever-increasing integration suggests itself not just for cost reasons.

Mechatronic modules can be used to combine actuators and sensors, their contacting and if necessary even an ECU into a single control system. The level of integration, or the scope, of a mechatronic module is dictated by the requirements of the relevant vehicle manufacturer (Figure 1).

Improvement Potential

Modular technology offers comprehensive potential for improvement in comparison with individual components (Figure 2).

Modules exhibit the following advantages over individual components:
- reduced space requirement,
- reduced mass,
- fewer individual parts,
- increased reliability,
- standardization of the components integrated in the module, and thus
- reduced costs.

These advantages are essentially provided by simplifications to the mechanical and electrical interfaces between the individual components and the transmission. There is in future still further potential for reducing costs in this field.

However, any further increase in module complexity could also lead to drawbacks; this is because a manufacturer of modules must demonstrate competence over the entire spectrum of the components contained therein. Since it is more difficult to separate the components in the event of a fault, there is the danger that increased follow-up costs may be incurred in the event of both assembly errors and repairs. Mechatronic systems therefore place strict demands on quality and reliability in the manufacturing process and necessitate precautionary measures to ensure that subsequent repairs can be carried out.

1　Variable scope of a mechatronic module

Electronic control unit	Sensors	Actuators
• Single unit • Mounting • Integrated • Combined	• Temperature • Pressure • Rotational speed • Position	• on /off • PWM • Pressure regulator • (Stepping motor)

Electronic module (EM)

Hydraulic module (HM)

Electrohydraulic module (EHM)

STS0291E

2 Improvement potential through modular system

Transmission manufacturer				Vehicle manufacturer
Development	Purchasing	Assembly	Final inspection	Purchasing/Assembly

Individual components / **Individual components**

- ECU → ECU → → ECU
- Wiring harness → Wiring harness → Wiring harness
- 4 to 8 actuators → 4 to 8 actuators → 4 to 8 actuators → Transmission → Transmission
- 4 to 8 sensors → 4 to 8 sensors → 4 to 8 sensors

Module / **Module**

- Electrohydraulic module → Electrohydraulic module → Electrohydraulic module → Transmission system → Transmission system

Improvement potential

- Reduced expenditure on electronics
- Optimized development
- Only one partner for development and supply
- Reduced logistics
- Reduced costs
- Simple assembly
- Lower wiring-harness scope
- Reduced costs
- Better adjustment
- Increased reliability

Improvement potential

- Reduced logistics
- No ECU mounting
- Lower wiring-harness scope
- Space saving

Module Types

3 HM 5R55 hydraulic module (example: Ford)

Hydraulic Modules (HM)

The hydraulic module (HM) (example, Figure 3) represents the first step towards the simplified assembly of modules. This module type comprises the following components (primarily sensors and actuators):

- pressure regulator (1),
- PWM valve (2),
- integrated switching valves (4),
- transmission connector matched to the vehicle type,
- temperature sensor,
- electrical connections, and
- joint filter seal between module housing and adapter plate with
- hydraulic channels (5).

Fig. 3
1 Pressure regulator
2 PWM valve
3 Transmission connector
4 Integrated switching valves
5 Adapter plate with hydraulic channels

Electronic Modules (EM)

Design

The next stage of development involves the fusion of mechanics and electronics into the mechatronic electronic module (EM), consisting of sensors and ECU. In many cases, it offers the possibility for system simplification and cost reduction.

Electronic modules already come in different types and versions. What they all have in common is their optimum adaptation to their environment, as the examples in Figures 4 to 6 demonstrate.

Bosch module for ZF transmission

The highly developed transmission functions with real-time calculations require a 32-bit microcontroller (Motorola MPC555). This means that up to 250 connections have to

4 Siemens-VDO electronic module

be routed, depending on the application. As well as sensitive signal conditioning, the electronic transmission-control system requires
- power electronics with a high-precision current regulator (1 A with 1% accuracy) for pressure control and
- semiconductor relays for currents of up to 8 A.

Further functions are:
- power supply,
- safety monitoring, and
- CAN bus system for data transfer.

The system also includes:
- electrically actuated pressure-control valves and solenoid valves,
- position sensor,
- rotational-speed sensors,
- temperature sensor, and
- transmission connector.

The environmental conditions which impact on the transmission pose a particular challenge.

The temperature temporarily reaches 140°C and acceleration values of up to 30 g occur. In addition, the electronic circuitry is completely immersed in transmission fluid, which contains contaminants, abrasion particles, and chemical additives (see also chapter entitled "Automatic transmissions AT" section "Automatic transmission luid/ATF").

5 Conti-TEMIC electronic module

6 Bosch electronic module for ZF 6HP26

Electrohydraulic Modules (EHM)

Expanding the components of the electronic module with the additional integration of actuators results in the electrohydraulic module (EHM).

An electrohydraulic module therefore comprises
- sensors,
- actuators, and
- ECU.

Figure 7 shows an electrohydraulic module:

A three-piece component carrier made of plastic holds the components integrated in the module. A pressed screen, located between the plastic components to provide protection against dirt and metal shavings, contains the electrical wiring. The components in the pressed screen are contacted by laser welding or special insulation-displacement connection technology.

7 Electrohydraulic module

1
4
2
5
3

UTS0223Y

Fig. 7
1 Transmission
 connector
2 Pressure-control
 valves
3 Rotational-speed
 sensor
4 Microhybrid-
 technology ECU
5 Position sensor

Antilock Braking System (ABS)

In hazardous driving conditions, it is possible for the wheels of a vehicle to lock up under braking. The possible causes include wet or slippery road surfaces, and abrupt reaction on the part of the driver (unexpected hazard). The vehicle can become uncontrollable as a result, and may go into a slip and/or leave the road. The antilock braking system (ABS) detects if one or more wheels are about to lock up under braking and if so makes sure that the brake pressure remains constant or is reduced. By so doing, it prevents the wheels from locking up and the vehicle remains steerable. As a consequence the vehicle can be braked or stopped quickly and safely.

System overview

The ABS braking system is based on the components of the conventional system. Those are
- the brake pedal (Fig. 1, 1),
- the brake booster (2),
- the master cylinder (3),
- the reservoir (4),
- the brake lines (5) and hoses (6), and
- the brakes and wheel-brake cylinders (7).

In addition there are also the following components:
- the wheel-speed sensors (8),
- the hydraulic modulator (9), and
- the ABS control unit (10).

The warning lamp (11) lights up if the ABS is switched off.

1 Braking system with ABS

Fig. 1
1 Brake pedal
2 Brake booster
3 Master cylinder
4 Reservoir
5 Brake line
6 Brake hose
7 Wheel brake
 with wheel-brake
 cylinder
8 Wheel-speed
 sensor
9 Hydraulic modulator
10 ABS control unit
 (in this case,
 attached unit fixed
 onto hydraulic
 modulator)
11 ABS warning lamp

Wheel-speed sensors

The speed of rotation of the wheels is an important input variable for the ABS control system. Wheel-speed sensors detect the speed of rotation of the wheels and pass the electrical signals to the control unit.

A car may have three or four wheel-speed sensors depending on which version of the ABS system is fitted (ABS system versions). The speed signals are used to calculate the degree of slip between the wheels and the road surface and therefore detect whether any of the wheels is about to lock up.

Electronic control unit (ECU)

The ECU processes the information received from the sensors according to defined mathematical procedures (control algorithms). The results of those calculations form the basis for the control signals sent to the hydraulic modulator.

Hydraulic modulator

The hydraulic modulator incorporates a series of solenoid valves that can open or close the hydraulic circuits between the master cylinder (Fig. 2, 1) and the brakes (4). In addition, it can connect the brakes to the return pump (6). Solenoid valves with two hydraulic connections and two valve positions are used (2/2 solenoid valves). The inlet valve (7) between the master cylinder and the brake controls pressure application, while the outlet valve (8) between the brake and the return pump controls pressure release. There is one such pair of solenoid valves for each brake.

Under normal conditions, the solenoid valves in the hydraulic modulator are at the "pressure application" setting. That means the inlet valve is open. The hydraulic modulator then forms a straight-through connection between the master cylinder and the brakes. Consequently, the brake pressure generated in the master cylinder when the brakes are applied is transmitted directly to the brakes at each wheel.

As the degree of brake slip increases due to braking on a slippery surface or heavy braking, the risk of the wheels locking up also increases. The solenoid valves are then switched to the "maintain pressure" setting. The connection between the master cylinder and the brakes is shut off (inlet valve is closed) so that any increase of pressure in the master cylinder does not lead to a pressure increase at the brakes.

If the degree of slip of any of the wheels increases further despite this action, the pressure in the brake(s) concerned must be reduced. To achieve this, the solenoid valves are switched to the "pressure release" setting. The inlet valve is still closed, and in addition, the outlet valve opens to allow the return pump integrated in the hydraulic modulator to draw brake fluid from the brake(s) concerned in a controlled manner. The pressure in the relevant brake(s) is thus reduced so that wheel lock-up does not occur.

2 Principle of hydraulic modulator with 2/2 solenoid valves (schematic)

Fig. 2
1 Master cylinder with reservoir
2 Brake booster
3 Brake pedal
4 Wheel brake with wheel-brake cylinder
Hydraulic modulator with
5 Damping chamber
6 Return pump
7 Inlet valve
8 Outlet valve
9 Brake-fluid accumulator

Inlet valve:
 shown in open setting
Outlet valve:
 shown in closed setting

Requirements placed on ABS

An ABS system must meet a comprehensive range of requirements, in particular all the safety requirements associated with dynamic braking response and the braking-system technology.

Handling stability and steerability

- The braking control system should be capable of ensuring that the car retains its handling stability and steerability on all types of road surface (from dry roads with good adhesion to black ice).
- An ABS system should utilize the available adhesion between the tires and the road surface under braking to the maximum possible degree, giving handling stability and steerability precedence over minimizing braking distance. It should not make any difference to the system whether the driver applies the brakes violently or gradually increases the braking force to the point at which the wheels would lock.
- The control system should be capable of adapting rapidly to changes in road-surface grip, e.g. on a dry road with occasional patches of ice, wheel lock-up on the ice must be restricted to such short time spans that handling stability and steerability are not impaired. At the same time, it should allow maximum utilization of adhesion where the road is dry.
- When braking under conditions where the amount of available grip differs between wheels (e.g. wheels on one side on ice, on dry tarmac on the other – referred to as "μ-split" conditions), the unavoidable yaw forces (turning forces around the vehicle's vertical axis which attempt to turn the vehicle sideways) should only be allowed to develop at a rate slow enough for the "average driver" to easily counteract them by steering in the opposite direction.
- When cornering, the vehicle should retain its handling stability and steerability under braking and be capable of being braked to a standstill as quickly as possible provided its speed is sufficiently below the corner's

limit speed (that is the absolute maximum speed at which a vehicle can successfully negotiate a bend of a given radius with the drive disengaged).

- The system should also be able to ensure that handling stability and steerability are maintained and the best possible braking effect obtained on a bumpy or uneven road surface, regardless of the force with which the driver applies the brakes.
- Finally, the braking control system should be able to detect aquaplaning (when the wheels "float" on a film of water) and respond appropriately to it. In so doing it must be able to maintain vehicle controllability and course.

Effective range

- The braking control system must be effective across the entire speed range of the vehicle right down to crawling speed (minimum speed limit approximately 2.5 km/h). If the wheels lock up below the minimum limit, the distance traveled before the vehicle comes to a halt is not critical.

Timing characteristics

- Adjustments to take account of braking-system hysteresis (delayed reaction to release of brakes) and the effects of the engine (when braking with the drive engaged) must take as little time as possible.
- Pitching of the vehicle in response to suspension vibration must be prevented.

Reliability

- There must be a monitoring circuit which continuously checks that the ABS is functioning correctly. If it detects a fault which could impair braking characteristics, the ABS should be switched off. A warning lamp must then indicate to the driver that the ABS is not functioning and only the basic braking system is available.

Dynamics of a braked wheel

Figs. 1 and 2 show the physical interdependencies during a braking sequence with ABS. The areas in which the ABS is active are shaded blue.

Curves 1, 2 and 4 in Fig. 1 relate to road conditions in which the level of adhesion and therefore the braking force increases up to a maximum limit as the brake pressure rises.

Increasing the brake pressure above that maximum adhesion limit on a car without ABS would constitute overbraking. When that happens, the "sliding" proportion of the tire footprint (the area of the tire in contact with the road) increases so much as the tire deforms that the static friction diminishes and the kinetic friction increases.

The brake slip λ is a measure of the proportion of kinetic friction, whereby $\lambda = 100\%$ is the level at which the wheel locks and only kinetic friction is present.

The degree of brake slip,

$$\lambda = \frac{(v_F - v_R)}{v_F} \cdot 100\%$$

indicates the degree to which the wheel's circumferential speed, v_R, lags behind the vehicle's linear speed (road speed), v_F.

From the progression shown in Fig. 1 of curves 1 (dry conditions), 2 (wet conditions) and 4 (black ice), it is evident that shorter braking distances are achieved with ABS than if the wheels are overbraked and lock up (brake slip $\lambda = 100\%$). Curve 3 (snow) shows a different pattern whereby a wedge of snow forms in front of the wheels when they lock up and helps to slow the vehicle down; in this scenario, the advantage of ABS is in its ability to maintain handling stability and steerability.

As the two curves for coefficient of friction, μ_{HF}, and lateral-force coefficient, μ_S, in Fig. 2 illustrate, the active range of the ABS has to be extended for the large lateral slip angle of $\alpha = 10°$ (i.e. high lateral force due to rapid

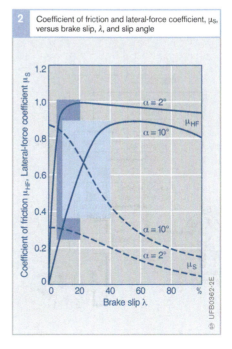

| 1 | Coefficient of friction, μ_{HF}, relative to brake slip, λ |

| 2 | Coefficient of friction and lateral-force coefficient, μ_S, versus brake slip, λ, and slip angle |

Fig. 1
1 Radial tire on dry concrete
2 Cross-ply tire on wet tarmac
3 Radial tire on loose snow
4 Radial tire on wet black ice
Blue shaded areas: ABS active zones

Fig. 2
μ_{HF} Coefficient of friction
μ_S Lateral-force coefficient
α Slip angle
Blue shaded areas: ABS active zones

lateral acceleration of the vehicle) as compared with the smaller lateral slip angle of $\alpha = 2°$, that is to say, if the vehicle is braked hard in a corner when the lateral acceleration is high, the ABS cuts in sooner and allows an initial level of brake slip of, say, 10%. At $\alpha = 10°$, an initial coefficient of friction of only $\mu_{HF} = 0.35$ is obtained, while the lateral-force coefficient is almost at its maximum at $\mu_S = 0.80$.

As the vehicle's speed and therefore its lateral acceleration are reduced by braking through the corner, the ABS is able to allow increasingly higher levels of brake slip so that the deceleration increases, while the lateral-force coefficient diminishes as the lateral acceleration reduces.

Braking while cornering causes the braking forces to rise so quickly that the overall braking distance is only marginally longer than when braking in a straight line under identical conditions.

ABS control loop

Overview
The ABS control loop (Fig. 1) consists of the following:

The controlled system
- The vehicle and its brakes
- The wheels and the friction pairing of tire and road surface

The external variables affecting the control loop:
- Changes in the adhesion between the tires and the road surface caused by different types of road surface and changes in the wheel loadings, e.g. when cornering
- Irregularities in the road surface causing the wheels and suspension to vibrate
- Lack of circularity of the tires, low tire pressure, worn tire tread, differences in circumference between wheels, (e.g. spare wheel)
- Brake hysteresis and fade
- Differences in master-cylinder pressure between the two brake circuits

The controllers
- The wheel-speed sensors
- The ABS control unit

The controlled variables
- The wheel speed and, derived from it, the wheel deceleration,
- The wheel acceleration and the brake slip

The reference variable:
- The foot pressure applied to the brake pedal by the driver, amplified by the brake booster, and generating the brake pressure in the braking system

The correcting variable:
- Braking pressure in the wheel-brake cylinder.

Fig. 1
1 Brake pedal
2 Brake booster
3 Master cylinder with reservoir
4 Wheel-brake cylinder
5 Wheel-speed sensor
6 Warning lamp

Controlled system

The data-processing operations performed by the ABS control unit are based on the following simplified controlled system:

- a non-driven wheel,
- a quarter of the vehicle's mass apportioned to that wheel,
- the brake on that wheel and, representing the friction pairing of tire and road surface,
- a theoretical curve for coefficient of friction versus brake slip (Fig. 2).

That curve is divided into a stable zone with a linear gradient and an unstable zone with a constant progression (μ_{HFmax}). As an additional simplification, there is also an assumed initial straight-line braking response that is equivalent to a panic braking reaction.

Fig. 3 shows the relationships between braking torque, M_B (the torque that can be generated by the brake through the tire), or road-surface frictional torque, M_R (torque that acts against the wheel through the friction pairing of tire and road surface), and time, t, as well as the relationships between the wheel deceleration (a) and time, t, whereby the braking torque increases in linear fashion over time. The road-surface frictional torque lags slightly behind the braking torque by the time delay, T, as long as the braking sequence is within the stable zone of the curve for friction coefficient versus brake slip. After about 130 ms, the maximum level (μ_{HFmax}) – and therefore the unstable zone – of the curve for friction coefficient versus brake slip is reached. From that point on, the curve for friction coefficient versus brake slip states that while the braking torque, M_B, continues to rise at an undiminished rate, the road-surface frictional torque, M_R, cannot increase any further and remains constant. In the period between 130 and 240 ms (this is when the wheel locks up), the minimal torque difference, $M_B - M_R$, that was present in the stable zone rises rapidly to a high figure. That torque difference is a precise measure of the wheel deceleration ($-a$) of the braked wheel (Fig. 3, bottom). In the stable

zone, the wheel deceleration is limited to a small rate, whereas in the unstable zone it increases rapidly. As a consequence, the curve for friction coefficient versus brake slip reveals opposite characteristics in the stable and unstable zones. The ABS exploits those opposing characteristics.

2 Ideal curve for friction coefficient versus slip

Fig. 2
a Stable zone
b Unstable zone
λ_K Optimum brake slip
μ_{HFmax} Maximum coefficient of friction

3 Initial braking phase, simplified

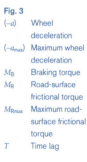

Fig. 3
($-a$) Wheel deceleration
($-a_{max}$) Maximum wheel deceleration
M_B Braking torque
M_R Road-surface frictional torque
M_{Rmax} Maximum road-surface frictional torque
T Time lag

Controlled variables

An essential factor in determining the effectiveness of an ABS control system is the choice of controlled variables. The basis for that choice are the wheel-speed sensor signals from which the ECU calculates the deceleration/acceleration of the wheel, brake slip, the reference speed and the vehicle deceleration. On their own, neither the wheel deceleration/acceleration nor the brake slip are suitable as controlled variables because, under braking, a driven wheel behaves entirely differently to a non-driven wheel. However, by combining those variables on the basis of appropriate logical relationships, good results can be obtained.

As brake slip is not directly measurable, the ECU calculates a quantity that approximates to it. The basis for the calculation is the reference speed, which represents the speed under ideal braking conditions (optimum degree of brake slip). So that speed can be determined, the wheel-speed sensors continuously transmit signals to the ECU for calculating the speed of the wheels. The ECU takes the signals from a pair of diagonally opposed wheels (e. g. right front and left rear) and calculates the reference speed from them. Under partial braking, the faster of the two diagonally opposite wheels generally determines the reference speed. If the ABS cuts in under emergency braking, the wheel speeds will be different from the vehicle's linear speed and can thus not be used for calculating the reference speed without adjustment. During the ABS control sequence, the ECU provides the reference speed based on the speed at the start of the control sequence and reduces it at a linear rate. The gradient of the reference-speed graph is determined by analyzing logical signals and relationships.

If, in addition to the wheel acceleration/deceleration and the brake slip, the vehicle's linear deceleration is brought into the equation as an additional quantity, and if the logical circuit in the ECU is modulated by computation results, then ideal brake control can be achieved. That concept has been realized in the Bosch Antilock Braking System (ABS).

Controlled variables for non-driven wheels

The wheel acceleration and deceleration are generally suitable as controlled variables for driven and non-driven wheels provided the driver brakes with the clutch disengaged. The reason can be found in the opposing characteristics of the controlled system in the stable and unstable zones of the curve for friction coefficient versus brake slip.

In the stable zone, the wheel deceleration is limited to relatively low rates so that when the driver presses harder on the brake pedal, the car brakes harder without the wheels locking up.

In the unstable zone, on the other hand, the driver only needs to apply slightly more pressure to the brake pedal to induce instantaneous wheel lock-up. This characteristic means that very often the wheel deceleration and acceleration can be used to determine the degree of brake slip for optimum braking.

A fixed wheel deceleration threshold for initiation of the ABS control sequence should only be fractionally above the maximum possible vehicle linear deceleration. This is particularly important if the driver initially only applies the brakes lightly but then increasingly applies more pressure to the pedal. If the threshold is set too high, the wheels could then progress too far along the curve for friction coefficient versus slip into the unstable zone before the ABS detects the imminent loss of control.

When the fixed threshold is initially reached under heavy braking, the brake pressure at the wheel(s) concerned should not automatically be reduced, because with modern tires, valuable braking distance would be lost on a surface with good grip, especially in cases where the initial speed is high.

Controlled variables for driven wheels

If first or second gear is engaged when the brakes are applied, the engine acts on the driven wheels and substantially increases their effective mass moment of inertia Θ_R, i.e. the wheels behave as if they were considerably heavier. The sensitivity with which the wheel deceleration responds to changes in the braking torque in the unstable zone of the curve for friction coefficient versus brake slip diminishes to an equal extent.

The starkly opposing characteristics displayed by non-driven wheels in the stable and unstable zone of the curve for friction coefficient versus brake slip are evened out to a substantial degree, so that in this situation the wheel deceleration is often insufficient as a controlled variable for identifying the degree of brake slip offering the greatest possible friction. It is necessary instead to introduce as an additional controlled variable a quantity that approximates to brake slip and to combine it in a suitable manner with the wheel deceleration.

Fig. 4 compares an initial braking sequence on a non-driven wheel and on a driving wheel that is connected to the drivetrain. In this example, the engine's inertia increases the effective wheel inertia by a factor of four. On the non-driven wheel, a specific threshold for deceleration $(-a)_1$ is exceeded very soon after leaving the stable zone of the curve for friction coefficient versus brake slip. Because it has a moment of inertia that is four times greater, the driven wheel requires a torque difference four times as big

$$\Delta M_2 = 4 \cdot \Delta M_1$$

to exceed the threshold $(-a)_2$. Consequently, the driven wheel may by then have progressed a long way into the unstable zone of the curve for friction coefficient versus brake slip, resulting in impaired vehicle handling stability.

Effectiveness of control

An efficient antilock braking system must meet the criteria listed below for the standard of control.
- Maintenance of handling stability by provision of adequate lateral forces at the rear wheels
- Maintenance of steerability by provision of adequate lateral forces at the front wheels
- Reduction of braking distances as compared with locked-wheel braking by optimum utilization of the available adhesion between tires and road surface
- Rapid adjustment of braking force to different friction coefficients, for instance when driving through puddles or over patches of ice or compacted snow.
- Ensuring application of low braking-torque control amplitudes to prevent suspension vibration
- High degree of user-friendliness due to minimal pedal feedback ("pedal judder") and low levels of noise from the actuators (solenoid valves and return pump in the hydraulic modulator).

The criteria listed can only be optimized collectively rather than individually. Nevertheless, vehicle handling stability and steerability are always among the top priorities.

4 Initial braking phase for a non-driven wheel and a driving wheel connected to the drivetrain

Fig. 4
Index 1: non-driven wheel
Index 2: driven wheel (in this example, the wheel moment of inertia is increased by a factor of 4)
$(-a)$ Threshold for wheel deceleration
M Torque difference $M_B - M_R$

Typical control cycles

Control cycle on surfaces with good grip (High coefficient of friction)

If the ABS sequence is activated on a road surface with good grip (high coefficient of friction), the subsequent pressure rise must be 5 to 10 times slower than in the initial braking phase in order to prevent undesirable suspension vibration. That requirement produces the control-cycle progression for high coefficients of friction illustrated in Fig. 1.

During the initial phase of braking, the brake pressure at the wheel and the rate of wheel deceleration (negative acceleration) rise. At the end of phase 1, the wheel deceleration passes the set threshold level $(-a)$.

As a result, the relevant solenoid valve switches to the "maintain pressure" setting. At this point the brake pressure must not be reduced, because the threshold $(-a)$ might be exceeded within the stable zone of the curve for friction coefficient versus brake slip and then potential braking distance would be "wasted". At the same time, the reference speed, v_{Ref}, reduces according to a defined linear gradient. The reference speed is used as the basis for determining the slip switching threshold, λ_1.

At the end of phase 2, the wheel speed, v_R, drops below the λ_1 threshold. At that point, the solenoid valves switch to the "reduce pressure" setting so that the pressure drops, and they remain at that setting as long as the

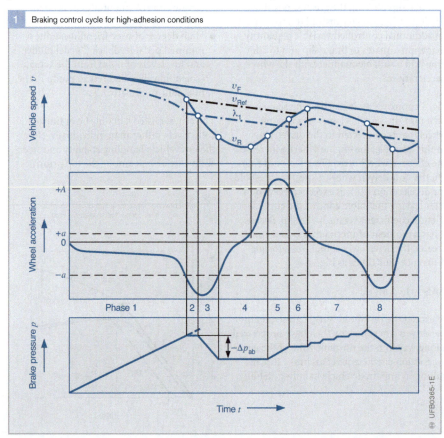

1 Braking control cycle for high-adhesion conditions

Vehicle speed v

v_F
v_{Ref}
λ_1
v_R

Wheel acceleration

$+A$
$+a$
0
$-a$

Phase 1 2 3 4 5 6 7 8

Brake pressure p

$-\Delta p_{ab}$

Time t

Fig. 1
v_F Vehicle speed
v_{Ref} Reference speed
v_R Wheel speed
λ_1 Slip switching
 threshold
Switching signals:
$+A$, $+a$ Thresholds for
 wheel acceleration
$-a$ Threshold for
 wheel deceleration
$-\Delta p_{ab}$ Brake-pressure
 drop

UFB0365-1E

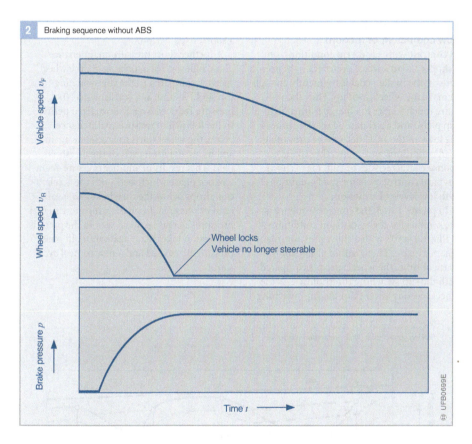

2 Braking sequence without ABS

Vehicle speed v_F

Wheel speed v_R

Wheel locks
Vehicle no longer steerable

Brake pressure p

Time t

UFB0699E

wheel deceleration is above the threshold $(-a)$.

At the end of phase 3, the deceleration rate falls below the threshold $(-a)$ again and a pressure-maintenance phase of a certain length follows. During that period, the wheel acceleration has increased so much that the threshold $(+a)$ is passed. The pressure continues to be maintained at a constant level.

At the end of phase 4, the wheel acceleration exceeds the relatively high threshold level $(+A)$. The brake pressure then increases for as long as the acceleration remains above the threshold $(+A)$.

In phase 6, the brake pressure is once again held constant because the acceleration is above the threshold $(+a)$. At the end of this phase, the wheel acceleration falls below the threshold $(+a)$. This is an indication that the wheel has returned to the stable zone of the curve for friction coefficient versus brake slip and is now slightly underbraked.

The brake pressure is now increased in stages (phase 7) until the wheel deceleration passes the threshold $(-a)$ (end of phase 7). This time, the brake pressure is reduced immediately without a λ_1 signal being generated.

By comparison, Fig. 2 shows the progressions for an emergency braking sequence without ABS.

Control cycle on slippery surfaces (low coefficient of friction)

In contrast with good grip conditions, on a slippery road surface even very light pressure on the brake pedal is frequently enough to make the wheels lock up. They then require much longer to emerge from a high-slip phase and accelerate again. The processing logic of the ECU detects the prevailing road conditions and adapts the ABS response characteristics to suit. Fig. 3 shows a typical control cycle for road conditions with low levels of adhesion.

In phases 1 to 3, the control sequence is the same as for high-adhesion conditions.

Phase 4 starts with a short pressure-maintenance phase. Then, within a very short space of time, the wheel speed is compared with the slip switching threshold λ_1. As the wheel speed is lower than the slip switching threshold, the brake pressure is reduced over a short, fixed period.

A further short pressure-maintenance phase follows. Then, once again, the wheel speed is compared to the slip switching threshold λ_1 and, as a consequence, the pressure reduced over a short, fixed period. In the following pressure-maintenance phase, the wheel starts to accelerate again and its acceleration exceeds the threshold $(+a)$. This results in another pressure-maintenance phase which lasts until the acceleration drops below the threshold $(+a)$ again (end of phase 5). In phase 6, the incremental pressure-increase pattern seen in the preceding section takes place again until, in phase 7, pressure is released and a new control cycle starts.

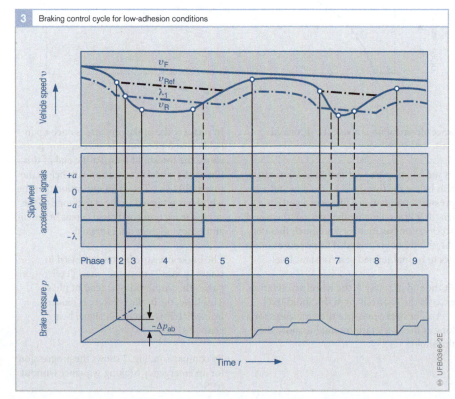

3 Braking control cycle for low-adhesion conditions

Fig. 3
v_F Vehicle speed
v_{Ref} Reference speed
v_R Wheel speed
λ_1 Slip switching
 threshold
Switching signals:
$+a$ Threshold for
 wheel acceleration
$-a$ Threshold for
 wheel deceleration
$-p_{ab}$ Brake-pressure
 drop

UFB0366-2E

In the cycle described above, the control logic detected that following pressure release – triggered by the signal (−a) – two more pressure-reduction stages were necessary to induce the wheel to accelerate again. The wheel remains in the higher-slip zone for a relatively long period, which is not ideal for handling stability and steerability.

In order to improve those two characteristics, this next control cycle and those that follow incorporate continual comparison of wheel speed with the slip switching threshold λ_1. As a consequence, the brake pressure is continuously reduced in phase 6 until, in phase 7, the wheel acceleration exceeds the threshold (+a). Because of that continuous pressure release, the wheel retains a high level of slip for only a short period so that vehicle handling and steerability are improved in comparison with the first control cycle.

Control cycle with yaw-moment buildup delay

When the brakes are applied in situations where the grip conditions differ significantly between individual wheels ("μ-split" conditions) – for example, if the wheels on one side of the car are on dry tarmac while those on the other side are on ice – vastly different braking forces will be produced at the front wheels (Fig. 4). That difference in braking force creates a turning force (yaw moment) around the vehicle's vertical axis. It also generates steering feedback effects of varying types dependent on the vehicle's kingpin offset. With a positive kingpin offset, corrective steering is made more difficult, while a negative kingpin offset has a stabilizing effect.

Heavy cars tend to have a relatively long wheelbase and a high level of inertia around the vertical axis. With vehicles of this type, the yaw effect develops slowly enough for the driver to react and take corrective steering action during ABS braking. Smaller cars with short wheelbases and lower levels of inertia, on the other hand, require an ABS system supplemented by a yaw-moment buildup delay (**GMA** system) to make them equally controllable under emergency braking in conditions where there are wide differences in grip between individual wheels. Development of the yaw moment can be inhibited by delayed pressure increase in the brake on the front wheel that is on the part of the road offering the higher level of adhesion (the "high" wheel).

Fig. 5 (overleaf) illustrates the principle of the yaw-moment buildup delay:
Curve 1 shows the brake pressure, p, in the master cylinder. Without yaw-moment buildup delay, the wheel on the tarmac quickly reaches the pressure p_{high} (curve 2) and the wheel on the ice, the pressure p_{low} (curve 5). Each wheel is braked with the specific maximum possible deceleration (individually controlled).

4 Yaw-moment build-up induced by areas of widely differing adhesion

$\mu_{HF1} = 0.8$ $\mu_{HF2} = 0.1$

F_{B1} F_{B2} M_{Yaw}

1 2

Fig. 4
M_{yaw} Yaw moment
F_B Braking force
1 "High" wheel
2 "Low" wheel

GMA 1 system

On vehicles with less extreme handling characteristics, the GMA 1 system is used. With this system, during the initial phases of braking (curve 3), as soon as the pressure is reduced for the first time at the "low" wheel because this shows a tendency to lock up, the brake pressure at the "high" wheel is increased in stages. Once the brake pressure at the "high" wheel reaches its lock-up point, it is no longer affected by the signals from the "low" wheel, and is individually controlled so that it is able to utilize the maximum possible braking force. This method gives the type of vehicle referred to satisfactory steering characteristics under emergency braking on surfaces offering unequal grip to individual wheels.
As the maximum brake pressure at the "high" wheel is reached within a relatively short time (750 ms), the braking distance is only marginally longer than for vehicles without a yaw-moment buildup delay facility.

GMA 2 system

The GMA 2 system is used on vehicles with more extreme handling characteristics. With this system, as soon as the brake pressure at the "low" wheel is reduced, the ABS solenoid valves for the "high" wheel are directed to maintain and then reduce the pressure for a specific period (Fig. 5, curve 4). Renewed pressure increase at the "low" wheel then triggers incremental increase of pressure at the "high" wheel, though with pressure-increase periods a certain amount longer than for the "low" wheel. This pressure metering takes place not only in the first control cycle but throughout the braking sequence.

The effect of the yaw moment on steering characteristics is all the more exaggerated the greater the speed of the vehicle when the brakes are first applied. The GMA 2 system divides the vehicle speed into four ranges. In each of those ranges, the yaw moment is inhibited to differing degrees. In the high speed ranges, the pressure-increase periods at the "high" wheel are progressively shortened, while the pressure-decrease periods at the "low" wheel are progressively lengthened in order to achieve effective inhibition of yaw-moment generation at high vehicle speeds. Fig. 5 below compares the steering angle progression necessary for maintaining a straight course under braking without GMA (curve 6) and with GMA (curve 7).
 Another important aspect with regard to GMA usage is the vehicle's response to braking in a corner. If the driver brakes when cornering at a high speed, the GMA increases the dynamic load on the front wheels while reducing it at the rear. As a result, the lateral forces acting on the front wheels become stronger while those acting on the rear wheels diminish. This generates a turning moment towards the inside of the bend as a result of which the vehicle slews off its line towards the inside of its intended course and is very difficult to bring back under control (Fig. 6a).

5 Progression of brake pressure/steering angle with yaw moment build-up delay

UFB0291-1E

Fig. 5

1 Pressure, p_{Hz}, in the master cylinder
2 Brake pressure, p_{high}, without GMA
3 Brake pressure, p_{high}, with GMA 1
4 Brake pressure, p_{high}, with GMA 2
5 Brake pressure, p_{low}
6 Steering angle, α, without GMA
7 Steering angle, α, with GMA

In order to avoid this critical response to braking, the GMA also takes the lateral acceleration into account. The GMA is deactivated at high lateral acceleration rates. As a result, a high braking force is generated at the outside front wheel during the initial phase of braking in a corner and creates a turning moment towards the outside of the bend. That turning moment balances out the turning moment acting in the opposite direction that is produced by the lateral forces, so that the vehicle slightly understeers and thus remains easily controllable (Fig. 6).

The ideal method of inhibiting yaw-moment buildup involves a compromise between good steering characteristics and suitably short braking distance and is developed by Bosch individually for a specific vehicle model through consultation with the manufacturer.

6 Response to braking when cornering at critical speeds with/without GMA

UFB0367-1Y

Fig. 6
a GMA activated (no individual control): vehicle oversteers
b GMA deactivated (individual control): vehicle slightly understeers
F_B Braking force
F_S Lateral force
M Torque

Control cycle for four-wheel-drive vehicles

The most important criteria for assessing the various types of four-wheel-drive configuration (Fig. 7) are traction, dynamic handling, and braking characteristics. As soon as differential locks are engaged, conditions are created that demand a different response from the ABS system.

When a rear-axle differential is locked, the rear wheels are rigidly interconnected, i.e. they always rotate at the same speed and respond to the braking forces (at each wheel) and friction levels (between each tire and the road surface) as if they were a single rigid body. The "select low" mode that would otherwise be adopted for the rear wheels (whereby the wheel with the lower degree of adhesion, μ_{HF}, determines the brake pressure for both rear wheels) is thus canceled, and both rear wheels utilize the maximum braking force. As soon as the inter-axle lock is engaged, the system forces the front wheels to assume the same average speed as the rear wheels. All four wheels are then dynamically interlinked and the engine drag (engine braking effect when backing off the throttle) and inertia act on all four wheels.

In order to ensure optimum ABS effectiveness under those conditions, additional features have to be incorporated according to the type of four-wheel-drive system (Fig. 7) in use.

7 Four-wheel-drive configurations

Fig. 7
a Four-wheel-drive system 1
b Four-wheel-drive system 2
c Four-wheel-drive system 3

1 Engine
2 Transmission
3 Freewheel and viscous clutch

Differential with
4 Manual lock or viscous lock
5 Proportional lock
6 Automatic clutch and lock
7 Automatic lock

UFB0368-1Y

Four-wheel-drive system 1

On four-wheel-drive system 1 with manual locks or permanently active viscous locks in the propeller shaft and the rear-axle differential, the rear wheels are rigidly interconnected and the average speed of the front wheels is the same as that of the rear wheels. As already mentioned, the rear differential lock results in deactivation of the "select low" mode for the rear wheels and utilization of the maximum braking force at each rear wheel. When braking on road surfaces with unequal levels of grip at the two rear wheels, this can generate a yaw moment with a potentially critical effect on vehicle handling stability. If the maximum braking force difference were also applied very quickly at the front wheels, it would not be possible to keep the vehicle on a stable course.

This type of four-wheel-drive arrangement therefore requires a GMA system for the front wheels in order to ensure handling stability and steerability are maintained in road conditions where there is a significant divergence between the levels of grip at the left and right wheels. In order to maintain ABS effectiveness on slippery surfaces, the engine drag effect – which of course acts on all four wheels on a four-wheel-drive vehicle – has to be reduced. This is done by using an engine drag-torque control system that applies just enough throttle to counteract the unwanted engine braking effect.

Another factor that demands refinement of the ABS control cycle in order to prevent wheel lock-up is the reduced sensitivity of the wheels to changes in road-surface adhesion on slippery surfaces that is caused by the effect of the engine inertia. The fact that all four wheels are dynamically linked to the engine's inertial mass therefore requires additional analysis and processing operations on the part of the ABS control unit. The vehicle longitudinal deceleration is calculated to detect smooth or slippery road surfaces where μ_{HF} is less than 0.3.

When analyzing braking on such surfaces, the response threshold $(-a)$ for wheel deceleration is halved and the diminishing rise in the reference speed is limited to specific, relatively low levels. As a result, imminent wheel lock-up can be detected early and "sensitively".

On four-wheel-drive vehicles, heavy application of the throttle on slippery road surfaces can cause all four wheels to spin. In such situations, special signal-processing methods ensure that the reference speed can only increase in response to the spinning wheels within the limits of the maximum possible vehicle acceleration. In a subsequent braking sequence, the initial ABS pressure reduction is triggered by a signal $(-a)$ and a specific, minimal difference in wheel speed.

Four-wheel-drive system 2

Because of the possibility of all four wheels spinning with four-wheel-drive system 2 (viscous lock with freewheel in the propeller shaft, proportional rear-axle differential lock), the same special procedures must be adopted for signal processing.

Other modifications for ensuring ABS effectiveness are not necessary as the freewheel disengages the wheels under braking. However, the system can be improved by the use of an engine drag-torque control system.

Four-wheel-drive system 3

As with the first two systems, four-wheel-drive system 3 (automatic differential/inter-axle locks) requires adoption of the signal processing procedures described above in the event of wheel-spin at all four wheels. In addition, automatic release of the differential locks whenever the brakes are applied is necessary. Other modifications for ensuring ABS effectiveness are not necessary.

Traction Control System (TCS)

Critical driving situations can occur not only while braking, but also whenever strong longitudinal forces should be transferred at the contact area between the tire and the ground. This is because the transferable lateral forces are reduced by this process. Critical situations can also occur when starting off and accelerating, particularly on a slippery road surface, on hills, and when cornering. These kinds of situations can overtax the driver not only causing him/her to react incorrectly but also causing the vehicle to become unstable. The traction control system (TCS) solves these problems, providing the vehicle remains within the physical limits.

Tasks

The antilock braking system (ABS) prevents the wheels from locking up when the brakes are applied by lowering the wheel brake pressures. The traction control system (TCS) prevents the wheels from spinning by reducing the drive torque at each driven wheel. TCS therefore provides a logical extension of ABS during acceleration.

In addition to this safety-relevant task of ensuring the stability and steerability of the vehicle when accelerating, TCS also improves the traction of the vehicle by regulating the optimum slip (see μ-slip curve in "Basic principles of vehicle dynamics"). The upper limit here is, of course, set by the traction requirement stipulated by the driver.

Function description

Unless otherwise stated, all the following descriptions refer to single-axle driven vehicles (Fig. 1). It makes no difference whether the vehicle is rear-wheel or front-wheel drive.

Drive slip and what causes it
If the driver presses the accelerator when the clutch is engaged, the engine torque will rise. The drive axle torque M_{Kar} also increases. This torque is distributed to both driven wheels in a ratio of 50:50 via the transversal differential (Fig. 1). If this increased torque can be transferred completely to the road surface, the vehicle will accelerate unhindered. However, if the drive torque $M_{Kar}/2$ at one driven wheel exceeds the maximum drive torque that can be transferred, the wheel will spin. The transferable motive force is therefore reduced and the vehicle becomes unstable due to the loss of lateral stability.

1 Drive concept of a single-axle driven vehicle with TCS

The TCS regulates the slip of the driven wheels as quickly as possible to the optimum level. To do this the system first determines a reference value for the slip. This value depends on a number of factors which are intended to represent the current driving situation as closely as possible. These factors include:
• the basic characteristic for TCS reference slip (based on the slip requirement of a tire during acceleration),
• effective coefficient of friction,
• external tractive resistance (deep snow, rough road, etc.),
• yaw velocity, lateral acceleration, and steering angle of the vehicle.

TCS interventions
The measured wheel speeds and the respective drive slip can be influenced by changing the torque balance M_{Ges} at each driven wheel. The torque balance M_{Ges} at each wheel results from the drive torque $M_{Kar}/2$ at this wheel, the respective braking torque M_{Br} and the road torque M_{Str} (Fig. 1).

$$M_{Ges} = M_{Kar}/2 + M_{Br} + M_{Str}$$
(M_{Br} and M_{Str} are negative here.)

This balance can obviously by influenced by the drive torque M_{Kar} provided by the engine as well as by the braking torque M_{Br}. Both these parameters are therefore correcting variables of the TCS which can be used to regulate the slip at each wheel to the reference slip level.
 In gasoline-engine vehicles, the drive torque M_{Kar} can be controlled using the following engine interventions:
• Throttle valve (throttle valve adjustment),
• Ignition system (ignition-timing advance),
• Fuel-injection system (phasing out individual injection pulses).

The last two interventions are rapid interventions, the first a slower means of intervention (Fig. 2). The availability of these interventions depends on the vehicle manufacturer and engine version.

In diesel-engine vehicles, the drive torque M_{Kar} is influenced by the electronic diesel control system (EDC) (reduction of the quantity of fuel injected).

The braking torque M_{Br} can be regulated for each wheel via the braking system. The TCS function requires the original ABS hydraulic system to be expanded because of the need for active pressure build-up (see also "Hydraulic modulator").

Fig. 2 compares the response times with various TCS interventions. The figure shows that exclusive drive torque regulation by means of the throttle valve can be unsatisfactory due to the relatively long response time.

2 Comparison of response times with various TCS interventions

Control variance ↑

1 2 3

TCS response time ⟶

UFB0382-1E

Fig. 2
1 Throttle-valve/wheel brake intervention
2 Throttle-valve intervention/ ignition adjustment
3 Throttle-valve intervention

Structure of traction control system (TCS)

The expanded ABS hydraulic system allows both symmetrical brake application (i.e. brake application at both driven wheels) as well as individual brake application. This is the key to further structuring of the TCS, i.e. structuring according to controlled variable rather than according to the actuator (engine/brake).

Drive axle speed controller

The drive axle speed v_{Kar} or the drive torque M_{Kar} can be influenced by means of engine interventions. Symmetric brake applications also influence the drive axle speed v_{Kar} and effect torque balance between the individual wheels in the same way as reducing the drive torque M_{Kar}. The *drive axle speed controller* is used to regulate the drive axle speed in this way.

Transversal differential lock controller

Asymmetric brake application (brake application at just one driven wheel) is used primarily to regulate the differential speed at the driven axle $v_{Dif} = v_L - v_R$. This task is carried out by the *differential speed controller*. Asymmetric brake application at just one

driven wheel is only noticeable at first in the torque balance of this wheel. The brake application has basically the same effect as an asymmetric distribution ratio of the transversal differential (but applied to a drive torque M_{Kar} that is reduced by the asymmetric braking torque). The differential speed controller is also referred to as the *transversal differential lock controller* because it can be used to influence to a certain extent the distribution ratio of the transversal differential, i.e. to mimic the effect of a differential lock.

Together the drive axle speed controller and transversal differential lock controller form the TCS system (Fig. 3). The drive axle speed controller uses the drive axle speed v_{Kar} to regulate the drive torque M_{Kar} provided by the engine. The transversal differential lock controller functions primarily like a controller that uses the differential speed v_{Dif} to regulate the distribution ratio M_L to M_R of the transversal differential and therefore the distribution of the drive torque M_{Kar} to the driven wheels.

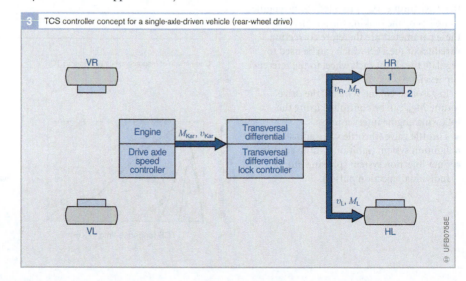

3 TCS controller concept for a single-axle-driven vehicle (rear-wheel drive)

Fig. 3
1 Wheel
2 Wheel brake

v_R, v_L Wheel speeds
v_{Kar} Drive axle speed
M_{Kar} Drive axle torque
V Front
H Rear
R Right
L Left

Typical control situations

Split: Transversal differential lock controller

Fig. 4 shows a typical situation ("μ-split") whereby the transversal differential lock controller of the TCS becomes active when the vehicle pulls away after being stationary. The left side of the vehicle is on a slippery road surface with a low coefficient of friction μ_l ("l" for low) and the right side of the vehicle is on dry asphalt with a considerably higher coefficient of friction μ_h ("h" for high).

Without brake application by the transversal differential lock controller only the drive force F_l could be transferred on both sides since the differential distributes the drive torque equally on both sides. A greater drive torque M_{Kar} causes the wheel on the side with μ_l to spin and leads to a differential speed of $v_{Dif} > 0$ (see also Fig. 5). In this case, the excess drive torque is lost as lost torque in the differential, engine, and transmission.

To prevent the wheel on the side of the vehicle with μ_l from spinning if the drive torque is too high, the braking force F_{Br} is applied (Fig. 4, see also Fig. 5). The differential can then transfer the force $F_{Br} + F_l$ to this side (or a torque corresponding to this force), whereby F_{Br} is diffused by the brake action. The drive force F_l remains as before. On the side of the vehicle with μ_h the force $F_{Br} + F_l$ is also transferred (characteristic of the differential). Since the brakes are not applied on this side, the entire force can be used as drive force $F_{Br}^* + F_l$ (F_{Br}^* results from F_{Br} taking the different effective radii into account). Overall the drive force transferred is increased by F_{Br}^* (the drive torque M_{Kar} must of course also be increased accordingly). This ability of the transversal differential lock controller to increase the traction is part of the traction control system (TCS).

The drive torque can be regulated to a maximum possible drive force. The value of μ_h represents a physical upper limit.

When both driven wheels run synchronously again ($v_{Dif} = 0$), the single-sided braking force F_{Br} or the corresponding braking torque M_{Br} is reduced again (Fig. 5).

The exact buildup and reduction of M_{Br} depends on the internal implementation of the transversal differential lock controller (the controller functions like a PI-controller).

Low: Drive axle speed controller

If both driven wheels are on a slippery road surface with a low coefficient of friction (e.g. the vehicle is standing on ice) when the vehicle pulls away, the drive axle speed controller of the TCS becomes active.

4 Differential lock effect due to asymmetric brake application

5 Typical wheel speed and braking torque curves when setting off on μ-split

Fig. 4
M_{Kar} Drive torque
F_{Br} Braking force
F_{Br}^* Braking force, based on effective radii
μ_l Low coefficient of friction
μ_h High coefficient of friction
F_l Transferable motive force on μ_l
F_h Transferable motive force on μ_h

Fig. 5
v Wheel speed
M_{Br} Braking torque
l Low-μ wheel
h High-μ wheel
v_{Fz} Vehicle speed
v_{Dif} Differential speed

If the driver increases the driver-specified torque $M_{FahVorga}$, the drive torque M_{Kar} increases almost simultaneously. Both driven wheels then spin at almost the same speed. The differential speed $v_{Dif} = v_L - v_R$ is approximately 0, while the drive axle speed $v_{Kar} = (v_L + v_R)/2 = v_L = v_R$ is considerably greater than a reasonable reference value v_{SoKar} determined by the TCS due to the spinning driven wheels. The drive axle speed controller reacts by reducing the drive torque M_{Kar} to a level below the torque specified by the driver $M_{FahVorga}$ and by initiating brief, symmetric brake application $M_{Br, Sym}$ (Fig. 6). As a result the drive axle speed v_{Kar} is reduced and thus the speed of the spinning wheels. The vehicle begins to accelerate. Since the optimum point of the μ-slip curve (see also "Basic principles of vehicle dynamics") would not be achieved without these TCS interventions, the acceleration would be slower while the wheels were spinning and considerably less lateral stability would be present.

The exact characteristic of M_{Kar} and $M_{Br, Sym}$ depends in turn on the internal implementation of the drive axle speed controller (the controller functions like a PID-controller).

Traction control system (TCS) for four wheel drive vehicles

In recent years four wheel drive vehicles have continued to increase in popularity. Amongst these types of vehicles, Sport Utility Vehicles, or SUVs, are the most popular of all. These are road vehicles with off-road characteristics.

If all four wheels of a vehicle are to be driven, both a second transversal differential and an additional longitudinal differential (also called a central differential) are required (Fig. 7). The first task of the longitudinal differential is to compensate for the differences between the drive axle speed of the front and rear axle $v_{Kar, VA}$ and $v_{Kar, HA}$, respectively. A rigid connection would result in tension between the front and rear axle. The second task is to achieve optimum distribution of the drive torque M_{Kar} to the two axles $M_{Kar, VA}$ and $M_{Kar, HA}$, respectively.

Less expensive SUVs are often equipped with a longitudinal differential with a preset distribution ratio. Unlike with a transversal differential, fixed distribution ratios other than 50:50 are useful – e.g. 60:40 for a design with emphasis on the rear-wheel drive of the vehicle. Brake application by the traction control system (TCS) can be used to mimic the behavior of a longitudinal differential lock.

By applying the brakes to eliminate part of $M_{Kar, VA}$, the distribution ratio $M_{Kar, HA}$ to $M_{Kar, VA}$ can be increased, or can be decreased by applying the brakes to eliminate part of $M_{Kar, HA}$. The principle is the same as previously described for the transversal differential lock or transversal differential. The only difference is that the braking torque of the TCS does not have to be asymmetric, (i.e. at one wheel of the driven axle) but can occur symmetrically at both wheels of a driven axle. Moreover, the longitudinal differential lock controller regards the two drive axle speeds $v_{Kar, VA}$ and $v_{Kar, HA}$ as input parameters rather than the speed difference of the left and right wheel of the driven axle (transversal differential lock controller, see above).

6 Typical wheel speed, engine and braking torque curves when setting off on low-μ

Fig. 6
v Wheel speed
v_{Fz} Vehicle speed
v_{Kar} Drive axle speed
v_{SoKar} Drive axle speed reference value
$M_{Br, Sym}$ Symmetric braking torque
$M_{FahVorga}$ Driver-specified drive torque (via accelerator pedal position)
L Left
R Right

Fig. 8 shows the expansion of the TCS concept from Fig. 3 for an all-wheel drive vehicle. As with a single-axle driven vehicle, the drive axle speed controller uses the drive axle speed v_{Kar} to regulate the drive torque M_{Kar} provided by the engine. As already described, the longitudinal differential lock controller distributes this torque to the front and rear axle ($M_{Kar, VA}$ and $M_{Kar, HA}$, respectively). The transversal differential lock controller uses the differential speed $v_{Dif, XA}$ to regulate the distribution of the drive torque $M_{Kar, XA}$ per axle to the driven wheels. This must now be carried out for both the front and rear axle ("X" = "V" (front) or "X" = "H" (rear)).

Electronic differential locks designed as part of the TCS software have the advantage that they do not require additional hardware. There are therefore very cost-efficient. They are used for road vehicles, which is usually the intended application of SUVs. When used in true off-road cross-country vehicles, electronic differential locks reach their limit in tough off-road conditions, at the latest

8 TCS controller concept for a four-wheel-drive vehicle

Fig. 8
1 Wheel
2 Wheel brake

v Wheel speed
v_{Kar} Drive axle speed
M_{Kar} Drive axle torque
A Axle
V Front
H Rear
R Right
L Left

7 Drive concept of a four-wheel-drive vehicle with TCS

Fig. 7
1 Engine with transmission
2 Wheel
3 Wheel brake
4 Transversal differential
5 Longitudinal differential
6 Control unit with TCS functionality
7 Transversal differential

Engine, transmission, gear ratios of differentials and losses are combined into one unit
v Wheel speed
v_{Kar} Drive axle speed
M_{Kar} Drive axle torque
M_{Br} Braking torque
R Right
L Left
V Front
H Rear
A Axle

when the brakes overheat. Vehicles for these conditions are therefore often fitted with mechanical locks (examples can be seen in Figs. 9 and 10). The lock controllers of the TCS software are then only used as a backup system and they do not intervene during normal operation.

9 Classic solution of a differential lock

UFB0762Y

10 Electronically controllable differential lock (Haldex coupling)

UFB0763Y

Fig. 10
1 Output shaft
2 Working piston
3 Lamella
4 Axial-piston pump
5 Control valve
6 Input shaft

Summary: Advantages of TCS

Below is a summary of the advantages of using TCS to prevent the driven wheels from spinning when starting off or accelerating on slippery road surfaces under one or both sides of the vehicle, when accelerating when cornering, and when starting off on an incline:

- Unstable vehicle conditions are avoided and therefore the driving safety is enhanced.
- Increased traction due to regulating the optimum slip.
- Mimicking the function of a transversal differential lock.
- Mimicking the function of a longitudinal differential lock with four-wheel-drive vehicles.
- Automatic control of the engine output.
- No "grinding" of the tires when driving around tight corners (unlike with mechanical differential locks).
- Reduction of tire wear.
- Reduction of wear to drive mechanism (transmission, differential, etc.) especially on μ-split or if a wheel suddenly starts to spin on a road surface providing good grip.
- Warning lamp informs the driver during situations close to the physical critical limits.
- Efficient double-use of existing ABS hydraulic components.
- Adoption of tasks of ESP vehicle dynamics control as subordinate wheel controller (see "Closed-loop control system").

▶ Basic principles of automotive control engineering

Many subsystems of a driving safety system (e. g. ESP) influence the driving dynamics of a vehicle by means of a controller i. e., they form a control loop together with the relevant components of the vehicle.

Control loop
A simple standard control loop consists of controllers and a controlled system. The objective is to influence the characteristic of the parameter y_{actual} (controlled variable) of the controlled system using the controller such that the parameter follows a reference characteristic y_{ref} as closely as possible. To do this the controlled variable is measured and passed to the controller. The actual value of the controlled variable is constantly compared with the current reference value by generating the control variance $e = y_{ref} - y_{actual}$.

The main task of the controller is to determine a suitable value for the correcting variable u for every control variance e so that the control variance is decreased, i. e., $y_{actual} = y_{ref}$ at least approximately.

This task may be made more difficult by unknown natural dynamics of the controlled system and other external factors z, which also influence the controlled system.

Example: TCS transversal differential lock controller
The principle of a control loop can be explained using the transversal differential lock controller of the TCS system as an example. The controlled variable $y_{actual} = v_{Dif}$ is the differential speed of the two wheels of a driven axle. The reference value v_{SoDif} is determined by the TCS and adapted to the current driving situation. When driving straight ahead, this value is typically 0. The asymmetric braking torque is used as a correcting variable to influence the controlled variable. The controlled system is the vehicle itself which is affected by external influences such as changing road surfaces.

▶ Standard control loop using TCS transversal differential lock controller as an example

y_{actual} Controlled variable
y_{ref} Reference variable
e Control variance
 $y_{ref} - y_{actual}$
u Correcting variable
z External disturbance
 values

Standard controller
Proportional, Integral and Differential elements are often used as controllers. The correcting variable u is determined by the current control variance e as shown below:

P-controller	Multiplication	$u(t) = K_P \cdot e(t)$
I-controller	Time integration	$u(t) = K_I \cdot \int e(t) dt$
D-controller	Time derivative	$u(t) = K_D \cdot de(t)/dt$

The degree of counter-reaction of these controllers increases the greater the control variance (P-controller), the longer the control variance lasts (I-controller) or the greater the tendency of the control variance to change (D-controller). Combining these basic controllers gives PI, PD and PID-controllers.

The transversal differential lock controller of the traction control system (TCS) is designed as a PI-controller that contains additional nonlinear elements.

Electronic Stability Program (ESP)

Human error is the cause for a large portion of road accidents. Due to external circumstances, such as an obstacle suddenly appearing on the road or driving at inappropriately high speeds, the vehicle can reach its critical limits and it becomes uncontrollable. The lateral acceleration forces acting on the vehicle reach values that overtax the driver. Electronic systems can make a major contribution towards increasing driving safety.

The Electronic Stability Program (ESP) is a closed-loop system designed to improve vehicle handling and braking response through programmed intervention in the braking system and/or drivetrain. The integrated functionality of the ABS prevents the wheels from locking when the brakes are applied, while TCS inhibits wheel spin during acceleration. In its role as an overall system, ESP applies a unified, synergistic concept to control the vehicle's tendency to "plow" instead of obeying the helm during attempted steering corrections; and at the same time it maintains stability to prevent the vehicle breaking away to the side, provided the vehicle remains within its physical limits.

Requirements

ESP enhances driving safety by providing the following assets:
- Enhanced vehicle stability; the system keeps the vehicle on track and improves directional stability under all operating conditions, including emergency stops, standard braking maneuvers, coasting, acceleration, trailing throttle (overrun), and load shift
- Increased vehicle stability at the limits of traction, such as during sharp steering maneuvers (panic response), to reduce the danger of skidding or breakaway.
- In a variety of different situations, further improvements in the exploitation of traction potential when ABS and TCS come into action, and when engine drag torque control is active, by automatically increasing engine speed to inhibit excessive engine braking. The ultimate effects are shorter braking distances and greater traction along with enhanced stability and higher levels of steering response.

Fig. 1
1 Driver steers, lateral-force buildup.
2 Incipient instability because side-slip angle is too large.
3 Countersteer, driver loses control of vehicle.
4 Vehicle becomes uncontrollable.

M_G Yaw moment
F_R Wheel forces
β Directional deviation from vehicle's longitudinal axis (side-slip angle)

1 Lateral dynamic response on passenger car without ESP

Tasks and method of operation

The electronic stability program is a system that relies on the vehicle's braking system as a tool for "steering" the vehicle. When the stability-control function assumes operation it shifts the priorities that govern the brake system. The basic function of the wheel brakes – to decelerate and/or stop the vehicle – assumes secondary importance as ESP intervenes to keep the vehicle stable and on course, regardless of the conditions.

Specific braking intervention is directed at individual wheels, such as the inner rear wheel to counter understeer, or the outer front wheel during oversteer, as shown in Fig. 2. For optimal implementation of stability objectives, ESP not only initiates braking intervention, but it can also intervene on the engine side to accelerate the driven wheels.

Because this "discriminatory" control concept relies on two *individual intervention* strategies, the system has two options for steering the vehicle: it can brake selected wheels (selective braking) or accelerate the driven wheels. Within the invariable limits imposed by the laws of physics, ESP keeps the vehicle on the road and reduces the risk of accident and overturning. The system enhances road safety by furnishing the driver with effective support.

Below are four examples comparing vehicles with and without ESP during operation "on the limit". Each of the portrayed driving maneuvers reflects actual operating conditions, and is based on simulation programs designed using data from vehicle testing. The results have been confirmed in subsequent road tests.

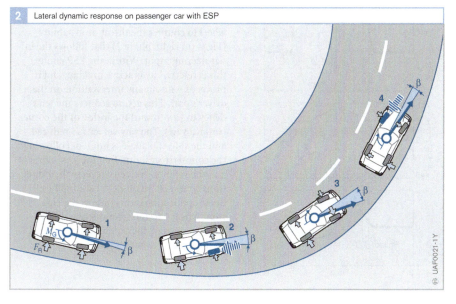

2 Lateral dynamic response on passenger car with ESP

Fig. 2

1 Driver steers, lateral-force buildup.
2 Incipient instability, ESP intervention at right front.
3 Vehicle remains under control.
4 Incipient instability, ESP intervention at left front, complete stabilization.

M_G Yaw moment
F_R Wheel forces
β Directional deviation from vehicle's longitudinal axis (side-slip angle)
⫘⫘ Increased braking force

Maneuvers

Rapid steering and countersteering

This maneuver is similar to lane changes or abrupt steering inputs such as might be expected for instance
- when a vehicle is moving too fast when it enters a series of consecutive S-bends,
- or which have to be initiated when, with oncoming traffic, an obstacle suddenly appears on a country road, or
- which are necessary when an overtaking maneuver on the highway or freeway suddenly has to be aborted.

Figs. 3 and 4 demonstrate the handling response of two vehicles (with and without ESP) negotiating a series of S-bends with rapid steering and countersteering inputs

- on a high-traction road-surface (coefficient of friction $\mu_{HF} = 1$),
- without the driver braking,
- with an initial speed of 144 km/h.

Initially, as they approach the S-bend, the conditions for both vehicles, and their reactions, are identical. Then come the first steering inputs from the drivers (phase 1).

Vehicle without ESP

As can be seen, in the period following the initial, abrupt steering input the vehicle without ESP is already threatening to become unstable (Fig. 4 on left, phase 2). Whereas the steering input has quickly generated substantial lateral forces at the front wheels, there is a delay before the rear wheels start to generate similar forces. The vehicle reacts with a clockwise movement around its vertical axis (inward yaw). The next stage is phase 3 with the second steering input. The vehicle without ESP fails to respond to the driver's attempt to countersteer and goes out of control. The yaw velocity and the side-slip angle rise radically, and the vehicle breaks into a skid (phase 4).

Vehicle with ESP

On this vehicle ESP brakes the left front wheel to counter the threat of instability (Fig. 4 on right, phase 2) that follows the initial steering input. Within the ESP context this is referred to as active braking, and it proceeds without any intervention on the driver's part. This action reduces the tendency to yaw toward the inside of the corner (inward yaw). The yaw velocity is reduced and the side-slip angle is limited. Following the countersteer input, first the yaw moment and then the yaw velocity reverse their directions (phase 3). In phase 4 a second brief brake application – this time at the right front wheel – restores complete stability. The vehicle remains on the course defined by the steering-wheel angle.

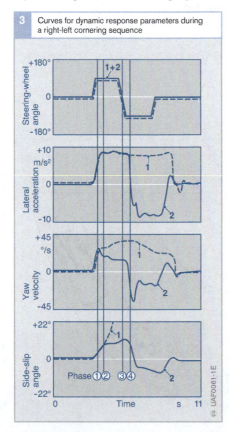

3 Curves for dynamic response parameters during a right-left cornering sequence

Fig. 3
1 Vehicle without ESP
2 Vehicle with ESP

4 Vehicle tracking during right-left cornering sequence

Vehicle without ESP

Vehicle with ESP

Phase ④
Phase ③
Phase ②
Phase ①

Phase ④
Phase ③
Phase ②
Phase ①

UAF0060-1E

5 Over and understeering behavior when cornering

a

b

3
2
1

3
2
1

with ESP
without ESP

with ESP
without ESP

UAF0086E

Fig. 4
⫸ Increased braking force
① Driver steers, lateral-force buildup.
② Incipient instability Right: ESP intervention at left front.
③ Countersteer Left: Driver loses control of vehicle; Right: Vehicle remains under control.
④ Left: Vehicle becomes uncontrollable, Right: ESP intervention at right front, complete stabilization.

Fig. 5
a Oversteering behavior.
1 The rear end of the vehicle breaks away.
2 ESP applies the brake at the outer front wheel and this reduces the risk of skidding.
3 The vehicle without ESP breaks into a slide.

b Understeering behavior
1 The front of the vehicle breaks away.
2 ESP applies the brake at the inner rear wheel and this reduces the risk of understeering.
3 The vehicle without ESP is understeered and leaves the road.

Lane change with emergency braking

When the last vehicle in a line of stopped traffic is hidden by a rise in the road, and thus invisible to traffic approaching from behind, drivers closing on the traffic jam cannot register the dangerous situation until the very last second. Stopping the vehicle without colliding with the stationary traffic will entail a lane change as well as braking.

Figs. 6 and 7 show the results of this evasive action as taken by two different vehicles:

- one equipped solely with the Antilock Braking System (ABS) and
- one that is also fitted with ESP. Both vehicles
- are traveling at 50 km/h and
- on a slippery road surface ($\mu_{HF} = 0.15$) at the moment when the driver registers the dangerous situation.

7 Curves for dynamic response parameters for lane change during an emergency stop at $v_0 = 50$ km/h and $\mu_{HF} = 0.15$

6 Lane change during emergency braking

Vehicle without ESP, with ABS Vehicle with ESP

Fig. 6
$v_0 = 50$ km/h
$\mu_{HF} = 0.15$

◀▦▶ Increased brake slip

Fig. 7
$v_0 = 50$ km/h
$\mu_{HF} = 0.15$

1 Vehicle without ESP
2 Vehicle with ESP

Vehicle with ABS but without ESP

Immediately after the initial steering input both the side-slip angle and the yaw velocity have increased to the point where driver intervention – in the form of countersteer – has become imperative (Fig. 6, on left). This driver action then generates a side-slip angle in the opposite direction (technically: with the opposite operational sign). This side-slip angle increases rapidly, and the driver must countersteer for a second time. Here the driver is able – but only just – to restabilize the vehicle and bring it to a safe halt.

Vehicle with ESP

Because ESP reduces yaw velocity and side-slip angle to easily controllable levels, this vehicle remains stable at all times. The driver is not confronted with unanticipated instability and can thus continue to devote full attention to keeping the vehicle on course. ESP substantially reduces

the complexity of the steering process and lowers the demands placed on the driver. Yet another asset is that the ESP vehicle stops in less distance than the vehicle equipped with ABS alone.

9 Oversteering and understeering when cornering

UAF0088Y

Fig. 9
1 Vehicle with ESP
2 Oversteered vehicle without ESP
3 Understeered vehicle without ESP

Fig. 8
Vehicle without ESP
1 Vehicle approaches an obstacle.
2 Vehicle breaks away and does not follow the driver's steering movements.
3 Vehicle slides uncontrolled off the road.

Vehicle with ESP
1 Vehicle approaches an obstacle.
2 Vehicle almost breaks away → ESP intervention, vehicle follows driver's steering movements.
3 Vehicle almost breaks away again when recentering the steering wheel → ESP intervention.
4 Vehicle is stabilized.

8 Critical obstacle-avoidance maneuver with and without ESP

with ESP

without ESP

UAF0087E

Extended steering and countersteering sequence with progressively greater input angles

A vehicle traversing a series of S-curves (for instance, on a snaking secondary road) is in a situation similar to that encountered on a slalom course. The way the ESP works can be clearly seen during this kind of dynamic maneuver when the steering wheel has to be turned to progressively greater angles to negotiate each turn.

Figs. 10 and 11 illustrate the handling response of two vehicles (one with and one without ESP) under these conditions

- on a snow-covered road ($\mu_{HF} = 0.45$),
- without the driver braking, and
- at a constant velocity of 72 km/h.

Vehicle without ESP

Engine output will have to be increased continually in order to maintain a constant road speed. This, in turn, will generate progressively greater slip at the driven wheels. A sequence of steering and countersteering maneuvers with a 40° steering-wheel angle can quickly increase drive slip to such levels that a vehicle without ESP becomes unstable. At some point in this alternating sequence the vehicle suddenly ceases to respond to steering inputs and breaks into a slide. While lateral acceleration remains virtually constant, both side-slip angle and yaw velocity rise radically.

Vehicle with ESP

The Electronic Stability Program (ESP) intervenes at an early stage in this sequence of steering and countersteering maneuvers to counter the instability that threatens right from the outset. ESP relies on engine intervention as well as individually controlled braking of all four wheels to maintain the vehicle's stability and steering response. Side-slip angle and yaw velocity are controlled so that the driver's steering demands can be complied with as far as possible considering the prevailing physical conditions.

10 Curves for dynamic response parameters for rapid steering and countersteering inputs with increasing steering-wheel angles

Fig. 10
1 Vehicle without ESP
2 Vehicle with ESP

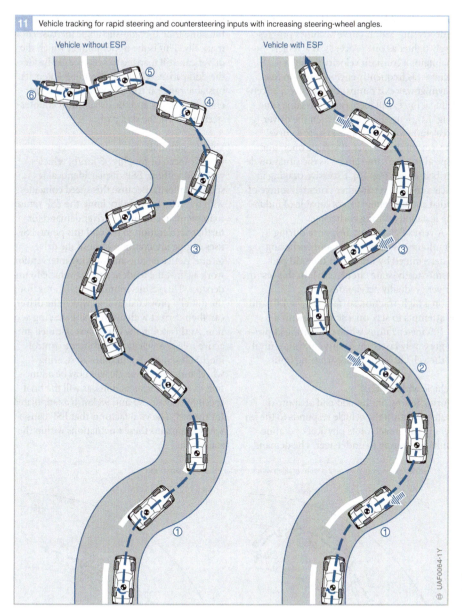

11 Vehicle tracking for rapid steering and countersteering inputs with increasing steering-wheel angles.

Vehicle without ESP

Vehicle with ESP

Fig. 11

◄▥▥ Increased braking
force

Acceleration/deceleration during cornering

A decreasing-radius curve becomes progressively tighter as one proceeds. If a vehicle maintains a constant velocity through such a curve (as frequently encountered on freeway/highway exit ramps) the outward, or centrifugal force, will increase at the same time (Fig. 12). This also applies when the driver accelerates too soon while exiting a curve. In terms of the physics of vehicle operation, it produces the same effects as the situation described above (Fig. 13). Excessive braking in such a curve is yet another potential source of radial and tangential forces capable of inducing instability during cornering.

A vehicle's handling response during simultaneous acceleration and cornering is determined by testing on the skid pad (semi-steady-state circulation). In this test the driver gradually accelerates

- on a high-traction surface ($\mu_{HF} = 1.0$) and
- attempts to stay on a skid pad with a 100 meter radius while circulating at progressively faster rates; this continues until the vehicle reaches its cornering limits.

Vehicle without ESP

During testing on the skid pad at approximately 95 km/h the vehicle responds to the impending approach of its physical operating limits by starting to understeer. The demand for steering input starts to rise rapidly, while at the same time the side-slip angle increases dramatically. This is the upper limit at which the driver can still manage to keep the car inside the skid pad. A vehicle without ESP enters its unstable range at roughly 98 km/h. The rear end breaks away and the driver must countersteer and leave the skid pad.

Vehicle with ESP

Up to a speed of roughly 95 km/h, vehicles with and without ESP display identical response patterns. Because this speed coincides with the vehicle's stability limit, the ESP refuses to implement continued driver demand for further acceleration to beyond this point. ESP uses engine intervention to limit the drive torque. Active engine and braking intervention work against the tendency of the vehicle to understeer. This results in minor deviations from the initially projected course, which the driver can then correct with appropriate steering action. At this point, the driver has assumed an active role as an element within the control loop. Subsequent fluctuations in steering-wheel and side-slip angle will now be a function of the driver's reactions, as will the final speed of between 95 and 98 km/h as stipulated for the test. The vital factor is that ESP consistently maintains these fluctuations within the stable range.

| 12 | Vehicle tracking when cornering while braking with a constant steering-wheel angle |

Fig. 12
◀▥ Increased braking
 force
◁ Decreased braking
 force

UAF0066-1Y

13 | Vehicle tracking when cornering while accelerating.

Vehicle without ESP Vehicle with ESP

Fig. 13
⊪⊪ Increased braking
force

14 | Comparison of cornering with vehicles with and without ESP

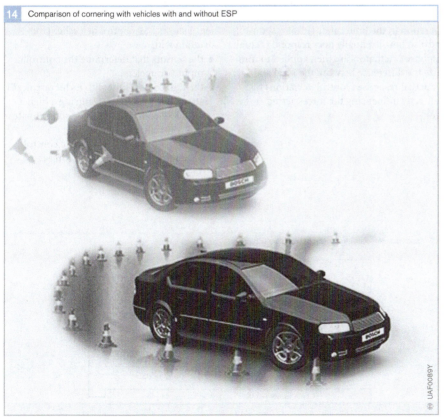

Closed-loop control system and controlled variables

Electronic stability program concept

Application of the ESP closed-loop stability control in the vehicle's limit situation as defined by the dynamics of vehicle motion is intended to prevent the
- linear (longitudinal) velocity, the
- lateral velocity and the
- rotational speed around the vertical axis (yaw velocity),

exceeding the ultimate control limits. Assuming appropriate operator inputs, driver demand is translated into dynamic vehicular response that is adapted to the characteristics of the road in an optimization process designed to ensure maximum safety. As shown in Fig. 1, the first step is to determine how the vehicle should respond to driver demand during operation in the limit range (ideal response), and also how it actually does respond (actual response). Actuators are then applied to minimize the difference between the ideal and the actual response (control deviation) by indirectly influencing the forces acting at the tires.

System and control structure

The Electronic Stability Program (ESP) embraces capabilities extending far beyond those of either ABS or ABS and TCS combined. Based on advanced versions of ABS and ABS/TCS system components, it allows active braking at all four wheels with a high level of dynamic sensitivity. Vehicle response is adopted as an element within the control loop. The system controls braking, propulsive and lateral forces so that the actual response converges with the ideal response under the given circumstances.

An engine-management system with CAN interface can vary engine output torque in order to adjust the driven-wheel slip rates. The advanced ESP system provides highly precise performance for selective adjustment of the dynamic longitudinal and lateral forces acting on each individual wheel.

Fig. 2 shows ESP control in a schematic diagram with
- the sensors that determine the controller input parameters,
- the ESP control unit with its hierarchically-structured controller, featuring a higher-level ESP controller and the subordinate slip controllers,
- the actuators used for ultimate control of braking, drive and lateral forces.

1 Block diagram of electronic stability program (ESP)

Controller hierarchy

Level 1 ESP controller

Task

The ESP controller is responsible for
- determining the current vehicle status based on the yaw velocity signal and the side-slip angle estimated by the "monitor" and then
- achieving maximum possible convergence between vehicle response in the limit range and its characteristics in the normal operating range (ideal response).

The following components register driver demand and the system processes their signals as the basis for defining ideal response:
- engine-management system (e. g. pressure on accelerator pedal),
- primary-pressure sensor (e. g. activation of brakes), or
- steering-wheel sensor (steering-wheel angle).

At this point the specified response is defined as driver demand. The coefficient of friction

and the vehicle speed are also included in the processing calculations as supplementary parameters. The "monitor" estimates these factors based on signals transmitted by the sensors for
- wheel speed,
- lateral acceleration,
- braking pressures, and
- yaw velocity.

The desired vehicle response is brought about by generating a yaw moment acting on the vehicle. In order to generate the desired yaw moment, the controller intervenes in the tire-slip rates to indirectly influence the longitudinal and lateral forces. The system influences the tire slip by varying the specifications for slip rate, which must then be executed by the subordinated ABS and TCS controllers.

The intervention process is designed to maintain the handling characteristics that the manufacturer intended the vehicle to have and to serve as the basis for ensuring consistently reliable control.

2 ESP control loop in vehicle

Fig. 2
1 Yaw-rate sensor with lateral-acceleration sensor
2 Steering-wheel-angle sensor
3 Primary-pressure sensor
4 Wheel-speed sensors
5 ESP control unit
6 Hydraulic modulator
7 Wheel brakes
8 Engine management ECU
9 Fuel injection

only for gasoline-engines:
10 Ignition-timing intervention
11 Throttle-valve intervention (ETC)

The ESP controller generates the specified yaw moment by relaying corresponding slip-modulation commands to the selected wheels.

The subordinate-level ABS and TCS controllers trigger the actuators governing the brake hydraulic system and the engine-management system using the data generated in the ESP controller.

Design
Fig. 3 is a simplified block diagram showing the design structure of the ESP controller. It portrays the signal paths for input and output parameters. Based on the

- yaw velocity (measured parameter),
- steering-wheel angle (measured parameter),
- lateral acceleration (measured parameter),
- vehicle's linear velocity (estimated parameter), and
- longitudinal tire forces and slip rates (estimated parameters)

the monitor determines the following:
- lateral forces acting on the wheel,
- slip angle,
- side-slip angle, and
- vehicle lateral speed.

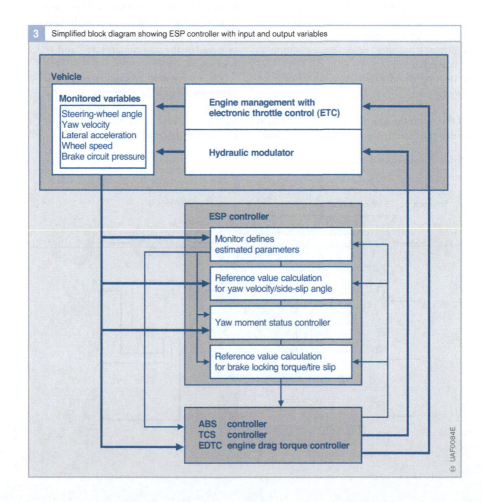

3 Simplified block diagram showing ESP controller with input and output variables

The specifications for side-slip angle and yaw velocity are determined on the basis of the following parameters, which may be either directly or indirectly defined by driver input:
- steering-wheel angle,
- estimated vehicle speed,
- coefficient of friction, which is determined on the basis of the longitudinal acceleration (estimated parameter) and lateral acceleration (measured parameter), and
- accelerator-pedal travel (engine torque) or brake-circuit pressure (force on brake pedal).

These processes also take into account the special characteristics related to vehicle dynamics, as well as unusual situations, such as a crowned road or μ-split surface (e.g. high traction on left of lane with right side slippery).

Method of Operation
The ESP controller governs the two status parameters "yaw velocity" and "side-slip angle" while calculating the yaw moment required to make the actual and desired-state parameters converge. As the side-slip angle increases, so does its significance for the controller.

The control program is based upon data concerning the maximum potential lateral acceleration and other data selected to reflect the vehicle's dynamic response patterns. These are determined for each vehicle in *steady-state skid-pad testing*. In subsequent steady-state vehicle operation, as well as during braking and acceleration, this data – defining how the steering angle and vehicle speed relate to the yaw velocity – serves as the basis for defining the desired vehicle motion. The required data (nominal yaw velocity) is stored in the program in the form of a single-track model.

The nominal yaw velocity must be limited in line with the friction coefficient so as to keep the vehicle on the predefined physically feasible track.

For example, if a vehicle breaks into over-steer while coasting into a right-handed

▶ **Single-track model**

Ranges of lateral acceleration

Passenger cars can achieve lateral accelerations up to 10 m/s². Lateral acceleration in the small signal range (0...0.5 m/s²) can be caused by road conditions such as ruts or by cross-winds.

The linear range extends from 0.5...4 m/s². Typical lateral maneuvers include changing lane or load-change reactions when cornering. The response of the vehicle in these situations can be described using the linear single-track model.

In the transition range (4...6 m/s²) some vehicles still respond linearly while others do not.

The critical range above 6 m/s² is only reached in extreme situations such as situations that almost result in an accident. In this case the vehicle response is highly non-linear.

Assumptions with the single-track model

The linear single-track model can be used to obtain important information about the lateral behavior of a vehicle. In the single-track model the lateral properties of an axle and its wheels are summarized into one effective wheel. In the simplest version, the properties taken into account are modeled using linear equations; thus the model is referred to as a linear single-track model.

The most important model assumptions are:

- Kinematics and elastokinematics of the axle are only modeled linearly.

- The lateral force buildup of the tire is linear and the aligning torque is ignored.

- The center of gravity is assumed to be at road level. The only rotational degree of freedom of the vehicle is therefore the yaw motion. Rolling, pitching and lifting (translational movements in z-axis) are not taken into account.

curve, and the specified yaw velocity is exceeded (the vehicle evinces a tendency to rotate too quickly around its vertical axis), ESP responds by braking the left front wheel to generate a defined brake slip which shifts the yaw moment toward greater counterclockwise rotation thus suppressing the vehicle's tendency to break away.

If a vehicle breaks into understeer while coasting into a right-handed curve, and the yaw velocity is below the specified yaw rate (the vehicle evinces a tendency to rotate too slowly around its vertical axis), ESP responds by braking the right rear wheel to generate a defined brake slip which shifts the yaw moment toward greater clockwise rotation thus suppressing the vehicle's tendency to push over the front axle.

ESP controller functions during ABS and TCS operation

The entire spectrum of monitored and estimated data is relayed to the subordinate controllers for continuous processing. This guarantees maximum exploitation of the traction available between tire and road surface for the basic ABS and TCS functions under all operating conditions.

During active ABS operation (with wheels tending to lock) the ESP controller provides the subordinate ABS controller with the following data:
• lateral vehicle velocity,
• yaw velocity,
• steering-wheel angle, and
• wheel speeds as the foundation for providing the desired ABS slip.

When TCS is active (wheels threatening to break into uncontrolled spin when moving off or during rolling acceleration) the ESP controller transmits the following offset values to the subordinate TCS controller:
• change in the specified value for the drive-slip,
• change in the slip tolerance range, and
• change in a value to influence the torque reduction.

Level 2 ABS controller
Task
The hierarchically subordinated ABS controller goes into operation whenever the desired slip rate is exceeded during braking, and it becomes necessary for ABS to intervene. During both ABS and "active" braking, the closed-loop control of wheel-slip rates as applied for various dynamic-intervention functions must be as precise as possible. The system needs precise data on slip as a precondition for dialing in the specified slip rates. It must be pointed out that the system does not measure the vehicle's longitudinal speed directly. Instead, this parameter is derived from the rotation rates of the wheels.

Design and method of operation
By briefly "underbraking" one wheel, the ABS controller performs an indirect measurement of vehicle speed. It interrupts the slip control to lower the current braking torque by a defined increment. The torque is then maintained at this level for a given period. Assuming that the wheel has stabilized and is turning freely with no slip at the end of this period, it can serve as a suitable source for determining (no-slip) wheel speed.

The calculated velocity at the center of gravity can be used to determine the effective (free-rolling) wheel speeds at all four wheels. These data, in turn, form the foundation for calculating the actual slip rates at the remaining three – controlled – wheels.

Level 2 engine drag-torque controller (EDTC)
Task
Following downshifts and when the accelerator is suddenly released, inertia in the engine's moving parts always exerts a degree of braking force at the drive wheels. Once this force and the corresponding reactive torque rise beyond a certain level, the tires will lose their ability to transfer the resulting loads to the road. Engine drag-torque control intervenes under these conditions (by "gently" accelerating the engine).

Design and method of operation
Factors such as variations in the character of the road surface can lead to conditions under which the engine-braking torque is suddenly too high. The result is a tendency for the wheels to lock. One available countermeasure is judicious throttle application. Here, the ECU transmits signals to trigger the corresponding actuators in the engine-management system (with ETC function) for an increase in drive torque. Intervention at the engine-management level is employed to regulate the driven wheel within the prescribed limits.

Level 2 TCS controller
Task
The hierarchically subordinated TCS controller is triggered in case of excessive slip during (for example) standing-start and during rolling acceleration when it becomes necessary for TCS to intervene. TCS is intended to prevent the driven wheels from breaking into a free spin. It functions by limiting engine torque to a level corresponding to the drive torque that the wheels can transfer to the road surface.

Intervention at the driven wheels is carried out by means of brake application or by including the engine management system in the process. In diesel-engine vehicles, the electronic diesel control system (EDC) reduces the engine torque by modifying the quantity of fuel injected. With gasoline-engines, the engine torque can be reduced by varying the throttle valve aperture (ETC), modifying the ignition timing, or suppressing individual injection pulses.

Active braking at non-driven wheels is governed directly by the ABS controller. Unlike with ABS, TCS receives values from the ESP controller for changing the specified slip and permissible slip difference of the driven axle(s). These changes take effect in the form of an offset applied to the basic values defined in the TCS.

Design
The specified propshaft speed and wheel-speed differential data are calculated from the specified slip values and speeds for the "coasting" wheels. The propshaft speed and wheel-speed differential controlled variables are derived from the driven wheel speeds.

Method of Operation
The TCS module calculates the desired braking torque for both driven wheels and the desired value for engine torque reduction to be implemented by the engine-management system.

Because the propshaft speed is affected by inertial forces originating from the drivetrain as a whole (engine, transmission, driven wheels, and the propshaft itself), a relatively large time constant is employed to describe its corresponding leisurely rate of dynamic response. In contrast, the time constant for the wheel-speed differential is relatively small, reflecting the fact that the wheels' own inertial forces are virtually the sole determining factor for their dynamic response. Another relevant factor is that the wheel-speed differential – unlike propshaft speed – is not affected by the engine.

The torques prescribed for the propshaft and wheel-speed differential are taken as the basis for defining the actuators' positioning forces. The system achieves the specified difference in braking torque between the left and right-side driven wheels by transmitting the appropriate control signals to valves in the hydraulic modulator.

Propshaft torques are regulated to the desired level using symmetrical braking as well as engine intervention. With a gasoline-engine, adjustments undertaken through the throttle valve are relatively slow to take effect (lag and the engine's transition response). Retarding the ignition timing and, as a further option, selective suppression of injection pulses are employed for rapid engine-based intervention, while symmetrical braking can be applied for brief transitional support of engine torque reduction.

Automatic brake functions

The possibilities of today's electronic brake systems go far beyond the tasks for which they were originally designed. Originally the antilock braking system (ABS) was only used to prevent the wheels of a vehicle from locking up and therefore to ensure the steerability of the vehicle even during emergency braking. Today, the brake system also controls the distribution of the braking-force. The electronic stability program (ESP), with its ability to build up brake pressure independently of the position of the brake pedal, offers a whole series of possibilities for active brake intervention. The ESP is intended to assist the driver by applying the brakes automatically and to therefore provide the driver with a higher level of comfort and convenience. Some functions, however, enhance the vehicle safety since automatic brake application during an emergency results in shorter braking distances.

Overview

The main function of the electronic brake system is the Electronic Braking-force Distribution (EBD) function which replaces the mechanical components for braking-force distribution between the front and rear axles. This function not only cuts costs, but also makes the electronic distribution of the braking force extremely flexible.

Additional functions are gradually being integrated into electronic brake systems. The following additional functions are currently available:

- *Hydraulic brake assist (HBA):*
 HBA detects emergency braking situations and shortens the braking distance by building up the brake pressure up to the wheel-lock limit.
- *Controlled deceleration for parking brake (CDP):*
 CDP brakes the vehicle until it is stationary when requested by the driver.

1 Block diagram

- *Hill hold control (HHC):*
 HHC intervenes in the brake system when pulling away on a hill and prevents the vehicle from rolling backwards.
- *Hill descent control (HDC):*
 HDC assists the driver when driving downhill on steep terrain by automatically applying the brakes.
- *Controlled deceleration for driver assistance systems (CDD):*
 CDD brakes the vehicle if required in combination with automatic vehicle-to-vehicle ranging.
- *Hydraulic fading compensation (HFC):*
 HFC intervenes if the maximum possible vehicle deceleration is not achieved even though the driver is forcefully pressing the brake pedal, e. g. due to high brake disk temperatures.
- *Hydraulic rear wheel boost (HRB):*
 HRB also increases the brake pressure in the rear wheels up to the wheel-lock limit during ABS brake application.

- *Brake disk wiping (BDW):*
 BDW removes splash water from the brake disks by briefly applying the brakes. This brake application is not noticed by the driver.

These functions work together with the Electronic Stability Program (ESP). Some of these functions may also be available with the Antilock Braking System (ABS) or the Traction Control System (TCS).

Most of the additional functions operate with the sensor technology of the existing electronic brake systems. Some functions, however, require additional sensors.

2 System layout of additional functions

Sensors and setpoint generators	ESP control unit	Hydraulic diagram

Sensor technology for HBA, CDP, HHC, HDC, CDD, HFC, HRB
- Master cylinder pressure sensor
- Stop lamp switch
- Wheel-speed sensor
- Ambient temperature sensor
- Windshield wiper signal
- Rain sensor

Additional sensor technology for HHC-S
- Pedal-travel sensor
- Parking brake switch
- Engine-speed sensor
- Engine torque
- Transmission information
- Lateral-acceleration sensor
- Tilt sensor

ESP control unit:
- HBA logic
- CDP logic
- HHC logic
- HDC logic
- CDD logic
- HFC logic
- HRB logic
- BDW logic

ESP logic

Safety logic

HSV SV
OV IV

UFB0777E

Standard function

Electronic braking-force distribution (EBD)

Requirements

Legal requirements demand that the braking systems of road vehicles are designed in such a way that they ensure a deceleration up to of $0.83\,g$[1] and provide stable driving behavior during all types of maneuvers so that the vehicle does not exhibit unstable handling characteristics (i. e. a tendency to skid).

Conventional braking-force distribution

On vehicles without ABS, this is achieved by a fixed braking-force distribution between the front and rear brakes or by the use of proportioning valves for the rear brakes (Fig. 1).

Curve 2 shows the pattern for a vehicle with a fixed braking-force distribution which, within the range $0...0.83\,g$ is below the ideal braking-force distribution curve (1l) for an unladen vehicle and fails to utilize the potential for higher rear-wheel braking forces. With a fully laden vehicle, (curve 1b), the utilization of potential is even lower. Curve 3 illustrates the characteristic obtained by using a proportioning valve, whereby a significant gain in rear-wheel braking force is obtained in an unladen vehicle. With the vehicle fully laden, however, the improvement is relatively small.

This latter characteristic can be improved by the use of load-dependent or deceleration-dependent proportioning valves, but at the cost of complex mechanics and hydraulics.

Electronic distribution

Electronic Braking-force Distribution (EBD) allows the distribution between front and rear brakes to be adjusted according to conditions. Handling response is continuously monitored and a greater proportion of the overall braking force can be applied to the rear brakes when conditions allow because the proportioning valve is dispensed with or else larger dimensioned rear wheel brakes are fitted. This releases additional braking potential for the front wheels which can be utilized on vehicles with a high forward weight bias in particular.

Design

The vehicle is designed in such a way that without a proportioning valve, the fixed braking-force distribution curve intersects the ideal braking-force distribution curve (curve 1) at a point P (Fig. 2) at a lower overall braking force, e.g. $0.5\,g$. The use of the existing ABS system's hydraulics, sensors and electronics, but with modified valves and software, allows the braking force at the rear wheels to be reduced at higher overall braking levels.

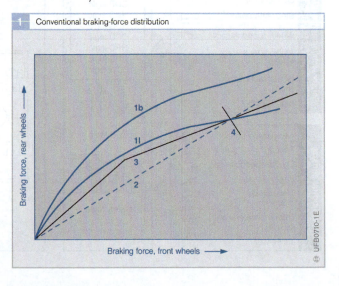

1 Conventional braking-force distribution

Braking force, rear wheels →

Braking force, front wheels →

UFB0710-1E

[1] Gravitational
 acceleration
 $g = 9.81$ m/s²

Fig. 1
1 Ideal braking-force
 distribution of a
 vehicle:
 1l Unladen vehicle
 1b Fully laden
 vehicle
2 Fixed braking-force
 distribution
3 Braking-force
 distribution with
 proportioning valve
4 Straight line for
 retardation of $0.83\,g$
 (g: gravitational
 acceleration)

Method of operation
The ECU continuously calculates the slip difference between the front and rear wheels in all driving situations. If the ratio of front to rear wheel slip exceeds a defined stable-handling threshold when braking, the ABS pressure inlet valve for the appropriate rear wheel is closed. This prevents further increase of brake pressure at that wheel.

If the driver then further increases the force applied to the brake pedal, and therefore the brake pressure, the degree of slip at the front wheels also increases. The difference between front and rear wheel slip diminishes and the pressure inlet valve is opened again so that the pressure at the rear wheel rises once more. This process may then be repeated a number of times depending on the brake pedal force and the maneuver being performed. The electronic braking-force distribution curve then takes on a staircase appearance (curve 3) which approximates to the ideal braking-force distribution curve.

For the Electronic Braking-force Distribution (EBD) function, only the rear-brake valves of the ABS system are activated, the return pump motor in the hydraulic modulator unit remains de-energized.

Advantages
The characteristics of the EBD system outlined above provide the following advantages:
● Optimized vehicle handling stability under all payload conditions, in all cornering situations, on uphill or downhill gradients, and in any drivetrain status (clutch engaged/disengaged, automatic transmission),
● No need for conventional proportioning valves or limiting valves,
● Reduced thermal stresses on the front brakes,
● Even wear between front and rear brake pads,
● Better vehicle deceleration with the same pedal force,
● Constant braking-force distribution patterns over the entire life of the vehicle,
● Only minor modifications to existing ABS components are required.

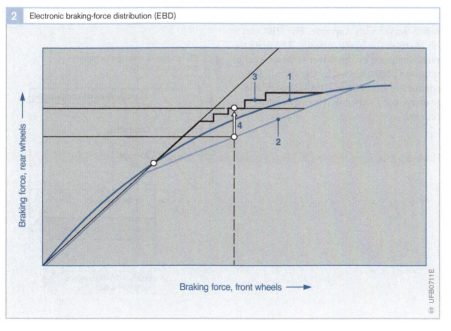

2 Electronic braking-force distribution (EBD)

Braking force, rear wheels →

Braking force, front wheels →

UFB0711E

Fig. 2
1 Ideal braking-force distribution
2 Design braking-force distribution
3 Electronic braking-force distribution
4 Gain in rear-wheel braking force

Additional functions

Hydraulic Brake Assist (HBA)

The main task of the hydraulic brake assist is to detect an emergency braking situation and as a result to automatically increase the vehicle deceleration. The vehicle deceleration is only limited by the intervention of ABS control and is therefore close to the optimum level possible within physical limits. A normal driver is therefore able to achieve the short braking distances that could previously only be achieved by specially trained drivers. If the driver reduces the desired level of braking, the vehicle deceleration is reduced in accordance with the force applied to the brake pedal. The driver can therefore precisely modulate the vehicle deceleration when the emergency braking situation has passed.

The driver's desired level of braking is determined by the force or pressure he/she applies to the brake pedal. The pedal pressure is derived from the measured master cylinder pressure taking into account the current hydraulic control.

The driver can intervene in the brake application at any time and can therefore directly influence the vehicle's response. The HBA can only increase the brake pressure. The primary pressure applied by the driver is therefore the minimum pressure used by the system. If a system error occurs, the HBA is shutoff and the driver is informed that an error has occurred.

Controlled Deceleration for Parking Brake CDP

The Electromechanical Parking Brake (EPB) is an automated parking brake system. It replaces the conventional hand brake lever or foot-operated parking brake lever with an electric motor. The disadvantage of the parking brake is that it only affects the rear axle and its braking force is limited in the event of emergency braking. The CDP function increases braking deceleration and simultaneously enables an ESP system controller to be used to ensure vehicle stability.

The CDP function is an additional function to actively increase the brake pressure in vehicles fitted with an hydraulic braking system and ESP system. On the driver's request, the CDP function automatically decelerates the vehicle until it is stationary. Once the vehicle has come to a standstill, the ESP hydraulics system briefly adopts all static parking brake processes.

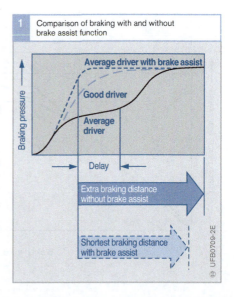

1 Comparison of braking with and without brake assist function

Braking pressure →

Average driver with brake assist

Good driver

Average driver

Delay

Extra braking distance without brake assist

Shortest braking distance with brake assist

UFB0709-2E

Hill Hold Control HHC

The Hill Hold Control (HHC) starting-off assistant is a comfort and convenience function that prevents the vehicle from rolling backwards when pulling away on hills and inclines. The gradient of the hill or slope is measured by a tilt sensor (longitudinal acceleration sensor). To operate, the starting-off assistant requires the brake pressure available when the vehicle is stationary that was built up when the driver pressed the foot brake.

The brake pressure specified by the driver when stopping the vehicle is maintained in the brake system when the system detects that the vehicle is stationary, even if the driver releases the brake pedal. The brake pressure is reduced after a pressure holding time of up to a maximum of two seconds. During this time the driver can press the accelerator pedal and pull away. The brake pressure is reduced when the system detects the driver's intention to pull away.

The system detects that the driver wishes to pull away if the engine torque is sufficient for the vehicle to move in the desired direction. This state can be triggered by the driver pressing the accelerator pedal and/or clutch as well as by the transmission outputting an engine torque (e. g. automatic/continuously variable transmission).

The brake pressure is not maintained in the brake system if there is already sufficient engine torque when the vehicle is stationary (e.g. due to the accelerative force of the automatic transmission).

If the driver presses the accelerator pedal during the vehicle-specific holding time, the holding time is extended until sufficient engine torque is available for the vehicle to pull away.

If the driver does not press the accelerator or the brake pedal, the function is deactivated at the latest after two seconds. The vehicle then starts rolling.

The HHC function is also designed as an additional function for the ESP system and it uses parts of this system. The function is activated automatically.

2 Function description of HHC

activated

1 3 5
2 4 R

deactivated

Activation
· Stop on an incline (> approx. 2%)
· Press brake pedal
· Engage 1st gear
 (if front of vehicle is facing uphill)
· Engage reverse gear
 (if rear of vehicle is facing uphill)

Deactivation
· Start off
· Shift to neutral position

UFB0713E

Hill Descent Control HDC

The Hill Descent Control (HDC) is a comfort and convenience function that assists the driver when driving down hill (on gradients of up to 50%) by automatically applying the brakes. Once this function is activated a predetermined, low desired-speed is introduced and regulated without any necessary intervention from the driver.

The driver must activate and deactivate the HDC function by pressing the HDC push-button.

If required the driver can vary the predetermined speed by pressing the brake and accelerator pedal or using the control buttons of a speed control system.

If the brake slip of the wheels is too high during HDC control, the ABS system intervenes automatically. If the wheels are on different road surfaces, the braking torque of the slipping wheels is automatically distributed to the wheels with a higher coefficient of friction.

If engine braking torque is available it is used automatically. Compared to the exclusive use of engine braking torque in order to maintain the vehicle speed, the HDC function has the advantages that if the wheels leave the surface of the road (loss of engine braking torque) the vehicle speed is maintained and sudden acceleration phases do not occur.

Another advantage of the HDC function is the variable distribution of the braking force which is coupled to the automatic driving direction recognition function. When reversing the rear axle is braked more forcefully to ensure optimum steerability even if the front axle is relieved of the load.

The level ground detection function integrated in the HDC only permits brake application by the HDC when driving downhill. If the vehicle is on the flat or driving uphill, the HDC switches to a standby mode and is reactivated automatically as soon as the system detects that the vehicle is traveling downhill.

To prevent the driver misusing this function, the HDC also switches to its standby mode if the accelerator pedal is pressed passed a certain threshold or if the vehicle exceeds a maximum control speed. The HDC function is deactivated if the vehicle accelerates beyond a preset cutoff-speed.

The status of the HDC function is indicated by the HDC indicator lamp. The brake lamp indicates when the HDC function applies the brakes.

3 Function description of HDC

Functions
- Default speed ≈ 8 km/h
 Speed range ≈ 5…35 km/h
 (can be adjusted by driver)
- Brake activation only when driving downhill
- Designed for off-road conditions
- Full functionality and power
 only in 1st gear or reverse gear

Activation
- Switch

Deactivation
- Switch
- High speed (> 60 km/h)

deactivated

activated

UFB0714E

Controlled Deceleration for Driver Assistance Systems CDD

The basic CDD function is an additional function for active brake application with the Adaptive Cruise Control (ACC), i.e. for automatic vehicle-to-vehicle ranging. The brakes are applied automatically without the driver pressing the brake pedal as soon as the distance to the vehicle in front falls below a predetermined distance. CDD is based on a hydraulic braking system and an ESP system.

The CDD function receives a request to decelerate the vehicle by a desired amount (input). CDD then sends the actual amount of deceleration (output) which is achieved by controlling the pressure using hydraulics. The requested amount of deceleration is specified by the cruise control system which is connected upstream.

Hydraulic Fading Compensation HFC

The Hydraulic Fading Compensation (HFC) function offers the driver additional brake servo assistance. The function is activated if the maximum possible vehicle deceleration is not achieved even if the driver forcefully presses the brake pedal to the point that would normally cause the lockup pressure to be reached (primary pressure over approx. 80 bar). This is the case, for example at high brake disk temperatures or if the brake pads have a considerably reduced coefficient of friction.

When the HFC is activated, the wheel pressures are increased until all wheels have reached the lockup pressure level and ABS control is initiated. The brake application is therefore at the physical optimum. The pressure in the wheel-brake cylinders can then exceed the pressure in the master cylinder, also during ABS control.

If the driver reduces the desired level of braking to a value below a particular threshold value, the vehicle deceleration is reduced in accordance with the force applied to the brake pedal. The driver can therefore precisely modulate the vehicle deceleration when the braking situation has passed. The HFC cuts off if the primary pressure or vehicle speed falls below the respective cutoff threshold.

4 Function description of CDD

Typical signal curve

5 Function description of HFC

Hydraulic Rear Wheel Boost HRB

The Hydraulic Rear wheel Boost (HRB) provides the driver with additional brake servo assistance for the rear wheels if the front wheels are controlled by the ABS system. This function was introduced because many drivers do not increase the pedal force at the start of ABS control even though the situation would require this. When the HRB function is active, the wheel pressures are increased at the rear wheels until they also reach the lockup pressure and ABS control is initiated. The brake application is therefore at the physical optimum. The pressure in the rear axle wheel-brake cylinders can then exceed the pressure in the master cylinder, also during ABS control.

The HRB cuts off when the wheels at the front axle are no longer under ABS control or when the primary pressure falls below the cutoff threshold.

Brake Disk Wiping BDW

The Brake Disk Wiping (BDW) function detects rain or a wet road by evaluating windshield wiper or rain sensor signals and then actively increases the brake pressure in the service brake. The brake pressure buildup is used to remove splash water from the disk brake to ensure minimum brake response times when driving in wet conditions. The pressure level when dry-braking is adjusted so that the vehicle deceleration cannot be perceived by the driver.

Dry-braking is repeated at a defined interval for as long as the system detects rain or a wet road. If required just the disks at the front axle can be wiped.

The BDW function interrupts the wiping procedure as soon as the driver applies the brakes.

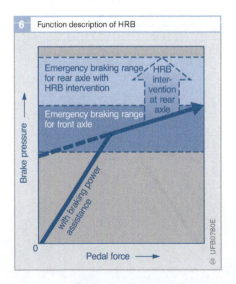

6 Function description of HRB

Emergency braking range for rear axle with HRB intervention

HRB intervention at rear axle

Emergency braking range for front axle

with braking power assistance

Brake pressure

0

Pedal force

UFB0780E

▶ Directional stability

Satisfactory handling is defined according to whether a vehicle maintains a path that accurately reflects the steering angle while at the same time remaining stable. To meet this stability criterion, the vehicle must remain consistently secure and sure-footed, without "plowing" or breaking away.

Dynamic lateral response is a critical factor in the overall equation. The response pattern is defined based on the vehicle's lateral motion (characterized by the side-slip angle) and its tendency to rotate around its vertical axis (yaw velocity) (Fig. 1).

Fig. 2 illustrates the dynamic lateral response of a vehicle being driven with a fixed steering angle (skid-pad circulation). Position 1 represents the instant when steering input is applied. Curve 2 is the vehicle's subsequent course on a high-grip road surface; this track is an accurate reflection of the steering angle. This is the case when the coefficient of friction is sufficient to transfer the lateral acceleration forces to the road surface. If the coefficient of friction is lower e.g. due to a slippery road sur-

face, the side-slip angle becomes excessive (curve 3). Although with a controlled yaw velocity the vehicle will rotate just as far around its vertical axis as in curve 2, the larger side-slip angle is now a potential source of instability. For this reason the electronic stability program controls the yaw velocity and limits the side-slip angle β (curve 4).

2 Dynamic lateral response

Fig. 2
1 Steering input, fixed steering-wheel angle
2 Vehicle path on high-grip surface
3 Vehicle path on low-grip surface with yaw velocity control
4 Vehicle path on low-grip surface with additional control of the side-slip angle β (ESP)

1 Vehicle travel directions

Vehicle vertical axis
Yaw velocity

Vehicle longitudinal axis

Vehicle horizontal axis
Lateral acceleration

Linear acceleration

Rolling

Steering movement

Hydraulic modulator

The hydraulic modulator forms the hydraulic connection between the master cylinder and the wheel-brake cylinders and is therefore the central component of electronic brake systems. It converts the control commands of the electronic control unit and uses solenoid valves to control the pressures in the wheel brakes.

A distinction is made between systems that modulate the brake pressure applied by the driver (Antilock Braking System, (ABS)) and systems that can build-up pressure automatically (Traction Control System (TCS) and Electronic Stability Program (ESP)).
All systems are only available as dual-circuit versions for legal reasons.

Development history

The transition from 3/3 to 2/2 solenoid valve was a milestone in the development of ABS. With 3/3 valves, which were used in generation 2, the control functions for building up, maintaining and reducing pressure could be carried out using just one valve. The valves had three hydraulic connections to carry out these functions. The disadvantages of this valve design were extremely expensive electrical control and great mechanical complexity. Control with the 2/2 valves of the current generations offers a less expensive solution. The following section describes how these valves work.

Generation 8, which was introduced onto the market in 2001, is designed as a fully modular system. The hydraulic system can therefore be tailored to meet the requirements of the respective vehicle manufacturer such as with regard to value added functions, comfort and convenience, vehicle segment (up to lightweight commercial vehicles), etc. Generation 8 is immersion-proof which means that the hydraulic modulator can withstand brief immersion in water.

1 ESP 8 hydraulic modulator

Fig. 1
1 Electronic control unit
2 Coil grid
3 Coils/solenoid group
4 Solenoid valves
5 Hydraulic block
6 DC motor
7 Plunger pump
8 Low-pressure reservoir
9 Pressure sensor

UFB0750-1Y

Design

Mechanical system

A hydraulic modulator for ABS/TCS/ESP consists of an aluminum block into which the hydraulic layout is drilled (Fig. 2). This block also houses the necessary hydraulic function elements (Fig. 1) presented below.

ABS hydraulic modulator

With a 3-channel ABS system, this block features one inlet valve and one outlet valve for each front wheel and one inlet valve and one outlet valve for the rear axle – a total of six valves. This system can only be used in vehicles featuring a dual brake-circuit configuration. The two wheels at the rear axle are not controlled individually, rather both wheels are controlled according to the select-low principle. This means that the wheel with the

greater slip determines the controllable pressure of the axle.

With a 4-channel ABS system (for dual or X-braking-force distribution), one inlet and one outlet valve is used for each wheel resulting in a total of eight valves. This system allows each wheel to be controlled individually.

Moreover, the system features one pump element (return pump) and one low-pressure reservoir for each brake circuit. Both pump elements are operated by a shared DC motor.

2 Hydraulic diagram of 4-channel ABS hydraulic modulator (X-brake-circuit configuration)

Fig. 2
HZ Master cylinder
RZ Wheel-brake
 cylinder
IV Inlet valve
OV Outlet valve
PE Return pump
M Pump motor
AC Low-pressure
 reservoir
V Front
H Rear
R Right
L Left

TCS hydraulic modulator

Unlike an ABS unit, a TCS system with II-brake-circuit configuration also has a switchover valve and an inlet valve at the rear axle (driven wheels) resulting in a total of 10 valves.

In TCS systems with X-brake-circuit configuration an additional switchover valve and inlet valve are required for each circuit (a total of 12 valves).

ESP hydraulic modulator

ESP systems require 12 valves irrespective of the brake-circuit configuration (Fig. 3). In these systems, the two inlet valves, as used in the TCS hydraulic modulator, are replaced by two high-pressure switching valves. The difference between the two types of valves is that the high-pressure switching valve can switch against higher differential pressures (> 0.1 MPa). With ESP it may be necessary to increase the brake pressure specified by the

driver in order to stabilize the vehicle. During such a maneuver (partially-active maneuver), the suction path of the pump must be opened despite the high primary pressure.

An integrated pressure sensor is used exclusively in ESP systems. The sensor detects the brake pressure in the master cylinder i.e. the driver command. This is also required for a partially-active ESP control maneuver since it is important to know the primary pressure applied by the driver pressing the brake pedal.

Since TCS/ESP systems are required to generate pressure automatically, the return pump is replaced by a self-priming pump in both these systems. An additional non-return valve with a specific closing pressure is required to prevent the pump from drawing unwanted media from the wheels.

Fig. 3

HZ	Master cylinder
RZ	Wheel-brake cylinder
EV	Inlet valve
AV	Outlet valve
USV	Switchover valve
SV	High-pressure switching valve
PE	Return pump
M	Pump motor
AC	Low-pressure reservoir
V	Front
H	Rear
R	Right
L	Left

3 Hydraulic diagram of ESP hydraulic modulator (X-brake-circuit configuration)

Evolution of the ABS system

Technological advances in the areas of
- solenoid-valve design and manufacturing,
- assembly and component integration,
- electronic circuitry (discrete components replaced by hybrid and integrated circuits with microcontrollers),
- testing methods and equipment (separate testing of electronic and hydraulic systems before combination in the hydraulic modulator), and
- sensor and relay technology

have enabled the weight and dimensions of ABS systems to be more than halved since the first-generation ABS2 in 1978. As a result, modern systems can now be accommodated even in vehicles with the tightest space restrictions. Those advances have also lowered the cost of ABS systems to the extent that it has now become standard equipment on all types of vehicle.

1 Evolution of ABS configurations

Hydraulic-modulator development

ABS 2 — 6,200 g
ABS 5.0 — 3,800 g
ABS 5.3/5.7 — 2,600 g
ABS 8 — <2,000 g

Development of electronics

ABS 2 — 1989
ABS 5.0 — 1993
ABS 5.3 — 1995
ABS 5.3/5.7
ABS 8 — 2001

UFB0700E

Fig. 1
Historical development of ABS showing technological advances: Decreasing weight accompanied by increasing processing power.

Pressure modulation

Modulation with ABS control

The pressure of an ABS/TCS/ESP system is modulated using solenoid valves. The outlet valves and the inlet and high-pressure switching valves (for TCS and ESP systems) are switching valves that are closed at zero current and can adopt two positions – closed or open.

In contrast, the inlet valves and switchover valves are both open at zero current and are used as control valves for the first time in generation 8. This is advantageous not only for instantaneous braking power and braking comfort but also for noise emissions. Pressures of up to 200 bar can be modulated using the standard valve set. Special systems for even higher pressures and also for greater throughput, usually required in the commercial vehicle sector, can be designed based on further developments of the generation 8 modular system.

All valves are controlled using coils which are controlled via the add-on control unit.

Pressure modulation with ABS hydraulic modulator

In the event of ABS brake application, the driver first generates the brake pressure at the wheel by pressing the brake pedal. This can be done without switching the valves since the inlet valve is open at zero current and the outlet valve is closed at zero current (Fig. 1a).

The *pressure maintenance* state is generated when the inlet valve is closed (Fig. 1b).

If a wheel locks up, the pressure from this wheel is released by opening the relevant outlet valve (Fig. 1c). The brake fluid can also escape from the wheel-brake cylinder to the relevant low-pressure reservoir. This chamber performs the role of a buffer. It can receive the brake fluid quickly and efficiently. The return pump in the circuit, which is driven by a shared motor via an eccentric element, reduces the pressure specified by the driver. The motor is actuated based on demand i.e. the motor is actuated according to the rotational speed. The ABS

system can of course control several wheels if they lock up simultaneously.

Fig. 1
a Pressure build-up
b Pressure
 maintenance
c Pressure reduction

IV Inlet valve
OV Outlet valve
PE Return pump
M Pump motor
AC Low-pressure
 reservoir
V Front
H Rear
R Right
L Left

1 Pressure modulation in ABS hydraulic modulator

Pressure modulation with ESP hydraulic modulators

Pressure is modulated with ESP control using ESP hydraulics in the same way as described for ABS. Unlike with ABS, however, the wheel-brake cylinder and master cylinder are also connected via a switchover valve that is open at zero current and a high-pressure

2 Pressure modulation in ESP hydraulic modulator

switching valve that is closed at zero current. These two valves are required for active/partially-active brake intervention (Fig. 2).

Pressure generation with ESP

The pressure generation chain consists of two self-priming pumps and a motor. Plunger pumps are used, as for ABS, but these pumps can generate pressure without requiring the primary pressure applied when the driver presses the brake pedal. These pumps are driven by a DC motor based on demand. The motor drives an eccentric bearing located on the shaft of the motor.

TCS/ESP pumps can build up pressure independently of the driver or increase the braking pressure already generated by the driver. These systems are therefore able to initiate brake application. To do this, the switchover valve is closed and the inlet valve or high-pressure switching valve is opened. Fluid can then be drawn from the brake fluid reservoir via the master cylinder and pressure can be built up in the wheel-brake cylinders (Fig. 2c). This is required not only for TCS/ESP functions, but also for many additional convenience functions (value added functions such as the brake assist (HBA)).

Demand-based control of the pump motor reduces noise emission during pressure generation and regulation. The pumps can be equipped with damping elements to satisfy vehicle manufacturer's strict low noise emission requirements.

Fig. 3

a Pressure build-up when braking

b Pressure reduction with ABS control

c Pressure build-up via self-priming pump due to TCS or ESP intervention

IV Inlet valve
OV Outlet valve
SV Switchover valve
HSV High-pressure switching valve
PE Return pump
M Pump motor
AC Low-pressure reservoir
V Front
H Rear
R Right
L Left

With ESP, there are basically three different applications:
- Passive – as previously described for ABS control.
- Partially-active – when the pressure specified by the driver is insufficient to stabilize the vehicle.
- Fully-active – when pressure is generated to stabilize the vehicle without the driver pressing the brake pedal.

Both pressure generation cases above, in addition to ESP control, are used for a whole series of additional functions such as adaptive cruise control and the brake assist.

Partially-active control

For partially-active control, the high-pressure switching valve must be able to open the suction path of the pump against high pressures. This is required since the driver has already generated a high pressure, but this pressure is insufficient to stabilize the vehicle.

The high-pressure switching valve is designed with two stages so that the valve can open against the high pressure. The first stage of the valve is opened via the magnetic force of the energized coil; the second stage via the hydraulic area difference.

If the ESP controller detects an unstable vehicle state, the switchover valves (open at zero current) are closed and the high-pressure switching valve (closed at zero current) is opened. The two pumps then generate additional pressure in order to stabilize the vehicle.

Once the vehicle is stabilized, the outlet valve is opened and the excess pressure in the controlled wheel escapes to the reservoir. As soon as the driver releases the brake pedal, the fluid is pumped from the reservoir back to the brake fluid reservoir.

Fully-active control

If the ESP controller detects an unstable vehicle state, the switchover valves are closed. This prevents the pumps from being hydraulically short-circuited via the switchover/high-pressure switching valve which would prevent pressure generation. The high-pressure switching valves are simultaneously opened. The self-priming pump now pumps brake fluid to the relevant wheel or wheels in order to build up pressure. If pressure generation is only required in one wheel, (for yaw rate compensation), the inlet valves of the other wheels are closed. To reduce the pressure, the outlet valves are opened and the high-pressure switching valves and switchover valves return to their original position. The brake fluid flows from the wheels to the reservoirs. The pumps then empty the reservoirs.

Development of hydraulic modulators

Control of hydraulic modulators
An electronic control unit processes the information received from the sensors and generates the control signals for the hydraulic modulator. The hydraulic modulator incorporates a series of solenoid valves that can open or close the hydraulic circuits between the master cylinder and the wheel-brake cylinders.

Hydraulic modulators with 3/3 solenoid valves
In 1978 version ABS2S was the first antilock braking system to go into series production. In this ABS system the electronic control unit switches the 3/3 solenoid valves of the hydraulic modulator to three different valve positions. There is a solenoid valve for each wheel-brake cylinder (Fig. 1a).
– The first (zero-current) position connects the master cylinder and the wheel-brake cylinder; the wheel brake pressure can rise.
– The second position (excitation at half the maximum current) separates the wheel brake from the master cylinder and return line so that the wheel brake pressure remains constant.
– The third position (excitation at the maximum current) separates the master cylinder and simultaneously connects the wheel brake and return line so that the wheel brake pressure decreases.

By applying these settings in the appropriate sequences, the brake pressure can thus be increased and reduced either continuously or incrementally (and therefore gradually).

Hydraulic modulators with 2/2 solenoid valves
While version ABS2S operates with 3/3 solenoid valves, the successor systems ABS5 and ABS8 feature 2/2 solenoid valves with two hydraulic connections and two valve positions. The inlet valve between the master cylinder and the wheel-brake cylinder controls pressure build-up, while the outlet valve between the wheel-brake cylinder and the return pump controls pressure release. There is a solenoid valve pair for each wheel-brake cylinder (Fig. 1b).
– In the "pressure build-up" position, the inlet valve connects the master cylinder to the wheel-brake cylinder so that the brake pressure built up in the master cylinder can be applied to the wheel-brake cylinder when the brakes are applied.
– In the "pressure maintenance" position, the inlet valve blocks the connection between the master cylinder and wheel-brake cylinder during rapid wheel deceleration (risk of locking up) and thus prevents the brake pressure from rising any higher. The outlet valve is also closed.
– If the wheel deceleration continues to increase, the inlet valve continues to block in the "decrease pressure" position. In addition the return pump pumps out the brake fluid via the open outlet valve so that the brake pressure in the wheel-brake cylinder drops.

1 Comparison of ABS systems

Fig. 1
a ABS2
b ABS5

1 Damping chamber
2 Constrictor
3 Return pump
4a 3/3 solenoid valve
4b 2/2 solenoid valves
5 Accumulator chamber

UFB0545-2Y

Sensotronic brake control (SBC)

Sensotronic brake control (SBC) is an electrohydraulic brake system that combines the functions of a brake servo unit and the ABS (antilock braking system) equipment, including ESP (electronic stability program). The mechanical operation of the brake pedal is redundantly measured by the actuator unit and transmitted to the control unit. There, control commands are calculated according to specific algorithms and passed to the hydraulic modulator where they are converted into pressure modulating operations for the brakes. If the electronics fail, a hydraulic fallback system is automatically available.

Purpose and function

By utilizing its "brake-by-wire" capabilities, SBC can control the hydraulic pressure in the wheel brake cylinders independently of driver input. As a result, functions beyond those performed by ABS (antilock braking system), TCS (traction control system) and ESP (electronic stability program) can be implemented. One example is the convenient method of brake application for ACC (adaptive cruise control).

Basic functions

As with a conventional braking system, sensotronic brake control must be capable of
- reducing the speed of the vehicle,
- bringing the vehicle to a halt, and
- keeping the vehicle stationary when it is stopped.

As an active braking system, it also performs the tasks of
- operating the brakes,
- amplifying the brake force, and
- modulating the brake force.

SBC is an electronic control system with hydraulic actuators. Braking force distribution takes place electronically to each wheel in response to driving conditions. A vacuum source for the brake servo function is no longer required. The self-diagnosis capability enables an early warning function for detection of possible system faults.

SBC uses hydraulic standard wheel brakes. Because of the fully electronic pressure control, SBC can be easily networked with vehicle handling systems. It thus meets all the demands made of future braking systems.

1 SBC components in the car

Fig. 1
1. Active wheel speed sensor with direction sensing
2. Engine management ECU
3. SBC ECU
4. Yaw rate and lateral acceleration sensor
5. Hydraulic modulator (for SBC, ABS, TCS and ESP)
6. Actuator unit with pedal travel sensor
7. Steering angle sensor

UFB0704-2Y

By using a high-pressure accumulator, SBC is capable of extremely rapid dynamic pressure increases and thus offers the potential for achieving short braking distances and excellent vehicle handling stability. Brake pressure modulation and active braking are silent and produce no brake pedal feedback. Consequently, SBC also satisfies demands for greater levels of comfort.

Braking characteristics can be adapted to the driving conditions (e.g. sharper response at high speeds or with more dynamic driving styles). "Duller" pedal characteristics allow the reduction of the braking effect, which is necessitated by physics, to be signaled to the driver before fading due to overheating occurs.

Additional SBC functions
The auxiliary functions provided by SBC make a significant extra contribution to safety and convenience.

Hill hold control
After hill hold control is activated by a significant increase in brake pressure while the vehicle is stationary, the vehicle remains braked without the need to keep the pedal down. The hill hold control is automatically released as soon as the driver has built up sufficient engine torque by depressing the accelerator. This allows the driver to start the car on a hill, for example, without activating the parking brake system. Likewise, in other situations in which the vehicle would roll out of position if not braked, the driver does not need to keep his or her foot on the brake at all times once hill hold control has been activated.

Enhanced brake assist function
If the driver abruptly releases the accelerator, an automatically regulated brake pressure build-up takes place that gently applies the brake pads. If panic braking follows, this allows the brake to "grab" more quickly and thus allows a shorter total braking distance.

If the system detects panic braking, the brake pressure is briefly increased until the optimum friction value is fully utilized. This results in a significant reduction of total braking distance for hesitant drivers. The highly dynamic braking force build-up of SBC exceeds that of conventional systems in this regard.

Soft stop assist
SBC provides comfortable braking that stops the car with no jerking by automatically reducing the pressure just before the car comes to a complete stop. If more deceleration is desired, this function is not activated and SBC minimizes the braking distance.

Traffic jam assist
When traffic jam assist is activated, SBC builds up a higher drag torque, which means that the driver does not have to constantly alternate between the accelerator and brake. The vehicle is automatically braked and, if necessary, brought to a complete stop and kept at a complete stop. This function can be activated at speeds of 50 - 60 km/h.

Brake wiping
"Brake wiping" is an operation whereby the film of water is regularly removed from the brake disks in wet weather. It results in shorter stopping distances in wet conditions. You can take the information for activating this function from the windshield wiper signal, for example.

2 Evolution of brake systems

Design

The sensotronic brake control system consists of the following components:
- Actuator unit (Fig. 1, Item 6)
- Vehicle dynamics sensors (1, 4 and 7)
- Discrete (separate) ECU (3)
- Hydraulic modulator with add-on ECU (5).

Those components are interconnected by electrical control leads and hydraulic pipes. Fig. 1 shows where they are fitted in a car.

Actuator unit

The actuator unit consists of:
- Master cylinder with expansion tank,
- Pedal travel simulator and
- Pedal travel sensor.

Pedal travel simulation

The pedal travel simulator makes it possible to realize a suitable force-distance curve and an appropriate damping of the brake pedal. In that way, the driver obtains the same "brake feel" with the sensotronic brake control system as with a very well designed conventional braking system.

Sensors of the SBC systems

The SBC sensors consist of the vehicle dynamics sensors familiar from ESP and the actual SBC sensors.

Vehicle dynamics sensors

The ESP sensors consist of four wheel speed sensors, the yaw rate sensor, a steering angle sensor and, where applicable, a lateral acceleration sensor. These sensors provide the ECU with data relating to the speed and movement status of the wheels and driving states such as cornering. Control functions such as ABS, TCS and ESP are executed in the familiar manner.

If the vehicle is fitted with ACC (adaptive cruise control), a radar system measures the distance to the vehicle in front. From these data, the SBC control unit calculates the brake pressure to be applied, which is built up without pedal feedback.

SBC sensors

Four pressure sensors measure the pressure individually for each wheel circuit (Fig. 3). A pressure sensor measures the storage pressure of the high-pressure accumulator. The driver's braking request is calculated by the pedal travel sensor attached to the actuator unit and a pressure sensor that detects the brake pressure applied by the driver.

The pedal travel sensor consists of two redundant, independent angle sensors. Together with the pressure sensor for the driver's brake pressure, they allow a threefold detection of the driver's request, and the system can continue to work without errors even if one of these systems fails.

Method of operation

Normal operation

Fig. 3 illustrates the components of SBC in the form of a block diagram. An electric motor drives a hydraulic pump. This charges a high-pressure accumulator to a pressure between approx. 90 and 130 bar, which is monitored by the accumulator pressure sensor. The four independent wheel pressure modulators are supplied by this accumulator and adjust the required pressure individually for each wheel. The pressure modulators themselves each consist of two valves with proportional control characteristics and a pressure sensor.

In normal operation, the isolating valves interrupt the connection to the actuator. The system is in "brake-by-wire" mode. It electronically detects the driver's braking request and transmits it "by wire" to the wheel pressure modulators. The interaction of the engine, valves and pressure sensors is controlled by the electronics, which are installed in the add-on ECU in hybrid technology. They have two microcontrollers that monitor each other. The essential feature of these electronics is their extensive self-diagnosis,

which monitors the plausibility of every system state at all times. In this way, the driver can be notified of any failures before critical states occur. If components fail, the system automatically provides the optimal remaining partial function to the driver. An extensive fault memory allows prompt diagnostics and repair in the event of a fault.

An intelligent interface with CAN bus establishes the connection to the discrete ECU. The following functions are integrated there:
- ESP (electronic stability program),
- TCS (traction control system),
- ABS (antilock braking system),
- Driver brake request calculation and
- SBC auxiliary functions (assist functions).

Braking in the event of system failure
For safety reasons, the SBC system is designed so that in the event of any serious errors (such as failure of the power supply), the system is switched to a state in which the vehicle can be braked even without active brake force support. When de-energized, the isolating valves establish a direct connection to the actuator (Fig. 3) and thus allow a direct hydraulic connection from the actuator unit to the wheel brake cylinders.

To maintain optimum function even if the system fails, the plunger pistons in the illustration serve as a medium separator between the active circuit of the SBC and the conventional front axle brake circuit. These prevent any gas that might escape from the high-pressure accumulator from reaching the brake circuit of the front wheels, which would diminish the instantaneous braking power in the event of a system failure.

3 Interaction of SBC functional modules

Overview of common-rail systems

The demands placed on diesel-engine fuel-injection systems are continuously increasing. Higher pressures, faster switching times, and a variable rate-of-discharge curve modified to the engine operating state have made the diesel engine economical, clean, and powerful. As a result, diesel engines have even entered the realm of luxury-performance sedans.

One of the advanced fuel-injection systems is the *common-rail (CR)* fuel-injection system. The main advantage of the common-rail system is its ability to vary injection pressure and timing over a broad scale.
This was achieved by separating pressure generation (in the high-pressure pump) from the fuel-injection system (injectors). The rail here acts as a pressure accumulator.

Areas of application

The common-rail fuel-injection system for engines with diesel direct injection (Direct Injection, DI) is used in the following vehicles:

- *Passenger cars* ranging from high-economy 3-cylinder engines with displacements of 800 *cc*, power outputs of 30 kW (41 HP), torques of 100 Nm, and a fuel consumption of 3.5 l/100 km through to 8-cylinder engines in luxury-performance sedans with displacements of approx. 4 *l*, power outputs of 180 kW (245 HP), and torques of 560 Nm.
- *Light-duty trucks* with engines producing up to 30 kW/cylinder, and
- *Heavy-duty trucks*, *railway locomotives*, and *ships* with engines producing up to approx. 200 kW/cylinder

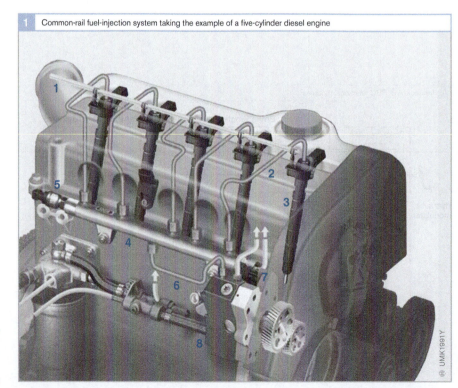

1 Common-rail fuel-injection system taking the example of a five-cylinder diesel engine

Fig. 1
1 Fuel return line
2 High-pressure fuel line to injector
3 Injector
4 Fuel rail
5 Rail-pressure sensor
6 High-pressure fuel line to rail
7 Fuel return line
8 High-pressure pump

The common-rail system is a highly flexible system for adapting fuel injection to the engine. This is achieved by:

- High injection pressure up to approx. 1,600 bar, in future up to 1,800 bar.
- Injection pressure adapted to the operating status (200...1,800 bar).
- Variable start of injection.
- Possibility of several pre-injection events and secondary injection events (even highly retarded secondary injection events).

In this way, the common-rail system helps to raise specific power output, lower fuel consumption, reduce noise emission, and decrease pollutant emission in diesel engines.

Today common rail has become the most commonly used fuel-injection system for modern, high-rev passenger-car direct-injection engines.

Design

The common-rail system consists of the following main component groups (Figs. 1 and 2):

- *The low-pressure stage,* comprising the fuel-supply system components.
- *The high-pressure system,* comprising components such as the high-pressure pump, fuel rail, injectors, and high-pressure fuel lines.
- *The electronic diesel control (EDC),* consisting of system modules, such as sensors, the electronic control unit, and actuators.

The key components of the common-rail system are the injectors. They are fitted with a rapid-action valve (solenoid valve or piezo-triggered actuator) which opens and closes the nozzle. This permits control of the injection process for each cylinder.

2 System modules of an engine control unit and a common-rail fuel-injection system

Electronic diesel control (EDC): engine management, sensors, interface

Fuel supply system (low-pressure stage)

Air-intake and exhaust-gas systems

Engine

High-pressure section

← Signals
← Diesel fuel

NMK1871E

Fig. 2
1 High-pressure pump
2 Fuel rail
3 Injectors

All the injectors are fed by a common fuel rail, this being the origin of the term "common rail".

One of the main features of the common-rail system is that system pressure is variable dependent on the engine operating point. Pressure is adjusted by the pressure-control valve or the metering unit (Fig. 3).

The modular design of the common-rail system simplifies modification of the system to different engines.

Operating concept

In the common-rail fuel-injection system, the functions of pressure generation and fuel injection are separate. The injection pressure is generated independent of the engine speed and the injected fuel quantity. The electronic diesel control (EDC) controls each of the components.

Pressure generation

Pressure generation and fuel injection are separated by means of an accumulator volume. Fuel under pressure is supplied to the accumulator volume of the common rail ready for injection.

A continuously operating high-pressure pump driven by the engine produces the desired injection pressure. Pressure in the fuel rail is maintained irrespective of engine speed or injected fuel quantity. Owing to the almost uniform injection pattern, the high-pressure pump design can be much smaller and its drive-system torque can be lower than conventional fuel-injection systems. This results in a much lower load on the pump drive.

The high-pressure pump is a radial-piston pump. On commercial vehicles, an in-line fuel-injection pump is sometimes fitted.

Pressure control

The pressure control method applied is largely dependent on the system.

Control on the high-pressure side

On passenger-car systems, the required rail pressure is controlled on the high-pressure side by a pressure-control valve (Fig. 3a, 4). Fuel not required for injection flows back to the low-pressure circuit via the pressure-control valve. This type of control loop allows rail pressure to react rapidly to changes in operating point (e.g. in the event of load changes).

Fig. 3

a Pressure control on the high-pressure side by means of pressure-control valve for passenger-car applications
b Pressure control on the suction side with a metering unit flanged to the high-pressure pump (for passenger cars and commercial vehicles)
c Pressure control on the suction side with a metering unit and additional control with a pressure-control valve (for passenger cars)

1 High-pressure pump
2 Fuel inlet
3 Fuel return
4 Pressure-control valve
5 Fuel rail
6 Rail-pressure sensor
7 Injector connection
8 Return fuel connection
9 Pressure-relief valve
10 Metering unit
11 Pressure-control valve

3 Examples of high-pressure control for common-rail systems

SMK1993Y

Control on the high-pressure side was adopted on the first common-rail systems. The pressure-control valve is mounted preferably on the fuel rail. In some applications, however, it is mounted directly on the high-pressure pump.

Fuel-delivery control on the suction side

Another way of controlling rail pressure is to control fuel delivery on the suction side (Fig. 3b). The metering unit (10) flanged on the high-pressure pump makes sure that the pump delivers exactly the right quantity of fuel to the fuel rail in order to maintain the injection pressure required by the system. In a fault situation, the pressure-relief valve (9) prevents rail pressure from exceeding a maximum.

Fuel-delivery control on the suction side reduces the quantity of fuel under high pressure and lowers the power input of the pump. This has a positive impact on fuel consumption. At the same time, the temperature of the fuel flowing back to the fuel tank is reduced in contrast to the control method on the high-pressure side.

Two-actuator system

The two-actuator system (Fig. 3c) combines pressure control on the suction side via the metering unit and control on the high-pressure side via the pressure-control valve, thus marrying the advantages of high-pressure-side control and suction-side fuel-delivery control (see the section on "Common-rail system for passenger cars").

Fuel injection

The injectors spray fuel directly into the engine's combustion chambers. They are supplied by short high-pressure fuel lines connected to the fuel rail. The engine control unit controls the switching valve integrated in the injector to open and close the injector nozzle.

The injector opening times and system pressure determine the quantity of fuel delivered. At a constant pressure, the fuel quantity delivered is proportional to the switching time of the solenoid valve. This is, therefore, independent of engine or pump speed (time-based fuel injection).

Potential hydraulic power

Separating the functions of *pressure generation* and *fuel injection* opens up further degrees of freedom in the combustion process compared with conventional fuel-injection systems; the injection pressure is more or less freely selectable within the program map. The maximum injection pressure at present is 1,600 bar; in future this will rise to 1,800 bar.

The common-rail system allows a further reduction in exhaust-gas emissions by introducing pre-injection events or multiple injection events and also attenuating combustion noise significantly. Multiple injection events of up to five per injection cycle can be generated by triggering the highly rapid-action switching valve several times. The nozzle-needle closing action is hydraulically assisted to ensure that the end of injection is rapid.

Control and regulation
Operating concept
The engine control unit detects the accelerator-pedal position and the current operating states of the engine and vehicle by means of sensors (see the section on "Electronic diesel control"). The data collected includes:
- Crankshaft speed and angle
- Fuel-rail pressure
- Charge-air pressure
- Intake air, coolant temperature, and fuel temperature
- Air-mass intake
- Road speed, etc.

The electronic control unit evaluates the input signals. In sync with combustion, it calculates the triggering signals for the pressure-control valve or the metering unit, the injectors, and the other actuators (e.g. the EGR valve, exhaust-gas turbocharger actuators, etc.).

The injector switching times, which need to be short, are achievable using the optimized high-pressure switching valves and a special control system.

The angle/time system compares injection timing, based on data from the crankshaft and camshaft sensors, with the engine state (time control). The electronic diesel control (EDC) permits a precise metering of the injected fuel quantity. In addition, EDC offers the potential for additional functions that can improve engine response and convenience.

Basic functions
The basic functions involve the precise control of diesel-fuel injection timing and fuel quantity at the reference pressure. In this way, they ensure that the diesel engine has low consumption and smooth running characteristics.

Correction functions for calculating fuel injection
A number of correction functions are available to compensate for tolerances between the fuel-injection system and the engine (see the section on "Electronic diesel control"):
- Injector delivery compensation
- Zero delivery calibration
- Fuel-balancing control
- Average delivery adaption

Additional functions
Additional open- and closed-loop control functions perform the tasks of reducing exhaust-gas emissions and fuel consumption, or providing added safety and convenience. Some examples are:
- Control of exhaust-gas recirculation
- Boost-pressure control
- Cruise control
- Electronic immobilizer, etc.

Integrating EDC in an overall vehicle system opens up a number of new opportunities, e.g. data exchange with transmission control or air-conditioning system.

A diagnosis interface permits analysis of stored system data when the vehicle is serviced.

Control unit configuration
As the engine control unit normally has a maximum of only eight output stages for the injectors, engines with more than eight cylinders are fitted with two engine control units. They are coupled within the "master/slave" network via an internal, high-speed CAN interface. As a result, there is also a higher microcontroller processing capacity available. Some functions are permanently allocated to a specific control unit (e.g. fuel-balancing control). Others can be dynamically allocated to one or other of the control units as situations demand (e.g. to detect sensor signals).

Diesel engine applications

At the start of automobile history, the spark-ignition engine (Otto cycle) was the drive unit for road vehicles. The first time a diesel engine was mounted on a truck was 1927. Passenger cars had to wait until 1936.

The diesel engine made strong headway in the truck sector due to its fuel economy and long service life. By contrast, the diesel engine in the car sector was long relegated to a fringe existence. It was only with the introduction of supercharged direct-injection diesel engines – the principle of direct injection was already used in the first truck diesel engines – that the diesel engine changed its image. Meanwhile, the percentage of diesel-engined passenger cars among new registrations is fast approaching 50% in Europe.

Features of the diesel engine

What is the reason for the boom in diesel engines in Europe?

Fuel economy

Firstly, fuel consumption compared to gasoline engines is still lower – this is due to the greater efficiency of the diesel engine. Secondly, diesel fuel is subject to lower taxes in most European countries. For people who travel a lot, therefore, diesel is the more economical alternative despite the higher purchase price.

Driving pleasure

Almost all diesel engines on the market are supercharged. This produces a high cylinder charge at low revs. The metered fuel quantity can also be high, and this produces high engine torque. The result is a torque curve that permits driving at high torque and low revs.

It is torque, and not engine performance, that is the decisive factor for engine power. Compared to a gasoline engine without supercharging, a driver can experience more "driving pleasure" with a diesel engine of lower performance. The image of the "stinking slowcoach" is simply no longer true for diesel-engined cars of the latest generation.

Environmental compatibility

The clouds of smoke that diesel-engined cars produced when driven at high loads are a thing of the past. This was brought about by improved fuel-injection systems and electronic diesel control (EDC). These systems can meter fuel quantity with high precision, adjusting it to the engine operating point and environmental conditions. This technology also meets prevailing exhaust-gas emission standards.

Oxidation-type catalytic converters, that remove carbon monoxide (CO) and hydrocarbons (HC) from exhaust gas, are standard equipment on diesel engines. Future, more stringent exhaust-gas emission standards, and even U.S. legislation, will be met by other exhaust-gas treatment systems, such as particulate filters and NO_x accumulator-type catalytic converters.

▶ Typical torque and power curves of a passenger-car diesel engine

⊕ SMK2023E

Common-rail system for passenger cars

Fuel supply

In common-rail systems for passenger cars, electric fuel pumps or gear pumps are used to deliver fuel to the high-pressure pump.

Systems with electric fuel pump

The electric fuel pump is either part of the in-tank unit (in the fuel tank) or is fitted in the fuel line (in-line). It intakes fuel via a pre-filter and delivers it to the high-pressure pump at a pressure of 6 bar (Fig. 3). The maximum delivery rate is 190 *l*/h. To ensure fast engine starting, the pump switches on as soon as the driver turns the ignition key. This builds up the necessary pressure in the low-pressure circuit when the engine starts.

The fuel filter (fine filter) is fitted in the supply line to the high-pressure pump.

Systems with gear pump

The gear pump is flanged to the high-pressure pump and is driven by its input shaft (Figs. 1 and 2). In this way, the gear pump starts delivery only after the engine has started. Delivery rate is dependent on the engine speed and reaches rates up to 400 *l*/h at pressures up to 7 bar.

A fuel pre-filter is fitted in the fuel tank. The fine filter is located in the supply line to the gear pump.

Combination systems

There are also applications where the two pump types are used. The electric fuel pump improves starting response, in particular for hot starts, since the delivery rate of the gear pump is lower when the fuel is hot, and therefore, thinner, and at low pump speeds.

High-pressure control

On first-generation common-rail systems, rail pressure is controlled by the pressure-control valve. The high-pressure pump (type CP1) generates the maximum delivery quantity, irrespective of fuel demand. The pressure-control valve returns excess fuel to the fuel tank.

Second-generation common-rail systems control rail pressure on the low-pressure side by means of the metering unit (Figs. 1 and 2). The high-pressure pump (types CP3 and CP1H) need only deliver the fuel quantity that the engine actually requires. This lowers the energy demand of the high-pressure pump and reduces fuel consumption.

Third-generation common-rail systems feature piezo-inline injectors (Fig. 3).

If pressure is only adjustable on the low-pressure side, it takes too long to lower the pressure in the fuel rail when rapid negative load changes occur. Adapting pressure to dynamic changes in load conditions is then too slow. This is particularly the case with piezo-inline injectors due to their very low internal leakage. For this reason, some common-rail systems are equipped with an additional pressure-control valve (Fig. 3) besides the high-pressure pump and metering unit. This two-actuator system combines the advantages of control on the low-pressure side with the dynamic response of control on the high-pressure side.

Another advantage compared with control on the low-pressure side only is that the high-pressure side is also controllable when the engine is cold. The high-pressure pump then delivers more fuel than is injected and pressure is controlled by the pressure-control valve. Compression heats the fuel, thus eliminating the need for an additional fuel heater.

1 | Example of a second-generation common-rail system for a 4-cylinder engine

Fig. 1
1 High-pressure pump (CP3) with fitted geared presupply pump and metering unit
2 Fuel filter with water separator and heater (optional)
3 Fuel tank
4 Pre-filter
5 Fuel rail
6 Rail-pressure sensor
7 Solenoid-valve injector
8 Pressure-relief valve

2 | Example of a second-generation common-rail system with two-actuator system for a V8 engine

Fig. 2
1 High-pressure pump (CP3) with fitted geared presupply pump and metering unit
2 Fuel filter with water separator and heater (optional)
3 Fuel tank
4 Pre-filter
5 Fuel rail
6 Rail-pressure sensor
7 Solenoid-valve injector
8 Pressure-control valve
9 Function module (distributor)

3 | Example of a third-generation common-rail system with two-actuator system for a 4-cylinder engine

Fig. 3
1 High-pressure pump (CP1H) with metering unit
2 Fuel filter with water separator and heater (optional)
3 Fuel tank
4 Pre-filter
5 Fuel rail
6 Rail-pressure sensor
7 Piezo-inline injector
8 Pressure-control valve
9 Electric fuel pump

System diagram for passenger cars
Fig. 4 shows all the components in a common-rail system for a fully equipped, 4-cylinder, passenger-car diesel engine. Depending on the type of vehicle and its application, some of the components may not be fitted.

The sensors and setpoint generators (A) are not depicted in their real installation position to simplify presentation. Exceptions are the exhaust-gas treatment sensors (F) and the rail-pressure sensor as their installation positions are required to understand the system.

Data exchange between the various sections takes place via the CAN bus in the "Interfaces" (B) section:
- Starter motor
- Alternator
- Electronic immobilizer
- Transmission control
- Traction Control System (TSC)
- Electronic Stability Program (ESP)

The instrument cluster (13) and the air-conditioning system (14) are also connectable to the CAN bus.

Two possible combined systems are described (a or b) for exhaust-gas treatment.

Fig. 4
Engine, engine management, and high-pressure fuel-injection components
17 High-pressure pump
18 Metering unit
25 Engine ECU
26 Fuel rail
27 Rail-pressure sensor
28 Pressure-control valve (DRV 2)
29 Injector
30 Glow plug
31 Diesel engine (DI)
M Torque

A Sensors and setpoint generators
1 Pedal-travel sensor
2 Clutch switch
3 Brake contacts (2)
4 Operator unit for vehicle-speed controller (cruise control)
5 Glow-plug and starter switch ("ignition switch")
6 Road-speed sensor
7 Crankshaft-speed sensor (inductive)
8 Camshaft-speed sensor (inductive or Hall sensor)
9 Engine-temperature sensor (in coolant circuit)
10 Intake-air temperature sensor
11 Boost-pressure sensor
12 Hot-film air-mass meter (intake air)

B Interfaces
13 Instrument cluster with displays for fuel consumption, engine speed, etc.
14 Air-conditioner compressor with operator unit
15 Diagnosis interface
16 Glow control unit
CAN Controller Area Network
(on-board serial data bus)

C Fuel-supply system (low-pressure stage)
19 Fuel filter with overflow valve
20 Fuel tank with pre-filter and Electric Fuel Pump, EFP (presupply pump)
21 Fuel-level sensor

D Additive system
22 Additive metering unit
23 Additive control unit
24 Additive tank

E Air supply
32 Exhaust-gas recirculation cooler
33 Boost-pressure actuator
34 Turbocharger (in this case with Variable Turbine Geometry (VTG))
35 Control flap
36 Exhaust-gas recirculation actuator
37 Vacuum pump

F Exhaust-gas treatment
38 Broadband lambda oxygen sensor, type LSU
39 Exhaust-gas temperature sensor
40 Oxidation-type catalytic converter
41 Particulate filter
42 Differential-pressure sensor
43 NO_x accumulator-type catalytic converter
44 Broadband lambda oxygen sensor, optional NO_x sensor

4 Common-rail diesel fuel-injection system for cars

▶ Overview of diesel fuel-injection systems

Areas of application

Diesel engines are characterized by high fuel economy. Since the first volume-production fuel-injection pump was introduced by Bosch in 1927, fuel-injection systems have experienced a process of continuous development.

Diesel engines are used in a wide variety of design for many different purposes (Fig. 1 and Table 1), for example
- To drive mobile power generators (up to approx. 10 kW/cylinder)
- As fast-running engines for cars and light-duty trucks (up to approx. 50 kW/cylinder)
- As engines for construction-industry and agricultural machinery (up to approx. 50 kW/cylinder)
- As engines for heavy trucks, omnibuses and tractor vehicles (up to approx. 80 kW/cylinder)
- To drive fixed installations such as emergency power generators (up to approx. 160 kW/cylinder)
- As engines for railway locomotives and ships (up to 1,000 kW/cylinder)

Requirements

Ever stricter statutory regulations on noise and exhaust-gas emissions and the desire for more economical fuel consumption continually place greater demands on the fuel-injection system of a diesel engine.

Basically, the fuel-injection system is required to inject a precisely metered amount of fuel at high pressure into the combustion chamber in such a way that it mixes effectively with the air in the cylinder as demanded by the type of engine (direct or indirect-injection) and its present operating status. The power output and speed of a diesel engine is controlled by means of the injected fuel quantity as it has no air intake throttle.

Mechanical control of diesel fuel-injection systems is being increasingly replaced by **E**lectronic **D**iesel **C**ontrol (EDC) systems. All new diesel-injection systems for cars and commercial vehicles are electronically controlled.

1 Applications for Bosch diesel fuel-injection systems

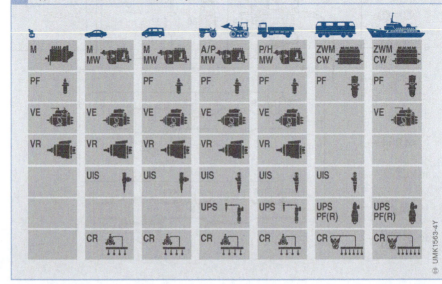

Fig. 1
M, MW,
A, P, H,
ZWM,
CW In-line fuel-injection
 pumps of increas-
 ing size
PF Discrete fuel-
 injection pumps
VE Axial-piston pumps
VR Radial-piston
 pumps
UIS Unit injector system
UPS Unit pump system
CR Common-rail
 system

Common-rail system for commercial vehicles

Fuel supply

Presupply

Common-rail systems for light-duty trucks differ very little from passenger-car systems. Electric fuel pumps or gear pumps are used for fuel presupply. On common-rail systems for heavy-duty trucks, only gear pumps are used to deliver fuel to the high-pressure pump (see the subsection "Gear-type supply pumps" in the section "Fuel supply in the low-pressure stage"). The presupply pump is normally flanged to the high-pressure pump (Figs. 1 and 2). In many applications, it is mounted on the engine.

Fuel filtering

As opposed to passenger-car systems, the fuel filter (fine filter) is fitted to the pressure side. For this reason, an exterior fuel inlet is required, in particular when the gear pump is flanged to the high-pressure pump.

1 Common-rail system for commercial vehicles with high-pressure pump (CP3)

Fig. 1
1 Fuel tank
2 Pre-filter
3 Fuel filter
4 Gear presupply pump
5 High-pressure pump (CP3.4)
6 Metering unit
7 Rail-pressure sensor
8 Fuel rail
9 Pressure-relief valve
10 Injector

2 Common-rail system for commercial vehicles with high-pressure pump (CPN2)

Fig. 2
1 Fuel tank
2 Pre-filter
3 Fuel filter
4 Gear presupply pump
5 High-pressure pump (CPN2.2)
6 Metering unit
7 Rail-pressure sensor
8 Fuel rail
9 Pressure-relief valve
10 Injector

System diagram for commercial vehicles

Fig. 3 shows all the components in a common-rail system for a 6-cylinder commercial-vehicle diesel engine. Depending on the type of vehicle and its application, some of the components may not be fitted.

Only the sensors and setpoint generators are depicted at their real position to simplify presentation, as their installation positions are required to understand the system.

Data exchange to the various sections takes place via the CAN bus in the "Interfaces" (B) section (e.g. transmission control, Traction Control System (TCS), Electronic Stability

Program (ESP), oil-grade sensor, trip recorder, **Active Cruise Control** (ACC), brake coordinator – up to 30 ECUs). The alternator (18) and the air-conditioning system (17) are also connectable to the CAN bus.

Three systems are described for exhaust-gas treatment: a purely DPF system (a) mainly for the U.S. market, a purely SCR system (b) mainly for the EU market, and a combined system (c).

Fig. 3

Engine, engine management, and high-pressure fuel-injection components
22 High-pressure pump
29 Engine ECU
30 Fuel rail
31 Rail-pressure sensor
32 Injector
33 Relay
34 Auxiliary equipment (e.g. retarder, exhaust flap for engine brake, starter motor, fan)
35 Diesel engine (DI)
36 Flame glow plug (alternatively grid heater)
M Torque

A Sensors and setpoint generators
1 Pedal-travel sensor
2 Clutch switch
3 Brake contacts (2)
4 Engine brake contact
5 Parking brake contact
6 Operating switch (e.g. vehicle-speed controller, intermediate-speed regulation, rpm- and torque reduction)
7 Starter switch ("ignition lock")
8 Turbocharger-speed sensor
9 Crankshaft-speed sensor (inductive)
10 Camshaft-speed sensor
11 Fuel-temperature sensor
12 Engine-temperature sensor (in coolant circuit)
13 Boost-air temperature sensor
14 Boost-pressure sensor
15 Fan-speed sensor
16 Air-filter differential-pressure sensor

B Interfaces
17 Air-conditioner compressor with operator unit
18 Alternator
19 Diagnosis interface

20 SCR control unit
21 Air compressor
CAN Controller Area Network (on-board serial data bus) (up to three data buses)

C Fuel-supply system (low-pressure stage)
23 Fuel presupply pump
24 Fuel filter with water-level and pressure sensors
25 Control unit cooler
26 Fuel tank with pre-filter
27 Pressure-relief valve
28 Fuel-level sensor

D Air intake
37 Exhaust-gas recirculation cooler
38 Control flap
39 Exhaust-gas recirculation positioner with exhaust-gas recirculation valve and position sensor
40 Intercooler with bypass for cold starting
41 Exhaust-gas turbocharger (in this case with variable turbine geometry) with position sensor
42 Boost-pressure actuator

E Exhaust-gas treatment
43 Exhaust-gas temperature sensor
44 Oxidation-type catalytic converter
45 Differential-pressure sensor
46 Catalyst-coated particulate filter (CSF)
47 Soot sensor
48 Level sensor
49 Reducing-agent tank
50 Reducing-agent pump
51 Reducing-agent injector
52 NO_x sensor
53 SCR catalytic converter
54 NH_3 sensor

3 Common-rail diesel fuel-injection system for commercial vehicles

SMK1820-1Y

High-pressure components of common-rail system

The high-pressure stage of the common-rail system is divided into three sections: pressure generation, pressure storage, and fuel metering. The high-pressure pump assumes the function of pressure generation. Pressure storage takes place in the fuel rail to which the rail-pressure sensor and the pressure-control and pressure-relief valves are fitted. The function of the injectors is correct timing and metering the quantity of

fuel injected. High-pressure fuel lines interconnect the three sections.

Overview

The main difference in the various generations of common-rail systems lie in the design of the high-pressure pump and the injectors, and in the system functions required (Table 1).

1	Overview of common-rail systems			
CR generation	Maximum pressure	Injector	High-pressure pump	
1st generation Pass. cars	1,350...1,450 bar	Solenoid-valve injector	CP1 Pressure control on high-pressure side by pressure-control valve	
1st generation Comm. veh.	1,400 bar	Solenoid-valve injector	CP2 Suction-side fuel-delivery control by two solenoid valves	
2nd generation Pass. cars and comm. veh.	1,600 bar	Solenoid-valve injector	CP3, CP1H Suction-side fuel-delivery control by metering unit	
3rd generation Pass. cars	1,600 bar (in future 1,800 bar)	Piezo-inline injector	CP3, CP1H Suction-side fuel-delivery control by metering unit	
3rd generation Comm. veh.	1,800 bar	Solenoid-valve injector	CP3.3NH Metering unit	

Table 1

1	Common-rail fuel-injection system taking the example of a four-cylinder diesel engine

Fig. 1
1 Hot-film air-mass meter
2 Engine ECU
3 High-pressure pump
4 High-pressure accumulator (fuel rail)
5 Injector
6 Crankshaft-speed sensor
7 Engine-temperature sensor
8 Fuel filter
9 Pedal-travel sensor

Cleanliness quality

The sharp rise in the performance of new assemblies, e.g. the common-rail system for high-pressure diesel fuel injection, requires extreme precision in machining, and ever tighter tolerances and fits. Particle residue from the production process may lead to increased wear, or even the total failure of the assembly. This results in high requirements and tight tolerances for cleanliness quality, with a continuous reduction in the permitted particle size.

The cleanliness quality of components is currently determined in the production process by light-microscope image-analysis systems. They supply information on particle-size distribution. Additional information, such as the nature of the particles and their chemical composition, is required to develop innovative cleaning processes. This information is obtained by electron microscopes.

Particle-analysis system (SEM)

Bosch uses a particle-analysis system based on a Scanning Electron Microscope (SEM). This system performs an automated analysis of particles adhering to a product. The results of the analysis show particle-size distribution, the chemical composition of the particles, and images of the individual particles. Using this information, the source of the particles are identifiable. Action can then be taken to avoid, reduce, or wash off certain particle types. In this way, solutions are not based on the increased use of cleaning techniques, but on avoiding and reducing residual soiling during the production process.

The automated particle-analysis system (SEM) provides the cleanliness process with an analysis system that produces important information on the type of residual soiling. The precise identification of particles and their sources is vital to developing new cleaning techniques.

▶ Principle of the particle-analysis system (SEM)

Evolution of particulate-analysis process
Particulate Microparticles

 up to < 1μm

Light Electron
microscope microscope

Growth in information content ▷

- Number of particles - EDX analysis
- Size distribution - Number of particles
 - Size distribution
 - High precision
 (focus depth)

The electron beam and its action

Backscatter electrons and X-rays from depth of several μm Secondary electrons from depth of several μm

 Heat

● Primary electron beam 20 kV
● Backscatter electrons to BSE detector (up to 20 keV)
● Secondary electrons to SE detector (several eV)
○ X-rays to EDX detector (up to 10 keV)

Detection of interaction between electron beam and sample

SE detector
Secondary electrons of sample surface are converted into image signals.
- Plastic image of surface (REM images).

BSE detector
Backscatter electrons are converted into image signals.
→ Phase composition
→ TOPO mode plastic image

EDX detector
Characteristic X-ray is converted into "energy-dispersive" spectrum.
→ Identification of chemical elements

SAN0169E

Injector

On a common-rail diesel injection system, the injectors are connected to the fuel rail by short high-pressure fuel lines. The injectors are sealed to the combustion chamber by a copper gasket. The injectors are fitted into the cylinder head by means of taper locks. Depending on the injection-nozzle design, common-rail injectors are intended for straight or inclined mounting in direct-injection diesel engines.

One of the system's features is that it generates an injection pressure irrespective of engine speed or injected fuel quantity. The start of injection and injected fuel quantity are controlled by the electrically triggered injector. The injection time is controlled by the angle/time system of the Electronic Diesel Control (EDC). This requires the use

of sensors to detect the crankshaft position and the camshaft position (phase detection).

An optimum mixture formation is required to reduce exhaust-gas emissions and comply with continuous demands to reduce the noise of diesel engines. This calls for injectors with very small pre-injection quantities and multiple injection events.

There are presently three different injector types in serial production:
- Solenoid-valve injectors with one-part armature
- Solenoid-valve injectors with two-part armature
- Injector with piezo actuator

Fig. 1
a Resting position
b Injector opens
c Injector closes

1 Fuel-return
2 Solenoid coil
3 Overstroke spring
4 Solenoid armature
5 Valve ball
6 Valve-control chamber
7 Nozzle spring
8 Pressure shoulder of nozzle needle
9 Chamber volume
10 Injection orifice
11 Solenoid-valve spring
12 Outlet restrictor
13 High-pressure connection
14 Inlet restrictor
15 Valve plunger (control plunger)
16 Nozzle needle

1 Solenoid-valve injector (functional schematic)

Solenoid-valve injector

Design

The injector can be subdivided into a number of function modules:

- The hole-type nozzle (see the section on "Injection nozzle")
- The hydraulic servo system
- The solenoid valve

Fuel is conveyed by the high-pressure connection (Fig. 1a, 13) via a supply passage to the injection nozzle and via an inlet restrictor (14) to the valve-control chamber (6). The valve-control chamber is connected to the fuel return (1) via the outlet restrictor (12) which can be opened by a solenoid valve.

Operating concept

The function of the injector can be subdivided into four operating states when the engine and the high-pressure pump are operating:

- Injector closed (with high pressure applied)
- Injector opens (start of injection)
- Injector fully open
- Injector closes (end of injection)

The operating states are caused by the balance of forces acting on the injector components. When the engine is not running and the fuel rail is not pressurized, the nozzle spring closes the injector.

Injector closed (resting position)

In its resting position, the injector is not triggered (Fig. 1a). The solenoid-valve spring (11) presses the valve ball (5) onto the seat of the outlet restrictor (12). Inside of the valve-control chamber, the pressure rises to the pressure in the fuel rail. The same pressure is also applied to the chamber volume (9) of the nozzle. The forces applied by the rail pressure to the end faces of the control plunger (15), and the force of the nozzle spring (7) retain the nozzle needle closed against the opening force applied to its pressure shoulders (8).

Injector opens (start of injection)

To begin with, the injector is in its resting position. The solenoid valve is triggered by the "pickup current". This makes the solenoid valve open very rapidly (Fig. 1b). The required rapid switching times are achieved by controlling solenoid-valve triggering in the ECU at high voltages and currents.

The magnetic force of the now triggered electromagnet exceeds the force of the valve spring. The armature raises the valve ball from the valve seat and opens the outlet restrictor. After a short time the increased pickup current is reduced to a lower holding current in the electromagnet. When the outlet restrictor opens, fuel flows from the valve-control chamber to the cavity above and then via the fuel-return line to the fuel tank. The inlet restrictor (14) prevents a complete pressure compensation. As a result, pressure in the valve-control chamber drops. Pressure in the valve-control chamber falls below the pressure in the nozzle chamber, which is still the same as the pressure in the fuel rail. The reduction in pressure in the valve-control chamber reduces the force acting on the control plunger and opens the nozzle needle. Fuel injection commences.

Injector fully open

The rate of movement of the nozzle needle is determined by the difference in the flow rates through the inlet and outlet restrictors. The control plunger reaches its upper stop and dwells there on a cushion of fuel (hydraulic stop). The cushion is created by the flow of fuel between the inlet and outlet restrictors. The injector nozzle is then fully open. Fuel is injected into the combustion chamber at a pressure approaching that in the fuel rail.

The balance of forces in the injector is similar to that during the opening phase. At a given system pressure, the fuel quantity injected is proportional to the length of time that the solenoid valve is open. This is entirely independent of the engine or pump speed (time-based injection system).

Injector closes (end of injection)
When the solenoid valve is no longer triggered, the valve spring presses the armature down and the valve ball closes the outlet restrictor (Fig. 1c). When the outlet restrictor closes, pressure in the control chamber rises again to that in the fuel rail via the inlet restrictor. The higher pressure exerts a greater force on the control plunger. The force on the valve-control chamber and the nozzle-spring force then exceed the force acting on the nozzle needle, and the nozzle needle closes. The flow rate of the inlet restrictor determines the closing speed of the nozzle needle. The fuel-injection cycle comes to an end when the nozzle needle is resting against its seat, thus closing off the injection orifices.

This indirect method is used to trigger the nozzle needle by means of a hydraulic servo system because the forces required to open the nozzle needle rapidly cannot be generated directly by the solenoid valve. The "control volume" required in addition to the injected fuel quantity reaches the fuel-return line via the restrictors in the control chamber.

In addition to the control volume, there are also leakage volumes through the nozzle-needle and valve-plunger guides. The control and leakage volumes are returned to the fuel tank via the fuel-return line and a collective line that comprises an overflow valve, high-pressure pump, and pressure-control valve.

Program-map variants
Program maps with fuel-quantity flat curve
With injectors, a distinction is made in the program map between ballistic and non-ballistic modes. The valve plunger/nozzle needle unit reaches the hydraulic stop if the triggering period in vehicle operation is of sufficient length (Fig. 2a). The section until the nozzle needle reaches its maximum stroke is termed ballistic mode. The ballistic and nonballistic sections in the fuel-quantity map, where the injected fuel quantity is applied for the triggering period (Fig. 2b), is separated by a kink in the program map.

Another feature of the fuel-quantity map is the flat curve that occurs with small triggering periods. The flat curve is caused by the solenoid armature rebounding on opening. In this section, the injected fuel quantity is independent on the triggering period. This allows small injected fuel quantities to be represented as stable. Only after the armature has stopped rebounding does the injected fuel quantity curve continue to rise linearly as the triggering period becomes longer.

Injection events with small injected fuel quantities (short triggering periods) are used as pre-injection in order to suppress noise. Secondary injection events help to enhance soot oxidation in selected sections of the operating curve.

Program maps without fuel-quantity flat curve
The increasing stringency of emission-control legislation has lead to the use of the two system functions: *injector delivery compensation* (IMA) and *zero delivery calibration* (NMK), as well as to short intervals in injection between pre-injection, main injection, and secondary injection events. With injectors that have no flat-curve section, IMA allows a precise adjustment of the pre-injection fuel quantity when new. NMK corrects fuel-quantity drifts over time in the low-pressure section. The key condition for deploying these two system functions is a constant, linear rise in quantity, i.e. there is no flat curve in the fuel-quantity map (Fig. 2c). If the valve plunger/nozzle needle unit is operated in nominal mode without lift-stop at the same time, this represents a fully ballistic operating mode of the valve plunger and there is no kink in the fuel-quantity map.

Injector variants
A distinction is made between two different solenoid-valve concepts with solenoid-valve injectors:
● Injectors with one-part armature (1-spring system)
● Injectors with two-part armature (2-spring system)

The short intervals between injection events are ensured when the armature can return to its resting position very rapidly on closing. This is best achieved by a two-part armature with an overstroke stop. During the closing process, the armature plate moves down by positive locking. The bottoming-out of the armature plate is limited by an overstroke stop. As a result, the armature reaches its resting position faster. Armature rebound on closing can end faster by decoupling the armature masses and adapting the setting parameters. This helps to achieve shorter intervals between two injection events with the two-part armature concept.

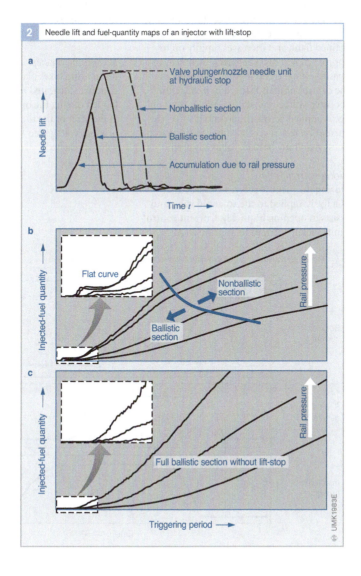

2 Needle lift and fuel-quantity maps of an injector with lift-stop

a
Needle lift ↑

Valve plunger/nozzle needle unit at hydraulic stop

Nonballistic section

Ballistic section

Accumulation due to rail pressure

Time t ⟶

b
Injected-fuel quantity ↑

Flat curve

Nonballistic section

Ballistic section

Rail pressure

c
Injected-fuel quantity ↑

Full ballistic section without lift-stop

Rail pressure

Triggering period ⟶

UMK1983E

Triggering the solenoid-valve injector

In its resting position, the injector's high-pressure solenoid valve is not triggered and is therefore closed. The injector injects when the solenoid valve opens.

Triggering the solenoid valve is divided into five phases (Figs. 3 and 4).

Opening phase

Initially, in order to ensure tight tolerances and high levels of reproducibility for the injected fuel quantity, the current for opening the solenoid valve features a steep, precisely defined flank and increases rapidly up to approx. 20 A. This is achieved by means of a *booster voltage* of up to 50 V. It is generated in the control unit and stored in a capacitor (booster-voltage capacitor). When this voltage is applied across the solenoid valve, the current increases several times faster than it does when only battery voltage is used.

Pickup-current phase

During the pickup-current phase, battery voltage is applied to the solenoid valve and assists in opening it quickly. Current control limits pickup current to approx. 20 A.

Holding-current phase

In order to reduce power loss in the ECU and injector, the current is dropped to approx. 13 A in the holding-current phase. The energy which becomes available when pickup current and holding current are reduced is routed to the booster-voltage capacitor.

Switchoff

When the current is switched off in order to close the solenoid valve, the surplus energy is also routed to the booster-voltage capacitor.

Recharging the step-up chopper

Recharging takes place by means of a step-up chopper integrated in the ECU. The energy tapped during the opening phase is recharged at the start of the pickup phase until the original voltage required to open the solenoid valve is reached.

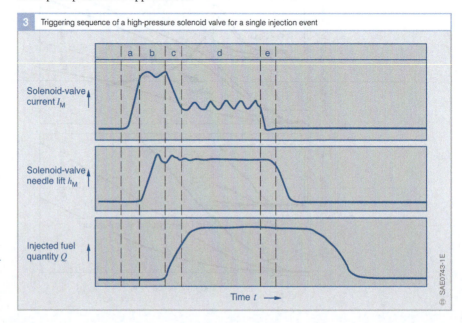

3 Triggering sequence of a high-pressure solenoid valve for a single injection event

Solenoid-valve current I_M

Solenoid-valve needle lift h_M

Injected fuel quantity Q

Time t →

SAE0743-1E

Fig. 3
a Opening phase
b Pickup-current phase
c Transition to holding-current phase
d Holding-current phase
e Switchoff

4 Common-rail system: Block diagram of the triggering phases for a cylinder group

a Opening phase

b Pickup-current phase

c Transition to holding-current phase

d Holding-current phase

e Switchoff

f Recharging the step-up chopper

DC/DC converter: recharging the energy accumulator (9)

Energy transfer (from 9 to 5)

Fig. 4
1 Battery
2 Current control
3 Solenoid windings of the high-pressure solenoid valves
4 Booster switch
5 Booster-voltage capacitor
6 Free-wheeling diodes for energy recovery and high-speed quenching
7 Cylinder selector switch
8 DC/DC switch
9 DC/DC coil
10 DC/DC diode
I Current flow

SMK1757-1E

Piezo-inline injector

Design and requirements

The design of the piezo-inline injector is divided into its main modules in the schematic (see Fig. 5):

- Actuator module (3)
- Hydraulic coupler or translator (4)
- Control or servo valve (5)
- Nozzle module (6)

The design of the injector took account of the high overall rigidity required within the actuator chain composed of actuator, hydraulic coupler, and control valve. Another design feature is the avoidance of mechanical forces acting on the nozzle needle. Such forces occurred as a result of the push rod used on previous solenoid-valve injectors. On aggregate, this design effectively reduces the moving masses and friction, thus enhancing injector stability and drift compared to conventional systems.

In addition, the fuel-injection system allows the implementation of very short intervals ("hydraulic zero") between injection events. The number and configuration of fuel-metering operations can represent up to five injection events per injection cycle in order to adapt the requirements to the engine operating points.

A direct response of the needle to actuator operation is achieved by coupling the servo valve (5) to the nozzle needle. The delay between the electric start of triggering and hydraulic response of the nozzle needle is about 150 microseconds. This meets the contradictory requirements of high needle speeds and extremely small reproducible injected fuel quantities.

As a result of this principle, the injector also includes small direct leakage points from the high-pressure section to the low-pressure circuit. The result is an increase in the hydraulic efficiency of the overall system.

Operating concept

Function of the 3/2-way servo valve in the CR injector

The nozzle needle on piezo-inline injector is controlled indirectly by a servo valve. The required injected fuel quantity is then controlled by the valve triggering period. In its non-triggered state, the actuator is in the starting position and the servo valve is closed (Fig. 6a), i.e. the high-pressure section is separated from the low-pressure section.

5 Construction of the piezo-inline injector

UMK1974-1Y

The nozzle is kept closed by the rail pressure exerted in the control chamber (3). When the piezo actuator is triggered, the servo valve opens and closes the bypass passage (Fig. 6b). The flow-rate ratio between the outlet restrictor (2) and the inlet restrictor (4) lowers pressure in the control chamber and the nozzle (5) opens. The control volume flows via the servo valve to the low-pressure circuit of the overall system.

To start the closing process, the actuator is discharged and the servo valve releases the bypass passage. The control chamber is then refilled by reversing the inlet and outlet restrictors, and pressure in the control chamber is raised. As soon as the required pressure is attained, the nozzle needle starts to move and the injection process ends.

The valve design described above and the greater dynamic design of the actuator system result in much shorter injection times compared to injectors of conventional design, i.e. push rod and 2/2-way valve. Ultimately, this has a positive impact on exhaust-gas emissions and engine performance. Due to requirements regarding the engine in EU 4, the injector program maps were optimized to apply corrective functions (injector delivery compensation (IMA) and zero delivery cali-

bration (NMK). The pre-injection quantity can then be selected at will, and IMA can minimize the quantity spread in the program map using full ballistic mode (see Fig. 7).

7 Injection-quantity program map of the piezo-inline injector

Fig. 7
Injected fuel quantities at different injection pressures
a 1,600 bar
b 1,200 bar
c 1,000 bar
d 800 bar
e 250 bar

6 Function of the servo valve

Rail pressure Leakage-oil pressure Control-chamber pressure

Fig. 6
a Start position
b Nozzle needle opens (bypass closed, normal function with outlet and inlet restrictors)
c Nozzle needle closes (bypass open, function with two inlet restrictors)

1 Servo valve (control valve)
2 Outlet restrictor
3 Control chamber
4 Inlet restrictor
5 Nozzle needle
6 Bypass

Function of the hydraulic coupler
Another key component in the piezo-inline injector is the hydraulic coupler (Fig. 8, 3) that implements the following functions:
- Translates and amplifies the actuator stroke.
- Compensates for any play between the actuator and the servo valve (e.g. caused by thermal expansion).
- Performs a failsafe function (automatic safety cutoff of fuel injection if electrical decontacting fails).

The actuator module and the hydraulic coupler are immersed in the diesel fuel flow at a pressure of about 10 bar. When the actuator is not triggered, pressure in the hydraulic coupler is in equilibrium with its surroundings. Changes in length caused by temperature are compensated by small leakage-fuel quantities via the guide clearances of the two plungers. This maintains the coupling of forces between actuator and switching valve at all times.

To generate an injection event, a voltage (110...150 V) is applied to the actuator until the equilibrium of forces between the switching valve and the actuator is exceeded. This increases the pressure in the coupler, and a small leakage volume flows out of the coupler via the piston guide clearances into the low-pressure circuit of the injector. The pressure drop caused in the coupler has no impact on injector function for a triggering period lasting several milliseconds.

At the end of the injection process, the quantity missing in the hydraulic coupler needs refilling. This takes place in the reverse direction via the guide clearances of the plungers as a result of the pressure difference between the hydraulic coupler and the low-pressure circuit of the injector. The guide clearances and the low-pressure level are matched to fill up the hydraulic coupler fully before the next injection cycle starts.

8 Function of the hydraulic coupler

Voltage →

Coupler pressure →

Volume change in coupler →

p_{System}

Recharging $p_K < p_{System}$

Leakage $p_K > p_{System}$

Time t →

Rail pressure
Coupler pressure
10 bar
1 bar

UMK1986E

Fig. 8
1 Low-pressure fuel rail with valve
2 Actuator
3 Hydraulic coupler (translator)

Triggering the common-rail piezo in-line injector

The injector is triggered by an engine control unit whose output stage was specially designed for these injectors. A reference triggering voltage is predetermined as a function of the rail pressure of the set operating point. The voltage signal is pulsed (Fig. 9) until there is a minimum deviation between the reference and the control voltage. The voltage rise is converted proportionally into a piezoelectric actuator stroke. The actuator stroke produces a pressure rise in the coupler by means of hydraulic translation until the equilibrium of forces is exceeded at the switching valve, and the valve opens. As soon as the switching valve reaches its end position, pressure in the control chamber starts to drop via the needle, and injection ends.

Benefits of the piezo-inline injector

- Multiple injection with flexible start of injection and intervals between individual injection events.
- Production of very small injected fuel quantities for pre-injection.
- Small size and low weight of injector (270 g compared to 490 g).
- Low noise (−3 dB [A]).
- Lower fuel consumption (−3%).
- Lower exhaust-gas emission (−20%).
- Increased engine performance (+7%).

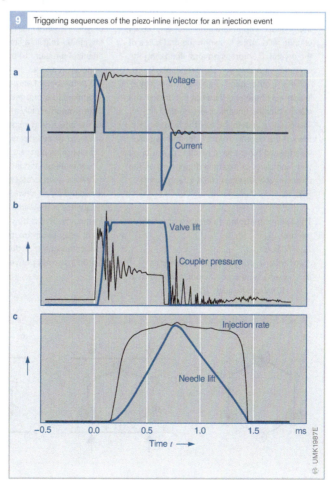

9 Triggering sequences of the piezo-inline injector for an injection event

a — Voltage — Current

b — Valve lift — Coupler pressure

c — Injection rate — Needle lift

−0.5 0.0 0.5 1.0 1.5 ms

Time t →

UMK1987E

Fig. 9

a Current and voltage curves for triggering the injector

b Valve-lift curve and coupler pressure

c Valve-lift curve and injection rate

▸ The piezoelectric effect

In 1880 Pierre Curie and his brother Jacques discovered a phenomenon that is still very little known today, but is present in the everyday lives of millions of people: the piezoelectric effect. For example, it keeps the pointers of a crystal clock operating in time.

Certain crystals (e.g. quartz and turmaline) are piezoelectric: Electric charges are induced on the crystal surface by exerting a compression or elongation force along certain crystal axes. This electrical polarization arises by shifting positive and negative ions in the crystal relative to each other by exerting force (see Fig., b). The shifted centers of charge gravity within the crystal compensate automatically, but an electric field forms between the end faces of the crystal. Compressing and elongating the crystal create inverse field directions.

On the other hand, if an electrical voltage is applied to the end faces of the crystal, the effect reverses (inverse piezoelectric effect): The positive ions in the electric field migrate toward the negative electrode, and negative ions toward the positive electrode. The crystal then contracts or expands depending on the direction of the electric field strength (see Fig., c).

The following applies to piezoelectric field strength E_p:

$$E_p = \delta \, \Delta x / x$$

$\Delta x/x$: relative compression or elongation

δ: piezoelectric coefficient, numeric value 10^9 V/cm through 10^{11} V/cm

The change in length Δx results from the following when a voltage U is applied:

$U / \delta = \Delta x$ (using quartz as an example: deformation of about 10^{-9} cm at $U = 10$ V)

The piezoelectric effect is not only used in quartz clocks and piezo-inline injectors, it has many other industrial applications, either as a direct or inverse effect:

Piezoelectric sensors are used for knock control in gasoline engines. For example, they detect high-frequency engine vibrations as a feature of combustion knock. Converting mechanical vibration to electric voltage is also used in the crystal audio pickup of a record player or crystal microphones. The piezoelectric igniter (e.g. in a firelighter) causes mechanical pressure to produce the voltage to generate a spark.

On the other hand, if an alternating voltage is applied to a piezoelectrical crystal, it vibrates mechanically at the same frequency as the alternating voltage. Oscillating crystals are used as stabilizers in electrical oscillating circuits or as piezoelectric acoustic sources to generate ultrasound.

When used in clocks, the oscillating quartz is excited by an alternating voltage whose frequency is the same as the quartz's natural frequency. This is how an extremely time-constant resonant frequency is generated. In a calibrated quartz, it deviates by only approx. 1/1,000 second per year.

Principle of the piezoelectric effect (represented as a unit cell)

a Quartz crystal SiO_2

b Piezoelectric effect: When the crystal is compressed, negative O^{2-} ions shift upward, positive Si^{4+} ions shift downward: Electric charges are induced at the crystal surface.

c Inverse piezoelectric effect: By applying an electrical voltage, O^{2-} ions shift upward, Si^{4+} ions shift downward: The crystal contracts.

This term actually goes back to the ancient Greeks. For them, the word "electron" meant amber. Its force of attraction on woollen threads or similar was known to Thales von Milet over 2,500 years ago.

Electrons, and therefore electronics as such, are extremely fast due to their very small mass and electric charge. The term "electronics" comes directly from the word "electron".

The mass of an electron has as little effect on a gram of any given substance as a 5 gram weight has on the total mass of our earth.

The word "electronics" was born in the 20th century. There is no evidence available as to when the word was used for the first time. It could be Sir John Ambrose Fleming, one of the inventors of the electron tube in about 1902.

Even the first "Electronic Engineer" already existed in the 19th century. Fleming was listed in the 1888 edition of "Who's Who", published during the reign of Queen Victoria. The official title was "Kelly's Handbook of Titled, Landed and Official Classes". The Electronic Engineer can be found under the title "Royal Warrant Holders", that is the list of persons who had been awarded a Royal Warrant.

What was this Electronic Engineer's job? He was responsible for the correct functioning and cleanliness of the gas lamps at court. And why did he have such a splendid title? Because he knew that "electrons" in ancient Greece stood for glitter, shine, and sparkle.

Source:
"Basic Electronic Terms" ("Grundbegriffe der Elektronik") – Bosch publication (reprint from the "Bosch Zünder" (Bosch Company Newspaper)), 1988.

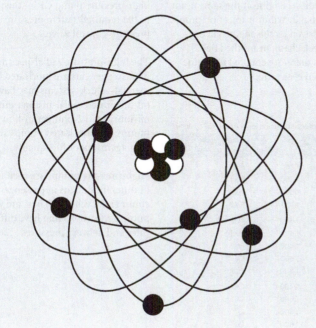

LAE0047Y

High-pressure pumps

Design and requirements

The high-pressure pump is the interface between the low-pressure and high-pressure stages. Its function is to make sure there is always sufficient fuel under pressure available in all engine operating conditions. At the same time it must operate for the entire service life of the vehicle. This includes providing a fuel reserve that is required for quick engine starting and rapid pressure rise in the fuel rail.

The high-pressure pump generates a constant system pressure for the high-pressure accumulator (fuel rail) independent of fuel injection. For this reason, fuel – compared to conventional fuel-injection systems – is not compressed during the injection process.

A 3-plunger radial-piston pump is used as the high-pressure pump to generate pressure in passenger-car systems. 2-plunger in-line fuel-injection pumps are also used on commercial vehicles. Preferably, the high-pressure pump is fitted to the diesel engine at the same point as a conventional distributor injection pump. The pump is driven by the engine via coupling, gearwheel, chain, or toothed belt. Pump speed is therefore coupled to engine speed via a fixed gear ratio.

The pump plunger inside of the high-pressure pump compresses the fuel. At three delivery strokes per revolution, the radial-piston pump produces overlapping delivery strokes (no interruption in delivery), low drive peak torques, and an even load on the pump drive.

On passenger-car systems, torque reaches 16 Nm, i.e. only 1/9th of the drive torque required for a comparable distributor injection pump. As a result, the common-rail system places fewer demands on the pump-drive system than conventional fuel-injection systems. The power required to drive the pump increases in proportion to the pressure in the fuel rail and the rotational speed of the pump (delivery quantity). On a 2-liter engine, the high-pressure pump draws a power of 3.8 kW at nominal speed and a pressure of 1,350 bar in the fuel rail (at a mechanical efficiency of approx. 90%). The higher power requirements of common-rail systems compared to conventional fuel-injection systems is caused by leakage and control volumes in the injector, and – on the high-pressure pump CP1 – the pressure drop to the required system pressure across the pressure-control valve.

The high-pressure radial-piston pumps used in passenger cars are lubricated by fuel. Commercial-vehicle systems may have fuel- or oil-lubricated radial-piston pumps, as well as oil-lubricated 2-plunger in-line fuel-injection pumps. Oil-lubricated pumps are more robust against poor fuel quality.

High-pressure pumps are used in a number of different designs in passenger cars and commercial vehicles. There are versions of pump generations that have different delivery rates and delivery pressures (Table 1).

1 Bosch high-pressure pumps for common-rail systems		
Pump	Pressure in bar	Lubrication
CP1	1,350	Fuel
CP1+	1,350	Fuel
CP1H	1,600	Fuel
CP1H-OHW	1,100	Fuel
CP3.2	1,600	Fuel
CP3.2+	1,600	Fuel
CP3.3	1,600	Fuel
CP3.4	1,600	Oil
CP3.4+	1,600	Fuel
CP2	1,400	Oil
CPN2.2	1,600	Oil
CPN2.2+	1,600	Oil
CPN2.4	1,600	Oil

Table 1
H Increased pressure
 section
+ Higher delivery rate
OHW Off-Highway

Radial-piston pump (CP1)

Design

The drive shaft in the housing of CP1 is mounted in a central bearing (Fig. 1, 1). The pump elements (3) are arranged radially with respect to the central bearing and offset by 120°. The eccenter (2) fitted to the drive shaft forces the pump plunger to move up and down.

Force is transmitted between the eccentric shaft and the delivery plunger by means of a drive roller, a sliding ring mounted on the shaft eccenter, and a plunger base plate attached to the plunger base plate.

Operating concept

Fuel delivery and compression

The presupply pump – an electric fuel pump or a mechanically driven gear pump – delivers fuel via a filter and water separator to the inlet of the high-pressure pump (6). The inlet is located inside of the pump on passenger-car systems with a gear pump flanged to the high-pressure pump. A safety valve is fitted behind the inlet. If the delivery pressure of the presupply pump exceeds the opening pressure (0.5 to 1.5 bar) of the safety valve, the fuel is pressed through the restriction bore of the safety valve into the lubrication and cooling circuit of the high-pressure pump. The drive shaft with its eccenter moves the pump plunger up and down to mimic the eccentric lift. Fuel passed through the high-pressure pump's inlet valve (4) into the element chamber and the pump plunger moves downward (inlet stroke).

When the bottom-dead center of the pump plunger is exceeded, the inlet valve closes, and the fuel in the element chamber can no longer escape. It can then be pressurized beyond the delivery pressure of the presupply pump. The rising pressure opens the outlet valve (5) as soon as pressure reaches the level in the fuel rail. The pressurized fuel then passes to the high-pressure circuit.

1 High-pressure pump (schematic, cross-section)

Fig. 1
1 Drive shaft
2 Eccenter
3 Pump element with pump plunger
4 Inlet valve
5 Outlet valve
6 Fuel inlet

The pump plunger continues to deliver fuel until it reaches its top-dead center position (delivery stroke). The pressure then drops so that the outlet valve closes. The remaining fuel is depressurized and the pump plunger moves downward.

When the pressure in the element chamber exceeds the pre-delivery pressure, the inlet valve reopens, and the process starts over.

Transmission ratio

The delivery quantity of a high-pressure pump is proportional to its rotational speed. In turn, the pump speed is dependent on the engine speed. The transmission ratio between the engine and the pump is determined in the process of adapting the fuel-injection system to the engine so as to limit the volume of excess fuel delivered. At the same time it makes sure that the engine's fuel demand at WOT is covered to the full extent. Possible gear ratios are 1:2 or 2:3 relative to the crankshaft.

Delivery rate

As the high-pressure pump is designed for high delivery quantities, there is a surplus of pressurized fuel when the engine is idling or running in part-load range. On first-generation systems with a CP1, excess fuel delivered is returned to the fuel tank by the pressure-control valve on the fuel rail. As the compressed fuel expands, the energy imparted by compression is lost; overall efficiency drops. Compressing and then expanding the fuel also heats the fuel.

2 High-pressure pump (CP1), variant with mounted pressure-control valve (3D view)

Fig. 2

1 Flange
2 Pump housing
3 Engine cylinder head
4 Inlet connection
5 High-pressure inlet
6 Return connection
7 Pressure-control valve
8 Barrel bolt
9 Shaft seal
10 Eccentric shaft

Radial-piston pump (CP1H)

Modifications

An improvement in energetic efficiency is possible by controlling fuel delivery by the high-pressure pump on the fuel-delivery side (suction side). Fuel flowing into the pump element is metered by an infinitely variable solenoid valve (metering unit, ZME). This valve adapts the fuel quantity delivered to the rail to system demand. This fuel-delivery control not only drops the performance demand of the high-pressure pump, it also reduces the maximum fuel temperature. This system designed for the CP1H was taken over by the CP3.

Compared to the high-pressure pump CP1, the CP1H is designed for higher pressures up to 1,600 bar. This was achieved by reinforcing the drive mechanism, modifying the valve units, and introducing measures to increase the strength of the pump housing.

The metering unit is mounted on the high-pressure pump (Fig. 3, 13).

Design of the metering unit (ZME)

Fig. 4 shows the design of the metering unit. The plunger operated by solenoid force frees up a metering orifice depending on its position.

The solenoid valve is triggered by a PWM signal.

4 Metering unit design

Fig. 4
1 Plug with electrical interface
2 Magnet housing
3 Bearing
4 Armature with tappet
5 Winding with coil body
6 Cup
7 Residual air-gap washer
8 Magnetic core
9 O-ring
10 Plunger with control slots
11 Spring
12 Safety element

3 High-pressure pump (CP1H) with metering unit (exploded view)

Fig. 3
1 Flange
2 Eccentric shaft
3 Bushing
4 Drive roller
5 Pump housing
6 Plate
7 Spring
8 Engine cylinder head
9 Return-flow connection
10 Overflow valve
11 Inlet connection
12 Filter
13 Metering unit
14 Cage
15 Pump plunger

Radial-piston pump (CP3)
Modifications

The CP3 is a high-pressure pump with suc-tion-side fuel-delivery control by means of a metering unit (ZME). This control was first used on the CP3 and was assumed later on the CP1H.

The principle design of the CP3 (Fig. 5) is similar to the CP1 and the CP1H. The main difference in features are:

- Monobloc housing: This construction reduces the number of leak points in the high-pressure section, and permits a higher delivery rate.
- Bucket tappets: Transverse forces arising from the transverse movement of the eccenter drive roller are not removed di-rectly by the pump plungers but by buckets on the housing wall. The pump then has greater stability under load and is capable of withstanding higher pressures. Poten-tially, it can withstand pressures up to 1,800 bar.

Variants

Pumps of the CP3 family are used in both passenger cars and commercial vehicles. A number of different variants are used depending on the delivery rate required. The size, and thus the delivery rate, increases from the CP3.2 to the CP3.4. The oil-lubricated CP3.4 is only used on heavy-duty trucks. On light-duty trucks and vans, pumps primarily designed for passenger cars may also be used.

A special feature of systems for medium-duty and heavy-duty trucks is the fuel filter located on the pressure side. It is situated between the gear pump and the high-pressure pump, and permits a greater filter storage capacity before requiring a change. The high-pressure pump requires an external connec-tion for the fuel inlet in any case, even if the gear pump is flanged onto the high-pressure pump.

| 5 | High-pressure pump CP3 with metering unit and mounted gear presupply pump |

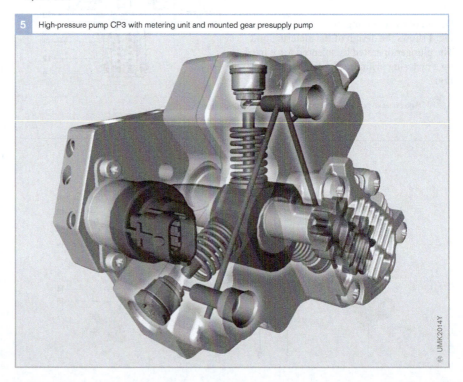

UMK2014Y

In-line piston pump (CP2)

Design

The oil-lubricated, quantity-controlled high-pressure pump (CP2) is only used on commercial vehicles. This is a 2-plunger pump with an in-line design, i.e. the two pump plungers are arranged adjacently (Fig. 6).

A gear pump with a high gear ratio is located on the camshaft extension. Its function is to draw fuel from the fuel tank and route it to the fine filter. From there, the fuel passes through another line to the metering unit located on the upper section of the high-pressure pump. The metering unit controls the fuel quantity delivered for compression dependent on actual demand in the same way as other common-rail high-pressure pumps of the recent generation.

Lube oil is supplied either directly via the mounting flange of the CP2 or a side-mounted inlet.

The drive gear ratio is 1:2. The CP2 is therefore mountable together with conventional in-line fuel-injection pumps.

Operating concept

Fuel enters the pump element and the compressed fuel is conveyed to the fuel rail via a combined inlet/outlet valve on the CP2.

6 High-pressure pump CP2

Fig. 6
1 Zero delivery restrictor
2 Metering unit
3 Internal gear
4 Pinion
5 Gear presupply pump
6 High-pressure connection
7 Two-part inlet/outlet valve
8 C-coated plunger
9 Plunger return spring
10 Oil inlet
11 C-coated roller bolt
12 Concave cam

Fuel rail (high-pressure accumulator)

Function

The function of the high-pressure accumulator (fuel rail) is to maintain the fuel at high pressure. In so doing, the accumulator volume has to dampen pressure fluctuations caused by fuel pulses delivered by the pump and the fuel-injection cycles. This ensures that, when the injector opens, the injection pressure remains constant. On the one hand, the accumulator volume must be large enough to meet this requirement. On the other hand, it must be small enough to ensure a fast enough pressure rise on engine start. Simulation calculations are conducted during the design phase to optimize the performance features.

Besides acting as a fuel accumulator, the fuel rail also distributes fuel to the injectors.

Design

The tube-shaped fuel rail (Fig. 1, 1) can have as many designs as there are engine mounting variants. It has mountings for the rail-pressure sensor (5) and a pressure-relief valve or pressure-control valve (2).

Operating concept

The pressurized fuel delivered by the high-pressure pump passes via a high-pressure fuel line to the fuel-rail inlet (4). From there, it is distributed to the individual injectors (hence the term "common rail").

The fuel pressure is measured by the rail-pressure sensor and controlled to the required value by the pressure-control valve. The pressure-relief valve is used as an alternative to the pressure-control valve – depending on system requirements – and its function is to limit fuel pressure in the fuel rail to the maximum permissible pressure. The highly compressed fuel is routed from the fuel rail to the injectors via high-pressure delivery lines.

The cavity inside the fuel rail is permanently filled with pressurized fuel. The compressibility of the fuel under high pressure is utilized to achieve an accumulator effect. When fuel is released from the fuel rail for injection, the pressure in the high-pressure accumulator remains virtually constant, even when large quantities of fuel are released.

1 Common rail with attached components

Fig. 1
1 Fuel rail
2 Pressure-control valve
3 Return line from fuel rail to fuel tank
4 Inlet from high-pressure pump
5 Rail-pressure sensor
6 Fuel line to injector

SMK1996Y

High-pressure sensors

Application
In automotive applications, high-pressure sensors are used for measuring the pressures of fuels and brake fluids.

Diesel rail-pressure sensor
In the diesel engine, the rail-pressure sensor measures the pressure in the fuel rail of the common-rail accumulator-type injection system. Maximum operating (nominal) pressure p_{max} is 160 MPa (1,600 bar). Fuel pressure is controlled by a closed control loop, and remains practically constant independent of load and engine speed. Any deviations from the setpoint pressure are compensated for by a pressure-control valve.

Gasoline rail-pressure sensor
As its name implies, this sensor measures the pressure in the fuel rail of the DI Motronic with gasoline direct injection. Pressure is a function of load and engine speed and is 5...12 MPa (50...120 bar), and is used as an actual (measured) value in the closed-loop rail-pressure control. The rpm and load-dependent setpoint value is stored in a map and is adjusted at the rail by a pressure control valve.

Brake-fluid pressure sensor
Installed in the hydraulic modulator of such driving-safety systems as ESP, this high-pressure sensor is used to measure the brake-fluid pressure which is usually 25 MPa (250 bar). Maximum pressure p_{max} can climb to as much as 35 MPa (350 bar). Pressure measurement and monitoring is triggered by the ECU which also evaluates the return signals.

Design and operating concept
The heart of the sensor is a steel diaphragm onto which deformation resistors have been vapor-deposited in the form of a bridge circuit (Fig. 1, 3). The sensor's pressure-measuring range depends on diaphragm thickness (thicker diaphragms for higher pressures and thinner ones for lower pressures). When the pressure is applied via the pressure connection (4) to one of the diaphragm faces, the resistances of the bridge resistors change due to diaphragm deformation (approx. 20 µm at 1,500 bar).

The 0...80 mV output voltage generated by the bridge is conducted to an evaluation circuit which amplifies it to 0...5 V. This is used as the input to the ECU which refers to a stored characteristic curve in calculating the pressure (Fig. 2).

1 High-pressure sensor

2 cm

Fig. 1
1 Electrical
 connection (socket)
2 Evaluation circuit
3 Steel diaphragm
 with deformation
 resistors
4 Pressure connection
5 Mounting thread

2 High-pressure sensor (curve, example)

Output voltage

Pressure

Pressure-control valve

Function
The function of the pressure-control valve is to adjust and maintain the pressure in the fuel rail as a factor of engine load, i.e.:
- It opens when the rail pressure is too high. Part of the fuel then returns from the fuel rail via a common line to the fuel tank.
- It closes when the rail pressure is too low, thus sealing the high-pressure side from the low-pressure side.

Design
The pressure-control valve (Fig. 1) has a mounting flange which attaches it to the high-pressure pump or the fuel rail. The armature (5) forces a valve ball (6) against the valve seat in order to seal the high-pressure stage from the low-pressure stage; this is achieved by the combined action of a valve spring (2) and an electromagnet (4) which force the armature downwards.

Fuel flows around the whole of the armature for lubrication and cooling purposes.

Operating concept
The pressure-control valve has two closed control loops:
- A slower, closed electrical control loop for setting a variable average pressure level in the fuel rail.
- A faster hydromechanical control loop for balancing out high-frequency pressure pulses.

Pressure-control valve not activated
The high pressure present in the fuel rail or at the high-pressure pump outlet is applied to the pressure-control valve via the high-pressure fuel supply. As the deenergized electromagnet exerts no force, the high-pressure force is greater than the spring force. The pressure-control valve opens to a greater or lesser extent depending on the delivery quantity. The spring is dimensioned to maintain a pressure of approx. 100 bar.

Pressure-control valve activated
When the pressure in the high-pressure circuit needs to be increased, the force of the electromagnet is added to that of the spring. The pressure-control valve is activated and closes until a state of equilibrium is reached between the high pressure and the combined force of the electromagnet and the spring. At this point, it remains in partly open position and maintains a constant pressure. Variations in the delivery quantity of the high-pressure pump and the withdrawal of fuel from the fuel rail by the injectors are compensated by varying the valve aperture. The magnetic force of the electromagnet is proportional to the control current. The control current is varied by pulse-width modulation. A pulse frequency of 1 kHz is sufficiently high to prevent adverse armature movement or pressure fluctuations in the fuel rail.

Designs
The pressure-control valve DRV1 is used in first-generation common-rail systems. Second- and third-generation CR systems operate using the two-actuator concept. Here, the rail pressure is adjusted by both a metering unit as well as a pressure-control valve. In this

1 Pressure-control valve DRV1 (section)

Fig. 1
1 Electrical connections
2 Valve spring
3 Armature
4 Valve housing
5 Solenoid coil
6 Valve ball
7 Support ring
8 O-ring
9 Filter
10 High-pressure fuel supply
11 Valve body
12 Drain to low-pressure circuit

case, either the pressure-control valve DRV2 is used or the DRV3 variant for higher pressures. This control strategy achieves lower fuel heating and eliminates the need for a fuel cooler.

The DRV2/3 (Fig. 2) differs from the DRV1 in the following features:
● Hard seal to the high-pressure interface (bite edge).
● Optimized magnetic circuit (lower power consumption).
● Flexible mounting concept (free plug orientation).

2 Pressure-control valve DRV2

3 Pressure-relief valve DBV4

Pressure-relief valve

Function
The pressure-relief valve has the same function as a pressure limiter. The latest version of the internal pressure-relief valve now has an integrated limp-home function. The pressure-relief valve limits pressure in the fuel rail by releasing a drain hole when pressure exceeds a certain limit. The limp-home function ensures that a certain pressure is maintained in the fuel rail to permit the vehicle to continue running without any restriction.

Design and operating concept
The pressure-relief valve (Fig. 3) is a mechanical component. It consists of the following parts:
● A housing with an external thread for screwing to the fuel rail.
● A connection to the fuel-return line to the fuel tank (3).
● A movable plunger (2).
● A plunger return spring (5).

At the end which is screwed to the fuel rail, there is a hole in the valve housing which is sealed by the tapered end of the plunger resting against the valve seat inside the valve housing. At normal operating pressure, a spring presses the plunger against the valve seat so that the fuel rail remains sealed. Only if the pressure rises above the maximum system pressure is the plunger forced back against the action of the spring by the pressure in the fuel rail so that the high-pressure fuel can escape. The fuel is routed through passages into a central bore of the plunger and returned to the fuel tank via a common line. As the valve opens, fuel can escape from the fuel rail to produce a reduction in fuel-rail pressure.

Fig. 2
1 Filter
2 Bite edge
3 Valve ball
4 O-ring
5 Union bolt with circlip
6 Armature
7 Solenoid coil
8 Electrical connection
9 Valve spring

Fig. 3
1 Valve insert
2 Valve plunger
3 Low-pressure section
4 Valve holder
5 Spring
6 Diaphragm disc

Electronic Diesel Control (EDC)

Electronic control of a diesel engine allows fuel-injection parameters to be varied precisely for different conditions. This is the only means by which a modern diesel engine is able to satisfy the many demands placed upon it. The EDC (Electronic Diesel Control) system is subdivided into three areas, "Sensors and desired-value generators", "Control unit", and "Actuators".

System overview

Requirements
Present-day development in the field of diesel technology is focused on lowering fuel consumption and exhaust-gas emissions (NO_x, CO, HC, particulate), while increasing engine performance and torque. In recent years this has led to an increase in the popularity of the direct-injection (DI) diesel engine, which uses much higher fuel-injection pressures than indirect-injection (IDI) engines with whirl or prechamber systems. Due to the more efficient mixture formation and the absence of flow-related losses between the whirl chamber/prechamber and the main combustion chamber, the fuel consumption of direct-injection engines is 10…20% lower than that achieved by indirect-injection designs.

In addition, diesel engine development has been influenced by the high levels of comfort and convenience demanded in modern cars. Noise levels, too, are subject to more and more stringent requirements.

As a result, the performance demanded of fuel-injection and engine-management systems has also increased, specifically with regard to:
- High fuel-injection pressures
- Rate-of-discharge curve control
- Pre-injection and, where applicable, secondary injection
- Variation of injected fuel quantity, charge-air pressure, and start of injection to suit operating conditions
- Temperature-dependent excess-fuel quantity for starting
- Control of idle speed independent of engine load
- Controlled exhaust-gas recirculation (cars)
- Cruise control
- Tight tolerances for injection duration and injected fuel quantity, and maintenance of high precision over the service life of the system (long-term performance)

Main components of EDC

Conventional mechanical governing of engine speed uses a number of adjusting mechanisms to adapt to different engine operating conditions and ensures a high mixture formation quality. Nevertheless, it is restricted to a simple engine-based control loop and there are a number of important influencing variables that it cannot take account of or cannot respond quickly enough to.

As demands have increased, what was originally a straightforward system using electric actuator shafts has developed into the present-day EDC, a complex electronic control system capable of processing large amounts of data in real time. It can form part of an overall electronic vehicle control system ("drive-by-wire"). And a result of the increasing integration of electronic components, the control-system circuitry can be accommodated in a very small space.

Operating concept

Electronic Diesel Control (EDC) is capable of meeting the requirements listed above as a result of microcontroller performance that has risen considerably in the last few years.

In contrast with diesel-engine vehicles with conventional mechanically controlled fuel-injection pumps, the driver of a vehicle equipped with EDC has no direct control over the injected fuel quantity via the accelerator pedal and cable. The injected fuel quantity is actually determined by a number of different influencing variables. They include:
- The vehicle response desired by the driver (accelerator-pedal position)
- The engine operating status
- The engine temperature
- Interventions by other systems (e.g. TCS)
- The effect on exhaust-gas emission levels, etc.

The control unit calculates the injected fuel quantity on the basis of all these influencing variables. Start of delivery can also be varied. This demands a comprehensive monitoring concept that detects inconsistencies and initiates appropriate actions in accordance with the effects (e.g. torque limitation or limp-

home mode in the idle-speed range). EDC, therefore incorporates a number of control loops.

Electronic diesel control allows data exchange with other electronic systems, such as the Traction Control System (TCS), Electronic Transmission Control (ETC), or Electronic Stability Program (ESP). As a result, the engine management system can be integrated in the vehicle's overall control system network, thereby enabling functions such as reduction of engine torque when the automatic transmission changes gear, regulation of engine torque to compensate for wheel spin, disabling of fuel injection by the engine immobilizer, etc.

The EDC system is fully integrated in the vehicle's diagnostic system. It meets all OBD (On-Board Diagnosis) and EOBD (European OBD) requirements.

System modules

Electronic Diesel Control (EDC) is divided into three system modules (Fig. 1):

1. *Sensors and setpoint generators* detect operating conditions (e.g. engine speed) and setpoint values (e.g. switch position). They convert physical variables into electrical signals.

2. The *electronic control unit* processes data from the sensors and setpoint generators based on specific open- and closed-loop control algorithms. It controls the actuators by means of electrical output signals. In addition, the control unit acts as an interface to other systems and to the vehicle diagnostic system.

3. *Actuators* convert electrical output signals from the control unit into mechanical parameters (e.g. the solenoid valve for the fuel-injection system).

Common-rail system for passenger cars

1 | Overview of EDC components for common-rail systems in passenger cars

Pedal-travel sensor with low-idle switch and kickdown switch

Engine speed (crankshaft)

Engine speed and cylinder detection (camshaft)

Rail pressure

Charge-air pressure

Charge-air temperature

Engine temperature (coolant)

Exhaust-gas temperature*

Air mass or exhaust-gas signal check-back

Road speed

Terminal 15

Terminal 50*

Clutch switch (alternative P/N contact on automatic transmission)

Multistage switch for maximum-speed limitation (HGB) and vehicle-speed controller (FGR)

Main relay

Sensors and setpoint generators

*optional

ECU
EDC 16 C / EDC 7

Signal inputs
Sensor evaluation
Atmospheric-pressure sensor

Functions:
- Idle-speed control
- Intermediate-speed regulation
- Smooth-running control
- Active-surge damper
- External torque intervention
- Vehicle immobilizer
- Fuel-delivery control and limitation
- Vehicle-speed controller (FGR)*
- Vehicle-speed limitation*
- Cylinder shutoff
- Fuel-quantity control
- Rail-pressure control
- Start-of-injection control
- Pre-injection control
- Secondary-injection control*
- Additional special adaptations*

Diagnostic functions:
- System diagnostics
- Substitute functions
- Engine diagnosis

Solenoid-valve output stages
Power output stages
Signal outputs
CAN communcation
Diagnosis communication
End-of-line programming
Power supply

+ 12V

BOSCH

Injectors
(max. 8 per ECU)

Rail-pressure control valve
High-pressure pump

Exhaust-gas recirculation actuator

Boost-pressure actuator*

Throttle-valve actuator

Auxiliary heating*

Intake-duct switchoff*

Radiator-fan control*

Starter motor*

Additional output stages*

Actuators

ISO interface
(e.g. diagnostics)

Glow-plug control unit

Engine-speed signal (TD)

Diagnosis lamp

CAN interface

Communication

SAE0750-1E

Common-rail system for commercial vehicles

2 Overview of EDC components for common-rail systems in commercial vehicles

Pedal-travel sensor with low-idle switch and kickdown switch

Engine speed (crankshaft)

Engine speed and cylinder detection (camshaft)

Rail pressure

Charge-air pressure

Oil pressure

Charge-air temperature

Engine temperature (coolant)

Fuel temperature

Differential pressure (particulate filter)

Exhaust-gas temperature*

Air mass or exhaust-gas signal check-back**

Terminal 15

Terminal 50*

Clutch switch (alternative P/N contact on automatic transmission)

Engine brake switch

Road speed

Multistage switch for maximum-speed limitation (HGB) and vehicle-speed controller (FGR)

Radiator-fan speed

VTG rpm*

Brake switch

Sensors and setpoint generators

*optional **light-duty only

ECU
EDC 16 / EDC 7

Signal inputs
Sensor evaluation

Functions:
- Idle-speed control
- Intermediate-speed regulation
- Smooth-running control
- Active-surge damper
- External torque intervention
- Vehicle immobilizer
- Fuel-delivery control and limitation
- Vehicle-speed controller (FGR)*
- Vehicle-speed limitation
- Cylinder shutoff
- Rail-pressure control
- Start-of-injection control
- Multiple-injection control
- Additional special adaptations*

Diagnostic functions:
On-board diagnosis (OBD)
Substitute functions
Engine diagnosis

Solenoid-valve output stages
Power output stages
Signal outputs
CAN communcation
Diagnosis communication
End-of-line programming

Power supply

12V / 24V
Main relay

BOSCH

Injectors
(max. 8 per ECU)

Metering unit
High-pressure pump

Exhaust-gas recirculation actuator

Boost-pressure actuator*/ VTG*

Auxiliary heating*

Cold-start system

Fuel heater*

Intake-duct switchoff*

Engine brake valve*/ retarder

Radiator-fan control

Radiator-fan coupling

Starter motor*

Additional output stages*

Actuators

K

L ISO interface (e.g. diagnostics)

Glow-plug control unit**

Engine-speed signal (TD)

Diagnosis lamp

CAN CAN interface

Communication

UAE0986E

Data processing

The main function of the electronic diesel control (EDC) is to control the injected fuel quantity and the injection timing. The common-rail fuel-injection system also controls injection pressure. Furthermore, on all systems, the engine ECU controls a number of actuators. For all components to operate efficiently, the EDC functions must be precisely matched to every vehicle and every engine. This is the only way to optimize component interaction (Fig. 2).

The control unit evaluates the signals sent by the sensors and limits them to the permitted voltage level. Some input signals are also checked for plausibility. Using these input data together with stored program maps, the microprocessor calculates injection timing and its duration. This information is then converted to a signal characteristic which is aligned to the engine's piston strokes. This calculation program is termed the "ECU software".

The required degree of accuracy together with the diesel engine's outstanding dynamic response requires high-level computing power. The output signals trigger output stages that supply sufficient power for the actuators (e.g. high-pressure solenoid valves for the fuel-injection system, exhaust-gas recirculation positioners, and boost-pressure actuators). Apart from this, a number of other auxiliary-function components (e.g. glow relay and air-conditioning system) are triggered.

Faulty signal characteristics are detected by output-stage diagnostic functions for the solenoid valves. Furthermore, signals are exchanged with other systems in the vehicle via the interfaces. The engine ECU monitors the complete fuel-injection system as part of a safety strategy.

1 Schematic using the example of a current regulator

2 Electronic Diesel Control (EDC): Basic sequence

Fuel control circuit 1 (fuel-injection components) Air control circuit
Fuel control circuit 2 (engine) Data and signal flow
"Diversion" via driver

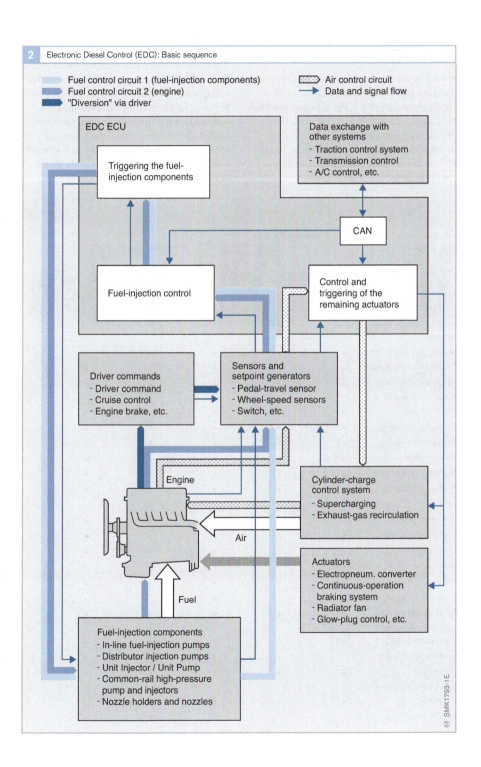

EDC ECU

Triggering the fuel-injection components

Data exchange with other systems
- Traction control system
- Transmission control
- A/C control, etc.

CAN

Fuel-injection control

Control and triggering of the remaining actuators

Driver commands
- Driver command
- Cruise control
- Engine brake, etc.

Sensors and setpoint generators
- Pedal-travel sensor
- Wheel-speed sensors
- Switch, etc.

Engine

Cylinder-charge control system
- Supercharging
- Exhaust-gas recirculation

Air

Fuel

Actuators
- Electropneum. converter
- Continuous-operation braking system
- Radiator fan
- Glow-plug control, etc.

Fuel-injection components
- In-line fuel-injection pumps
- Distributor injection pumps
- Unit Injector / Unit Pump
- Common-rail high-pressure pump and injectors
- Nozzle holders and nozzles

SMK1793-1E

Fuel-injection control

An overview of the various control functions which are possible with the EDC control units is given in Table 1. Fig. 1 opposite shows the sequence of fuel-injection calculations with all functions, a number of which are special options. These can be activated in the ECU by the workshop when retrofit equipment is installed.

In order that the engine can run with optimal combustion under all operating conditions, the ECU calculates exactly the right injected fuel quantity for all conditions. Here, a number of parameters must be taken into account. On a number of solenoid-valve-controlled distributor pumps, the sol-enoid valves for injected fuel quantity and start of injection are triggered by a separate pump ECU (PSG).

1 EDC variants for road vehicles: Overview of functions

Fuel-injection system	In-line injection pumps	Helix-controlled distributor injection pumps	Solenoid-valve-controlled distributor injection pumps	Unit Injector System and Unit Pump System	Common Rail System
	PE	VE-EDC	VE-M, VR-M	UIS, UPS	CR
Function					
Injected-fuel-quantity limitation	•	•	•	•	•
External torque intervention	•[3]	•	•	•	•
Vehicle-speed limitation	•[3]	•	•	•	•
Vehicle-speed control (Cruise Control)	•	•	•	•	•
Altitude compensation	•	•	•	•	•
Boost-pressure control	•	•	•	•	•
Idle-speed control	•	•	•	•	•
Intermediate-speed control	•[3]	•	•	•	•
Active surge damping	•[2]	•	•	•	•
BIP control	–	–	•	•	–
Intake-tract switch-off	–	–	•	•[2]	•
Electronic immobilizer	•[2]	•	•	•	•
Controlled pilot injection	–	–	•	•[2]	•
Glow control	•[2]	•	•	•[2]	•
A/C switch-off	•[2]	•	•	•	•
Auxiliary coolant heating	•[2]	•	•	–	•
Cylinder-balance control	•[2]	•	•	•	•
Control of injected fuel quantity compensation	•[2]	–	•	•	•
Fan (blower) triggering	–	•	•	•	•
EGR control	•[2]	•	•	•[2]	•
Start-of-injection control with sensor	•[1,3]	•	•	–	–
Cylinder shutoff	–	–	•[3]	•[3]	•[3]

Table 1

1 Only control-sleeve in-line injection pumps
2 Passenger cars only
3 Commercial vehicles only

1 Calculation of fuel-injection process in the ECU

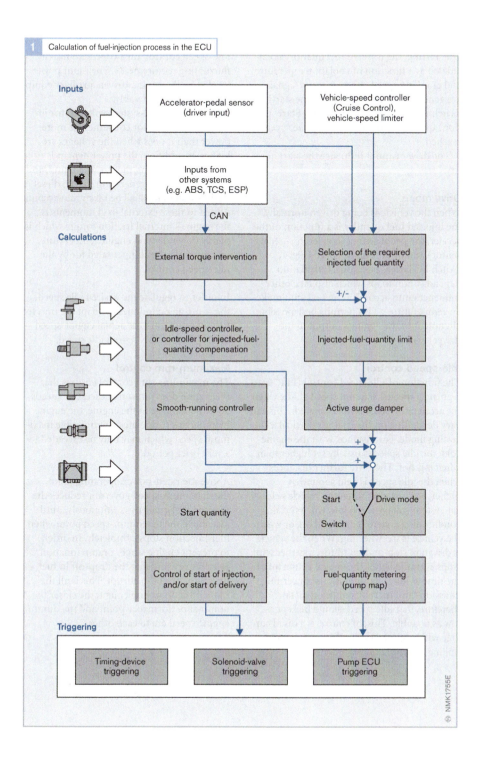

Inputs

Accelerator-pedal sensor
(driver input)

Vehicle-speed controller
(Cruise Control),
vehicle-speed limiter

Inputs from
other systems
(e.g. ABS, TCS, ESP)

CAN

Calculations

External torque intervention

Selection of the required
injected fuel quantity

+/-

Idle-speed controller,
or controller for injected-fuel-
quantity compensation

Injected-fuel-quantity limit

Smooth-running controller

Active surge damper

+

+

Start quantity

Start Drive mode

Switch

Control of start of injection,
and/or start of delivery

Fuel-quantity metering
(pump map)

Triggering

Timing-device
triggering

Solenoid-valve
triggering

Pump ECU
triggering

NMK1755E

Start quantity

For starting, the injected fuel quantity is calculated as a function of coolant temperature and cranking speed. Start-quantity signals are generated from the moment the starting switch is turned (Fig. 1, switch in "Start" position) until a given minimum engine speed is reached.

The driver cannot influence the start quantity.

Drive mode

When the vehicle is being driven normally, the injected fuel quantity is a function of the accelerator-pedal setting (accelerator-pedal sensor) and of the engine speed (Fig. 1, switch in "Drive" position). Calculation depends upon maps which also take other influences into account (e. g. fuel and intake-air temperature). This permits best-possible alignment of the engine's output to the driver's wishes.

Idle-speed control

The function of idle speed control (LLR) is to regulate a specific setpoint speed at idle when the accelerator pedal is not operated. This can vary depending on the engine's particular operating mode. For instance, with the engine cold, the idle speed is usually set higher than when it is hot. There are further instances when the idle speed is held somewhat higher. For instance, when the vehicle's electrical-system voltage is too low, when the air-conditioning system is switched on, or when the vehicle is freewheeling. When the vehicle is driven in stop-and-go traffic, together with stops at traffic lights, the engine runs a lot of the time at idle. Considerations concerning emissions and fuel consumption dictate, therefore, that idle speed should be kept as low as possible. This, of course, is a disadvantage with respect to smooth-running and pulling away.

When adjusting the stipulated idle speed, the idle-speed control must cope with heavily fluctuating requirements. The input power needed by the engine-driven auxiliary equipment varies considerably.

At low electrical-system voltages, for instance, the alternator consumes far more power than it does when the voltages are higher. In addition, the power demands from the A/C compressor, the steering pump, and the high-pressure generation for the diesel injection system must all be taken into account. Added to these external load moments is the engine's internal friction torque which is highly dependent on engine temperature, and must also be compensated for by the idle-speed control.

In order to regulate the desired idle speed, the controller continues to adapt the injected fuel quantity until the actual engine speed corresponds to the desired idle speed.

Maximum-rpm control

The maximum-rpm control ensures that the engine does not run at excessive speeds. To avoid damage to the engine, the engine manufacturer stipulates a permissible maximum speed which may only be exceeded for a very brief period.

Above the rated-power operating point, the maximum-speed governor reduces the injected fuel quantity continuously, until just above the maximum-speed point when fuel-injection stops completely. In order to prevent engine surge, a ramp function is used to ensure that the drop-off in fuel injection is not too abrupt. This is all the more difficult to implement, the closer the nominal performance point and maximum engine speed are to each other.

Intermediate-speed control

Intermediate-speed control (ZDR) is used on commercial vehicles and light-duty trucks with power take-offs, e.g. crane), or for special vehicles (e.g. ambulances with a power generator). With the control in operation, the engine is regulated to a load-independent intermediate speed.

With the vehicle stationary, the intermediate-speed control is activated via the cruise-control operator unit. A fixed rotational speed can be called up from the data store at the push of a button. In addition, this operator unit can be used for preselecting specific engine speeds. The intermediate-speed control is also applied on passenger cars with automated transmissions (e.g. Tiptronic) to control the engine speed during gearshifts.

Vehicle-speed controller (cruise control)

Cruise control allows the vehicle to be driven at a constant speed. It controls the vehicle speed to the speed selected by the driver without him/her needing to press the accelerator pedal. The driver can set the required speed either by operating a lever or by pressing buttons on the steering wheel. The injected fuel quantity is either increased or decreased until the desired (set) speed is reached.

On some cruise-control applications, the vehicle can be accelerated beyond the current set speed by pressing the accelerator pedal. As soon as the accelerator pedal is released, cruise control regulates the speed back down to the previously set speed.

If the driver depresses the clutch or brake pedal while cruise control is activated, control is terminated. On some applications, the control can be switched off by the accelerator pedal.

If cruise control has been switched off, the driver only needs to shift the lever to the restore position to reselect the last speed setting.

The operator controls can also be used for a step-by-step change of the selected speed.

Vehicle-speed limiter
Variable limitation

Vehicle-speed limitation (FGB, also called the limiter) limits the maximum speed to a set value, even if the driver continues to depress the accelerator pedal. On very quiet vehicles, where the engine can hardly be heard, this is a particular help for the driver who can no longer exceed speed limits inadvertently.

The vehicle-speed limiter keeps the injected fuel quantity down to a limit corresponding to the selected maximum speed. It can be deactivated by pressing the lever or depressing the kickdown switch. In order to reselect the last speed setting, the driver only needs to press the lever to the restore position. The operator controls can also be used for a step-by-step change of the selected speed.

Fixed limitation

In a number of countries, fixed maximum speeds are mandatory for certain classes of vehicles (for instance, for heavy trucks). Vehicle manufacturers also limit the maximum speeds of their heavy vehicles by installing a fixed speed limit which cannot be deactivated.

In the case of special vehicles, the driver can also select from a range of fixed, programmed speed limits (for instance, when workers are standing on the platform of a garbage truck).

Active-surge damping

Sudden engine-torque changes excite the vehicle's drivetrain, which, as a result, goes into bucking oscillation. These oscillations are perceived by the vehicle's occupants as unpleasant periodic changes in acceleration (Fig. 2, a). The function of the active-surge damper (ARD) is to reduce these changes in acceleration (b).

Two different methods are used:
- In case of sudden changes in the torque required by the driver (through the accelerator pedal), a precisely matched filter function reduces drivetrain excitation (1).
- The speed signals are used to detect drivetrain oscillations which are then damped by an active control. In order to counteract the drivetrain oscillations (2), the active control reduces the injected fuel quantity when rotational speed increases, and increases it when speed drops.

Smooth-running control (SRC)/ Control of injected-fuel-quantity compensation (MAR)

Presuming a constant injection duration, not all of the engine's cylinders generate the same torque. This can be due to differences in cylinder-head sealing, as well as differences in cylinder friction, and hydraulic-injection components. These differences in torque output lead to rough engine running and an increase in exhaust-gas emissions.

Smooth-running control (LRR) or fuel-balancing control (MAR) have the function of detecting these differences based on the resulting fluctuations in engine speed, and to compensate by adjusting the injected fuel quantity in the cylinder affected. Here, the rotational speed at a given cylinder after injection is compared to a mean speed. If the particular cylinder's speed is too low, the injected fuel quantity is increased; if it is too high, the fuel quantity is reduced (Fig. 3).

Fig. 2
a Without active-surge
 damper
b With active-surge
 damper
1 Filter function
2 Active correction

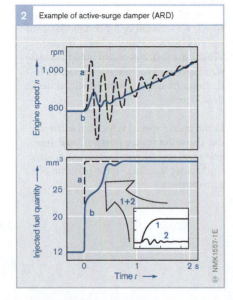

2 Example of active-surge damper (ARD)

3 Example of smooth-running control (LRR)

Smooth-running control is a convenience feature. Its primary object is to ensure that the engine runs smoothly at near-idle. The injected-fuel-quantity compensation function is aimed at not only improving comfort at idle, but also at reducing exhaust-gas emissions in the medium-speed ranges by ensuring identical injected fuel quantities for all cylinders.

On commercial vehicles, smooth-running control is also known as the AZG (adaptive cylinder equalization).

Injected-fuel-quantity limit

There are a number of reasons why the fuel quantity actually required by the driver, or that which is physically possible, should not always be injected. The injection of such fuel quantities could have the following effects:

- Excessive exhaust-gas emissions
- Excessive soot
- Mechanical overloading due to high torque or excessive engine speed
- Thermal overloading due to excessive temperatures of the exhaust gas, coolant, oil, or turbocharger
- Thermal overloading of the solenoid valves if they are triggered too long

To avoid these negative effects, a number of input variables (for instance, intake-air quantity, engine speed, and coolant temperature) are used to generate this limitation figure. The result is that the maximum injected fuel quantity is limited and with it the maximum torque.

Engine-brake function

When a truck's engine brake is applied, the injected fuel quantity is either reduced to zero, or the idle fuel quantity is injected. For this purpose, the ECU detects the position of the engine-brake switch.

Altitude compensation

Atmospheric pressure drops as altitude increases so that the cylinder is charged with less combustion air. This means that the injected fuel quantity must be reduced accordingly, otherwise excessive soot will be emitted. In order that the injected fuel quantity can be reduced at high altitudes, the atmospheric pressure is measured by the ambient-pressure sensor in the ECU. This reduces the injected fuel quantity at higher elevations. Atmospheric pressure also has an effect on boost-pressure control and torque limitation.

Cylinder shutoff

If less torque is required at high engine speeds, very little fuel needs to be injected. As an alternative, cylinder shutoff can be applied to reduce torque. Here, half of the injectors are switched off (commercial-vehicle UIS, UPS, and CRS). The remaining injectors then inject correspondingly more fuel which can be metered with even higher precision.

When the injectors are switched on and off, special software algorithms ensure smooth transitions without noticeable torque changes.

Injector delivery compensation

New functions are added to common-rail (CR) and UIS/UPS systems to enhance the high precision of the fuel-injection system further, and ensure them for the service life of the vehicle.

With injector delivery compensation (IMA), a mass of measuring data is detected for each injector during the injector manufacturing process. The data is then affixed to the injector in the form of a data-matrix code. With piezo-inline injectors, data on lift response is included. This data is transferred to the ECU during vehicle production. While the engine is running, these values are used to compensate for deviations in metering and switching response.

Zero delivery calibration

The reliable mastery of small pre-injection events for the service life of the vehicle is vitally important to achieve the required level of comfort (through reduced noise) and exhaust-gas emission targets. There must be some form of compensation for fuel-quantity drifts in the injectors. For this reason, a small quantity of fuel is injected in one cylinder in overrun conditions in second- and third-generation CR systems.

The wheel-speed sensor detects the resulting torque increase as a minor dynamic change in engine speed. This increase in torque, which remains imperceptible to the driver, is clearly linked to the injected fuel quantity. The process is then repeated for all cylinders and at various operating points. A teach-in algorithm detects minor changes in pre-injection quantity and corrects the injector triggering period accordingly for all pre-injection events.

Average delivery adaption

The deviation of the actually injected fuel quantity from the setpoint value is required to adapt exhaust-gas recirculation and charge-air pressure correctly. The average delivery adaption (MMA), therefore determines the average value of the injected fuel quantity for all cylinders from the signals received from the lambda oxygen sensor and the air-mass sensor. Correction values are then calculated from the setpoint and actual values (see "Lambda closed-loop control for passenger-car diesel engines").

The MMA teach-in function ensures a constant level of favorable exhaust-gas emission values in the lower part-load range for the service life of the vehicle.

Pressure-wave correction

Injection events trigger pressure waves in the line between the nozzle and the fuel rail in all CR systems. These pressure pulses affect systematically the injected fuel quantity of later injection events (pre-injection/main injection/secondary injection) within a combustion cycle. The deviations of later injection events are dependent on the fuel quantity previously injected, the time interval between injection events, rail pressure, and fuel temperature. The control unit can calculate a correction factor by including these parameters in suitable compensation algorithms.

However, extremely high application resources are required for this correction function. The benefit is the possibility of flexibly adjusting the interval between pre-injection and main injection, for example, in order to optimize combustion.

▶ Injector delivery compensation

Functional description

Injector delivery compensation (IMA) is a software function to make fuel quantity metering more precise and increase injector efficiency on the engine. The feature has the function of correcting injected fuel quantity to the setpoint value over the entire program map individually for every injector in a CR system. This reduces system tolerances and exhaust-gas emission spread. The compensation values required for IMA represent the difference from the setpoint value of each factory test point, and are inscribed on each injector in encoded form.

The entire engine environment is corrected by means of a correction program map that uses compensation values to calculate a correction quantity. At the end of the line of the car assembly plant, the EDC compensation values belonging to the injectors fitted and their cylinder assignment are programmed in the electronic control unit using EOL programming. The compensation values are also re-programmed when an injector is replaced at the customer service workshop.

Necessity for this function

The technical resources required for a further restriction of the manufacturing tolerances for injectors rise exponentially and appear to be financially unprofitable. IMA is a viable solution to increase efficiency, enhance the metering precision of fuel quantity injected in the engine, and reduce exhaust-gas emissions.

Measured values in testing

The end-of-line test measures every injector at several points that are representative for the spread of the particular injector type. Deviations from setpoint values at these points (compensation values) are calculated and then inscribed on the injector head.

▶ EMI characteristic curve without IMA

Fig. 1
Curves of various injectors as a function of rail pressure. IMA reduces curve spread.
EMI Injected-fuel-quantity indicator

▶ Considering the matrix in the injection calculation

Fig. 2
Calculation of injector triggering period based on setpoint quantity, rail pressure, and correction values

▶ Process chain

Fig. 3
Schematic of process chain from injector delivery compensation at Bosch through to end-of-line programming at the vehicle manufacturer's plant

Lambda closed-loop control for passenger-car diesel engines

Application

The lawmakers are continually increasing the severity of legislation governing exhaust-gas emission limits for cars powered by diesel engines. Apart from the measures taken to optimize the engine's internal combustion, the open and closed-loop control of functions related to exhaust-gas emissions are continuing to gain in importance. Introduction of lambda closed-loop control offers major potential for reducing emission-value spread in diesel engines.

A broadband lambda oxygen sensor in the exhaust pipe (Fig. 1, 7) measures the residual oxygen content in the exhaust gas. This is an indicator of the A/F ratio (excess-air-factor lambda λ). The lambda oxygen-sensor signal is adapted while the engine is running. This ensures a high level of signal accuracy throughout the sensor's service life. The lambda oxygen-sensor signal is used as the basis for a number of lambda functions, which will be described in more detail in the following.

Lambda closed-loop control circuits are used to regenerate NO_X accumulator-type catalytic converters.

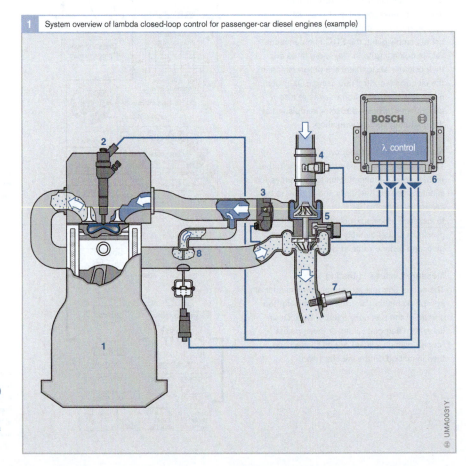

1 System overview of lambda closed-loop control for passenger-car diesel engines (example)

Fig. 1
1 Diesel engine
2 Diesel injection
 component
 (here, common-
 rail injector)
3 Control flap
4 Hot-film air-mass
 meter
5 Exhaust-gas
 turbocharger
 (here, VTG version)
6 Engine ECU for
 EDC
7 Broadband lambda
 oxygen sensor
8 EGR valve

Lambda closed-loop control is designed for all passenger-car fuel-injection systems with engine control units dating dating from the EDC16 generation.

Basic functions

Pressure compensation

The unprocessed lambda oxygen-sensor signal is dependent on the oxygen concentration in the exhaust gas and the exhaust-gas pressure at the sensor installation point. The influence of pressure on the sensor signal must, therefore, be compensated.

The *pressure-compensation* function incorporates two program maps, one for exhaust-gas pressure, and one for pressure dependence of the lambda oxygen-sensor output signal. These two maps are used to correct the sensor output signal with reference to the particular operating point.

Adaption

In overrun mode (trailing throttle), lambda oxygen-sensor adaption takes into account the deviation of the measured oxygen concentration from the fresh-air oxygen concentration (approx. 21%). As a result, the system "learns" a correction value which is used at every engine operating point to correct the measured oxygen concentration. This leads to a precise, drift-compensated lambda output signal for the service life of the lambda oxygen sensor.

Lambda-based EGR control

Compared with air-mass-based exhaust-gas recirculation, detecting oxygen concentration in the exhaust gas allows tighter emission tolerance bands for an automotive manufacturer's entire vehicle fleet. For future limits, an emission advantage of approx. 10 to 20% can be gained in this way for the exhaust-gas test.

2 Operating concept of average delivery adaption in "indirect control" mode

Average delivery adaption

Average delivery adaption supplies a precise injection quantity signal to form the setpoint for the exhaust-gas-related closed control loop. Correction of exhaust-gas recirculation plays a major role in emissions here. Average delivery adaption operates in the lower part-load range and determines the average deviation in the injected fuel quantity of all cylinders.

Fig. 2 (previous page) shows the basic structure of average delivery adaption and its influence on the exhaust-gas-related closed control loops.

The lambda oxygen-sensor signal and the air-mass signal are used to calculate the actually injected fuel mass, which is then compared to the desired injected fuel mass. Differences are stored in an adaption map in defined "learning points". This procedure ensures that, when the operating point requires an injected fuel quantity correction, it can be implemented without delay even during dynamic changes of state.

These correction quantities are stored in the EEPROM of the ECU and are available immediately the engine is started.

Basically speaking, there are two average-delivery adaption operating modes. They differ in the way they apply detected deviations in injected fuel quantity:

Operating mode: Indirect Control
In *Indirect Control* mode (Fig. 2), a precise injection quantity setpoint is used as the input variable in various exhaust-gas-related reference program maps. The injected fuel quantity is not corrected during the fuel-metering process.

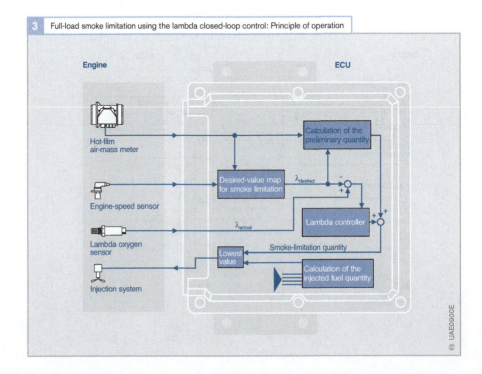

3 Full-load smoke limitation using the lambda closed-loop control: Principle of operation

UAE0900E

Operating mode: Direct Control

In *Direct Control* mode, the quantity deviation is used in the metering process to correct the injected fuel quantity so that the actual fuel quantity injected coincides more precisely with the reference injected fuel quantity. In this case, this is (more or less) a closed quantity control loop.

Full-load smoke limitation

Fig. 3 shows the block diagram of the control structure for full-load smoke limitation using a lambda oxygen sensor. The objective here is to determine the maximum fuel quantity which may be injected without exceeding a given smoke-emission value.

The signals from the air-mass meter and the engine-speed sensor are applied together with a smoke-limitation map to determine the desired air/fuel ratio value $\lambda_{desired}$. This, in turn, is applied together with the air mass to calculate the precontrol value for the maximum permissible injected fuel quantity.

This form of control is already in serial production, and has a lambda closed-loop control imposed on it. The lambda controller calculates a correction fuel quantity from the difference between the desired air/fuel ratio $\lambda_{desired}$ and the actual air/fuel ratio value λ_{actual}. The maximum full-load injected fuel quantity is the total of the pilot-control quantity and the correction quantity.

This control architecture permits a high level of dynamic response due to pilot control, and improved precision due to the superimposed lambda control loop.

Detection of undesirable combustion

The lambda oxygen sensor signal helps to detect the occurrence of undesirable combustion in overrun mode. It is detected if the lambda oxygen-sensor signal drops below a calculated threshold. In this case, the engine can be switched off by closing a control flap and the EGR valve. The detection of undesirable combustion represents an additional engine safeguard function.

Summary

A lambda-based exhaust-gas recirculation system can substantially reduce emission-value spread over a manufacturer's vehicle fleet due to production tolerances or aging drift. This is achieved by using average delivery adaption.

Average delivery adaption supplies a precise injection quantity signal to form the setpoint for the exhaust-gas-related closed control loop. The precision of these control loops is increased as a result. Correction of exhaust-gas recirculation plays the major role on emissions here.

In addition, the application of lambda closed-loop control permits the precise metering of the full-load smoke quantity and detection of undesirable combustion in overrun (trailing throttle) mode.

Furthermore, the lambda oxygen sensor's high-precision signal can be used in a lambda closed control loop to regenerate NO_X catalytic converters.

► Closed-loop and open-loop control

Application

The *closed-loop* and *open-loop* control applications are of vital importance for various on-board systems.

The term *(open-loop) control* is used in many cases, not only for the process of controlling, but also for the entire system in which control takes place (for this reason, the general term *"control unit"* is used, although it may perform a closed-loop control function). Accordingly, arithmetic processes run in control units to calculate both closed-loop and open-loop functions.

Closed-loop control

Closed-loop control is a process in which a parameter (controlled variable x) is detected continuously, compared to another parameter (reference variable w_1), and adapted to the reference variable in an adjustment process depending on the result of the comparison. The resulting action takes place in a closed circuit (closed control loop).

Closed-loop control has the function of adjusting the value of controlled variables to a value specified by a reference variable, despite any disturbance influences that may occur.

The *closed control loop* (Fig. 1a) is a closed-loop control circuit with a discrete action. Controlled variable x acts within a loop configuration in a form of negative feedback. Contrary to open-loop control, closed-loop control considers the impact of all disturbance values (z_1, z_2) occurring within the control loop. Examples of closed-loop systems in a vehicle:

- Lambda closed-loop control
- Idle-speed control
- ABS/TSC/ESP control
- Air conditioning (interior temperature)

Open-loop control

Open-loop control is the process within a system in which one or several parameters act as input variables affecting other parameters due to intrinsic laws governing the system. A feature of open-loop control is the open action sequence across an individual transfer element or the open control loop.

An *open control loop* (Fig. 1b) is an arrangement of elements that interact on each other in a loop structure. It may interact in any possible way with other systems as an entity within a higher-level system. The open control loop can only counter the impact of a disturbance value measured by the control unit (e.g. z_1); other disturbance values (e.g. z_2) may act unimpaired. Examples of open-loop systems in a vehicle:

- Electronic Transmission Control (ETC)
- Injector delivery compensation and pressure-wave correction for calculating injected fuel quantity

Fig. 1
a Closed control loop
b Open control loop
c Block diagram of a digital closed-control loop

w Reference variable
x Controlled variable (closed loop)
x_A Controlled variable (open loop)
y Manipulated variable
z_1, z_2 Disturbance values

T Sampling time
* Digital signal values
A Analog
D Digital

1 Closed-loop and open-loop control applications

Torque-controlled EDC systems

The engine-management system is continually being integrated more closely into the overall vehicle system. Through the CAN bus, vehicle dynamics systems such as TCS, and comfort and convenience systems such as cruise control, have a direct influence on the Electronic Diesel Control (EDC). Apart from this, much of the information registered and/or calculated in or by the engine management system must be passed on to other ECUs through the CAN bus.

In order to be able to incorporate the EDC even more efficiently in a functional alliance with other ECUs, and implement other changes rapidly and effectively, it was necessary to make far-reaching changes to the newest-generation controls. These changes resulted in the torque-controlled EDC which was introduced with the EDC16. The main feature is the changeover of the module interfaces to the parameters, as commonly encountered in practice in the vehicle.

Engine parameters

Essentially, an IC engine's output can be defined using the three parameters: power P, engine speed n, and torque M.

For 2 diesel engines. Fig. 1 compares typical curves of torque and power as a function of engine speed. Basically speaking, the following equation applies:

$$P = 2 \cdot \pi \cdot n \cdot M$$

In other words, it suffices to use the torque as the reference (command) variable. Engine power then results from the above equation. Since power output cannot be measured directly, torque has turned out to be a suitable reference (command) variable for engine management.

Torque control

When accelerating, the driver uses the accelerator pedal (sensor) to directly demand a given torque from the engine. At the same time, but independent of the driver's requirements, via the interfaces other vehicle systems submit torque demands resulting from the power requirements of the particular component (e.g. air conditioner, alternator). Using these torque-requirement inputs, the engine management calculates the output torque to be generated by the engine and controls the fuel-injection and air-system actuators accordingly. This method has the following advantages:

- No single system (for instance, boost pressure, fuel injection, pre-glow) has a direct effect on engine management. This enables the engine management to also take into account higher-level optimization criteria (such as exhaust-gas emissions and fuel consumption) when processing external requirements, and thus control the engine in the most efficient manner,
- Many of the functions which do not directly concern the engine management can be designed to function identically for diesel and gasoline engines.
- Extensions to the system can be implemented quickly.

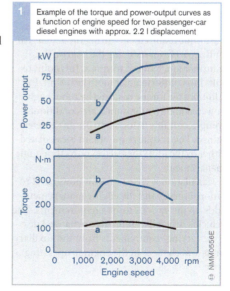

1 Example of the torque and power-output curves as a function of engine speed for two passenger-car diesel engines with approx. 2.2 l displacement

NMM0556E

Fig. 1
a Year of manufacture 1968
b Year of manufacture 1998

Engine-management sequence

Fig. 2 shows (schematically) the processing of the setpoint inputs in the engine ECU. In order to be able to fulfill their assignments efficiently, the engine management's control functions all require a wide range of sensor signals and information from other ECUs in the vehicle.

Propulsion torque

The driver's input (that is, the signal from the accelerator-pedal sensor) is interpreted by the engine management as the request for a propulsive torque. The inputs from the cruise control and the vehicle speed limiter are processed in exactly the same manner.

Following this selection of the desired propulsive torque, should the situation arise, the vehicle-dynamics system (TCS, ESP) increases the desired torque value when there is the danger of wheel lockup and decreases it when the wheels show a tendency to spin.

Further external torque demands

The drivetrain's torque adaptation must be taken into account (drivetrain transmission ratio). This is defined for the most part by the ratio of the particular gear, or by the torque-converter efficiency in the case of automatic transmissions. On vehicles with an automatic-gearbox, the transmission control stipulates the torque requirement during the actual gear shift. Apart from reducing the load on the transmission, reduced torque at this point results in a comfortable, smooth gear shift. In addition, the torque required by other engine-powered units (for instance, air-conditioner compressor, alternator, servo pump) is determined. This torque requirement is calculated either by the units themselves or by the engine management.

Calculation is based on unit power and rotational speed, and the engine management adds up the various torque requirements. The vehicle's drivability remains unchanged despite varying requirements from the auxiliary units and changes in the engine's operating state.

Internal torque demands

At this stage, the idle-speed control and the active surge damper intervene.

For instance, if demanded by the situation, in order to prevent mechanical damage, or excessive smoke due to the injection of too much fuel, the torque limitation reduces the internal torque requirement.

In contrast to the previous engine-management systems, limitations are no longer only applied to the injected fuel-quantity, but instead, depending upon the required effects, also to the particular physical quantity involved.

The engine's losses are also taken into account (e.g. friction, drive for the high-pressure pump). The torque represents the engine's measurable effects to the outside. The engine management, though, can only generate these effects in conjunction with the correct fuel injection together with the correct injection point, and the necessary marginal conditions as apply to the air-intake system (e.g. boost pressure and EGR rate). The required injected fuel quantity is determined using the current combustion efficiency. The calculated fuel quantity is limited by a protective function (for instance, protection against overheating), and if necessary can be varied by smooth-running control (SRC). During engine start, the injected fuel quantity is not determined by external inputs such as those from the driver, but rather by the separate "start-quantity control" function.

Actuator triggering

Finally, the desired values for the injected fuel quantity are used to generate the triggering data for the injection pump and/or the injectors, and for defining the optimum operating point for the intake-air system.

2 Engine-management sequence for torque-controlled diesel injection

Sensor
signals

Data
exchange

Propulsion torque:

Driver input:
- Accelerator-pedal
 sensor
- Vehicle-speed control
 (cruise control)
- Vehicle-speed limitation

Selection of the
desired propulsion
torque

Input from the vehicle-
dynamics systems:
- TCS
- ESP

Coordination of the
propulsion torque

Further external torque demands

Drivetrain transm.

Input from the
transmission ECU

Engine loading due
to auxiliary units

Coordination of the
drivetrain torque

Internal torque requirements

Idle-speed control

Active-surge damping

Torque limitation

Control of the
engine torque
(internal functions)

Start quantity

Engine efficiency

Fuel-quantity limit

Smooth-running control

Start Drive
mode
Fuel-
quantity
input

Actuator triggering

Inputs:
- Boost pressure
- EGR rate
- ...

Input:
- Start of delivery
- Timing device
- Rail pressure
- ... (depending on system)

☐ External
 inputs

☐ Internal
 sequences

--- Data trans-
 mission
 possible
 through CAN

Intake-air system
- Turbocharger
- EGR ...

Injection system
- Fuel-injection
 pump
- Injectors ...

NMK1763E

Data exchange with other systems

Fuel-consumption signal

The engine ECU (Fig. 1, 3) determines fuel consumption and sends this signal via CAN to the instrument cluster or a separate onboard computer (6), where the driver is informed of current fuel consumption and/or the range that can be covered with the remaining fuel in the tank. Older systems used pulse-width modulation (PWM) for the fuel-consumption signal.

Starter control

The starter motor (8) can be triggered from the engine ECU. This ensures that the driver cannot operate the starter motor with the engine already running. The starter motor only turns long enough to allow the engine to reach a self-sustaining speed reliably. This function leads to a lighter, and thus lower-priced, starter motor.

Glow control unit

The glow control unit (GZS, 5) receives information from the engine ECU to control glow start and duration. It then triggers the glow plugs accordingly and monitors the glow process, and reports back to the engine ECU on any faults (diagnostic function). The preglow indicator lamp is usually triggered from the engine ECU.

Electronic immobilizer

To prevent unauthorized starting and drive-off, the engine cannot be started before a special immobilizer (7) ECU removes the block from the engine ECU.

The driver can signal the immobilizer ECU that he/she is authorized to use the vehicle, either by remote control or by means of the glow-plug and starter switch ("Ignition" key). The immobilizer ECU then removes the block on the engine ECU to allow engine start and normal operation.

1 Possible components involved in the exchange of data with the Electronic Diesel Control (EDC)

Fig. 1

1 ESP ECU
 (with ABS and TCS)
2 ECU for transmission-shift control
3 Engine ECU (EDC)
4 A/C ECU
5 Glow control unit
6 Instrument cluster
 with onboard
 computer
7 Immobilizer ECU
8 Starter motor
9 Alternator
10 A/C compressor

External torque intervention

In the case of external torque intervention, the injected fuel quantity is influenced by another (external) ECU (for instance, for transmission-shift control, or TCS). This informs the engine ECU whether the engine torque is to be changed, and if so, by how much (this defines the injected fuel quantity).

Alternator control

By means of a standard serial interface, the EDC can control and monitor the alternator (9) remotely. The regulator voltage can be controlled, just the same as the complete alternator assembly can be switched off. In case of low battery power, for instance, the alternator's charging curve can be improved by increasing the idle speed. It is also possible to perform simple alternator diagnosis through this interface.

Air conditioner

In order to maintain comfortable temperatures inside the vehicle when the ambient temperature is high, the air conditioner (A/C) cools down cabin air with the help of an A/C compressor (10). Depending on the engine and operating conditions, the A/C compressor may draw as much as 30% of the engine's output power.

Immediately the driver hits the accelerator pedal (in other words he/she wishes maximum torque), the compressor can be switched off briefly by the engine ECU to concentrate all of the engine's power to the wheels. Since the compressor is only switched off very briefly, this has no noticeable effect on interior temperature.

Serial data transmission (CAN)

Modern-day vehicles are equipped with a constantly increasing number of electronic systems. Along with their need for extensive exchange of data and information in order to operate efficiently, the data volumes and speeds are also increasing at a rapid rate.

Although CAN (Controller Area Network) is a linear bus system (Fig. 1) specifically designed for automotive applications, it has already been introduced in other sectors (for instance, in building automation).

Data is relayed in serial form, that is, one after another on a common bus line. All CAN stations have access to this bus, and via a CAN interface in the ECUs, they can receive and transmit data over the CAN bus line. Since a considerable amount of data can be exchanged and repeatedly accessed on a single bus line, this network results in far fewer lines required. On conventional systems, data exchange takes place point to point over individually assigned data lines.

1 Linear bus topology

Transmission-shift control
Station 1

Engine management
Station 2

CAN

ABS/TCS/ESP
Station 3

Instrument cluster
Station 4

UAE0283-2E

Active steering

The development of vehicle steering systems is characterized by the consistent introduction of hydraulic servo assist and the replacement of ball-and-nut-type steering in the car by the easier and more inexpensive rack-and-pinion steering. Recently, electromechanical power steering has been displacing hydraulic power steering in small and lightweight cars. By law, however, pure "steer-by-wire" technology is not yet permitted in motor vehicles. European Union safety regulations still require a mechanical connection between the steering wheel and the wheels of the vehicle.

All of these developments have the goal of making vehicle handling as easy as possible and to limit steering forces to a logical amount. The best possible feedback about the contact of the tires to the road is to be ensured. This has a decisive impact on the driver's ability to manage his or her task in the control loop between driver, vehicle and environment.

Purpose

The newly developed active steering can affect the steering forces and the steering angle set by the driver. It fulfills the wish for a direct steering ratio to improve handling at low speeds. It also meets requirements for ensuring comfort, drivability and straight-running stability at high speeds. Active steering is an initial step towards a "steer-by-wire" function. Although it does not let the car drive itself, it provides correction functions and added comfort and convenience.

Design

The primary difference between active steering and a "steer-by-wire" system is the fact that the steering train, and thus the driver's mechanical control of the steered front wheels, is maintained during active steering.

Fig. 1
Changing the ratio between the steering wheel angle and the median angle of the front wheels → Reducing the steering effort at low speeds and providing stability at high speeds

1 | Variable steering ratio

Steering wheel angle: $\Omega_S = 320°$ $\Omega_S = 406°$ $\Omega_S = 512°$

Motor angle: $\Omega_M = 7000°$ $\Omega_M = 0°$ $\Omega_M = -9000°$

$\Omega_F = 32°$ $\Omega_F = 32°$ $\Omega_F = 32°$

Front wheel angle

i_V
$i_{V\,max}$
$i_{V\,mechanical}$
$i_{V\,min}$
direct
indirect
0
0 60 100 km/h

Vehicle speed v_x

SFL0033E

Mechanical system
The steering train consists, as usual, of the steering wheel, steering column, steering gear and tie rods. The special feature of the new active steering is a differential gearbox (Fig. 2). For this purpose, a planetary gear (6) is integrated with two input shafts and one output shaft in the steering gear. One input shaft is connected to the steering wheel, and the other drives an electric motor (4) via a worm gear pair (3) as a reduction stage. The connected ECU processes the necessary sensor signals, controls the electric motor and monitors the entire steering system.

The electric motor and differential gearbox allow steering intervention at the front axle to take place independently of the driver. At low speeds, the effective steering angle at the wheels is greater than the angle set on the steering wheel, as the system adds a part that is proportional to the steering angle. At high speeds, it subtracts a corresponding amount so that the wheel angle is

smaller than that set by the driver (Fig. 1). When the electric motor is inactive, the steering wheel controls the vehicle's wheels directly, as with conventional steering systems.

Hydraulic system
The principle of differential steering usually requires a hydraulic servo assist to limit the forces applied by hand to a logical amount. This is accomplished using an "open center" steering valve specially adapted to the high performance requirements. The vector superimposition of the positioning rates of the drivers and engine can, in certain cases, cause significantly higher rack-and-pinion speeds than those of conventional steering systems. The geometric flow of the vane-type power steering pump with flow rate controller is designed for the theoretical maximum positioning rate. Regulation at the output side provides a highly dynamic and quiet energy supply for the active steering system.

2 Actuators

3 Components of active steering

Other vehicle sensors (e.g. ESP):
• Steering wheel angle
• Wheel speeds
• Reference speed

Direction of travel

Fig. 2
1 Servotronic II valve
2 Electromagnetic block
3 Worm
4 Electric motor
5 Steering rack
6 Planetary gear
7 Worm gear

Fig. 3
1 Electronic control unit
2 Pinion angle sensor
3 Substructure
4 Block
5 Servotronic II valve
6 Actuator module
7 Motor angle sensor

Method of operation

Activation concept

The actuator adjustment at the front axle, which takes place independently of the driver, requires a complex activation concept that is implemented in an ECU with two processors that communicate with each other. One processor is responsible for activating the servomotor, the other for calculating the correct control angle. Both processors monitor each other to ensure that they are functioning properly. The movement status of the steering gear is detected using one angle sensor each on the steering pinion and servomotor. The steering wheel angle signal is also used as the setpoint specification from the driver. Sensors for yaw velocity, lateral acceleration and wheel speed, which are already in place for the driving stability systems (ESP), provide additional input signals for the active steering.

The ECU is networked to the system using powertrain CAN and the new chassis CAN at the required high data rate (see Fig. 3). Approximately 100 times a second, the necessary data are collected using sensors and evaluated by the ECU. The ECU then decides whether, and by what amount, the steering angle needs to be changed. At low speeds, it adds an amount proportional to the steering angle; at high speeds, it sub-

tracts a corresponding amount. Therefore, from the driver's point of view, the impression given is of a steering ratio that is variable over the driving speed. The steering effort remains largely constant over a wide speed range. Except at extremely low speeds, a significantly greater steering angle than 180° is required.

Driving stabilization

To calculate the stabilizing steering intervention, the vehicle movement variables of yaw angular acceleration and lateral acceleration are returned and compared in the stabilization controller with the setpoint specified by the driver, which is made up of the steering wheel angle and driving speed (Fig. 4).

Setpoint value

The setpoint for the control angle to be set, which is calculated in the ECU of the active steering, can be separated into a part with open-loop control and a part with closed-loop control. The open-loop control part, known simply as the variable steering ratio, by definition is calculated from the reference variable only, the driver's steering angle. Additional information from the closed-loop control system known as the "vehicle" provides the basis for the closed-loop part. The partial setpoints meet at a convergence point. They modify the response of the vehicle closed-loop control system to the driver's steering input. These steering interventions are usually continuous and thus do not disturb the driver.

4 Yaw moment compensation

- Different braking forces generate a yaw moment

- Required steering intervention is carried out automatically

➡ Braking forces
➡ Lateral forces

Fig. 4
Example: braking on non-uniform road surface

Cooperation with driving stability systems

Compared to conventional driving stability systems based on wheel slip control, driving stabilization via steering intervention at the front axle has different characteristics:

- The steering intervention is considerably less noticeable to the driver than the clearly audible braking intervention.
- The steering intervention is faster than a radial braking intervention, which requires a certain pressure build-up time.
- The braking intervention provides superior stabilization performance than the steering.

By combining active steering (steering intervention) and wheel slip control (braking intervention), optimum driving stabilization is attained.

Safety concept

If the servomotor has to be switched off due to a fault, this path is mechanically blocked. The planetary gear then rolls off internally while the worm gear is blocked, and the vehicle remains steerable with no restrictions and at a constant gear ratio. Thus active steering is possible even if the mechanical connection fails. This is a great advantage over pure "steer-by-wire" systems. In active steering, all relevant input signals are safeguarded by redundant sensors or measurements. Two different processors calculate the setpoint signal in the ECU. Although the conversion of the setpoint signal in the electromechanical converter takes place in one channel only, no unwanted actuation in the region relevant to vehicle dynamics is possible due to the selection of a BLDC motor.

The safety concept is expanded by adding an adapted shutoff concept. Function shutoffs range from temporary or long-term disabling of the driving stabilization to limited vehicle operation with substitute values all the way to complete shutoff of the entire system. The state transitions here are comparable to familiar, manageable disturbances such as those caused by crosswinds or rail grooves in the road.

Active steering requires no additional control elements because all of its subfunctions are activated automatically when the engine is started. If the internal combustion engine is not working (while the vehicle is being towed), the active steering, like conventional power steering, is disabled. Such situations are indicated by a lamp in the instrument cluster.

Benefits of active steering for the driver

- Driving errors are corrected or compensated for in such a way that the driver is not surprised by the vehicle's reaction.
- The steering ratio, which depends on the driving situation, makes it easier to maneuver the car at very low speeds, as fewer complete turns of the steering wheel are required with the same amount of effort.
- Greater convenience and comfort at high speeds, as the driver no longer needs to fear losing control of the car by accidentally applying too much steering force.
- The steering lead is another comfort and convenience feature. It allows more agile activation in response to the steering command.

Drive and adjustment systems

Power windows

Application

Power windows have mechanisms that are driven by electric motors. There are two types of systems in use (Fig. 1). The available installation space assumes a prominent place among the criteria applied in determining which system to install.

- *Rod linkage regulator mechanism*: The drive motor pinion engages with a quadrant gear, which is connected to a rod linkage. The use of this type of window regulator mechanism is decreasing.
- *Flexible cable mechanism*: The drive motor turns a cable reel, which operates a flexible cable mechanism.

Power window motors

Space limitations inside the door make narrow drive units imperative (flat motors, Fig. 2). The reduction gearing is a self-inhibiting worm gear design. This prevents the window from opening inadvertently of its own accord or being opened by force.

Dampers integrated in the gearing mechanism provide good damping characteristics at the end positions.

Power window control

The window is controlled manually by means of a rocker switch. For greater convenience, power windows can be linked to a central or decentralized closing system which, when the vehicle is locked, closes the windows automatically or sets them to a predefined partially open position for ventilation purposes.

When the windows are closed, a force limitation device comes into effect. The device serves to prevent human limbs from being caught by the closing window. In Germany, paragraph 30 of the StVZO (Road Traffic Licensing Regulations) stipulates that the force limitation device must be effective when the window is moving upwards within a range of 200 to 4 mm from the upper edge of the window aperture.

The power window drive units include integral Hall-effect sensors to monitor motor speed during operation. If a reduction in speed is detected, the motor's direction of rotation is immediately reversed. The window closing force must not exceed 100 N at a spring rate of 10 N/mm. The unit automatically overrides the force limitation function

Fig. 1

a System with rod linkage
b System with flexible cable

1 Electric gearmotor
2 Guide rail
3 Coupling
4 Rod linkage
5 Drive cable

Fig. 2

1 Reduction gear (worm gear pair)
2 Flat motor

1 Power windows

2 Power window motor with integrated electronics (force limiter)

immediately before the window enters the door seal, allowing the motor to run to its end position and permitting complete closure of the window. The window position is monitored over its entire range of movement.

Electronic control (Fig. 3) may be concentrated in a central control unit, or the control elements may be dispersed among the individual power window motors in order to reduce the complexity of the wiring. Future decentralized electronics will be networked via bus interfaces (LIN bus/CAN bus). The advantages offered are fault diagnosis for the electronics and further reduction of the amount of wiring.

Power sunroofs

Application

Power sunroofs can combine tilting and sliding sunroof functions. This requires

3 Power window control unit featuring electronic force limitation device

4 Power sunroof drive unit with electromechanical control of opening, closing and tilt functions

the use of special control systems that may be electronic or electromechanical in design.

Open-loop control

With electromechanical control (Fig. 4), mechanical interlocks on limit switches a) and b) ensure that the roof can either slide or tilt open from the closed position, depending on the polarity applied to Terminals 1 and 2. Once the sunroof has been tilted or opened, a polarity changeover will initiate the lowering or closing process. An electronic control unit featuring integral force limitation allows the sliding/tilting sunroof to be incorporated in a central locking system. The electronic control system includes a microcomputer responsible for evaluating incoming signals and monitoring sliding sunroof position. The closed and end positions for the sliding action of the sunroof are monitored with the aid of microswitches or Hall-effect sensors. Possible auxiliary functions include:

- Preset position control
- Closing via rain sensor
- Motor speed control
- Electronic motor protection

Drive systems

Drive for the roof is provided by Bowden cables or other torsion and pressure-resistant control cables. The drive motor is usually installed in the roof or at the rear of the vehicle (e.g. in the trunk). Permanently energized worm drive motors with outputs of approximately 30 W are used as the drive motors. They are protected against thermal overload by a thermostatic switch (phasing out) or (most frequently) by a software thermal protection switch.

Provision must also be made for ensuring that the roof can be closed using simple on-board tools (such as a crank) if the electrical system fails.

Fig. 3
1 Microcomputer
2 Relay output stage
3 Setpoint commands
4 Multiplex bus
5 Hall-effect sensors

Seat and steering column adjustment

Application

Electric power seats (Fig. 1) are still primarily installed in passenger cars in the mid-price and high-price ranges. The primary function of these systems is to enhance comfort, but multiple adjustment options combined with space restrictions and difficult access to the controls limit the practicality of manually operated mechanisms. Up to seven motors control the following functions:

- Seat cushion height adjustment, front/rear
- Fore/aft seat adjustment
- Seat cushion depth adjustment
- Backrest tilt adjustment
- Lumbar support adjustment, height/curvature
- Shoulder support tilt angle (top third of backrest)
- Head restraint height adjustment

Adjustment systems

A common seat bottom frame includes four motors, which drive the height adjustment gearing or the combined fore/aft and height adjustment gearing. The unit for adjusting the seat height is not included on simpler seat designs. Another concept features three identical motor-and-gear assemblies with four height and two fore/aft adjustment gear sets. The motor-and-gear assemblies drive the gear sets via flexible shafts. This is a highly universal system and can be installed on any seat design.

Modern seats (especially for sportscars) do not merely fasten the lap belt to the seat frame, they also attach the shoulder strap, including its height adjustment, inertia reel and seat belt pretensioner mechanism, to the backrest. This type of seat design ensures optimized seat belt positioning for a wide range of different occupant sizes at all available seat positions, and makes an important contribution to occupant safety. The seat frame must be reinforced for this type of design, while both the gear set components and their connections to the frame must be strengthened.

The optional programmable electric seat adjustment ("memory seat") can recall several preset seat positions. Potentiometers or Hall-effect sensors provide seat position feedback. On two-door vehicles, a feature designed to slide the front seat to its extreme forward position can be incorporated to help rear passengers enter the vehicle.

Electrically adjustable steering columns are also on the increase as another means of enhancing driver comfort. The adjustment mechanism, consisting of a single electric motor and self-inhibiting gear set for each adjustment plane, forms an integral part of the steering column. The gear set for fore/aft adjustment must be capable of absorbing all impact forces (crash forces) that might be applied to the steering column. The adjustment can be triggered in either of two ways, by the manual position switch or by the programmable seat adjustment. The column can also be tilted upward when the ignition is switched off to help the driver enter or exit the vehicle.

Fig. 1
Positioning units for
1 Curvature of the backrest
2 Angle adjustment of the backrest
3 Depth adjustment of the seat cushions
4 Height adjustment of the head restraint
5 Height adjustment of the seat
6 Fore/aft seat adjustment

1 Electromechanical seat adjustment (example)

Heating, ventilation and air conditioning

The vehicle's heating, ventilation and air conditioning systems have the following tasks:
- Providing a pleasant climate for all occupants of the car at different outside temperatures (Fig. 1).
- Ensuring good visibility through all windows and the windshield.
- Creating an environment calculated to minimize driver stress and fatigue.
- More recent units use filters to remove particulate matter (pollen, dust) and even odors from the air.

Electronic heater control

Function
In many countries, the performance of the heater unit is governed by legal requirements, with emphasis on the defroster's ability to maintain clear windows and windshields (such as EEC Directive 78/317 within the European Union, and MVSS 103 in the USA).

Method of operation
Variations in outside air temperature and vehicle speed cause fluctuations in the temperature of the passenger compartment. Without a control system, the heater controls must be constantly readjusted by hand in order to maintain an agreeable temperature. Electronic heater control dispenses with this requirement by automatically maintaining the interior temperature of the vehicle at the desired level.

With heater units featuring water-side control, sensors monitor the temperatures of the vehicle's interior and of the emerging air. The control unit processes this information and compares it with the preselected temperature. Meanwhile, a solenoid valve installed in the cooling circuit opens and closes at a given frequency in response to the signals which it receives from the control unit. Adjustments in open/close ratio in the cycle periods regulate the flow rate from the closed position up to the maximum.

An adjustment flap, actuated by an electric gearmotor, is the usual means of providing infinitely variable temperature control in air-side systems (pneumatic linear adjusters are also occasionally used). Sophisticated systems allowing separate adjustment for the left and right sides of the vehicle are also available.

Electronically controlled air conditioning system

Function
The heater unit alone is not capable of providing a comfortable environment at all times. When outside temperatures exceed 20°C, the required inside temperatures can be attained only by cooling the air using compressor refrigeration systems and the refrigerant R 134 (Fig. 2, next page). The moisture contained in the cooled air is also extracted as condensate, thus reducing the air's humidity to the desired level.

Automatic climate control is particularly useful for vehicles in which both air conditioner and heater are installed, because the constant monitoring and adjustment required to maintain a temperate climate presents the occupants with a complicated task. This rule applies to bus drivers in particular, as they can only feel the temperature at the front of the vehicle.

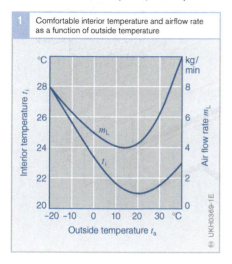

1 Comfortable interior temperature and airflow rate as a function of outside temperature

The task of automatic climate control systems incorporating a preselection feature is to automatically maintain the correct temperature, air flow, and air distribution in the passenger compartment. These parameters are mutually interdependent, and changes to one will affect the others.

Method of operation

The compressor, powered by the motor (Fig. 2, Item 1) compresses and heats the gaseous coolant. It then cools in the condenser (3) and becomes a liquid. The generated heat dissipates into the environment. An expansion valve (14) injects the cooled refrigerant into the evaporator (11), where it evaporates, and the fresh air inflow removes the required evaporation heat. The moisture is taken out of the cooled air as condensed water, dehumidifying the air.

At the center of the system is a temperature control circuit for interior temperature. The setpoint temperature (as described in "Electronic heater control") determined by the control unit is achieved by means of

water-side or air-side adjustments (Fig. 3). The fresh air (a) taken in by the fan (Item 1) is, depending on the temperature, either cooled by the evaporator (2) or heated by the heater core (4) and then, depending on the flap position, guided into the desired areas of the interior (b, c, f).

By means of various temperature sensors (3, 5, 7) the ECU (8) collects all important influencing and disturbance values and the temperature selected by the occupants on the setpoint controller (6) and uses all of this information to continuously calculate the setpoint value. This setpoint value is compared to the actual temperature, and the calculated difference generates reference variables in the ECU for regulating the heating (4, 11), cooling (2, 10) and air flow (1). Another function controls the position of the air distribution flaps (b, c, d, e, f) depending on the program the occupants have selected. Meanwhile, all control circuits continue to respond to manual inputs.

Infinitely variable or stepped blower control is used to adjust the air flow to the speci-

Fig. 2
1 Compressor
2 Solenoid clutch
 (for switching the
 compressor on/off)
3 Condenser
4 Auxiliary fan
5 High-pressure
 switch
6 Fluid reservoir with
 desiccant insert
7 Low-pressure switch
8 Thermostatic switch
 or on/off control (for
 compressor on/off)
9 Temperature sensor
10 Condensate drip
 pan
11 Evaporator
12 Evaporator fan
13 Fan switch
14 Expansion valve

2 Refrigerant circuit of an air conditioning system

▬▬▬ High pressure, liquid
▬▪▬ High pressure, gaseous
ⅢⅢ Suction pressure, liquid
═══ Suction pressure, gaseous

UKH0371-1E

fied level. Generally, the control system does not process the actual value in this operation.

This type of arrangement is inadequate for dealing with increases in flow rate caused by aerodynamic pressure at high driving speeds. Here, a special control function can compensate by responding to increasing vehicle speeds, initially by reducing the blower speed to zero, and then, if the flow continues to rise, by using a restriction flap to throttle the flow of incoming air. Air distribution to the three levels – defroster (b), upper compartment (c), and footwell (f) – is controlled manually, with preselection, or with a fully automatic program. Most popular are units featuring program control buttons, in which each button provides a specific air distribution pattern for the three levels.

The defroster mode ("DEF" setting) is a special case. In order to clear the windows as rapidly as possible, the temperature control must revert to maximum heat and maximum blower speed, while the air flow is directed through the upper defroster outlets. On systems with program switches and fully automatic units, this operating mode is selected with a single button; at temperatures above 0°C, the refrigeration unit is also activated to extract humidity from the air. To prevent unheated air from causing drafts, the blower is switched off electronically after cold starts in winter, except when "DEF" and the cooling function are in operation.

The variations described above are used in both passenger cars and trucks. Buses require more complicated layouts. The passenger compartment can be divided into several control zones, in which the temperature is controlled by electronically regulating the speed of the zone's individual water pump.

3 Air conditioner with electronic water-side control (schematic)

© UKH0370-2Y

Fig. 3

1 Fan
2 Evaporator
3 Evaporator temperature sensor
4 Heater core
5 Air exit temperature sensor
6 Setpoint controller
7 Interior sensor (ventilated)
8 ECU
9 Water drainage
10 Compressor
11 Solenoid valve

a Fresh air
b Defroster
c Ventilation
d Recirculated air
e Bypass
f Footwell

Vehicle security systems

Acoustic signaling devices

Application

The internationally applicable ECE Regulation No. 28 specifies that acoustic signals produced by motor vehicles must maintain a uniform sound quality with no perceptible frequency fluctuations during operation. Operation is permitted only as a hazard warning function. In countries where this regulation does not apply, the acoustic signaling device is a wearing part subject to heavy use. The use of sirens, signal bells and the like is not permitted, nor is playing melodies by triggering sound generators in a given sequence. Signaling devices must be installed facing forward in the vehicle and must produce the signal at a sufficient loudness level at a distance of 2 m. They are subjected to temperatures ranging from −40°C to +90°C and must be designed to be resistant to moisture, salt spray as well as mechanical shock and vibration. Elastic couplings must be used to decouple electric horns and fanfare horns from the vehicle body, as the horn would otherwise induce sympathetic oscillations in the adjacent bodywork. The resulting feedback would diminish both loudness level and tone quality. Both electric and electropneumatic horns and fanfare horns are sensitive to series resistors in their control circuits. When horns are installed in pairs, they should be triggered by relays. The warning produced by the impact horn is more conspicuous; therefore, it is the better choice for vehicles that are frequently used for long-distance overland truck traffic instead of fanfare horns. Alternatively, the fanfare horn is superior for city driving, as pedestrians often find the standard horn excessively loud and unpleasant. Both sets of requirements can be fulfilled by installing the two horn types together in a single system with a selector switch for city or overland traffic. Standard and fanfare horns both operate at standardized frequencies. High and low tones can be combined to produce a harmonious dual-tone sound.

Horns

The mass of the armature together with the flexible diaphragm forms an oscillating system within the *standard horn*. When voltage is applied to the solenoid coil controlled by a contact breaker, the armature impacts against the magnetic core at the horn's basic frequency. A fixed tone disk, directly attached to the armature, responds to these intense periodic impacts by radiating harmonic waves. Legal regulations stipulate that the maximum sound energy must lie within a band of 1.8-3.55 kHz. This frequency explains the horn's relatively piercing sound. This sound, which is essentially emitted along the horn's axis and outwards to the front, can be heard above the background noise in traffic, even over longer distances. The horn's size is one of the factors which determine the basic frequency and loudness level.

In addition to larger diameters, *supertone horns* have a higher electrical power input. Their warning signals can therefore still be heard under extreme conditions (e.g. truck driver's cab).

Fanfare horns

The electropneumatic fanfare horn employs the same basic actuating system as its standard counterpart. The main difference is that the armature is allowed to vibrate freely in front of the coil with no impact occurring. The oscillating diaphragm induces vibration of an air column within a tube. The resonant frequency of the diaphragm and the air column are tuned to each other and determine the signal pitch. The tube is in the shape of a funnel, with a wide opening to enhance the efficiency of sound propagation. The funnel tube is generally coiled to minimize the unit's size.

The presence of upper harmonics in the lower range of the frequency spectrum provides the fanfare horn with a rich, melodious sound. The tone is less penetrating than that of a standard horn due to the more consistent distribution of sound energy over a wide spectrum.

Central locking system

Application

The central locking and unlocking of the vehicle doors, trunk and fuel filler flap are an integral part of the convenience system in motor vehicles. Electric or pneumatic actuators operate the mechanisms in the individual locking devices.

Locking systems

In pneumatic systems, a central electric dual-pressure pump (vacuum/gauge pressure) and dimensionally stable lines leading to the diaphragms of the individual actuators provide the opening and closing movement of the locking system.

The electromechanical central locking system (Fig. 1) features an electric servomotor to drive the locking function at each point in the system. This electric motor with a reduction gear drive unit actuates a mechanical connection (rod, lever) responsible for the locking device (Fig. 2).

Open-loop control

All central locking systems feature a control unit for signal processing. The control unit is integrated in the housing in pneumatic dual-pressure pump systems. In electromechanical systems, it is located centrally, or to minimize wiring, decentrally in the door modules in connection with multiplex systems. The lines associated with decentral or distributed door multiplex systems have

multiple functions, i.e. they are also used for the power window drive units and door mirror adjustment. Electromechanical and pneumatic central locking systems are actuated via electrical contact switches in the actuators or locking cylinders in the driver's door, passenger's door, trunk lid, and operating switches in the vehicle interior. Infrared and radio remote control systems enhance operating convenience.

In newer comfort and convenience systems, the central locking system is locked on starting off (vehicle speed signal) to prevent the vehicle from being opened from the outside. The system is unlocked in the event of an accident (crash sensor).

A distinction is made between opening, closing and locking. In connection with the opening and closing functions, manual operation from the inside (e.g. occupants) is possible at any time. When the system is

1 Central locking system with electric motor

Fig. 1
1 Central switch
2 Contacts in door
 lock mechanisms
3 Control unit
4 Servomotors

2 Central locking mechanism servo unit

Fig. 2
1 Wiring connection
2 Flexible end position
 coupling
3 Gearing
4 Electric motor
5 Adjusting lever
h Stroke

locked, the locking mechanism is blocked as theft deterrence and can only be operated with the vehicle key from the outside or with the remote control.

Locking systems

Application

The function of locking systems is to ensure and maintain access authorization at all times. Although lock mechanisms in motor vehicles are subject to substantial wear and tear as a result of prolonged cold, wet and dirty conditions over several years, their main design feature must be to reliably protect the vehicle and its occupants while ensuring smooth and efficient operation.

Structure

The structure of a locking system is made up of the following components (Fig. 1):
● Locking bars on the body pillars
● Side doors, door lock mechanisms, and the associated mechanical and electrical parts
● Electrical components that make up the access authorization facility and radio remote control (frequently assigned to locking systems).

Depending on their location in the vehicle, a further distinction is made between the following assemblies (with differing scope of functions):
● Side door assembly
● Trunk assembly
● Hood assembly

The most important component of the locking system is the door lock mechanism (Fig. 2). Its primary functions include:
● Transmitting the structural forces between the door and the vehicle body
● Reliably opening and closing the locking mechanism
● Evaluating the mechanical or electrical commands
● Storing the logic statuses (mechanical computer).

Method of operation

Accordingly, the lock mechanism is made up of assemblies that perform the following functions:

The locking mechanism is responsible for transmitting forces to initiate the opening and closing functions. It consists of a catch, pawl and latch.

Fig. 1
■ Mechanical connection
— Electrical signals

1 Structure (example: side door)

Side door

| Lock barrel (key) | | Locking element (knob) |

Exterior operating element (handle) | Interior operating element

Door lock | Electronics (option)

Actuators

Mechanical logic unit

Lock stirrup (B-pillar) | Data bus | Power supply

UKT0068-1E

Fig. 2
1 Backplate with locking mechanism
2 Catch
3 Electrical interface
4 Bowden cable to inner handle

2 Mechanical door lock

4 1 2 3

UKT0069-1Y

The corresponding door lock striker is mounted on the body pillar. During the closing procedure, the door lock striker engages in the latch. This centers the door, while the catch holds it in the closed position (main catch). In turn, the pawl then positively locks the catch in position.

This lock status is initially canceled to initiate the opening procedure. For this purpose, the actuating forces applied at the inner or outer door handles are transmitted to the pawl. The pawl now releases the catch to allow the door to be opened. The door lock striker is fixed in position while the catch swivels into the open position.

Mechanical locking system
Logic statuses of the lock mechanism
The logic unit of the lock mechanism permits the mechanical decoupling or isolation of the outer door handle, the inner opener and the locking element. For instance, with the door locked, the outer door handle can be freely pulled with no response, and it is consequently not possible to open the door from the outside.

In addition, the inner opener (opening from inside) and the locking element (unlocking from inside) are inoperative in connection with a theft deterrence system.

The rear door locks cannot be opened from the inside when the childproof lock is engaged. However, the occupants are able to unlock the door from the inside to allow a person providing assistance to open the door from the outside. (Note the difference compared to the theft deterrence system!) The following status matrix (Table 1) illustrates the basic functions of one rear door variant.

1 Basic function of a rear door variant ("single-stroke ejection" type) Table 1

	ODH	BUS	CLC	BUTTON	IDH	
	Outer door handle	LIN or CAN	Central locking switch	Locking button	Inner door handle	
Command	Open	–	–	–	Open	
	–	Unlock	Unlock	–	Unlock (single-stroke ejector)	
	–	Lock	Lock	–	–	
	–	Activate theft deterrent	–	–	–	
Status						DS Theft deterrent activated
						VR locked
						ER unlocked
			OPENED			

Design features

The lock mechanism is located in the wet area inside the door. The adverse effects of water and dust, the considerable impact load that occurs when the door is slammed, and the theft deterrence requirements necessitate a sturdy, tough design.

Currently, the typical locking parts and lever mechanisms (Fig. 3) are made of steel materials (precision punched parts, springs, riveted connections). Plastic components are being used to an ever greater extent. However, they must fulfill their function even after a side impact or in the event of a vehicle fire. Corrosion, friction and wear dictate that the metal components undergo a comprehensive surface treatment process (vibratory grinding, galvanizing, coating).

The parts of the locking mechanism are sheathed in plastic materials to provide effective sound deadening in order to satisfy requirements to minimize closing noise. Higher-grade lock mechanisms normally feature extensive sound deadening measures incorporated in the lever mechanisms and actuators.

In the past, operation of automotive locks was purely mechanical. Central locking was then introduced in the form of flange-mounted locking actuators (electrically and pneumatically operated control elements).

These actuators are now integrated in the housing of more recent lock mechanisms. Miniature motors carry out the electrical commands from the access authorization facility in the form of mechanical actuating movements. The locking status is stored mechanically in the lever mechanisms.

The output from the high-speed drive units is converted in worm gear, multistage spur gear or planetary gear mechanisms (plastic materials). Spindle drive systems are also used.

Sensors in the lock mechanism permit the electrical evaluation of the lock status. Hall effect sensors and microswitches have proven particularly suitable for this purpose.

Open-by-wire

The widespread use of electrical systems in motor vehicle applications initially gave rise to the central locking system, which was soon followed by the theft deterrence system and radio remote control. The childproof lock is now switched electrically. Some door closing and locking systems employ motors to actuate the catch or the door lock striker (servo closing).

The obvious next step, therefore, is to use an electric motor for the opening function, i.e. to swivel the pawl ("open-by-wire", Fig. 4).

Consequently, development of the "x-by-wire" function for lock mechanisms has been consistently pursued.

The main advantage of the "open-by-wire" function is its interaction with access authorization, particularly in connection with the use of "passive entry" systems, as they require either extremely fast unlocking actuators or "overtake" solutions.

Based on the "overtake principle", the actuator begins the opening procedure in the locking mechanism before the lock logic has been completely released. This function is

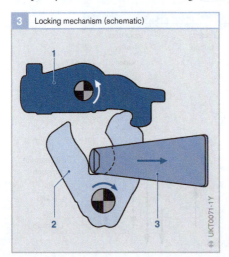

3 Locking mechanism (schematic)

Fig. 3
1 Pawl
2 Catch in lock
3 Door lock striker on
 B-pillar or C-pillar

intended to avoid waiting times for the user. Corresponding design measures reliably rule out the possibility of being locked in.

Electrical locking system

Demands for quality and reliability as well as the pressures of rising costs render it necessary in the future to minimize the number of mechanical parts in the locking system and replace them by electrical components. The end point of the development is the electrical locking system (Fig. 5).

The lock then consists of only the locking mechanism, opening actuator and the associated electronics. The door handles and other operating elements are equipped with sensors. Electrical wires replace the mechanical connections to the lock.

The advantages of an electrical locking system are considerable:
- Reduced size and weight of the lock
- Symmetrical design
- Only one lock variant per vehicle (variant encoding at the end of the assembly line)
- Door handles no longer move (or can be dispensed with completely)

Auxiliary functions such as interior lighting, status indicators and a whole range of other functions will then be easy to implement in electrical locking systems, as the lock is equipped with electronics that can perform these functions.

Communication with the locks, access authorization and power supply all take place via the data bus interfaces. Prognoses using fault tree analysis indicate that the electrical locking system is just as reliable as conventional systems. Economically feasible systems can be expected with the introduction of the 42-V vehicle electrical system at the latest (active redundant power supply).

4 Open-by-wire (schematic)

UKT0072-1Y

Fig. 4
1 Electric motor
2 Pawl
3 Catch
4 Door lock striker

5 Electromechanical door lock (schematic)

UKT0073-1Y

Fig. 5
1 Locking mechanism
2 Electronics
3 Electric motor
4 Gearing
5 Electrical interface

Biometric systems

Function

"Biometric systems" have the task of determining or confirming the identity of persons on the basis of "biometric" features. Currently, approximately ten biometric methods are known, including fingerprint identification, face recognition, iris recognition and voice recognition.

Biometric systems in motor vehicles

Fingerprint identification is becoming established in automotive applications, as criminology has developed a fundamental understanding of the techniques of fingerprint distinction.

As fingerprint identification will in the future replace passwords on computers, in time management and access control systems, and on mobile phones as well, the development of sensors and algorithms has already become widespread. Due to the confined space available on board a motor vehicle, there is also limited space for installing the sensor.

Biometric features

Biometric features are features that are uniquely characteristic to a particular person. They may either be physiological features (such as the pattern of a fingerprint) or behavioral characteristics (such as the way a person walks or writes his/her signature). The advantage of biometric systems is, firstly, for the greater convenience of the user. Since biometric features are part of a person's individuality, they are carried by the person at all times and cannot be forgotten or lost. There is no need to carry a key or a transponder card. Secondly, they offer the advantage of an attainable enhancement in security. Since biometric features are intrinsic and unique to a particular individual, they can neither be deliberately given to others nor stolen in the traditional sense of the word. Therefore, a high level of reliance can be placed on whether the vehicle really is being driven by a known driver.

Enrollment

In order to be able to recognize a user, a biometric system must first of all learn the fingerprint pattern of that person. This process is known as "enrollment". It involves the user placing his/her finger on a fingerprint sensor that generates a grayscale image (typical size: 64,000 to 96,000 pixels at a resolution of 8 bits/pixel, Fig. 1).

Using signal-processing algorithms, a processor calculates the characteristics features within that image, e.g. branches or nodal points in the line pattern. The biometric system then stores these features – though not as an actual fingerprint pattern – in a permanent database, typically in an EEPROM memory (capacity requirement: 250 to 600 bytes per fingerprint).

When the biometric system checks a person's fingerprint pattern the next time, it again identifies the characteristic features of the fingerprint and then searches the database for a matching set of features. If it finds a match, the person is recognized as an authorized user.

Examples of application

Many vehicles already have programmable seat adjustment functions. The control panel typically has three numbered memory buttons, which can be assigned to three different drivers, and a teach-in button. The drivers of the vehicle will have to select a memory button for each driver and then remember which button contains their own settings. The number of drivers that can use the system is limited by the number of memory buttons.

The use of a biometric system significantly enhances the level of personalized convenience. In this system, a fingerprint sensor replaces the memory buttons for the programmable seat adjustment system (Fig. 2). Instead of pressing a memory button to recall his/her personal settings, the driver places a finger on the sensor. Since biometric features are intrinsic to a particular individ-

ual, the biometric system can always identify a particular driver reliably and apply the correct personal settings.

If a previously unknown driver places his/her finger on the sensor to store his/her personal settings, the biometric system automatically performs an enrollment sequence and then memorizes the current settings. Once a driver is known to the biometric system, only the current settings are stored.

The advantage for the user lies in the simplified human-machine interface. Users no longer have to agree on the allocation of memory buttons, nor do they have to remember which button is theirs. In addition, the maximum number of users is not limited to the number of memory buttons available. Ultimately, the number of users is limited by the biometric system's memory capacity for user identities and personal settings.

Personalization can be extended beyond seat adjustment, mirror adjustment and steering column adjustment functions. Theoretically, all configurable vehicle systems can be connected to the biometric personalization system. It could be conceivable to personalize air conditioning and automatic transmission settings (sports/economy). For the radio, a set of personal favorite stations and preferred volume and tone settings could be selected, while there could be list of favorite destinations for the navigation system. Similarly, if the vehicle has configurable displays, a driver's personal preferences for the display elements could be recalled.

In the case of self-teaching assistance systems which adapt to the driver's driving patterns, the "last known settings" regarding driver characteristics could be retrieved by the personalization system before starting off.

Biometric systems can also be used even before starting the engine as a means of implementing a vehicle immobilizer and for vehicle access. Vehicles equipped with this technology could then be driven entirely without the use of keys. With regard to the use as a vehicle immobilizer, it has not yet been definitively demonstrated that the same degree of security is attainable as with the present transponder-based immobilizers.

For vehicle access applications, the sensors for identifying the fingerprint pattern would have to be integrated in the vehicle's outer skin. The technical challenge is then to design a sufficiently resilient sensor system that is adequately protected against external conditions so that effective usability is guaranteed at all times – including in winter.

1 Fingerprint sensor based on the direct optical scanning (DiOS™) method (example)

2 Seat adjustment via fingerprint sensor

Electromagnetic compatibility (EMC) and interference suppression

Electromagnetic compatibility (EMC) consists of two elements. One is understood as the ability of a device to continue providing reliable service when exposed to electromagnetism from external sources. The second aspect focuses on electromagnetic fields generated by the same device; these should remain minimal in order to avoid creating interference that would impinge upon the quality of radio reception, etc. in the vicinity.

Today's vehicle contains a multitude of systems whose functions are implemented by electrical or electronic components. These same functions, if present at all, used to be carried out fully or predominantly mechanically. With the exception of two-way transceivers used in special-purpose applications, a car's radio was once the only device in which signal reception was of any importance.

In contrast, today's vehicles feature a host of devices that rely on the reception of electromagnetic radio waves. Mobile phones, navigation systems, theft-deterrent systems with radio remote control, and integrated fax and PC units are now being installed and used in vehicles. This trend has led to a commensurate increase in the importance of suppressing interference and ensuring reception in the vehicle of the operationally vital radio waves.

EMC ranges

Transmitter and receiver

All vehicle systems must remain impervious to electromagnetic radiation emitted from such external sources as extremely powerful radio transmitters. In other words, there must be no threat to the motor vehicle's operational integrity, and malfunctions representing a potential source of driver irritation are unacceptable. Another consideration is that stationary receivers should remain unaffected by passing traffic. Both considerations are governed by national and international codes (EU Directives, German StVZO).

Electrical and electronic components

Motor vehicles contain an extensive array of electrical and electronic components including servo and fan motors, solenoid valves, electronic sensors and control units with microprocessors. These devices must rely on a single vehicle electrical system for their power supply. It is thus vital to avoid mutual interference and feedback phenomena generated by one or several systems so that these do not cause malfunction.

On-board electronic systems

Mobile communications equipment – such as the radio – also exists within an interlinked environment including all of the vehicle's electronic systems. Every device is powered via the same vehicle electrical system, and its reception antennas are located in the immediate vicinity of potential radio-interference sources. These considerations make it imperative that strict limits are imposed on the levels of interference emitted by on-board electronic systems. Compliance with official regulations is essential, and it is important to maintain interference-free reception in the vehicle even when conditions are well below optimal.

EMC between different systems in the vehicle

Shared vehicle electrical system

All of the motor vehicle's electrical systems rely on a single shared vehicle electrical system. Because the electric lines leading to the individual systems are frequently combined within a single wiring harness, feedback pulses can easily travel from one system to the I/O ports of its neighbor (Fig. 1).

This transfer of interference can be in the form of signal pulses (abrupt, steep jumps in current and voltage) generated during the switching on and off of various electrical components such as electric motors, solenoid valves and actuators. Yet another source is the ignition system's high-tension circuit. Similar to other interference signals (for instance the ripple on the power supply), these signal pulses can propagate through the wiring harness. These interference pulses then proceed

to the I/O ports of adjacent systems either directly, through shared conductors such as the power supply (galvanic coupling), or indirectly, through capacitive and inductive coupling stemming from electromagnetic emissions.

Galvanic coupling

Currents for two different electric circuits (such as a solenoid valve's trigger loop and the electric circuit for assessing sensor data) flowing through a single conductive path (common ground through the vehicle chassis, etc.) will both generate a voltage owing to the consistent resistance in the shared conductor (Fig. 2a, next page). Continuing with this sample scenario, the voltage produced by radio-interference source u_1 has the effect of a supplementary signal voltage in signal circuit 2, and could lead to erroneous interpretation of the sensor signal. One remedy is to use separate return lines for each electric circuit (Fig. 2b, next page).

1 Mutual interference between two systems as transmitted through the shared vehicle electrical system (A) and wiring harnesses (B and C)

UAE0240-1Y

Fig. 1

System I:
1 Control unit
2 Actuator
3 Sensor

System II:
4 Control unit
5 Actuator
6 Sensor

Capacitive coupling

Capacitive coupling allows variable-peri-odicity signals such as pulse voltages and sinusoidal alternating voltages to produce interference and crosstalk in adjacent electric circuits, even without the exis-tence of a direct physical link (Fig. 3). The potential level of capacitive (interfer-ence) voltage is proportional to such fac-tors as the closeness of the neighboring conductor paths and the rise rate of the pulse-shaped voltage shifts (or the fre-quency of the alternating voltage). The first step is to separate the conductive paths while at the same time extending the ramp periods during which signals rise and fall (or to limit the frequency of the alternating voltages to the absolute minimum required for the function).

Inductive coupling

Currents recurring with variable periodicity in one conductor can induce voltage pulses in adjoining circuits. These voltage pulses then generate current in the secondary cir-cuit (Fig. 4). This is the inductive principle exploited in transformer design. One prime factor defining susceptibility to overcou-pling is the signal's rise and fall time (or al-ternating voltage frequency), reflecting the situation encountered with capacitive cou-pling. Also significant is the effective mutual inductance, as determined by such factors as the size of the wires and their relative routing. Strategies for avoiding inductive coupling include minimizing the dimensions of circuit wires, keeping critical circuits as far apart as possible from each other, and the avoidance of parallel conductive paths. The tendency toward inductive interference is especially pronounced in circuits carrying low-frequency signals (e.g. coupling into loudspeaker wiring).

Fig. 2
a Electric circuit
 with shared return
 conductor path
b Electric circuit with
 separate return
 lines

u_1, u_2 Voltage source
Z_i Internal resistor
Z_a Terminal resistor

Fig. 3
1 Electric circuit 1
2 Electric circuit 2

u_1 Voltage source
Z_i Internal resistor
R_E Input resistance
C_E Input capacitance
$C_{1,2}$ Capacitance
 between two
 conductor paths
u_s Radio interference
 voltage

Fig. 4
1 Electric circuit 1
2 Electric circuit 2

u_1 Voltage source
u_2 Voltage source
Z_i Internal resistor
Z_a Terminal resistor
L_1, L_2 Inductance of
 conductors
$M_{1,2}$ Inductive coupling
u_s Radio interference
 voltage

2 Galvanic coupling of interference signals

3 Capacitive coupling of interference signals

4 Inductive coupling of interference signals

Pulsation in the vehicle electrical system

One strategy for dealing with interference pulses in the vehicle electrical system entails limiting the amplitude of the interference emanating from the radio-interference sources. On the other hand, the susceptible devices (affected electronic components) are designed for insensitivity to pulses of specific shapes and amplitudes. The initial step was to list and classify the pulses encountered within vehicular electrical systems (Table 1). Special-purpose pulse generators can be used to generate the test patterns defined in Table 1 as a basis for assessing resistance to these interference waves in susceptible devices. Both the test pulses and the test technology are codified in standards (DIN 40 839, Section 1; ISO 7637, Section 1) including definitions of the measurement technology for evaluating emissions of pulse-pattern interference. Classifications based on pulse amplitude levels facilitate effective definition of radio-interference sources and the susceptible devices (interference receptors) within each vehicle. It would thus be possible to specify Class II for all of the radio-interference sources

within a vehicle while designing all the susceptible devices (e.g. control units) to comply with and exceed – by a certain safety margin – Class III. If suppressing interference at the source proves to be cheaper or to involve less technical complexity than reducing sensitivity at the receptors, the logical response might be to shift the definitions to Classes I/II. If the scenario is inverted, with the shielding of potential susceptible devices as the cheaper and simpler solution, then a move to Classes III/IV is warranted.

Because numerous wires are combined within a single wiring harness, each individual conductor is potentially susceptible to inductive and capacitive interference. Although reduced in intensity, the resulting voltage pulses in adjacent supply lines can then appear as spurious signals at the input ports and control outputs in neighboring systems. The test technology for simulating crosstalk interference within wiring harnesses (as defined in DIN 40 839, Section 3 and ISO 7637, Section 3) uses a standardized substitute wiring layout (capacitive clip) with a defined wiring capacitance. Test pulses are fed into this

| **1** Mutual interference within power supply |||||||||
|---|---|---|---|---|---|---|---|
| **Test pulses as defined in DIN 40 839, Section 1** |||| **Max. pulse amplitude classes** ||||
| Pulse pattern | This is caused by | Internal resistance | Pulse duration | I | II | III | IV[1] |
| 1 | Shutdown of inductive consumers | 10 Ω | 2 ms | −25 V | −50 V | −75 V | −100 V |
| 2 | Shutdown of motorized consumers | 10 Ω | 50 µs | +25 V | +50 V | +75 V | +100 V |
| 3a | Steep over-voltages | 50 Ω | 0.1 µs | −40 V | −75 V | −110 V | −150 V |
| 3 b | | | | +25 V | +50 V | +75 V | +100 V |
| 4 | Voltage curve during starting operation | 10 mΩ | to 20 s | 12 V −3 V | 12 V −5 V | 12 V −6 V | 12 V −7 V |
| 5 | Alternator load dump [1] | 1 Ω | to 400 ms | +35 V | +50 V | +80 V | +120 V |

Table 1

[1] "Load dump", i.e. when the alternator is feeding high current to the battery and the connection to the battery is suddenly interrupted.

layout and through the specimen's wiring harness to produce overcoupling in the signal and control lines. The effects of low-frequency oscillations within vehicular electrical systems can be simulated by producing the desired signals with a signal generator and projecting these into the wiring harness through a current clamp. This process reflects the procedure described above by serving as the basis for the correct balance between the amplitude of radiated interference pulses and the resistance to interference of the potentially susceptible devices (receptors).

Effects on mobile radio reception of high-frequency signal feedback in the vehicle electrical system
Along with the pulses and other interference signals described above, high-frequency signals constitute yet another source of undesired interference within the vehicle electrical system. These signals can stem from periodic switching operations, as found in high-intensity ignition systems, at the commutators in DC motors and from the CPU clock signals that are generated in microprocessor-equipped control units. The interference these sig-

Fig. 5
a As a function
 of time
b As a function
 of frequency

T Period duration
T_r Rise time
T_i Pulse duration
$f_0 = T^{-1}$ Fundamental
 wave
f_g Fundamental
 frequencies
f_{min} Periodic minima
H Envelope curve

5 Voltage amplitude

Fig. 6
a Wide-band
 interference
b Narrow-band
 interference

6 Interference-signal spectrum

nals can induce in mobile communications receiver equipment may impair reception or even render it impossible.

Spectrum
Pulsation within on-board electrical systems is usually based on the analysis of current and voltage progressions as function of time (Fig. 5a). Amplitudes at specific frequencies are generally viewed as the prime criterion for evaluating interference signals affecting radio reception (Fig. 5b). Under standard conditions, the interference signals encountered in the automotive environment are rarely isolated sinusoidal waves with single amplitudes recurring at consistent frequencies. Much more common are superimpositions consisting of numerous oscillation components with a variety of amplitudes and frequencies. The "spectrum" for an interference signal is a portrayal of amplitude versus frequency designed to facilitate evaluation of interference potential in individual wavebands (Figs. 6a and 6b). Table 4 (refer to section "Interference-suppression categories") lists the most significant wavebands for automotive applications.

Interference test technology divides interference signals into wide-band and narrow-band interference: a spectrum with a continuous, progressive curve (Fig. 6a) is indicative of wide-band interference, and identifies the corresponding radio-interference source as a wide-band interferer. A contrasting pattern composed of isolated spikes forming a so-called line spectrum indicates narrow-band interference stemming from narrow-band emitters (Fig. 6b).

Initially, the classifications in either category are arbitrary: among the factors that determine whether interference is emanating from a "wide-band" or "narrow-band" radio-interference source are the reception properties of the potentially susceptible devices as reflected in the characteristics of the test equipment employed to monitor the emissions.

In the case of radio interference this equipment is in the form of a selective laboratory receiver or spectral analysis unit. Similar to a radio tuner therefore, the tester thus measures the signal amplitude only within a specific narrow frequency band (receiver bandwidth). The test progresses through the entire relevant frequency range using a procedure analogous to the station search function in standard radios: while maintaining the initial bandwidth, the laboratory receiver proceeds through a range of frequencies, either at graduated intervals or in a continuous progression.

Interference signals recurring at frequencies lower than the test bandwidth produce the continuous signal pattern indicative of wide-band interference. If the frequency is higher than the test bandwidth, the test monitor will also pick up gaps in the spectrum, and the result will be the line spectrum that indicates narrow-band interference.

Electric motors are a typical source of wide-band interference. Commutation frequencies in electric motors are located around just a few 100 Hz, with the exact figure depending on the motor's number of poles and its speed. At a test bandwidth of 120 kHz (bandwidth corresponding to an FM radio tuner receiver bandwidth) this produces a continuous spectrum. At the same test bandwidth, a 2 MHz cycle frequency (of the kind that might be encountered in a microprocessor-equipped control unit) will generate a completely different spectrum, this time with the spiked line spectrum typical of narrow-band interference (interference signal recurs at frequency greater than test bandwidth).

While all electric motors – fans, windshield wipers, servo units and fuel-supply pump, etc. – join the alternator as typical emitters of wide-band interference, yet another potential emitter is the high-tension ignition circuit. In addition, low-frequency clock signals stemming from devices such as switching elements can produce wide-

band interference. The list of narrow-band interference emitters includes the microprocessors in control units as well as all other devices using high-frequency control signals.

Interference signals picked up by radio receivers may also be of the so-called conducted interference type, that is they may propagate through wiring and conductors (such as the radio's power supply) or they can enter the receiver's I/O ports via inductive and capacitive radiation in the wiring harness. However, interference usually enters through the antenna, either through coupling in the antenna cable, or because the antenna picks up the electromagnetic fields generated by radio-interference sources. One particularly effective transmission antenna is the wiring harness. Other factors influencing the amplitude of the interference signals picked-up by the receiver are the vehicle-body structure and the type and location of the antenna.

Fig. 7
Terminals:
P-B Test
A-B Power supply
M-B Radio-interference monitor
S Switch
B Reference ground (metal sheet, shielding for simulated vehicle electrical system)

Measuring incident interference

Test techniques (DIN 57 879/VDE 0879, Sections 2 and 3; CISPR 25) are stipulated for monitoring interference transmitted through wiring and antennas. Research on individual components is carried out using

7 Schematic diagram of simulated automotive electrical circuit

UAE0686-1Y

2 Interference-suppression levels: permissible radio-interference voltage limits for individual frequency ranges in dBµV for wide-band (B) and narrow-band (S) interference as defined in CISPR 25 (DIN/VDE 0879-2)

Interference-suppression levels	Interference-suppression levels									
	0.15 to 0.3 MHz (LW)		0.53 to 2.0 MHz (MW)		5.9 to 6.2 MHz (SW)		30 to 54 MHz		70 to 108 MHz (VHF)	
	B	S	B	S	B	S	B	S	B	S
1	100	90	82	66	64	57	64	52	48	42
2	90	80	74	58	58	51	58	46	42	36
3	80	70	66	50	52	45	52	40	36	30
4	70	60	58	42	46	39	46	34	30	24
5	60	50	50	34	40	33	40	28	24	18

Table 2

3 Permissible radio interference-field strength in dBµV/m for interference-suppression levels in individual frequency ranges according to DIN/VDE 0879, Section 2, or CISPR 25 for wide-band interferers, measured with quasi-peak detector (B), and narrow-band interference, measured with peak detector (S).

Interference-suppression level	Interference-field strength level																	
	0.15 to 0.3 MHz (LW)		0.53 to 2.0 MHz (MW)		5.9 to 6.2 MHz (SW)		30 to 54 MHz		68 to 87 MHz		76 to 108 MHz (VHF)		142 to 175 MHz		380 to 512 MHz		820 to 960 MHz	
	B	S	B	S	B	S	B	S	B	S	B	S	B	S	B	S	B	S
1	83	61	70	50	47	46	47	46	36	36	36	42	36	36	43	43	49	49
2	73	51	62	42	41	40	41	40	30	30	30	36	30	30	37	37	43	43
3	63	41	54	34	35	34	35	34	24	24	24	30	24	24	31	31	37	37
4	53	31	46	26	29	28	29	28	18	18	18	24	18	18	25	25	31	31
5	43	21	38	18	23	22	23	22	12	12	12	18	12	12	19	19	25	25

Table 3

laboratory setups and standardized test setups in shielded chambers. Interference is monitored with a laboratory receiver. Line lengths and other geometric dimensions are specified. The power supply must be provided by a precisely defined vehicle electrical system. Voltage sources are also precisely defined, with laboratory specimens being powered from simulated vehicle electrical systems (Fig. 7).

Interference-suppression categories

Similar to pulsation in the vehicle electrical system, narrow- and wide-band interference factors are classified by interference-suppression category in order to facilitate selection and design for specific applications. Within this classification system, demands on sporadically active radio-interference sources are less stringent than the requirements placed on components in continuous operation, such as the alternator. The permissible CISPR 25 (DIN/VDE 0879-2) radio interference levels are listed in Table 2. Table 3 indicates the permissible interference-field strengths for radiation measurements with antennas.

Narrow-band interference of the kind generated by CPU clock signals in control units poses an especially acute problem for radio reception. These interference signals are always present (the control

units are usually switched on along with the ignition). Also of significance is the fact that the radio receiver cannot distinguish these spurious signals from useful signals, making reception of weak transmissions impossible. This situation is reflected in the definition of interference-suppression categories. In a given interference-suppression category, narrow-band interferences are assigned lower permissible interference levels than wide-band interferences.

Because individual vehicle configuration also has a substantial effect on the quality of broadcast reception, the interference-suppression data derived from laboratory testing must be confirmed in a practical automotive environment. The test involves measuring antenna voltage at the end of the antenna cable to which the radio receiver will subsequently be connected.

CISPR 25 also prescribes limits for radio interference voltage as determined using this procedure (Table 4). It contains voltage levels defined to reflect the unfavorable reception conditions as encountered in motor vehicles, where the strengths of useful signals not only fail to exceed just a few mV, but also suffer from substantial fluctuation owing to vehicle motion and the multiple reception paths generated by signal reflections.

Table 4
QP-B Quasi-peak detector relays aural impression produced by interference
B Wide-band interference emitter with peak detector indicates maximum level
S Narrow-band interference emitter with peak detector indicates maximum level
* Limit values for high-voltage ignition

4 Limits defined for permissible radio interference voltage at vehicle antenna in dBµV

Frequency range	Frequency	Continuous wide-band interference		Sporadic wide-band interference		Narrow-band interference
	MHz	QP-B	b	QP-B	b	S
LW	0.14 to 0.30	9	22	15	28	6
MW	0.53 to 2.0	6	19	15	28	0
SW	5.9 to 6.2	6	19	6	19	0
2-way transceivers	30 to 54	6 (15*)	28	15	28	0
2-way transceivers	70 to 87	6 (15*)	28	15	28	0
VHF	87 to 108	6 (15*)	28	15	28	6
2-way transceivers	144 to 172	6 (15*)	28	15	28	0
C-network car phone	420 to 512	6 (15*)	28	15	28	0
D-network car phone	800 to 1000	6 (15*)	28	15	28	0

Interference from electrostatic charges

ESD, or electrostatic discharge, can damage or even destroy electronic components. The voltages produced in the event of this discharging can be anywhere up to several volts, which in turn also generate very high, pulsed currents. For this reason, suitable measures are required to prevent the destructive effect or, better still, the discharges themselves. Electronic components that come into contact with vehicle passengers are particularly endangered.

ISO TR 10605 defines test procedures for examining the effects of electrostatic discharge and the immunity to interference that components exhibit both in the laboratory and when installed in the vehicle. The test is conducted with a suitable ESD test-pulse generator, usually in the shape of a pistol. The generator produces the high-voltage pulses that are then coupled into the test specimen.

EMC between the vehicle and its surroundings

Since early 1996, an official regulation has placed mandatory limits on radiation by the vehicle of interference affecting stationary radio reception, while at the same time defining suppression requirements during exposure to external electromagnetic fields. This ordinance (European Directive 95/54/EC) superseded an earlier directive of more limited scope, which only limited interference emissions while defining the homologation certification procedure governing EMC in motor vehicles.

Emitted interference

Electromagnetic emissions radiated by motor vehicles must remain within the limits for narrow and wide-band signals (Fig. 1) to ensure that they do not interfere with radio and television broadcasts or private transmissions. These limits are specified in the Directive 95/54/EC cited above as well as in VDE 0879, Section 1 and the CISPR 12 standard. These regulations contain detailed descriptions of the test procedure, which employs antennas set up at defined distances from the vehicle (10 meters, 3 meters).

The specifics of the test procedure are described in the regulations quoted.

In practice, it is the ignition system that is usually responsible for the greatest amount of interference radiation. However, since a comprehensive range of measures have been taken to ensure good radio reception in the vehicle, the emitted interference is limited to such an extent that compliance with the legal limit values is usually comfortably achieved.

Incident radiation

As a vehicle travels through the high-intensity electromagnetic field (near field) immediately surrounding a high-intensity emissions source, radiation penetrates through gaps and apertures in the vehicle

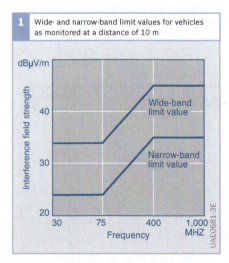

1 Wide- and narrow-band limit values for vehicles as monitored at a distance of 10 m

body and interferes with on-board electrical systems. The intensity of this effect (incident radiation) varies according to component locations, body configuration and the wiring harness.

Vehicle measurements

At one time the procedure for verifying that various electronic systems would continue to provide satisfactory performance under exposure to high-intensity fields involved driving the motor vehicle to a number of different broadcast transmitter locations. Now, special test chambers are available for this purpose.

These chambers must be surrounded (shielded) with a metallic casing to prevent the electromagnetic field generated during testing from radiating to the outside. In addition, the chambers must be equipped with absorption materials to inhibit formation of standing waves (nodes and antinodes), which would lead to major variations in field strength between the various measuring points.

The anechoic chamber is used for investigating the overall performance of vehicle electrical systems under real-world operating conditions. The Bosch anechoic chamber (Fig. 2) is equipped with devices capable of generating high-frequency fields within a frequency range extending from 10 kHz to 18 GHz, with a maximum field strength of E_{max} = 200 V/m. Since these field strengths represent a health hazard, the test vehicle is operated via remote control from within a shielded control room, while testing is monitored by video cameras. The chamber is screened with metallic sheeting, and non-conductive materials (wood and plastic) line the interior, as metallic substances could foster spurious readings. To prevent reflections and inhibit the formation of standing waves, walls and ceiling are covered by pyramid-shaped absorption elements made of graphite-filled polyurethane foam.

Vehicle testing proceeds on a chassis dynamometer (rolling road) capable of accommodating simulated speeds of up to

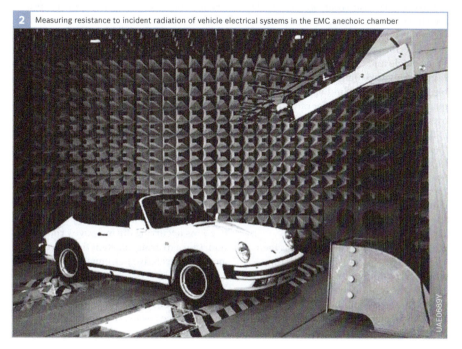

2 Measuring resistance to incident radiation of vehicle electrical systems in the EMC anechoic chamber

UAE0689Y

200 km/h. A fan is also present to direct up to 40,000 m³/h (corresponding to a wind speed of 80 km/h) of air over the vehicle.

Among the advantages that distinguish indoor testing from the old open-air tests near transmitters is that the former method allows latitude for considerable variations in both frequency and field strength. This facilitates evaluation of vehicular resistance to incident radiation under a wide range of conditions, and not at just a few available points. The ability to gradually increase field strengths right up to each electronic device's operational limits furnishes information on safety margins.

Specifics defining the test procedures used in determining radiation resistance in the overall vehicle are described in DIN 40 839, Section 4. ISO 11 452, Sections 1 – 4 contain similar descriptions as elements within a more extensive compilation of special-purpose test procedures.

Laboratory measurements

Although the data garnered from assessments of incident radiation on the vehicle as a whole are invaluable, the disadvantage of this type of testing is that it can only be performed at an advanced stage in the vehicle and electronic system design process. Only very limited latitude is available for responding to indications of inadequate resistance to incident radiation uncovered at this stage.

This explains the necessity to determine at an early stage how an electronic system will operate during subsequent use in its automotive environment: prompt information allows effective recourse to remedial action when needed. Various test procedures have emerged for obtaining the required information.

In the first of the three tests described below, interference waves propagating through a conductor are coupled into the wiring harness of the system under investigation. As the suitability of these test arrangements for evaluating the frequency range beyond > 400 MHz is restricted,

a different procedure is used for this range. The alternate procedure relies on antennas to project electromagnetic fields into standardized bench setups.

The specifics of the methods outlined below are defined in DIN 40 839, Section 4, and ISO 11 452, Sections 1 – 7 (which also contains additional process descriptions detailing less widespread methods).

All these methods furnish a precise picture of a system's resistance to incident radiation. This can then be used to implement improvements while the development phase is still in progress. Thanks to this inestimable advantage it is now impossible to imagine development projects without these laboratory measurements as flanking measures.

It must be taken into consideration that the design of the electrical system in itself is not the only important factor. Because installation conditions in the vehicle and the routing of the wiring harness can also exercise a decisive influence on ultimate levels of interference resistance, all earlier results still need to be confirmed on a production vehicle in the anechoic chamber.

Stripline method (Figs. 3 and 4)
The designation, "stripline", refers to a conductor in the form of a strip. This conductor has a length of 4.1 m and a width of 0.74 m. It is arranged with a clearance of 0.15 m above the conductive sheet (counter-electrode). A high-frequency alternator serves as the source of a transverse magnetic wave generated between the stripline and the counter-electrode, continuing until it reaches the terminal resistor. The stripline dimensions have been selected to minimize the likelihood of reflections occurring during wave propagation, thus ensuring constant field-strength amplitudes relative to frequency.

A typical system setup might include a control unit, wiring harness and peripheral devices (sensors and actuators). These are set up at a height half way be-

tween the base plate and the stripline, with the wiring harness in alignment with the waves' direction of propagation.

The field strength between the planes at a given frequency is then increased until the system malfunctions or until a maximum specified level is reached. If the increments are small enough, repeating this process at progressively altered frequencies will produce a detailed diagram portraying resistance to incident radiation as a function of frequency (Fig. 4).

Bulk current injection method (Fig. 5)
Bulk current injection (BCI) is an immunity test commonly used in the automotive and aircraft industry. As with the stripline method, the system being tested is arranged above a conductive sheet (counter-electrode). A current clamp attached to the wiring harness injects transformer current into its individual wires. The vectorial sum of these currents corresponds to the total current radiated by the clamp. While testing using the stripline method proceeds through a range of increasing field strengths, BCI relies on progressively higher current flows as it continues until the system malfunctions or until current reaches the specified terminal value.

TEM cell (Fig. 6)
As with the stripline method, a transverse electromagnetic field (TEM) is generated here between a strip conductor and a counter-electrode. The TEM counter-electrode, however, is an enclosed housing instead of a metal plate. This leads to another distinguishing feature: TEM test benches do not need to be set up within specially shielded chambers.

3 Stripline method

G

UAE0246-1Y

4 Resistance to incident radiation

V·m⁻¹

Resistance to incident radiation →

Frequency → MHz

UAE0249-1E

5 BCI Method

G

UAE0247-1Y

Fig. 3
1 High-frequency alternator
2 Resistor
3 Stripline
4 Counter-electrode (conductive sheet or cell)
5 System under test
6 Wiring harness
7 Peripherals (sensors, actuators)

Fig. 4
Determined using stripline, BCI or TEM cell method

Fig. 5
1 High-frequency alternator
2 Resistor
3 Counter-electrode (conductive sheet or cell)
4 System under test
5 Wiring harness
6 Peripherals (sensors, actuators)
7 Current clamp

A further difference lies in the fact that only the test specimen per se (for example: a control unit) is subject to electromagnetic exposure. The peripheral equipment is located outside the TEM cell and is con-

wiring harness arranged at right angles to the direction of propagation of the electromagnetic waves.

The measurement procedure is consistent with that of the stripline method. Here, too, the field strength is increased until the system malfunctions or until a maximum specified level is reached.

Incident radiation via antenna
Once again similarities with the stripline method are encountered, as the test specimen, including control unit, wiring harness and peripherals, is set up on a base plate. The wiring harness is laid out at a defined distance from this base plate. An antenna located at a specified distance from this assembly generates an electromagnetic field for projection into the entire test setup. Again, field strength is raised until the system malfunctions or until field strength reaches the maximum value specified for testing.

Guarantee of immunity and interference suppression

As early as the planning and conception phases, EMC requirements for immunity to and suppression of radio interference have to be considered as part of every development project focusing on any electronic system or component. The finished products must incorporate EMC measures as an integral design element.

EMC in the electronic control unit (ECU)

The first EMC requirement reflected in any ECU is to select microprocessor cycle frequencies that are as low as possible while simultaneously making every effort to keep the signal transit curves as flat as possible. EMC characteristics join standard operational aspects as an essential criterion governing the selection of components (integrated circuits). Whereas the components' resistance to interference must be maximal, its interference emissions should be minimal. On printed-circuit boards designed for optimal EMC performance, all components with pronounced sensitivity to interference, and all of those with the potential to cause it,

6 TEM-cell test setup

Fig. 6
1 High-frequency
 alternator
2 Resistor
3 Stripline
4 Counter-electrode
 (conductive sheet
 or cell)
5 System under test
6 Peripherals
 (sensors, actuators)

UAE0248-1Y

are decoupled from the connected wiring harness. This is done by locating the components as far as possible from the plug connections.

Suppression components, usually capacitors designed for high-frequency applications, place the required restrictions on potentially disruptive interference. These interference-suppression capacitors are located either directly within the integrated circuits or in the plug-connector area. The suppression elements in the plug-connector area combine with housings featuring optimal conductive properties (shielded casing) to prevent high-frequency radiation in the immediate vicinity from reaching the inside of the ECU. This ensures that signals produced outside the unit do not cause interference inside the unit. On the other side of the coin, high-frequency signals produced inside the unit do not cause undesirable interference in the surrounding environment.

Electric motors and other electromechanical components

Standard practice in the design of electric motors reflects the procedures used for ECUs and sensors by incorporating interference suppression in the design process right from the initial development stages. A sample problem is the interference produced by the brush arcing during current reversal in motors with commutator control. This can severely disrupt radio reception. Suitable suppression elements (capacitors and throttles) limit the interference. Motors are designed with optimal operation of these interference suppressors in mind.

The voltage pulses produced during switching operations in electromagnetic actuators are kept to an acceptable level by suitable elements, such as quench resistors, in the circuit.

High-voltage ignition

The ignition system's high-voltage circuit represents a potential source of substantial interference to radio reception. One standard solution is to use spark plugs and cable sockets with noise-suppression resistors. While the resistor element is usually located at the end of the ignition cables, in modern (multiple-coil) systems it can be integrated within the dedicated ignition coils plugged directly onto the spark plugs for the individual cylinders. In using resistors, the object is to find a balance between the demand for ignition energy and the required level of interference suppression.

Retrofit interference suppression

As made clear in the above sections, EMC design considerations must be carefully aligned and balanced against operational requirements. Subsequent remedial measures superimposed on the original design at a later date usually entail substantial effort, and should be avoided in production vehicles.

Should the interference-suppression measures incorporated in electrical components prove inadequate for dealing with the conditions encountered in specific applications (in special-purpose emergency vehicles, etc.), supplementary action can be implemented to deal with the problem. Options include installation of filters as well as supplementary shielding for components and wiring.

These options should be exercised with great care, as field modifications to electronic components can lead to operational malfunctions.

Fault diagnostics

The rise in the sheer amount of electronics in the automobile, the use of software to control the vehicle, and the increased complexity of modern fuel-injection systems place high demands on the diagnostic concept, monitoring during vehicle operation (on-board diagnosis), and workshop diagnostics (Fig. 1). The workshop diagnostics is based on a guided troubleshooting procedure that links the many possibilities of on-board and offboard test procedures and test equipment. As emission-control legislation becomes more and more stringent and continuous monitoring is now called for, lawmakers have now acknowledged on-board diagnosis as an aid to monitoring exhaust-gas emissions, and have produced manufacturer-independent standardization. This additional system is termed the *on-board diagnostic system*.

Monitoring during vehicle operation (on-board diagnosis)

Overview

ECU-integrated diagnostics belong to the basic scope of electronic engine-management systems. Besides a self-test of the control unit, input and output signals, and control-unit intercommunication are monitored.

On-board diagnosis of an electronic system is the capability of a control unit to interpret and perform self-monitoring using "software intelligence", i.e. detect, store, and diagnostically interpret errors and faults. On-board diagnosis runs without the use of any additional equipment.

Monitoring algorithms check input and output signals during vehicle operation, and check the entire system and all its functions for malfunctions and disturbances. Any errors or faults detected are stored in the control-unit fault memory. Stored fault information can be read out via a serial interface.

1 Diagnostic system

Diagnostic tester

Offboard tester

Test lines

Offboard Onboard

Control units

CAN

UWT0104E

Input-signal monitoring

Sensors, plug connectors, and connecting lines (signal path) to the control unit (Fig. 2) are monitored by evaluating the input signal. This monitoring strategy is capable of detecting sensor errors, short-circuits in the battery-power circuit U_{Batt} and vehicle-ground circuit, and line breaks. The following methods are applied:

- Monitoring sensor supply voltage (if applicable).
- Monitoring detected values for permissible value ranges (e.g. 0.5....4.5 V).
- If additional information is available, a plausibility check is conducted using the detected value (e.g. comparison of crankshaft speed and camshaft speed).
- Critical sensors (e.g. pedal-travel sensor) are fitted in redundant configuration, which means that their signals can be directly compared with each other.

Output-signal monitoring

Actuators triggered by a control unit via output stages (Fig. 2) are monitored. The monitoring functions detects line breaks and short-circuits in addition to actuator faults. The following methods are applied:

- Monitoring an output signal by the output stage. The electric circuit is monitored for short-circuits to battery voltage U_{Batt}, to vehicle ground, and for open circuit.
- Impacts on the system by the actuator are detected directly or indirectly by a function or plausibility monitor. System actuators, e.g. exhaust-gas recirculation valves, throttle valves, or swirl flaps, are monitored indirectly via closed-control loops (e.g. continuous control variance), and also partly by means of position sensors (e.g. position of turbine geometry in the exhaust-gas turbocharger).

2 System chart of an electronic system (example: common rail)

Sensors and setpoint generators	ECU	Actuators
Pedal-travel sensor with low-idle switch and kickdown switch	Sensor evaluation	Injectors
Wheel-speed sensors (crankshaft, camshaft)	Signal processing	Rail-pressure control valve / High-pressure pump
Pressure sensors (charge-air pressure, rail pressure)		Exhaust-gas recirculation actuator
		Boost-pressure actuator
Temperature sensors (charge-air, exhaust-gas, and engine temperature)	Fault diagnosis	Throttle-valve actuator
		Additional output stages
Air-mass meter	Output stages	**Communication** — ISO interface (e.g. diagnostics) K / L
		Glow-plug control unit
Switch inputs (clutch switch, terminal 15, vehicle-speed controller)		Diagnosis lamp
		CAN interface

SAE0750-2E

Monitoring internal ECU functions

Monitoring functions are implemented in control-unit hardware (e.g. "intelligent" output-stage modules) and software to ensure that the control unit functions correctly at all times. The monitoring functions check each of the control-unit components (e.g. microcontroller, flash EPROM, RAM).

Many tests are conducted immediately after startup. Other monitoring functions are performed during normal operation and repeated at regular intervals in order to detect component failure during operation. Test runs that require intensive CPU capacity, or that cannot be performed during vehicle operation for other reasons, are conducted in after-run more when the engine is switched off. This method ensures that the other functions are not interfered with. In the common-rail system for diesel engines, functions such as the injector switchoff paths are tested during engine runup or after-run. With a spark-ignition engine, functions such as the flash EPROM are tested in engine after-run.

Monitoring ECU communication

As a rule, communication with other ECUs takes place over the CAN bus (**C**ontroller **A**rea **N**etwork). The CAN protocol contains control mechanisms to detect malfunctions. As a result, transmission errors are even detectable at CAN-module level. A number of other checks are also performed in the ECU. Since the majority of CAN messages are sent at regular intervals by the individual control units, the failure of a CAN controller in a control unit is detectable by testing at regular intervals. In addition, when redundant information is available in the ECU, the received signals are checked in the same way as all input signals.

Error handling

Error detection

A signal path is categorized at finally defective if an error occurs over a definite period of time. Until the defect is categorized, the system uses the last valid value detected. When the defect is categorized, a standby function is triggered (e.g. engine-temperature substitute value $T = 90°C$).

Most errors can be rectified or detected as intact during vehicle operation, provided the signal path remains intact for a definite period of time.

Fault storage

Each fault is stored as a fault code in the non-volatile area of the data memory. The fault code also describes the fault type (e.g. short-circuit, line break, plausibility, value range exceeded). Each fault-code input is accompanied by additional information, e.g. the operating and environmental conditions (freeze frame) at the time of fault occurrence (e.g. engine speed, engine temperature).

Limp-home function

If a fault is detected, limp-home strategies can be triggered in addition to substitute values (e.g. engine output power or speed limited). These strategies help to:
- Maintain driving safety
- Avoid consequential damage
- Minimize exhaust-gas emissions

On-board diagnosis system for passenger cars and light-duty trucks

The engine system and its components must be constantly monitored in order to comply with exhaust-gas emission limits specified by law in everyday driving situations. For this reason, regulations have come into force to monitor exhaust-gas systems and components, e.g. in California. This has standardized and expanded manufacturer-specific on-board diagnosis with respect to the monitoring of emission-related components and systems.

Legislation
OBD I (CARB)
1988 marked the coming into force of OBD I in California, that is, the first stage of CARB legislation (California Air Resources Board). The first OBD stage makes the following requirements:
- Monitoring emission-related electrical components (short-circuits, line breaks) and storage of faults in the control-unit fault memory.
- A Malfunction Indicator Lamp (MIL) that alerts the driver to the malfunction.
- The defective component must be displayed by means of on-board equipment (e.g. blink code using a diagnosis lamp).

OBD II (CARB)
The second stage of the diagnosis legislation (OBD II) came into force in California in 1994. OBD II became mandatory for diesel-engine cars with effect from 1996. In addition to the scope of OBD I, system functionality was now monitored (e.g. plausibility check of sensor signals).

OBD II stipulates that all emission-related systems and components must be monitored if they cause an increase in toxic exhaust-gas emissions in the event of a malfunction (by exceeding the OBD limits). Moreover, all components must be monitored if they are used to monitor emission-related components or if they can affect the diagnosis results.

Normally, the diagnostic functions for all components and systems under surveillance must run at least once during the exhaust-gas test cycle (e.g. FTP 75). A further stipulation is that all diagnostic functions must run with sufficient frequency during daily driving mode. For many monitoring functions, the law defines a monitoring frequency (In Use Monitor Performance Ratio) in daily operation starting model year 2005.

Since the introduction of OBD II, the law has been revised in several stages (updates). The last update came into force in model year 2004. Further updates have been announced.

OBD (EPA)
Since 1994 the laws of the EPA (Environmental Protection Agency) have been in force in the remaining U.S. states. The scope of these diagnostics comply for the most part with the CARB legislation (OBD II).

The OBD regulations for CARB and EPA apply to all passenger cars with up to 12 seats and to light-duty trucks weighing up to 14,000 lbs (6.35 t).

EOBD (EU)
The OBD attuned to European conditions is termed EOBD and is based on the EPA-OBD.

EOBD has been valid for all passenger cars and light-duty trucks equipped with gasoline engines and weighing up to 3.5 t with up to 9 seats since January 2000. Since January 2003 the EOBD also applies to passenger cars and light-duty trucks with diesel engines.

Other countries
A number of other countries have already adopted or are planning to adopt EU or US-OBD legislation.

Requirements of the OBD system

The ECU must use suitable measures to monitor all on-board systems and components whose malfunction may cause a deterioration in exhaust-gas test specifications stipulated by law. The Malfunction Indicator Lamp (MIL) must alert the driver to a malfunction if the malfunction could cause an overshoot in OBD emission limits.

Emission limits

The U.S. OBD II (CARB and EPA) prescribes thresholds that are defined relative to emission limits. Accordingly, there are different permissible OBD emission limits for the various exhaust-gas categories that are applied during vehicle certification (e.g. TIER, LEV, ULEV). Absolute limits apply in Europe (Table 1).

Malfunction Indicator Lamp (MIL)

The Malfunction Indicator Lamp (MIL) alerts the driver that a component has malfunctioned. When a malfunction is detected, CARB and EPA stipulate that the MIL must light up no later than after two driving cycles of its occurrence. Within the scope of EOBD, the MIL must light up no later than in the third driving cycle with a detected malfunction.

If the malfunction disappears (e.g. loose contact), the malfunction remains entered in the fault memory for 40 trips (warmup cycles). The MIL goes out after three fault-free driving cycles.

Communication with scan tool

OBD legislation prescribes standardization of the fault-memory information and access to the information (connector, communication interface) compliant with ISO 15031 and the corresponding SAE standards (Society of Automotive Engineers). This permits the readout of the fault memory using standardized, commercially available testers (scan tools, see Fig. 1).

Depending on their scope of application, various communication protocols are used throughout the world: The most important are:

- ISO 9141-2 for European passenger cars.
- SAE J 1850 for U.S. passenger cars.
- ISO 14 230-4 (KWP 2000) for European passenger cars and commercial vehicles.
- SAE J 1708 for U.S. commercial vehicles.

These serial interfaces operate at a bit rate (baud rate) of 5 to 10 k baud. They are designed as a single-wire interface with a common wire for both transmission and reception, or as a two-wire interface with a separate *data line* (K-line) and *initiate line* (L-line). Several electronic control units (such as Motronic, ESP, or EDC, and transmission-shift control, etc.) can be combined on one diagnosis connector.

Communication between the tester and ECU is set up in three phases:

- Initiate the ECU.
- Detect and generate baud rate.
- Read key bytes which identify the transmission protocol.

1	OBD limits for passenger cars and light-duty trucks			
	Otto passenger cars		**Diesel passenger cars**	
CARB	– Relative emission limits – Mostly 1.5 times the limit of a specific exhaust-gas category		– Relative emission limits – Mostly 1.5 times the limit of a specific exhaust-gas category	
EPA	– Relative emission limits – Mostly 1.5 times the limit of a specific exhaust-gas category		– Relative emission limits (U.S. Federal) – Mostly 1.5 times the limit of a specific exhaust-gas category	
EOBD	2000 CO: 3.2 g/km HC: 0.4 g/km NO$_X$: 0.6 g/km	2005 (draft) CO: 1.9 g/km HC: 0.3 g/km NO$_X$: 0.53 g/km	2003 CO: 3.2 g/km HC: 0.4 g/km NO$_X$: 1.2 g/km PM: 0.18 g/km	2005 (draft) CO: 3.2 g/km HC: 0.4 g/km NO$_X$: 1.2 g/km PM: 0.18 g/km

Table 1

Evaluation is performed subsequently. The following functions are possible:
- Identify the ECU
- Read the fault memory
- Erase the fault memory
- Read the actual values

In future, the CAN bus (ISO 15 765-4) will be used increasingly to handle communication between ECUs and the tester. Starting 2008 diagnostics will only be permitted over this interface in the U.S.A.

To make it easier to read out the ECU fault-memory information, a standardized diagnosis socket will be fitted at an easily accessible place in every car (easy to reach from the driver's seat). The socket is used to connect the scan tool (Fig. 2).

Reading the fault information

Any workshop can use the scan tool to read out emission-relevant fault information from the ECU (Fig. 3). In this way, workshops not franchized to a particular manufacturer are also able to carry out repairs. Manufacturers are obliged to make the required tools and information available (on the internet) in return for a reasonable payment, to allow this.

Vehicle recall

If vehicles fail to comply with OBD requirements by law, the authorities may demand the vehicle manufacturer to start a recall at his own cost.

1 OBD system

MIL ⊗

ISO 15031 SAE…

Sensors
Actuators → Diagnostic manager
Systems

Communication

Generic scan tool

OBD socket Repair

Onboard Offboard

2 Pin allocation of the OBD socket

8 ■ 16 ■
7 ■ 15 ■
6 ■ 14 ■
5 ■ 13 ■
4 ■ 12 ■
3 ■ 11 ■
2 ■ 10 ■
1 ■ 9 ■

Pins 7 and 15:
Data transmission to DIN ISO 9141-2
Pins 2 and 10:
Data transmission to SAE J 1850
Pins 1, 3, 8, 9, 11, 12, 13:
Not assigned by OBD
Pin 4: Vehicle ground
Pin 5: Signal ground
Pin 6: CAN high
Pin 14: CAN low
Pin 16: Battery +

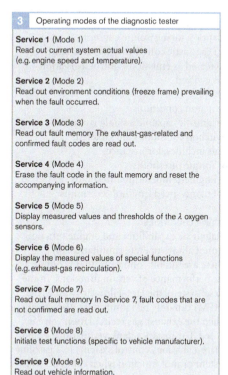

3 Operating modes of the diagnostic tester

Service 1 (Mode 1)
Read out current system actual values (e.g. engine speed and temperature).

Service 2 (Mode 2)
Read out environment conditions (freeze frame) prevailing when the fault occurred.

Service 3 (Mode 3)
Read out fault memory The exhaust-gas-related and confirmed fault codes are read out.

Service 4 (Mode 4)
Erase the fault code in the fault memory and reset the accompanying information.

Service 5 (Mode 5)
Display measured values and thresholds of the λ oxygen sensors.

Service 6 (Mode 6)
Display the measured values of special functions (e.g. exhaust-gas recirculation).

Service 7 (Mode 7)
Read out fault memory In Service 7, fault codes that are not confirmed are read out.

Service 8 (Mode 8)
Initiate test functions (specific to vehicle manufacturer).

Service 9 (Mode 9)
Read out vehicle information.

Functional requirements

Overview

Just as for on-board diagnosis, all ECU input and output signals, as well as the components themselves, must be monitored.

Legislation demands the monitoring of electrical functions (short-circuit, line breaks), a plausibility check for sensors, and a function monitoring for actuators.

The pollutant concentration expected as the result of a component failure (empirical values), and the monitoring mode partly required by law determine the type of diagnostics. A simple functional test (black/white test) only checks system or component operability (e.g. swirl flap opens and closes). The extensive functional test provides more detailed information about system operability. As a result, the limits of adaption must be monitored when monitoring adaptive fuel-injection functions (e.g. zero delivery calibration for a diesel engine, lambda adaption for a gasoline engine).

Diagnostic complexity has constantly increased as emission-control legislation has evolved.

Switchon conditions

Diagnostic capabilities only run if the switchon conditions are satisfied.
This includes, for instance:
- Torque thresholds
- Engine-temperature thresholds
- Engine-speed thresholds or limits

Inhibit conditions

Diagnostic capabilities and engine functions cannot always operate simultaneously. There are inhibit conditions that prohibit the performance of certain functions. In the diesel system, the Hot-Film Air-Mass Meter (HFM) can only be monitored satisfactorily when the exhaust-gas recirculation valve is closed. For instance, tank ventilation (evaporative-emissions control system) in a gasoline system cannot function when catalytic-converter diagnosis is in operation.

Temporary disabling of diagnostic functions

Diagnostic capabilities may only be disabled under certain conditions in order to prevent false diagnosis. Examples include:
- Height too large.
- Low ambient temperature at engine switchon.
- Low battery voltage.

Readiness code

When the fault memory is checked, it is important to know that the diagnostic capabilities ran at least once. This can be checked by reading out the readiness code over the diagnostic interface. After erasing the fault memory in service, the readiness codes must be reset after checking the functions.

Diagnostic System Management (DSM)

The diagnostic capabilities for all components and systems checked must normally run in driving mode, but at least once during the exhaust-gas test cycle (e.g. FTP 75, NEDC). Diagnostic System Management (DSM) can change the sequence for running the diagnostic capabilities dynamically, depending on driving conditions.

The DSM comprises the following three components (Fig. 4):

Diagnosis Fault Path Management (DFPM)
The primary role of DFPM is to store fault states that are detected in the system. In addition to faults, it also stores other information, such as environmental conditions (freeze frame).

Diagnostic Function SCHEDuler (DSCHED)
DSCHED is responsible for coordinating
assigned engine capabilities (MF) and diag-
nostic capabilities (DF). It obtains informa-
tion from DVAL and DFPM to carry this out.
In addition, it reports functions that require
release by DSCHED to perform their readi-
ness, after which the current system state is
checked.

Diagnosis VALidator (DVAL)
The DVAL (only installed in gasoline systems
to date) uses current fault-memory entries
and additionally stored information to decide
for each detected fault whether it is the actual
cause, or a consequence of the fault. As a re-
sult, validation provides stored information
to the diagnostic tester for use in reading out
the fault memory.

 In this way, diagnostic capabilities can
be released in any sequence. All released
diagnoses and their results are evaluated sub-
sequently.

4 Diagnostic system management

OBD functions

Overview

Whereas EOBD only contains detailed moni-
toring specifications for individual compo-
nents, the requirements in CARB OBD II are
much more detailed. The list below shows
the current state of CARB requirements
for gasoline-engined and diesel-engined
passenger cars. Requirements that are also
described in detail in the EOBD legislation
are marked by (E):

- Catalytic converter (E), heated catalytic
 converter
- Combustion (ignition) misfire
 (E, for diesel system not for EOBD)
- Evaporation reduction system
 (fuel-tank leak diagnosis, only for
 gasoline system)
- Secondary-air injection
- Fuel system
- Lambda oxygen sensors (E)
- Exhaust-gas recirculation
- Crankcase ventilation
- Engine cooling system
- Cold-start emission reduction system
 (presently only for gasoline system)
- Air conditioner (components)
- Variable valve timing (presently only in
 use with gasoline systems)
- Direct ozone reduction system (presently
 only in use with gasoline systems)
- Particulate filter (particulate filter, only for
 diesel system) (E)
- Comprehensive components (E)
- Other emission-related
 components/systems (E)

"Other emission-related components/sys-
tems" refer to components and systems not
mentioned in this last and that may exceed
OBD emission limits, or block other diagnos-
tic functions if they malfunction.

Catalytic converter diagnosis

In the diesel system, carbon monoxide (CO) and unburned hydrocarbons (HC) are oxidized in the oxidation-type catalytic converter. Work is ongoing on diagnostic capabilities for monitoring the operation of oxidation-type catalytic converters relating to temperature and differential pressure. One approach focuses on active secondary injection (intrusive operation). Here, heat is generated in the oxidation-type catalytic converter by an exothermic HC reaction. The temperature is measured and compared with calculated model values. These are used to derive the functionality of the catalytic converter.

Equally, work is ongoing on monitoring functions for the storage and regeneration capabilities of NO_x accumulator-type catalytic converters that will also be installed in the diesel system in future. The monitoring functions run based on loading and regeneration models, and the measured regeneration duration. This requires the use of lambda or NO_x sensors.

Combustion-miss detection

Incorrect fuel injection or loss of compression result in impaired combustion, and thus to changes in emission values. The misfire detector evaluates the time expired (segment time) from one combustion cycle to the next for each cylinder. This time is derived from the speed-sensor signal. A segment time that is longer than for the other cylinders indicates a misfire or loss of compression.

In the diesel system, diagnosis of combustion misses is only required and performed when the engine is at idle speed.

Fuel system diagnosis

In the common-rail system, a fuel-system diagnosis includes the electrical monitoring of injectors and rail-pressure control (high-pressure control). In the Unit Injector System, it also includes monitoring of the injector switching time. Special functions of the fuel-injection system that increase injected-fuel-quantity precision are also monitored.

Examples of this include zero-fuel-quantity calibration, quantity-mean-value adaptation, and the ASMOD observer function (air-system model observer). The two last functions use information from the lambda oxygen sensor as input signals. From the models, they calculate fluctuations between setpoint and actual quantities.

Lambda-oxygen-sensor diagnosis

Modern diesel systems are fitted with broadband oxygen sensors. They require a different diagnostic procedure than two-stage sensors since their settings may deviate from $\lambda = 1$. They are monitored electrically (short-circuit, line interruption) and for plausibility. The heater element of the sensor heater is tested electrically and for permanent governor deviation.

Exhaust-gas recirculation system diagnosis

In the exhaust-gas recirculation system, the EGR valve and – if fitted – the exhaust-gas cooler are monitored.

The exhaust-gas recirculation valve is monitored for its electrical and functional operability. Functional monitoring is performed by air-mass regulators and position controllers. They check for permanent control variances.

If the exhaust-gas recirculation system has a cooler fitted, its function must also be monitored, i.e. an additional temperature measurement takes place downstream of the cooler. The temperature measured is compared with a setpoint value calculated from a model. If a fault occurs, it is detected by measuring the deviation between setpoint and actual values.

Crankcase ventilation diagnosis

Faults in crankcase ventilation are detected by the air-mass sensor, depending on the system. The legislation requires no monitoring if the crankcase ventilation has a "rugged" design.

On-board diagnosis system for passenger cars and light-duty trucks

Engine cooling system diagnosis

The cooling system comprises a thermostat and a coolant-termperature sensor. If the thermostat is defective, for instance, the engine temperature can only rise slowly and, consequently, the exhaust emission rates may increase. The diagnostic function for the thermostat uses the coolant-termperature sensor to check that a nominal temperature has been reached. A temperature model is also used for monitoring.

The coolant-termperature sensor is monitored to ensure that a minimum temperature has been reached in addition to monitoring for electrical faults by means of a dynamic plausibility function. Dynamic plausibility is performed as the engine cools down. These functions can monitor the sensor for "sticking" in both low and high temperature ranges.

Air-conditioner diagnosis

The engine can be operated at a different operating point in order to cover the air-conditioner's electrical load requirements. The required diagnosis must therefore monitor all electronic components in the air conditioner that may cause an increase in emissions if they malfunction.

Particulate-filter diagnosis

The particulate filter is currently monitored for filter breakage, removal, or blockage. A differential-pressure sensor is used to measure differential pressure (exhaust-gas backpressure downstream and upstream of the filter) at a specific volumetric flow. The measured value can be used to decide whether the filter is defective.

Comprehensive components

On-board diagnosis legislation requires that all sensors (e.g. air-mass meter, wheel-speed sensors, temperature sensors) and actuators (e.g. throttle valve, high-pressure pump, glow plugs) must be monitored if they either have an impact on emissions, or are used to monitor other components or systems (and consequently, may block other diagnoses).

Sensors monitor the following faults (Fig. 5):
- Electrical faults, i.e. short-circuits and line breaks *(signal-range check)*.
- Range faults *(out-of-range check)*, i.e. undercutting or exceeding voltage limits defined by the physical measurement range of a sensor.
- Plausibility faults *(rationality check)*; these are faults that are inherent in the components themselves (e.g. drift), or which may be caused by shunts, for instance. Monitoring is carried out by a plausibility check on the sensor signals, either by using a model, or directly using other sensors.

Actuators must be monitored for electrical faults and – if technically possible – also for function. Functional monitoring means that, when a control command (setpoint value) is given, it is monitored by observing or measuring (e.g. by a position sensor) the system reaction (actual value) in a suitable way by using information from the system.

Besides all output stages, the following actuators are monitored:
- throttle valve
- exhaust-gas recirculation valve
- variable turbine geometry of the exhaust-gas turbocharger
- swirl flap
- glow plugs

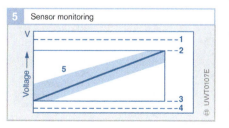

5 Sensor monitoring

Fig. 5
1 Upper threshold for *signal-range check*
2 Upper threshold for *out-of-range check*
3 Lower threshold for *out-of-range check*
4 Lower threshold for *signal-range check*
5 Plausibility range *rationality check*

On-board diagnosis system for heavy-duty trucks

In Europe and the U.S.A., there exist draft laws that have not yet been adopted; they are based closely on legislation for passenger cars.

Legislation

In the EU, there are plans to introduce new type approvals in October 2005 (coinciding with Euro 4 emission-control legislation). With effect from October 2006 an OBD system will become obligatory for every commercial vehicle. In the U.S.A., the draft of the Californian CARB provides for the introduction of an OBD system for Model Year (MY) 2007. It is probable that EPA (U.S. Federal) will also follow with a draft in 2004 for subsequent introduction in MY 2007. Besides that, there are initiatives to promote worldwide harmonization (World Wide Harmonized (WWH) OBD). However, this is not expected until 2012. Japan is planning to introduce an OBD system in 2005.

EOBD for trucks and buses > 3.5 t

European OBD legislation provides for a two-stage introduction. Stage 1 (2005) requires monitoring:
- of the fuel-injection system for closed electrical circuit and total failure.
- of emission-related engine components or systems for compliance with OBD emission limits (Table 1).
- of the exhaust-gas treatment system for major functional faults (e.g. damaged catalytic converter, urea deficit in the SCR system).

Stage 2 (2008) requires:
- Monitoring of the exhaust-gas treatment system for emission limits.
- The OBD emission limits must be adapted to the prevailing state of the art (availability of exhaust-gas sensors).

Protocols for scan-tool communication over CAN have been approved using either ISO 15765 or SAE J1939.

CARB OBD for HD trucks > 14,000 lb. (6.35 t)

The present draft law is very close to passenger-car legislation in its function requirements, and also provides for a two-stage introduction:
- MY 2007: Monitoring for functional faults.
- MY 2010: Monitoring for OBD emission limits (Table 1).

The main changes compared with present passenger-car legislation are as follows:
- Erasing the OBD fault memory by scan tool is no longer possible. This is only possible by self-healing (e.g. after repair).
- SAE J1939 has also been approved as an alternative to CAN diagnostic communication to ISO 15765 (as for passenger cars).

1	OBD emission limits for heavy-duty trucks (draft)	
CARB	2007 – Functional check no limits	2010 – Relative limit – 1.5 times the value of each exhaust-gas category – Exception: catalytic converter, factor 1.75
EPA	– to be defined	– to be defined
EU	2005 – Absolute limit NO_x: 7.0 g/kWh PM: 0.1 g/kWh – Functional check for exhaust-gas treatment system	2008 – Absolute limit NO_x: 7.0 g/kWh PM: 0.1 g/kWh – Subject to review by EU Commission

Table 1

"Once you have driven an automobile, you will soon realize that there is something unbelievably tiresome about horses (…). But you do require a conscientious mechanic for the automobile (…)".

Robert Bosch wrote these words to his friend Paul Reusch in 1906. In those days, it was indeed the case that breakdowns could be repaired on the road or at home by an employed chauffeur or mechanic. However, with the growing number of motorists driving their own cars after the First World War, the need for workshops offering repair services in-

creased rapidly. In the 1920s Robert Bosch started to systematically create a nationwide customer-service organization. In 1926 all the repair centers were uniformly named "Bosch Service" and the name was registered as a trademark.

Today's Bosch Service agencies retain the same name. They are equipped with the latest electronic equipment in order to meet the demands of 21st-century automotive technology and the quality expectations of the customers.

1 A repair shop in 1925 (photo: Bosch)

2 A Bosch car service workshop in the 21st century carried out with the very latest electronic testing equipment

▶ Index

CPI Antony Rowe
Eastbourne, UK
April 09, 2019